Plant Metal and Metalloid Transporters

Kundan Kumar • Sudhakar Srivastava
Editors

Plant Metal and Metalloid Transporters

Springer

Editors
Kundan Kumar
Department of Biological Sciences
Birla Institute of Technology & Science
Pilani, K. K. Birla Goa Campus
Goa, India

Sudhakar Srivastava
Institute of Environment & Sustainable Development
Banaras Hindu University
Varanasi, Uttar Pradesh, India

ISBN 978-981-19-6102-1   ISBN 978-981-19-6103-8 (eBook)
https://doi.org/10.1007/978-981-19-6103-8

This Springer imprint is published by the registered company Springer Nature Singapore Pte Ltd.
The registered company address is: 152 Beach Road, #21-01/04 Gateway East, Singapore 189721, Singapore

Preface

Metals and metalloids are essential for normal plant growth and development due to their direct or indirect involvement in every biological process. However, metal (loid)s become toxic when present in excess. Environmental contamination of metal (loid)s is a major problem threatening the growth and development of plants and also impacts the yield of crops. Metal(loid) uptake, accumulation, translocation and distribution in different tissues and cells of plants are highly regulated phenomena involving the role of several transporters, enzymes, metabolites, transcription factors and post-translational modifications. Transporters play a central role in the uptake and transport and tissue-specific localization of metal(loid)s. Toxic metals also gain entry into plants via transporters of essential metals and distribute inside various cells, tissues and organs of plants.

Research over the years has identified several transporters constituting various groups. These include ATP-binding cassette (ABC), natural resistance-associated macrophage proteins (NRAMPs), cation diffusion facilitator (CDF), iron-regulated transporter (IRT), zinc-iron permeases (ZIP), zinc transporters (Znt), yellow strip like (YSL), heavy metal ATPases (HMAs), aquaglyceroporins (AQPs), copper transporter (COPT), aluminium-activated malate transporter (ALMT), multidrug and toxic compound extrusion (MATE) and metal tolerance protein (MTP). A lot of information about genes and proteins, mechanism of their functioning and their regulation has emerged. Further, their physiological functions in metal(loid) transport have become increasingly evident with the ongoing research.

The present book includes a total of 19 chapters dealing with various transporters of metal and metalloids, nanoparticles, co-transport mechanisms and the role of transporters in phytoremediation and crop biofortification. Chapter 1 of the book presents an overview of all the metal(loid) transporters present in plants, which are involved in the accumulation and transport of metals and metalloids. Chapter 2 deals with the physiological and toxicological effects of toxic metal(loid)s on plants. Chapters 3 and 4 discuss ABC transporters and cadmium transporter family, respectively. Chapter 5 deals with the structural and functional significance of the NRAMP family of transporters in metal ion transportation and homeostasis. The role of heavy metal ATPases (HMAs) in zinc and cadmium transport in plants is documented in Chapter 6. The transportation of metalloids such as As, Si, B and Se by aquaglyceroporins or major intrinsic protein family is discussed in Chapter 7.

Chapter 8 discusses the role of the multidrug and toxic compound extrusion (MATE) transporter family in regulating the transport of various metals such as iron and aluminium. Chapter 9 deals with aluminium tolerance mechanism of a plant at the molecular level. Chapter 10 discusses zinc-regulated iron-regulated transporter-like protein (ZIP) family members, which are actively involved in zinc uptake and transport in plants. Potassium (K^+), a vital macronutrient, regulates different physiological processes. Chapter 11 deals with HAK and AKT transporters regulating K^+ homeostasis in plants. Silicon (Si) is the most abundant metalloid of Earth's crust and transported in plants via various channel proteins. Chapter 12 discusses the transporters, which are involved in Si transport in plants. Chapter 13 deals with copper transport mechanisms in plants. The role of plant metal tolerance proteins (MTPs) in metal transport and storage is discussed in Chapter 14. Many metals and metalloids transport themselves through more than one transporter. Chapter 15 deals with the co-transport mechanism of various transporters in plants. Chapter 16 discusses both positive and negative impacts of metal nanoparticles, their transport, detection methods, and understanding the possible molecular interactions of plant-metal nanoparticles. Transcriptional factors (TFs) are identified as regulators of metal stress signal transduction pathways. The involvement of various TFs in relation to different heavy metal stresses is discussed in Chapter 17. At the end, Chapters 18 and 19 deal with approaches involved in phytoremediation for the removal of harmful heavy metal(loid)s in non-edible parts as well as biofortification of essential nutrients into the edible parts of plants towards large-scale sustainable production of nutritious and safe food.

This book stands as a one place knowledge hub for plant metal(loid) transporters. The book comprehensively covers the holistic aspects of metal(loid) transporters involved in the uptake and translocation of essential as well as toxic metal(loid)s. This book is an excellent and comprehensive reference material for teachers, researchers, doctoral and graduate students working in the area of plant physiology, environmental biotechnology, plant biotechnology, phytoremediation and crop biofortification. The chapters are written by internationally reputed scientists actively working on the topic. The book has clear depictive figures and descriptive tables as standalone representative in most of the chapters.

Goa, India — Kundan Kumar
Varanasi, Uttar Pradesh, India — Sudhakar Srivastava

Contents

Editors and Contributors

About the Editors

Kundan Kumar is presently working as an Associate Professor at the Department of Biological Sciences, Birla Institute of Technology & Science, Pilani, K. K. Birla Goa Campus. He completed his Ph.D. in Plant Molecular Biology at National Institute of Plant Genome Research, New Delhi, India. He did his postdoctoral studies at the University of Massachusetts Amherst, USA, and McGill University, Canada, from 2009 to 2012. He has more than fifteen years of research experience in the area of Plant Molecular Biology and Stress Physiology with focus on the role of plant aquaporins in metalloid transport. He has published more than 45 publications in international and national journals of repute. Dr. Kumar was a recipient of Young Scientist Award from the Science and Engineering Research Board (SERB) in 2012 and life member of several societies. He is an elected member of Plant Tissue Culture Association of India (PTCAI). He was a reviewer for various international journals and serves as an Associate Editor for two international journals.

Sudhakar Srivastava is presently working as an Assistant Professor at the Institute of Environment and Sustainable Development (IESD), Banaras Hindu University (BHU). He completed his M.Sc. and Ph.D. in Botany at the University of Lucknow. He also served as a Scientific Officer at Bhabha Atomic Research Centre (BARC), Mumbai, from 2009 to 2014. He has more than nineteen years of research experience in the field of plant-metal interactions, with a major focus on arsenic stress responses in plants. He has published 85 research articles and 40 review and letter articles in high-impact international and national journals. He has also received several awards for his contributions, including a Fellow of International Society of Environmental Botanists, 2020, Young Scientist Award from the National Academy of Sciences, India (NASI), Allahabad, in 2011, Young Scientist Award from Uttar Pradesh Council of Science & Technology (UPCST) in 2013–2014, and Young Scientist Award from the Science and Engineering Research Board (SERB) in 2015. Dr. Srivastava is also an Associate of the National Academy of Agricultural Sciences

(NAAS) and a life member of several societies as well as non-governmental organizations and serves as an Editor for respected international journals. He was included among the top 2% of world scientists in 2020.

Contributors

Tariq Aftab Department of Botany, Aligarh Muslim University, Aligarh, India

Parinita Agarwal Division of Plant Omics, CSIR-Central Salt and Marine Chemicals Research Institute (CSIR-CSMCRI), Council of Scientific and Industrial Research (CSIR), Gijubhai Badheka Marg, Bhavnagar, Gujarat, India

Pradeep K. Agarwal Division of Plant Omics, CSIR-Central Salt and Marine Chemicals Research Institute (CSIR-CSMCRI), Council of Scientific and Industrial Research (CSIR), Gijubhai Badheka Marg, Bhavnagar, Gujarat, India

Arzoo Ahad Department of Plant Biotechnology, Atta-ur-Rahman School of Applied Biosciences, National University of Sciences and Technology, Islamabad, Pakistan

Namrah Ahmad Department of Plant Biotechnology, Atta-ur-Rahman School of Applied Biosciences, National University of Sciences and Technology, Islamabad, Pakistan

Shatha N. Alameeri Department of Applied Biology, College of Sciences, University of Sharjah, Sharjah, United Arab Emirates

Muna A. Ali Department of Applied Biology, College of Sciences, University of Sharjah, Sharjah, United Arab Emirates

Mani Anitha Ion Channel Biology Laboratory, AU-KBC Research Centre, MIT Campus of Anna University, Chennai, Tamil Nadu, India

Aya Awad Department of Applied Biology, College of Sciences, University of Sharjah, Sharjah, United Arab Emirates

Tuba Sharf Batool Department of Plant Biotechnology, Atta-ur-Rahman School of Applied Biosciences, National University of Sciences and Technology, Islamabad, Pakistan

Lavanya Bhagavatula Department of Biological Sciences, Indian Institute of Science Education and Research (IISER) Bhopal, Bhopal, Madhya Pradesh, India

Surajit Bhattacharya Amity Institute of Biotechnology, Amity University Kolkata, Kolkata, India

S. Antony Ceasar Division of Plant Molecular Biology and Biotechnology, Department of Biosciences, Rajagiri College of Social Sciences, Kochi, Kerala, India

Moumita Chatterjee V. Sivaram Research Foundation, Bangalore, Karnataka, India

Anjali Chittora Division of Plant Omics, CSIR-Central Salt and Marine Chemicals Research Institute (CSIR-CSMCRI), Council of Scientific and Industrial Research (CSIR), Gijubhai Badheka Marg, Bhavnagar, Gujarat, India

Sadaf Choudhary Department of Botany, Aligarh Muslim University, Aligarh, India

Binayak Das Amity Institute of Biotechnology, Amity University Kolkata, Kolkata, India

Natasha Das Department of Biotechnology, Indian Institute of Technology Kharagpur, Kharagpur, India

Soumi Datta Department of Pharmacognosy and Phytochemistry, School of Pharmaceutical Education and Research, Jamia Hamdard, New Delhi, India

Sourav Datta Department of Biological Sciences, Indian Institute of Science Education and Research (IISER) Bhopal, Bhopal, Madhya Pradesh, India

Priyanka Devi Department of Agronomy, School of Agriculture, Lovely Professional University, Jalandhar, Punjab, India

Shipa Rani Dey Department of Agronomy, School of Agriculture, Lovely Professional University, Jalandhar, Punjab, India

Om Parkash Dhankher Stockbridge School of Agriculture, University of Massachusetts Amherst, Amherst, MA, USA

Arka Dutta Department of Biological Sciences, Indian Institute of Science Education and Research (IISER) Bhopal, Bhopal, Madhya Pradesh, India

Padmanabh Dwivedi Department of Plant Physiology, Institute of Agricultural Sciences, Banaras Hindu University, Varanasi, India

Doddabhimappa R. Gangapur Division of Plant Omics, CSIR-Central Salt and Marine Chemicals Research Institute (CSIR-CSMCRI), Council of Scientific and Industrial Research (CSIR), Gijubhai Badheka Marg, Bhavnagar, Gujarat, India

Sharmistha Ganguly University Department of Botany, Ranchi University, Ranchi, Jharkhand, India

Pallabi Ghosh ICMR, National Instiute of Pathology, New Delhi, India

E. Lokesh Goud Department of Agronomy, School of Agriculture, Lovely Professional University, Jalandhar, Punjab, India

Alvina Gul Department of Plant Biotechnology, Atta-ur-Rahman School of Applied Biosciences, National University of Sciences and Technology, Islamabad, Pakistan

Namra Haq Atta-ur-Rahman School of Applied Biosciences, National University of Sciences & Technology, Islamabad, Pakistan

S. Ignacimuthu Xavier Research Foundation, St Xavier's College, Palayamkottai, Tirunelveli, Tamil Nadu, India

Mahnoor Ilyas Department of Plant Biotechnology, Atta-ur-Rahman School of Applied Biosciences, National University of Sciences and Technology, Islamabad, Pakistan

Siddhi Kashinath Jalmi Department of Botany, School of Biological Sciences and Biotechnology, Goa University, Taleigao, Goa, India

Debojyoti Kar Department of Biological Sciences, Indian Institute of Science Education and Research (IISER) Bhopal, Bhopal, Madhya Pradesh, India

Suhas Balasaheb Karle Department of Biological Sciences, Birla Institute of Technology & Science Pilani, K. K. Birla Goa Campus, Goa, India

T. P. Ajeesh Krishna Division of Plant Molecular Biology and Biotechnology, Department of Biosciences, Rajagiri College of Social Sciences, Kochi, Kerala, India

Kundan Kumar Department of Biological Sciences, Birla Institute of Technology & Science Pilani, K. K. Birla Goa Campus, Goa, India

Prasann Kumar Department of Agronomy, School of Agriculture, Lovely Professional University, Jalandhar, Punjab, India

Varun Kumar Plant Transgenic Laboratory, Molecular Biology and Biotechnology Division, CSIR-National Botanical Research Institute, Lucknow, India

T. Maharajan Division of Plant Molecular Biology and Biotechnology, Department of Biosciences, Rajagiri College of Social Sciences, Kochi, Kerala, India

Mrinal K. Maiti Department of Biotechnology, Indian Institute of Technology Kharagpur, Kharagpur, India

Shafia Maryam Department of Plant Biotechnology, Atta-ur-Rahman School of Applied Biosciences, National University of Sciences and Technology, Islamabad, Pakistan

Kareem A. Mosa Department of Applied Biology, College of Sciences, University of Sharjah, Sharjah, United Arab Emirates
Department of Biotechnology, Faculty of Agriculture, Al-Azhar University, Cairo, Egypt

Suchetana Mukherjee Department of Botany, Sripat Singh College, West Bengal, India

M. Naeem Department of Botany, Aligarh Muslim University, Aligarh, India

Ajar Anupam Pradhan Department of Biological Sciences, Indian Institute of Science Education and Research (IISER) Bhopal, Bhopal, Madhya Pradesh, India

Praveenya Tirunagari Department of Biotechnology, Indian Institute of Technology Kharagpur, Kharagpur, India

Khola Rafique Agriculture Department, Pest Warning and Quality Control of Pesticides, Rawalpindi, Punjab, Pakistan

Kalidoss Ramamoorthy Department of Applied Biology, College of Sciences, University of Sharjah, Sharjah, United Arab Emirates

S. Saharsha Reddy Department of Agronomy, School of Agriculture, Lovely Professional University, Jalandhar, Punjab, India

Sanket Roy Amity Institute of Biotechnology, Amity University Kolkata, Kolkata, India

Kavitha Sankaranarayanan Ion Channel Biology Laboratory, AU-KBC Research Centre, MIT Campus of Anna University, Chennai, Tamil Nadu, India

Indraneel Sanyal Plant Transgenic Laboratory, Molecular Biology and Biotechnology Division, CSIR-National Botanical Research Institute, Lucknow, India
Academy of Scientific and Innovative Research (AcSIR), Ghaziabad, India

Anuska Sen Amity Institute of Biotechnology, Amity University Kolkata, Kolkata, India

Supatra Sen Department of Botany, Asutosh College, Kolkata, West Bengal, India

Dwaipayan Sinha Department of Botany, Government General Degree College, Mohanpur, Paschim Medinipur, West Bengal, India

Gyan Prakash Srivastava Department of Botany, University of Allahabad, Prayagraj, Uttar Pradesh, India

Shobhit Kumar Srivastava Department of Botany, Asian International University, Imphal, West Manipur, India

Pramod Kumar Tandon Department of Botany, University of Lucknow, Lucknow, Uttar Pradesh, India

Kaiser Iqbal Wani Department of Botany, Aligarh Muslim University, Aligarh, India

Rand Alhaj Yahya Department of Applied Biology, College of Sciences, University of Sharjah, Sharjah, United Arab Emirates

Andleeb Zehra Department of Botany, Aligarh Muslim University, Aligarh, India

Plant Metal and Metalloid Transporters 1

Arzoo Ahad, Namrah Ahmad, Mahnoor Ilyas, Tuba Sharf Batool, and Alvina Gul

Abstract

In the developing countries, approximately 3.1 billion people reside in remote areas, and 2.5 billion of these people rely on farming for subsistence, which provides 30% to productivity expansion due to the GDP produced from agriculture. Plants' physiochemical functions, for instance, pigment (chlorophyll) production, photosynthesis, nucleic acid synthesis, peptide modifications, oxidation-reduction reaction, carbohydrate metabolism, and nitrification, depend greatly on micro- and macronutrients. Metal and metalloid toxic effects are becoming more prevalent around the world, owing mostly to human sources. Soil pollution is one of the most significant variables because it impacts agricultural output allowing the metals and metalloid ions to permeate the food chain and experience bioaccumulation, resulting in repercussions on health and environmental changes. Throughout evolution, plants have evolved various ways to deal with biotic and abiotic stressors. Plants employ various metal and metalloid transporters to maintain intraorganellar homeostasis in order to provide resilience toward their toxicity. These includes NRAMP, CDF, ZIP, ABC, HMAs, NIP, BOR, and Lsi2 transporters. This chapter provides a potential understanding of several putative transporters that are currently believed to be involved in the accumulation and transport of metals and metalloids in plants. Apart from the specific structure, this chapter focuses on the properties of several transporter families, with a specific attention on transportation of metals.

Keywords

Metal · Metalloid · Transporters · Toxicity · Plants

A. Ahad · N. Ahmad · M. Ilyas · T. S. Batool · A. Gul (✉)
Department of Plant Biotechnology, Atta-ur-Rahman School of Applied Biosciences, National University of Sciences and Technology, Islamabad, Pakistan
e-mail: alvina_gul@asab.nust.edu.pk

K. Kumar, S. Srivastava (eds.), *Plant Metal and Metalloid Transporters*,
https://doi.org/10.1007/978-981-19-6103-8_1

1.1 Introduction

Plants require different metal elements for their survival and reproduction, which should be mobilized and absorbed as metal ions from the subsurface into the root system. Metal ions are absorbed directly by the roots and distributed to various areas of the plant through the vascular tissues. All plants require 14 essential mineral elements in addition to H_2O, oxygen (O), and CO_2. Among them, potassium (K), calcium (Ca), sulfur (S), phosphorus (P), magnesium (Mg), and nitrogen (N) are required in high amount and are designated as macronutrients. In contrast, boron (B), molybdenum (Mo), iron (Fe), and copper (Cu) are required in minute amount and are known as microelements (DalCorso et al. 2014). Metal and metalloid ions are naturally present on Earth's crust in different layers. However, increasing anthropic inputs, industrialization, and modern cultivating approaches have escalated metal toxicity. Millions of hectares have been contaminated as a result of chemicals, fertilizers, maniple wastes, and heavy metals that penetrate into the soil through foundries, combustion, metalliferous mining, smelters, and agriculture (Singh et al. 2016; Nagajyoti et al. 2010).

Plants have developed adaptive mechanisms not only to absorb enough amounts of necessary macronutrients and micronutrients but also to prevent toxic accumulations. Excessive levels of metal ions generate ROS and other metal ions within metal proteins leaving them inactive. Moreover, plants also have to deal with harmful metals such as arsenic (As), cadmium (Cd), mercury (Hg), and uranium (U) by activating their homeostatic systems that not only allow metal uptake and distribution within tissues but also avoid accretion and toxicity of nonessential metal elements (DalCorso et al. 2014). Being micronutrients, heavy metals and metalloids are reported to have a prominent function in plant development because they contribute to several metabolic reactions (Fig. 1.1) (Rahman and Singh 2019; Angulo-Bejarano et al. 2021). When their concentration exceeds a certain threshold level, they are considered as toxic to plants, for example, aluminum toxicity in acidic land. The deleterious effect of a metallic ion on plants is based on its ability to interact with normally encountered ions, which are main cofactors for fundamental enzymes in primary and secondary metabolism. When the metal ion interacts with sulfhydryl group, it creates imbalance in protein function and upregulates the oxidative state of plants. Furthermore, they also replace the significant elements existing in the cell wall; for example, Zn, Al, Pb, and Cu bind quickly with cell wall pectins than calcium (Ca) (Angulo-Bejarano et al. 2021).

The major aspect to characterize heavy metals is their density. Fifty-three naturally occurring elements out of ninety are classified as heavy metals (Rahman and Singh 2019; Fryzova et al. 2018). This chapter highlights the significant role and effects of metals and metalloid on plants. Furthermore, potential plant transporters for metal and metalloid uptake are also discussed.

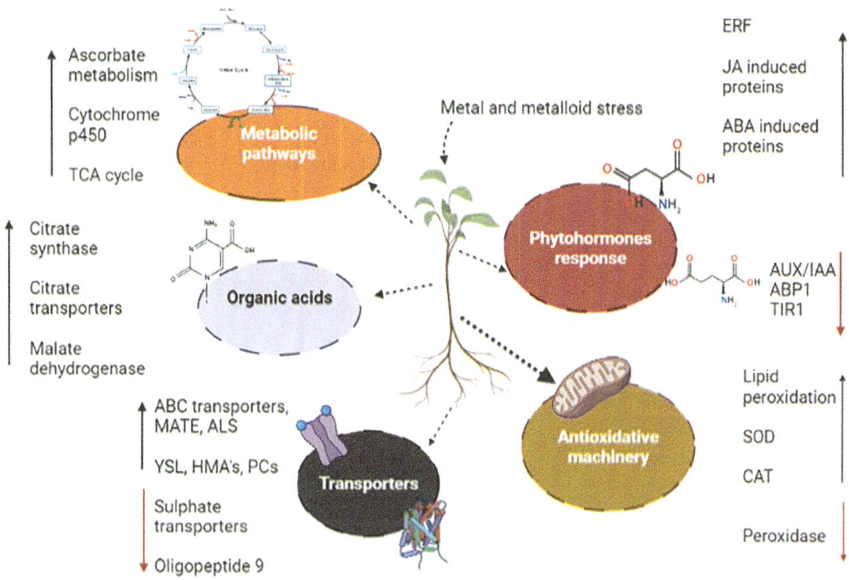

Fig. 1.1 Metals' and metalloids' role in plants

1.2 Metals and Their Significance in Plants

Plant nourishment requires metal cations. Several enzymatic processes require metal cofactors, for instance, manganese (Mn), zinc (Zn), copper (Cu), and other metals which serve crucial structural functions in proteins. Moreover, metal cations have recently been considered to be significant in signaling in both animals and plants (Table 1.1). Plants, on the other hand, must guard against unnecessary buildup of vital cations and harmful metals such as cadmium (Cd^{2+}), arsenic (As), and mercury (Hg) (Thomine et al. 2000a, b). Fe is a vital nutrient that is not easily accessible to plants from topsoil. It is available in the form of hydroxides and insoluble complexes thus limiting its availability to plants. Plants have evolved reductive strategies to mobilize Fe III present in rhizospheric region. Zn is essentially required for RNA polymerase, Cu/Zn superoxide dismutase, alcohol dehydrogenase, and zinc finger domains (Mitra et al. 2014). Mg is an essential metal element that has a physiological function in photosynthesis being central atom of chlorophyll. It is also involved in RUBISCO activation. Therefore its deficiency results in reducing the biomass, initiation of reactive oxygen species, and production of antioxidant enzymes in plants (Hauer-Jákli and Tränkner 2019). Ni is a micronutrient playing a key role in vegetative growth and development. Being an integral component of urease and as glyoxalase-I, it also performs many biological functions in plant's higher metabolism, nitrogen assimilation, and defense against biotic (pathogen, herbivore) as well as abiotic (drought, salinity) stresses (Shahzad et al. 2018).

Table 1.1 Impacts of metals on plants' mechanism and growth

Plant metals	Function in plant	Effects on plants	References
Copper	• Component of enzymes • Role in photosynthetic mechanisms • Cofactor of proteins • Play significant role in growth and development • Plasma membrane permeability	• Disturbance of plant photosynthesis and developmental and reproductive pathways • Reduce the surface area of the thylakoid	Laghlimi et al. (2015), Printz et al. (2016), Yruela (2005)
Zinc	• Involved in determining productivity of crops • Resistance against biotic and abiotic stresses • Nitrogen fixation in legumes • Intracellular second messenger	• Reduces Ni toxicity and seed development • Defense against herbivores and pathogens	Laghlimi et al. (2015), Cabot et al. (2019)
Cadmium		• Reduce lipid content • Inhibit seed germination, root elongation, photosynthesis, and plant growth • Interrupt enzymatic events and leaf gaseous exchange • Hamper symbiosis between microbes and plants	Laghlimi et al. (2015), Haider et al. (2021)
Nickel	• Component of enzymes • Activation of urease • Nodule formation in legumes • Activation of biological processes	• Decrease in protein, chlorophyll, and enzyme production • Accretion of photosynthetic pigments and dry matter	Shahzad et al. (2018), Laghlimi et al. (2015)
Lead		• Reduces chlorophyll contend and causes chlorosis • Blackening of root system • Inhibit enzyme activities • Reduce seed germination and root and shoot growth • Less biomass production	Laghlimi et al. (2015), Nas and Ali (2018)

1.3 Metalloids and Their Significance in Plants

Elements whose features are mediators among metals and nonmetals are referred as metalloids. Some prominent metalloids are tellurium (Te), silicon (Si), arsenic (As), germanium (Ge), antimony (Sb), and boron (B). Usually, soil solutions include sufficient concentrations of boric acid [B(OH)3], silicic acid [$Si(OH)_4$], and trivalent arsenious acid [$As(OH)_3$] or pentavalent As acid [$AsO(OH)_3$] (Lombi and Holm 2010). They have varying impacts on plant growth, depending on whether they are necessary (B), helpful (Si), or poisonous (As). As a result, plants have developed mechanisms to use or distill metalloids based on their functionality (Marschner 2011). For the normal functioning of vascular plants, boron has crucial importance (Uluisik et al. 2018). Its main physiological function is to polymerize two chains of pectin at the cell wall's rhamnogalacturonan II (Funakawa and Miwa 2015). B requirements vary by plant type, with gramineous plants requiring 5–10 mg/kg dry matter and most dicots requiring 20–70 mg/kg dry matter. These changes are linked to the amount of pectin in the cell wall. Whenever B is present predominantly, however, it causes oxidative injury. Additionally, the gap between B insufficiency and oxidative stress is razor thin. Plants have evolved a complex strategy to cope with B in the milieu of external changes by carefully controlling transporters caught up in its systemic absorption (Uluisik et al. 2018). However, due to a shortage of knowledge for its role in cellular metabolism except for diatom and *Equisetum*, silicon is still to be identified as a critical element in plants. Because Si is an extremely prevalent mineral element in soil, it is deposited in all terrestrial flora, albeit the quantity of Si deposited varies substantially between plant species (Marschner 2011).

Si content in terrestrial plant soils varies from 0.1% to 10% by dry weight, whereas leguminous plants, for example, rice (*Oryza sativa*), increase overall Si levels 100–1000-fold stronger than B. Si is absorbed via roots and subsequently accumulated at the surface of plant and primarily in silica cells as amorphous silica (nSiO2 nH2O), functioning as a protective border (Sun et al. 2020). This barrier is critical for safeguarding plants against biotic and abiotic stressors. Because of its chemical inertness and the moderate solubility (2 mM) of $Si(OH)_4$, Si does not normally cause excessive toxicity in plants (Coskun et al. 2019). As is an unimportant and lethal element for plants, albeit some species may hyper accumulate it without toxicity (Sun et al. 2019). $As(OH)_3$ is the prevalent inorganic form of arsenic in anaerobic paddy fields. Monomethylarsonic acid (MMA) and dimethylarsinic acid (DMA), both synthesized by soil bacteria or algae, are found in some soils (DiTusa et al. 2016). Accretion of As in plants is primarily influenced by soil conditions, albeit it is typically not more than 1 mg/kg dry matter in uncontaminated territory, which is substantially less than B and Si accumulation (Sun et al. 2020). Despite the fact that B, Si, and As in soil solution are noncharged molecules at pH 9, due to their identical pKa values, plants utilize both comparable and distinct transport pathways for these metalloids ($B(OH)_3$, 9.24; $Si(OH)_4$, 9.86; $As(OH)_3$, 9.2). Furthermore, diverse regulatory systems for these metalloids have evolved in

Table 1.2 Phytotoxic impacts of metalloids on plants' mechanism and growth

Metalloids	Targeted mechanism	Impact on plants	Plant species	References
Boron	• Disturbs cell cycle, division, elongation, and biosynthesis of cellular components • Affects metabolic processes of cell	• Stunted growth of the plant • Chlorosis and necrosis at tips and margins	• Sweet wormwood (*Artemisia*) • Barley	Aftab et al. (2012)
Arsenic	• Disrupts photosystem • Increased deposition of arsenic in roots	• Inhibition of chlorophyll biosynthesis • Stunted root growth	• Pea • Rice	Kaur et al. (2012)
Antimony	• Imbalance plant nutrient concentrations • Block photochemical efficiency and chlorophyll biosynthesis	• Suppress biomass production (aerial and underground parts) • Alter photosynthetic capacity	• Wheat • Corn	Ma et al. (2019)
Selenium	• Disrupt plant vascular system by restricting water flux	• Decreased water content in leaf	• Common bean	Aggarwal et al. (2011)
Germanium	• Increased concentration in plant root cells	• Stunted growth at the juvenile stage	• Lettuce	Seo et al. (2010)

plants to prevent phytotoxic threats posed by these metalloids on metabolic mechanism of plants as mentioned in Table 1.2 (Yamaji and Ma 2021).

1.4 Metal Transporters

Transition metals such as Zn, Mn, Fe, and Cu are vital micronutrients for typical plant growth and development, but their excessive quantity can be harmful. Thus, for nutritious plant growth, various transition metals must be provided to soil, distributed around the plant aerial parts and roots, and their levels are carefully maintained within different cells and tissues. These include NRAMPs, CDF family, the ZIP family, and HMAs (Hall and Williams 2003). Expression analysis of these metal ions in the plant under consideration indicated elevated tissue-specific expressions of the transcripts being utilized (Migeon et al. 2010).

1.4.1 NRAMP Transporters

NRAMPs are the natural macrophage proteins associated with the plant resistance which are metal ion carriers in animals and plants whereas *NRAMP1* gene acts as metal carrier across the phagocytic cell membrane, while inoperative *NRAMP1* confers sensitivity to numerous pathogenic microorganisms inside the cell (Nevo and Nelson 2006). Normal metal uptake is necessary to maintain cellular homeostasis and cell membrane functioning along with signaling pathways' exogenous stimuli playing a vital role in this regulatory mechanism (Radisky and Kaplan 1999). Plant cells must have some system to resist toxic heavy metal ions, and *Nramp* gene from *A. thaliana* such as *AtNramp3* and *AtNramp4* expressing in leaves, stem, and roots showed homology to the yeast mutants having the ability to uptake iron and cadmium (Thomine et al. 2000a, b). In an experiment conducted on *B. napus* metal transport system, computational analysis and RNA-seq methods were utilized to identify and annotate 22 *NRAMP* genes, and high expression of *BnNRAMP1b* gene showed its function retrieval in zinc, cadmium, and manganese transport (Meng et al. 2017). *Nicotiana tabacum* from Solanaceae family contains 21 *NRAMP* genes involved in phytoremediation, and significant expression pattern of *NtNRAMP3* gene was observed in the excess of Zn metal ions (Papierniak et al. 2018). Higher concentration of CsNRAMPs protein transcripts in aerial parts and roots was found to evaluate their effect on lead treatment. The findings disclosed that *CsNRAMPs* genes were upregulated in a distinct manner, and they might be critical factors for Pb transportation as *CsNRAMP2* and *CsNRAMP5* were residing in plasma membrane as depicted by subcellular localization studies (Li et al. 2021).

1.4.2 CDF Transporters

CDF family encodes transporter proteins which facilitate in cation diffusion of Zn, Cd, and Co which comprise of six transmembrane spanning domains, conserved signature sequence at N-terminus, and positive ion binding domain at C-terminus (Paulsen and Saier Jr 1997). There are two groups of metal tolerance proteins (MTP), that is, groups 1 and 3, present in plants out of four phylogenetically distinct groups in eukaryotes involved in phytoaccumulation (Krämer et al. 2007). It is a renowned fact that CDFs are the fundamental components in photoperiodism and flower formation in *Arabidopsis* by directing expression framework of vital regulators such as CONSTANS promoting blooming of flower and FLOWERING LOCUS T involved in long-day plants' growth (Fornara et al. 2009; Renau-Morata et al. 2020). In an investigation conducted to assess the role of *TuMTP* gene expression pattern in yeast, MTP_1 and $MTP_{1.1}$ of *T. urartu* showed tolerance to Co and Zn ions, but the same was not the case with other metals although $TuMTP_8$, $TuMTP_{11}$, $TuMTP_{8.1}$, and $TuMTP_{11.1}$ exhibited endurance to manganese (Mn). $TuMTP_1$ proteins in *Arabidopsis* were confined to vacuolar membrane and considerably increased Co and Zn metal tolerance (Wang et al. 2021).

1.4.3 ZIP Transporters

Initially, *AtIRT1*ZIP metal transporter was discovered in *A. thaliana*, which is considered as the primary carrier that potentially uptakes Fe from the growing medium (Weber et al. 2004). Regardless of the IRT1 which was first recognized as an Fe metal transporter, IRT1 can also translocate bivalent cations of Cd, Mn, Ni, and Zn as elevated levels of *AtIRT1* accumulate iron under inadequate micronutrient conditions (Shanmugam et al. 2011). It has been observed that Zn deficiency upregulates the expression pattern of *OsZIP9* in the endodermis and exodermis of mature root cells and helps in potential influx of Zn ion improving Zn content in rice (Huang et al. 2020). Functional characterization of ZDREs (Zn deficiency response elements) within the promoter sequence of *TaZIPs* indicates a sustained procedure in response to limited Zn. Due to this, *TabZIPF1-7DL* and *TabZIPF4-7AL* offered a high degree of support to the hypersensitive double mutant (*bzip19 bzip23*) of *Arabidopsis* genes due to Zn insufficiency (Evens et al. 2017). *OsZIP3* was found in xylem transfer cells and parenchymatic cells, and its knockdown downregulated the accumulation of Zn in shoot and upregulated its efflux through transpiration (Sasaki et al. 2015).

1.4.4 ABC Transporters

ATP binding cassette transporters represent a large group of carriers, whose representatives are present in plants, microorganisms, and animals. Fundamentally, they have 1–2 copies of two basic structural characteristics 4–6 transmembrane folds in a highly hydrophobic domain and a cytosolic ATP binding domain (Theodoulou 2000; Thomas et al. 2020). These transporters play leading roles in translocation of phytohormones as the ABCG-type proteins, namely, ABCG14, ABCG25, and ABCG40 are potential ABA shippers while ABCG14 was potentially expressed in cytokinin (CK) production pathway. Pleiotropic drug resistance 1 from a plant *Petunia hybrida* transfers strigolactones (SLs) outside the rhizosphere and from root to shoot as mycorrhizal fungi are attracted toward SLs (Borghi et al. 2015). *Phytolacca acinosa* naturally accumulates high potassium content in shoots and roots as a result of upregulation of 45 ABC transporters belonging to subtypes of ABCB, ABCC, ABCF, and ABCG. This implied that ABC transporters might be essential for potassium influx and shuttling under limited K^+ concentration (Xie et al. 2020).

1.4.5 Heavy Metal ATPases (HMAs)

HP locus is conserved only in plant *HMAs* and absent in two divalent cation transporters of algae, that is, *CrHMA1* and *CmHMA2*, that along with a nearby glutamate help in nucleotide coordination and also in catalytic activity as suggested by an experiment done on *E. coli* Znta (*EcZntA*) mutants (Okkeri et al. 2002;

Hanikenne et al. 2005). Heavy metals such as Cu, Mn, Fe, Ni, Zn, and Co are essential minerals for plant growth and development although an excess of these elements along with nonessential elements like Cd, Ag, Se, Pb, and Hg can cause phytotoxicity which can be the consequence of blockage or displacement of crucial biological molecules or enzymes that regulate the metabolic pathways. *Zea mays* cultivar also hyper accumulates chromium from the industrial wastewater into roots and young leaves when toxicity tests were conducted for tannery leftover water (Calheiros et al. 2008). *Arabidopsis HMA1* localized in green tissues (chloroplast envelope) contributes in the transport of copper ions, and it is analyzed by experiment on the deletion mutants of *HMA1* that absence of N-terminal histidine domain somewhat affects this transport, and these mutants are also characterized to have lower Cu^{+2} ion concentration and complete blockage of superoxide dismutase activity in the chloroplast (Seigneurin-Berny et al. 2006).

HMA6/PAA1 has the same catalytic activity as that of *HMA1* and mediates the import of copper ions in chloroplast envelope which are essential micronutrients and activate cell-damaging free radicals hence playing an important functional role in the diverse pathway of Cu^{+2} homeostasis, and *paa1* mutants recovered their phenotypes when provided with copper supplements (Boutigny et al. 2014). *HMA2* gene of *Triticum aestivum* (*TaHMA2*) when overexpressed in rice plants showed an improved level of Zn and Cd translocation (Tan et al. 2013). *AtHMA5* has been characterized as a transporter of Cu that delivers it in the secretory pathway with the help of ATX1-like metallochaperone, and its mutants after T-DNA insertion are hypersensitive to Cu but not to other metals defining their regulatory role in Cu accumulation in roots, and this regulation is due to binding of chaperons with strictly conserved amino acid residues (Andrés-Colás et al. 2006; Kobayashi et al. 2008). *OsHMA9* is involved in zinc transport as a metal ion efflux ATPase and can be utilized to increase levels of essential micronutrients such as Zn and Fe as a medium for biofortification in wheat and rice by QTL (quantitative trait loci) mapping and high-throughput genotyping, and in this way, candidate genes can be identified and allelic variations can be investigated (Lee et al. 2007; Tong et al. 2020).

1.5 Metalloid Transporters

Plants are affected differently by metalloids such as silicon, boron, and arsenic. A combination of (NOD26-like intrinsic proteins—metabolite exchange protein) NIP channels (Liu et al. 2009) and unique efflux transporters with discrete polar localizations mediates the transference of such metalloids in plants, including their absorption and dispersion. Plants have also evolved a robust network for controlling such transporters at the levels of transcription, translation, and translational modifications in rapidly changing environments. Such patterns of metalloid regulation are peculiar for different species since these mechanisms ensure their specific metalloid requisites to accurately and consistently carry out physiological functions (Yamaji and Ma 2021).

1.5.1 Diversity of Plant Metalloid Transporters

NIPs are specific to leguminous plants (due to the association of legume plants with nitrogen-fixing bacteria in the soil) (Liu and Zhu 2010). Such transporters facilitate absorption and two-way transportation of antimony, arsenic, boron (boric acid), and silicon. While NIPs are specific to legumes, similarly PIPs (plasma membrane intrinsic proteins), SIPs (small basic intrinsic proteins), and TIPs (alpha-tonoplast intrinsic protein) are involved in similar transportation mechanisms of metalloids (Kapilan et al. 2018). Such transporters are mostly found in association with membranes due to the requisites of their biological functions in the plant cell. Selenium uptake in the plant is facilitated by sulfate transporters (SULTRs), while it is also absorbed through aquaporins and phosphate transporters (Wang et al. 2020; Trippe and Pilon-Smits 2021). The primary function of phosphate transporters (PHT1) is the uptake of arsenic, particularly As(V), Glucose transporters (GLUT) are a multifaceted protein involved in the transport (uptake and detoxification) of antimony Sb(III) and arsenic As(III) (Singh et al. 2021). As explained before, BOR is responsible for boron transport and absorption in the form of boric acid. Such different transporter systems are depicted in Fig. 1.2.

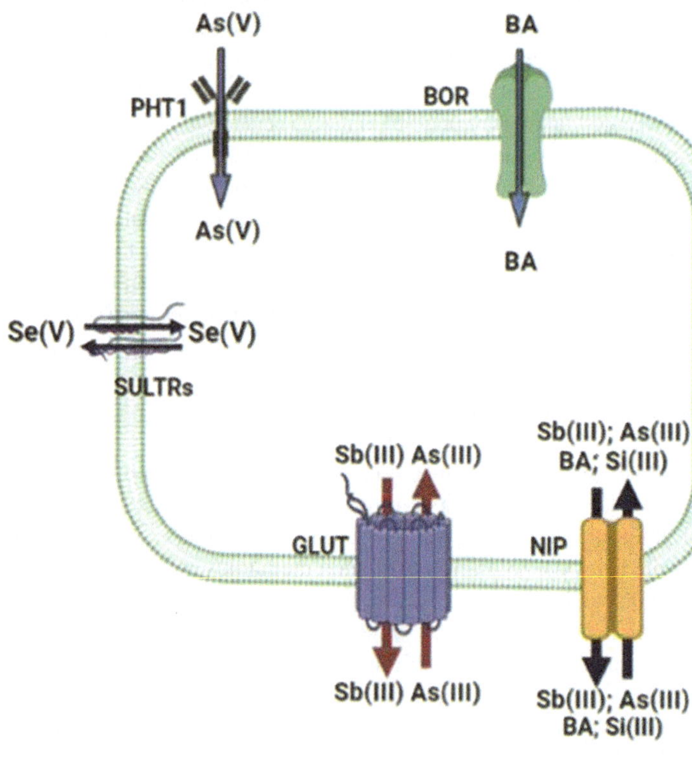

Fig. 1.2 Plant transporters for metalloid transportation and uptake

1.5.2 Metalloid Absorption Channels

Each metalloid's absorption, like that of other minerals, mandates the expression of transporters in the roots. Influx transporters are employed for transferring minerals, through channels specifically in the scenario of metalloids, and through utilizing efflux transporters, these minerals are transported toward the xylem from the root system (Che et al. 2018). Similar channels and mechanisms, that is, circular transport patterns, are used to supply minerals toward root stele from the soil solution source. Metalloid absorption channels are a type of route that allows metalloids to enter into biological systems (Sasaki et al. 2016). NIP from the aquaporin subfamily, that is, channel-type transporters, is employed for the mobility of metalloid from the soil to plant root system. Aquaporin is a membrane-crossing water channel with six helical structures at transmembrane interlinked by two termini and five loops that expose the cytoplasm (Bienert and Bienert 2017). NIP is a subfamily of aquaporin; it is plant-specific and performs the transfer of noncharged low molecular weight compounds such as metalloid oxoacids. With few exceptions, the NIPs are classified into three classes with significant sequence homology, NIPs I–III. Xenopus oocytes having an assay system of heterozygous nature display varying specificities for transport substrates (He et al. 2016).

1.5.3 Metalloid Channel Transporters and Their Specificity

NIP I is impervious to $As(OH)_3$ but not to $Si(OH)_4$. However, $As(OH)_3$, $B(OH)_3$, and $Si(OH)_4$ did not affect the NIP III subgroup. The aromatic/arginine (ar/R), which is a thinnest portion of the channel orifice acting as selectivity filter and creates the diameter limitation obstacle which grants preference to specific surfaces, is assumed to be the cause of the differences in transport materials. The ar/R detector is organized into four scattered amino acid residues, the most prevalent of which are WVAR, AIGR, and GSGR for NIP I, NIP II, and NIP III (Pommerrenig et al. 2020). However, understanding the specific process of determining the specificity of materials would require elucidating structural patterns of NIP protein members (Ma and Yamaji 2015). Ever since the scientific breakthrough about $Si(OH)_4$, being a permeable channel, has happened. It has been reported in genomes of angiosperms along monocot and dicot that had broad range of metalloid depositions. This has been due to (OsLsi1)/OsNIP2;1 and the $B(OH)_3$ permeable channel in rice that assists in reduction of metalloid efflux, that is, Si. While a homologous role was reported for AtNIP5;1 in *Arabidopsis thaliana* that pertain to NIPs III and II, perhaps NIP III in *Arabidopsis* has minimal similarity with any other channel proteins (and most likely Brassicaceae). This is in line with the fact that it has a low Si content (Deshmukh et al. 2020). Proteins of NIP II are mostly engaged in B absorption, whereas proteins of NIP III are prominently featured in Si absorption, according to physiological characterization in plants. Moreover, there seems to be some cross talk among proteins of NIPs II and III (Shao et al. 2018). For instance, the rice transporter OsLsi1 (NIP III) is a key transporter for B absorption, whereas the barley transporter

HvLsi1/HvNIP2;1 is engaged in B toxicity. Unlike NIPs II and III, the particular biological functioning of the majority of NIP I proteins is unclear (Bienert and Bienert 2017).

1.6 Metalloid Transporter Types

1.6.1 Aquaporin Transporters

1.6.1.1 NIP Transporters

Aquaporins in plants have a subfamily characterized as X-intrinsic proteins, which are lost during evolution in monocots and dicots including *Arabidopsis* as well. Thus, it can be concluded that it could be engaged in B transport in conjunction with NIP (Bienert et al. 2019). Although its physiological role in tobacco is unknown, NtXIP1;1 expression is mediated by the AtNIP5;1 promoter which will enhance *Arabidopsis* nip5;1 mutant's low-B sensitivity. Proteins of all NIP subgroups contribute to arsenite absorption in distinct ways. Following a lateral gene transfer incident, current findings postulated that plant NIPs developed from an unidentified bacterial AqpN-aquaporin group for As detoxification (Grégoire et al. 2012). Diverse transporters maintain the absorption of various As species, for example, arsenate, MMA, and DMA. Inorganic arsenate transporters absorb arsenate due to analogy in their chemical characteristic features. Although parts of such processes are indeed assisted by OsLsi1 in rice, the absorption streamlines for MMA and DMA in most plants are yet unknown (Li et al. 2016).

1.6.2 Metalloid Efflux Transporters in Plants

1.6.2.1 BOR Transporters

Metalloid uptake efflux transporters infiltrate metalloid into root cells being regulated by NIP proteins. The outflow of these metalloids from root cells, on the other hand, is governed by a variety of transporters that bear little resemblance to one another (Yoshinari and Takano 2017). High B requiring (BOR) family transporter regulates efflux of B; these transporters are common to plants and yeast while these have identical characteristics to the anion exchanger carrier protein in humans. Crystallographic protein structure of BOR1 in *A. thaliana*, namely AtBOR1, validates its morphological and functional nature for diverse roles, that is, "elevator transport mechanism" (Thurtle-Schmidt and Stroud 2016). Negative membrane potential in plant cells is the driving force for functioning of uniporter based on anion borate; such predictions are reported in limited plant species only (Nagarajan et al. 2016). Function of transporters for boron absorption has been demonstrated in *Arabidopsis*, rice, maize (*Zea mays*), and rapeseed as AtBOR1, OsBOR1, ZmRTE, and BnaC4.BOR1 (Yoshinari and Takano 2017).

1.6.2.2 Lsi2 Transporters

Lsi2-like transporters, which are unrelated to BOR, are responsible for Si efflux from root cells. A peculiar transporter family abundantly prevalent on land plants constitutes homologs of Lsi2. It has five homologs in *O. sativa* and 1 in *Arabidopsis*. Rice OsLsi2 serves as an antiporter between proton inflow and $Si(OH)_4$ outflow (Zhang et al. 2017). In pumpkin, horsetail, cucumber, maize, and barley Lsi2 functions as an efflux transporter of $Si(OH)_4$. Plant Lsi2 homologs share sequence similarities with arsenical resistance operon B (ArsB), a bacterial tolerance gene that functions as an arsenite efflux transporter. Rice OsLsi2 is also reportedly engaged in arsenite mobility, through efflux mechanism that has caused a large decrease in As intake (Tang and Zhao 2020).

1.7 Directional Transport Systems for Metalloid Uptake

1.7.1 Polar Localization of Metalloid Transporters in Plants

Current studies suggest that crucial elements involved in composing mineral concentrations are being absorbed by roots; thus, it needs the assistance of both inflow and out-flow transporters with polar localization (Nagarajan et al. 2016). Same scenario is validated for metalloids as well. Because NIP channels have bidirectional permeability, directed transport across concentration gradients necessitates coordination with active efflux transporters that are polarly localized (Che et al. 2018). The partnership of AtNIP5;1 and AtBOR1 in *Arabidopsis* facilitates boron uptake. The epidermis and endodermis of root have AtNIP5;1 at the plasma membrane, which is polarized on the distal side (soil side), whereas AtBOR1 is polarized on the proximal side (stele side), in the mature root endodermis, and numerous additional cells near the root tip. These transporters' polar localization results in an effective directed flow of B from the soil to the stele (Shao et al. 2018). Because the apoplastic route is obstructed by Casparian strips, which is necessary for endo- and exodermal layer of cells, that is, *Oryza sativa.*

OsLsi1 and OsLsi2, which are positioned on the distal and proximal polar ends of the endodermis and exodermis tissues, respectively, benefit rice roots (Sun et al. 2020). When OsLsi1 and OsLsi2 are knocked out, Si absorption is dramatically decreased (Sasaki et al. 2016). Radial Si transfer through different channels requires assistance among both transporters from each cell (Yamaji and Ma 2021). Structural orientation of these transporters at the polar ends on the both cellular types favors their functionality, hence proved through mathematical modeling as well (Ma and Yamaji 2015). Several plant species have been shown to have specific tissue locations and polarity of Si transporters. In barley and maize, for example, the Lsi1-type Si(OH)4 channel may be found in majority of tissue layers, that is, cortex and endodermis, with distal side polarity (Sun et al. 2020). The Lsi2-type Si(OH)4 efflux transporter is located solely inside the endoplasmic reticulum of root tissues. Moreover, the Lsi1-Lsi2 pair is essential for significant Si concentration buildup (Mitani-Ueno and Ma 2021).

Tomato roots, for example, have a working Lsi1 but not a working Lsi2, resulting in a low level of Si absorption (Sun et al. 2020). The Si transporter polarity, level of gene expressions, and localization are transporters in the cell of all such facets which are associated with Si accumulation (Mitani-Ueno and Ma 2021). Lsi1 and Lsi2 in rice facilitate absorption of arsenite but not arsenate. Rice mutants carrying the lsi1 and lsi2 genes absorb less arsenite. Besides, an ardent effect of Si on arsenite absorption was observed in wild-type (WT) mice (Tang and Zhao 2020). This shows that the OsLsi1 and OsLsi2 proteins collaborate to produce a major Si and As absorption pathway. This might account for why rice accumulates more Si than other plant species (Lindsay and Maathuis 2017).

1.8 Distribution of Metalloids by Transporters

1.8.1 Transporters for B Distribution

Even though no efflux transporters seem to be participating in B dispersion in *Arabidopsis*, BnaC4.BOR1;1c, a near relative of AtBOR1, participates in B dispersion in rapeseed shoots. OsNIP3;1 and OsBOR1 assist rice B dispersion. OsNIP3;1 is orientated in the vascular tissue of vasculature-oriented expanded vascular bundles (VB; EVBs), that is, xylems, in rice nodes, whereas OsBOR1 is oriented in the bundle sheath of EVBs (Shao et al. 2018). Throughout the reproductive development stage, the mutants' dissemination of boron to the panicle was reduced, while its transference to the main leaf was augmented (Shao et al. 2021; Leonard et al. 2014). Both mutants displayed the same phenotype under B restriction: substantial development defects in vegetative tissues, notably procreative structures (tassel and ear). Even though particular mobility mechanisms have yet to be investigated, TLS1 and RTE are found to be associated with boron dissemination in corn, similar to rice. The HvLsi6 $Si(OH)_4$ channel in barley, which is a comparable homolog of the Lsi1 that facilitates $Si(OH)_4$ channeling, was recently discovered as significant QTL genes connected to boron concentration in grain genetic variation (Jia et al. 2021). Rice requires both OsNIP3;1 and OsBOR1 for localized boron distribution. In the juvenile leaf stage, OsNIP3;1 is orientated at the xylem parenchyma cells exposing the xylem vessel, whereas OsBOR1 is orientated at the mestome sheath and the distal side of the xylem parenchyma cells (Shao et al. 2018). Because of certain polarity difference between OsNIP3;1 and OsBOR1, B flow to immature leaf tissues from the xylem may be guided, allowing for increased B needs for leaf development. OsBOR1 in tissue localizations for sheath cells of elderly leaf at the proximal side of cells facilitates the prevention of excessive boron from entering into growing tissues (Uluisik et al. 2018).

1.8.2 Transporters for Si Distribution

SI distribution, uptake, and absorption have been facilitated by mainly NIP III members and Lsi2 homologs. Silicon transport and efflux toward leaf from the xylem were assisted by Lsi2 and NIP2 transporters in rice and barley. Polarization of these transporters occurs at parenchyma cells of the xylem while their terminal sides oppose the vessels (similar to OsNIP3;1 B channel) (Shao et al. 2018). Silicon efflux transporters working with OsLsi6 or HvLsi6 have not been reported yet. Because substantial Si concentration in the husk is essential for grain fertility in rice, Si is preferentially transported to the panicles during the reproductive stage. Three transporters OsLsi6, OsLsi2, and OsLsi3 (a homolog of Lsi2) are excessively translated in nodal cells of rice and mediate such biased dissemination of silicon to the panicle (Di Giorgio et al. 2016). OsLsi6 is polarized in the xylem parenchyma cells, particularly in EVBs' XTCs, which face the xylem vessels. OsLsi6 is not translated in the phloem of DVBs, in contrast to OsNIP3;1 for B distribution (Mitani-Ueno and Ma 2021). These transporters provide a directed conduit for Si transfer from the EVBs' xylem to DVBs' xylem between vascular tissue systems, with Si eventually reaching the panicles (Ma and Yamaji 2015). Knocking out OsLsi6, OsLsi2, or OsLsi3 causes a lower distribution of Si in the panicles but a higher division of silicon in the flag leaf. Mathematical modeling confirms their contribution to the distribution of Si in the panicles. Similar to rice, in barley nodes, HvLsi6 and HvLsi2 were localized. In dicots, on the other hand, the distribution mechanism of Si is yet unclear (Jia et al. 2021).

1.8.3 Transporters for As Distribution

In rice node tissues, arsenic dissemination is reportedly mediated by two transporters (OsLsi2 and OsABCC1). OsLsi2 is an $Si(OH)_4$/arsenite transporter involved in efflux; it has been reported previously (Chen et al. 2015). Arsenite is the primary arsenic form in the xylem, but unlike Si, the majority of arsenite is maintained in node vacuoles (Moore et al. 2014). OsABCC1 is abundant in the phloem of nodal vascular tissues, which aids in As vacuolar sequestration (Tang and Zhao 2020). Nodes in rice plants have inhibition mechanisms for As translocation to the grain, as evidenced by the 13-fold increase in As accumulation in the grains when this gene was knocked out. Two transporters in *Arabidopsis*, inositol transporters 2 and 4 (AtINT2 and AtINT4), facilitate arsenite uptake by phloem (Duan et al. 2015). When this loading was disrupted, concentrations of arsenic have reportedly reduced in seed and phloem exudates, whereas homologs of these transporters with functional homology, i.e., assisting phloem loading with arsenic, have not been discovered in other plant species yet (Li et al. 2016).

1.9 Conclusions and Future Perspectives

Plants having efficient root system absorb metal and metalloid ions from the soil and transport them to aerial parts. During development, plants utilize Zn, Cd, Hg, Cu, As, B, Si, and As in a variety of ways as required. Experiments have been conducted all over the globe on phytotoxic effects of these metal ions in numerous plants. *NRAMPs*, *CDF*, *ZIP*, *ABC*, and *HMAs* families are the metal ion shuttles that accumulate ions into root cells and aerial parts of plants. Plants also use specialist transporters, for example, BOR1 and Lsi2, to efflux metalloids from cellular compartments. They have evolved a complicated mechanism for mobilization of metalloids. Such systems involve diverse types of transporters based on their functionality, that is, regulatory transporters and influx and efflux carriers. *Arabidopsis* and rice are expected to utilize metalloids in distinct ways. Despite having a higher B requirement than other plants, *Arabidopsis* (and presumably Brassicaceae) lacks the NIP III $Si(OH)_4$ channel. On the other hand, Poaceae, specifically rice, may gather a lot of Si constituting approximately 10% of its dry weight, whereas it requires significant amounts of B as well. As a result, *Arabidopsis* has evolved a system for carefully adjusting B absorption and distribution that are independent of Si transporters. Rice along with other legume-based crops possess a unique Si distribution mechanism in the node, which is defined by Si transfer through intervascular tissue layers, that is, xylem to xylem. The intravascular transfer, on the other hand, is impossible to conduct this technique in dicots. Recombinant technology can be utilized to overcome this barrier.

References

Aftab T, Khan MMA, Naeem M, Idrees M, Moinuddin T, da Silva JA, Ram M (2012) Exogenous nitric oxide donor protects Artemisia annua from oxidative stress generated by boron and aluminium toxicity. Ecotoxicol Environ Saf 80:60–68. https://doi.org/10.1016/j.ecoenv.2012.02.007

Aggarwal M, Sharma S, Kaur N, Pathania D, Bhandhari K, Kaushal N, Kaur R, Singh K, Srivastava A, Nayyar H (2011) Exogenous proline application reduces phytotoxic effects of selenium by minimising oxidative stress and improves growth in bean (Phaseolus vulgaris L.) seedlings. Biol Trace Elem Res 140(3):354–367. https://doi.org/10.1007/s12011-010-8699-9

Andrés-Colás N, Sancenón V, Rodríguez-Navarro S, Mayo S, Thiele DJ, Ecker JR, Puig S, Peñarrubia L (2006) The Arabidopsis heavy metal P-type ATPase HMA5 interacts with metallochaperones and functions in copper detoxification of roots. Plant J 45(2):225–236

Angulo-Bejarano PI, Puente-Rivera J, Cruz-Ortega R (2021) Metal and metalloid toxicity in plants: an overview on molecular aspects. Plan Theory 10(4):635. https://doi.org/10.3390/plants10040635

Bienert MD, Bienert GP (2017) Plant aquaporins and metalloids. In: Plant aquaporins. Springer, New York, pp 297–332

Bienert MD, Muries B, Crappe D, Chaumont F, Bienert GP (2019) Overexpression of X Intrinsic Protein 1; 1 in Nicotiana tabacum and Arabidopsis reduces boron allocation to shoot sink tissues. Plant Direct 3(6):e00143

Borghi L, Kang J, Ko D, Lee Y, Martinoia E (2015) The role of ABCG-type ABC transporters in phytohormone transport. Biochem Soc Trans 43(5):924–930. https://doi.org/10.1042/BST20150106

Boutigny S, Sautron E, Finazzi G, Rivasseau C, Frelet-Barrand A, Pilon M, Rolland N, Seigneurin-Berny D (2014) HMA1 and PAA1, two chloroplast-envelope PIB-ATPases, play distinct roles in chloroplast copper homeostasis. J Exp Bot 65(6):1529–1540

Cabot C, Martos S, Llugany M, Gallego B, Tolrà R, Poschenrieder C (2019) A role for zinc in plant defense against pathogens and herbivores. Front Plant Sci 10:1171. https://doi.org/10.3389/fpls.2019.01171

Calheiros CS, Rangel AO, Castro PM (2008) The effects of tannery wastewater on the development of different plant species and chromium accumulation in Phragmites australis. Arch Environ Contam Toxicol 55(3):404–414

Che J, Yamaji N, Ma JF (2018) Efficient and flexible uptake system for mineral elements in plants. New Phytol 219(2):513–517

Chen Y, Moore KL, Miller AJ, McGrath SP, Ma JF, Zhao F-J (2015) The role of nodes in arsenic storage and distribution in rice. J Exp Bot 66(13):3717–3724

Coskun D, Deshmukh R, Sonah H, Menzies JG, Reynolds O, Ma JF, Kronzucker HJ, Bélanger RR (2019) The controversies of silicon's role in plant biology. New Phytol 221(1):67–85

DalCorso G, Manara A, Piasentin S, Furini A (2014) Nutrient metal elements in plants. Metallomics 6(10):1770–1788. https://doi.org/10.1039/c4mt00173g

Deshmukh R, Sonah H, Belanger RR (2020) New evidence defining the evolutionary path of aquaporins regulating silicon uptake in land plants. J Exp Bot 71(21):6775–6788

Di Giorgio JAP, Bienert GP, Ayub ND, Yaneff A, Barberini ML, Mecchia MA, Amodeo G, Soto GC, Muschietti JP (2016) Pollen-specific aquaporins NIP4; 1 and NIP4; 2 are required for pollen development and pollination in Arabidopsis thaliana. Plant Cell 28(5):1053–1077

DiTusa SF, Fontenot EB, Wallace RW, Silvers MA, Steele TN, Elnagar AH, Dearman KM, Smith AP (2016) A member of the phosphate transporter 1 (Pht1) family from the arsenic-hyperaccumulating fern Pteris vittata is a high-affinity arsenate transporter. New Phytol 209(2):762–772

Duan G-L, Hu Y, Schneider S, McDermott J, Chen J, Sauer N, Rosen BP, Daus B, Liu Z, Zhu Y-G (2015) Inositol transporters AtINT2 and AtINT4 regulate arsenic accumulation in Arabidopsis seeds. Nat Plants 2(1):1–6

Evens NP, Buchner P, Williams LE, Hawkesford MJ (2017) The role of ZIP transporters and group F bZIP transcription factors in the Zn-deficiency response of wheat (Triticum aestivum). Plant J 92(2):291–304. https://doi.org/10.1111/tpj.13655

Fornara F, Panigrahi KC, Gissot L, Sauerbrunn N, Rühl M, Jarillo JA, Coupland G (2009) Arabidopsis DOF transcription factors act redundantly to reduce CONSTANS expression and are essential for a photoperiodic flowering response. Dev Cell 17(1):75–86. https://doi.org/10.1016/j.devcel.2009.06.015

Fryzova R, Pohanka M, Martinkova P, Cihlarova H, Brtnicky M, Hladky J, Kynicky J (2018) Oxidative stress and heavy metals in plants. Rev Environ Contam Toxicol 245:129–156. https://doi.org/10.1007/398_2017_7

Funakawa H, Miwa K (2015) Synthesis of borate cross-linked rhamnogalacturonan II. Front Plant Sci 6:223

Grégoire C, Rémus-Borel W, Vivancos J, Labbé C, Belzile F, Bélanger RR (2012) Discovery of a multigene family of aquaporin silicon transporters in the primitive plant Equisetum arvense. Plant J 72(2):320–330

Haider FU, Liqun C, Coulter JA, Cheema SA, Wu J, Zhang R, Wenjun M, Farooq M (2021) Cadmium toxicity in plants: impacts and remediation strategies. Ecotoxicol Environ Saf 211:111887. https://doi.org/10.1016/j.ecoenv.2020.111887

Hall JL, Williams LE (2003) Transition metal transporters in plants. J Exp Bot 54(393):2601–2613. https://doi.org/10.1093/jxb/erg303

Hanikenne M, Krämer U, Demoulin V, Baurain D (2005) A comparative inventory of metal transporters in the green alga Chlamydomonas reinhardtii and the red alga Cyanidioschyzon merolae. Plant Physiol 137(2):428–446

Hauer-Jákli M, Tränkner M (2019) Critical leaf magnesium thresholds and the impact of magnesium on plant growth and photo-oxidative defense: a systematic review and meta-analysis from 70 years of research. Front Plant Sci 10:766. https://doi.org/10.3389/fpls.2019.00766

He Z, Yan H, Chen Y, Shen H, Xu W, Zhang H, Shi L, Zhu YG, Ma M (2016) An aquaporin Pv TIP 4; 1 from Pteris vittata may mediate arsenite uptake. New Phytol 209(2):746–761

Huang S, Sasaki A, Yamaji N, Okada H, Mitani-Ueno N, Ma JF (2020) The ZIP transporter family member OsZIP9 contributes to root zinc uptake in rice under zinc-limited conditions. Plant Physiol 183(3):1224–1234. https://doi.org/10.1104/pp.20.00125

Jia Z, Bienert MD, von Wirén N, Bienert GP (2021) Genome-wide association mapping identifies HvNIP2; 2/HvLsi6 accounting for efficient boron transport in barley. Physiol Plant 171(4): 809–822

Kapilan R, Vaziri M, Zwiazek JJ (2018) Regulation of aquaporins in plants under stress. Biol Res 51(1):4. https://doi.org/10.1186/s40659-018-0152-0

Kaur S, Singh HP, Batish DR, Negi A, Mahajan P, Rana S, Kohli RK (2012) Arsenic (As) inhibits radicle emergence and elongation in Phaseolus aureus by altering starch-metabolizing enzymes vis-à-vis disruption of oxidative metabolism. Biol Trace Elem Res 146(3):360–368

Kobayashi Y, Kuroda K, Kimura K, Southron-Francis JL, Furuzawa A, Kimura K, Iuchi S, Kobayashi M, Taylor GJ, Koyama H (2008) Amino acid polymorphisms in strictly conserved domains of a P-type ATPase HMA5 are involved in the mechanism of copper tolerance variation in Arabidopsis. Plant Physiol 148(2):969–980

Krämer U, Talke IN, Hanikenne M (2007) Transition metal transport. FEBS Lett 581(12): 2263–2272. https://doi.org/10.1016/j.febslet.2007.04.010

Laghlimi M, Baghdad B, El Hadi H, Bouabdli A (2015) Phytoremediation mechanisms of heavy metal contaminated soils: a review. Open J Ecol 5:375–388. https://doi.org/10.4236/oje.2015.58031

Lee S, Kim Y-Y, Lee Y, An G (2007) Rice P1B-type heavy-metal ATPase, *OsHMA9*, is a metal efflux protein. Plant Physiol 145(3):831–842

Leonard A, Holloway B, Guo M, Rupe M, Yu G, Beatty M, Zastrow-Hayes G, Meeley R, Llaca V, Butler K (2014) tassel-less1 encodes a boron channel protein required for inflorescence development in maize. Plant Cell Physiol 55(6):1044–1054

Li N, Wang J, Song W-Y (2016) Arsenic uptake and translocation in plants. Plant Cell Physiol 57(1):4–13

Li J, Duan Y, Han Z, Shang X, Zhang K, Zou Z, Ma Y, Li F, Fang W, Zhu X (2021) Genome-wide identification and expression analysis of the NRAMP family genes in tea plant (Camellia sinensis). Plan Theory 10(6):1055. https://doi.org/10.3390/plants10061055

Lindsay ER, Maathuis FJ (2017) New molecular mechanisms to reduce arsenic in crops. Trends Plant Sci 22(12):1016–1026

Liu Q, Zhu Z (2010) Functional divergence of the NIP III subgroup proteins involved altered selective constraints and positive selection. BMC Plant Biol 10(1):256. https://doi.org/10.1186/1471-2229-10-256

Liu Q, Wang H, Zhang Z, Wu J, Feng Y, Zhu Z (2009) Divergence in function and expression of the NOD26-like intrinsic proteins in plants. BMC Genomics 10(1):313. https://doi.org/10.1186/1471-2164-10-313

Lombi E, Holm PE (2010) Metalloids, soil chemistry and the environment. In: MIPs their role in the exchange of metalloids. Springer, New York, pp 33–44

Ma JF, Yamaji N (2015) A cooperative system of silicon transport in plants. Trends Plant Sci 20(7): 435–442

Ma C, He M, Zhong Q, Ouyang W, Lin C, Liu X (2019) Uptake, translocation and phytotoxicity of antimonite in wheat (*Triticum aestivum*). Sci Total Environ 669:421–430. https://doi.org/10.1016/j.scitotenv.2019.03.145

Marschner H (2011) Marschner's mineral nutrition of higher plants. Academic Press, New York

Meng JG, Zhang XD, Tan SK, Zhao KX, Yang ZM (2017) Genome-wide identification of Cd-responsive NRAMP transporter genes and analyzing expression of NRAMP 1 mediated by miR167 in Brassica napus. Biometals 30(6):917–931. https://doi.org/10.1007/s10534-017-0057-3

Migeon A, Blaudez D, Wilkins O, Montanini B, Campbell MM, Richaud P, Thomine S, Chalot M (2010) Genome-wide analysis of plant metal transporters, with an emphasis on poplar. Cell Mol Life Sci 67(22):3763–3784. https://doi.org/10.1007/s00018-010-0445-0

Mitani-Ueno N, Ma JF (2021) Linking transport system of silicon with its accumulation in different plant species. Soil Sci Plant Nutr 67(1):10–17

Mitra A, Chatterjee S, Datta S, Sharma S, Veer V, Razafindrabe BHM, Walther C, Gupta DK (2014) Mechanism of metal transporters in plants. In: Gupta DK, Chatterjee S (eds) Heavy metal remediation. Nova Science Publishers, New York

Moore KL, Chen Y, van de Meene AM, Hughes L, Liu W, Geraki T, Mosselmans F, McGrath SP, Grovenor C, Zhao FJ (2014) Combined NanoSIMS and synchrotron X-ray fluorescence reveal distinct cellular and subcellular distribution patterns of trace elements in rice tissues. New Phytol 201(1):104–115

Nagajyoti PC, Lee KD, Sreekanth TVM (2010) Heavy metals, occurrence and toxicity for plants: a review. Environ Chem Lett 8(3):199–216. https://doi.org/10.1007/s10311-010-0297-8

Nagarajan Y, Rongala J, Luang S, Singh A, Shadiac N, Hayes J, Sutton T, Gilliham M, Tyerman SD, McPhee G (2016) A barley efflux transporter operates in a Na+-dependent manner, as revealed by a multidisciplinary platform. Plant Cell 28(1):202–218

Nas FS, Ali M (2018) The effect of lead on plants in terms of growing and biochemical parameters: a review. Eco Environ Sci 3(4):265–268. https://doi.org/10.15406/mojes.2018.03.00098

Nevo Y, Nelson N (2006) The NRAMP family of metal-ion transporters. Biochim Biophys Acta 1763(7):609–620. https://doi.org/10.1016/j.bbamcr.2006.05.007

Okkeri J, Bencomo E, Pietilä M, Haltia T (2002) Introducing Wilson disease mutations into the zinc-transporting P-type ATPase of Escherichia coli: the mutation P634L in the 'hinge'motif (GDGXNDXP) perturbs the formation of the E2P state. Eur J Biochem 269(5):1579–1586

Papierniak A, Kozak K, Kendziorek M, Barabasz A, Palusińska M, Tiuryn J, Paterczyk B, Williams LE, Antosiewicz DM (2018) Contribution of NtZIP1-Like to the regulation of Zn homeostasis. Front Plant Sci 9:185. https://doi.org/10.3389/fpls.2018.00185

Paulsen I, Saier M Jr (1997) A novel family of ubiquitous heavy metal ion transport proteins. J Membr Biol 156(2):99–103. https://doi.org/10.1007/s002329900192

Pommerrenig B, Diehn TA, Bernhardt N, Bienert MD, Mitani-Ueno N, Fuge J, Bieber A, Spitzer C, Bräutigam A, Ma JF (2020) Functional evolution of nodulin 26-like intrinsic proteins: from bacterial arsenic detoxification to plant nutrient transport. New Phytol 225(3):1383–1396

Printz B, Lutts S, Hausman J-F, Sergeant K (2016) Copper trafficking in plants and its implication on cell wall dynamics. Front Plant Sci 7:601. https://doi.org/10.3389/fpls.2016.00601

Radisky D, Kaplan J (1999) Regulation of transition metal transport across the yeast plasma membrane. J Biol Chem 274(8):4481–4484. https://doi.org/10.1074/jbc.274.8.4481

Rahman Z, Singh VP (2019) The relative impact of toxic heavy metals (THMs) (arsenic (As), cadmium (Cd), chromium (Cr)(VI), mercury (Hg), and lead (Pb)) on the total environment: an overview. Environ Monit Assess 191(7):419. https://doi.org/10.1007/s10661-019-7528-7

Renau-Morata B, Carrillo L, Dominguez-Figueroa J, Vicente-Carbajosa J, Molina RV, Nebauer SG, Medina J (2020) CDF transcription factors: plant regulators to deal with extreme environmental conditions. J Exp Bot 71(13):3803–3815. https://doi.org/10.1093/jxb/eraa088

Sasaki A, Yamaji N, Mitani-Ueno N, Kashino M, Ma JF (2015) A node-localized transporter Os ZIP 3 is responsible for the preferential distribution of Zn to developing tissues in rice. Plant J 84(2):374–384. https://doi.org/10.1111/tpj.13005

Sasaki A, Yamaji N, Ma JF (2016) Transporters involved in mineral nutrient uptake in rice. J Exp Bot 67(12):3645–3653

Seigneurin-Berny D, Gravot A, Auroy P, Mazard C, Kraut A, Finazzi G, Grunwald D, Rappaport F, Vavasseur A, Joyard J (2006) HMA1, a new Cu-atpase of the chloroplast envelope, is essential for growth under adverse light conditions. J Biol Chem 281(5):2882–2892

Seo D-C, Cheon Y-S, Park S-K, Park J-H, Kim A-R, Lee W-G, Lee S-T, Lee Y-H, Cho J-S, Heo J-S (2010) Applications of different types of germanium compounds on rice plant growth and its Ge uptake. Korean J Soil Sci Fertil 43(2):166–173

Shahzad B, Tanveer M, Rehman A, Cheema SA, Fahad S, Rehman S, Sharma A (2018) Nickel; whether toxic or essential for plants and environment - a review. Plant Physiol Biochem 132: 641–651. https://doi.org/10.1016/j.plaphy.2018.10.014

Shanmugam V, Lo JC, Wu CL, Wang SL, Lai CC, Connolly EL, Huang JL, Yeh KC (2011) Differential expression and regulation of iron-regulated metal transporters in Arabidopsis halleri and Arabidopsis thaliana–the role in zinc tolerance. New Phytol 190(1):125–137. https://doi.org/10.1111/j.1469-8137.2010.03606.x

Shao JF, Yamaji N, Liu XW, Yokosho K, Shen RF, Ma JF (2018) Preferential distribution of boron to developing tissues is mediated by the intrinsic protein OsNIP3. Plant Physiol 176(2): 1739–1750

Shao JF, Yamaji N, Huang S, Ma JF (2021) Fine regulation system for distribution of boron to different tissues in rice. New Phytol 230(2):656–668

Singh S, Parihar P, Singh R, Singh VP, Prasad SM (2016) Heavy metal tolerance in plants: role of transcriptomics, proteomics, metabolomics, and ionomics. Front Plant Sci 6:1143. https://doi.org/10.3389/fpls.2015.01143

Singh S, Kumar V, Datta S, Dhanjal DS, Singh S, Kumar S, Kapoor D, Prasad R, Singh J (2021) Physiological responses, tolerance, and remediation strategies in plants exposed to metalloids. Environ Sci Pollut Res 28(30):40233–40248. https://doi.org/10.1007/s11356-020-10293-2

Sun D, Feng H, Li X, Ai H, Sun S, Chen Y, Xu G, Rathinasabapathi B, Cao Y, Ma LQ (2019) Expression of new Pteris vittata phosphate transporter PvPht1; 4 reduces arsenic translocation from the roots to shoots in tobacco plants. Environ Sci Technol 54(2):1045–1053

Sun H, Duan Y, Mitani-Ueno N, Che J, Jia J, Liu J, Guo J, Ma JF, Gong H (2020) Tomato roots have a functional silicon influx transporter but not a functional silicon efflux transporter. Plant Cell Environ 43(3):732–744

Tan J, Wang J, Chai T, Zhang Y, Feng S, Li Y, Zhao H, Liu H, Chai X (2013) Functional analyses of T a HMA 2, a P 1B-type ATP ase in wheat. Plant Biotechnol J 11(4):420–431

Tang Z, Zhao F-J (2020) The roles of membrane transporters in arsenic uptake, translocation and detoxification in plants. Crit Rev Environ Sci Technol 2020:1–36

Theodoulou FL (2000) Plant ABC transporters. Biochim Biophys Acta 1465(1-2):79–103. https://doi.org/10.1016/S0005-2736(00)00132-2

Thomas C, Aller SG, Beis K, Carpenter EP, Chang G, Chen L, Dassa E, Dean M, Duong Van Hoa F, Ekiert D (2020) Structural and functional diversity calls for a new classification of ABC transporters. FEBS Lett 594(23):3767–3775. https://doi.org/10.1002/1873-3468.13935

Thomine S, Wang R, Ward JM, Crawford NM, Schroeder JI (2000a) Cadmium and iron transport by members of a plant metal transporter family in Arabidopsis with homology to Nramp genes. Proc Natl Acad Sci U S A 97(9):4991–4996. https://doi.org/10.1073/pnas.97.9.4991

Thomine S, Wang R, Ward JM, Crawford NM, Schroeder JI (2000b) Cadmium and iron transport by members of a plant metal transporter family in Arabidopsis with homology to Nramp genes. Proc Natl Acad Sci 97(9):4991–4996. https://doi.org/10.1073/pnas.97.9.4991

Thurtle-Schmidt BH, Stroud RM (2016) Structure of Bor1 supports an elevator transport mechanism for SLC4 anion exchangers. Proc Natl Acad Sci 113(38):10542–10546

Tong J, Sun M, Wang Y, Zhang Y, Rasheed A, Li M, Xia X, He Z, Hao Y (2020) Dissection of molecular processes and genetic architecture underlying iron and zinc homeostasis for biofortification: from model plants to common wheat. Int J Mol Sci 21(23):9280

Trippe RC, Pilon-Smits EAH (2021) Selenium transport and metabolism in plants: Phytoremediation and biofortification implications. J Hazard Mater 404:124178. https://doi.org/10.1016/j.jhazmat.2020.124178

Uluisik I, Karakaya HC, Koc A (2018) The importance of boron in biological systems. J Trace Elem Med Biol 45:156–162

Wang K, Wang Y, Li K, Wan Y, Wang Q, Zhuang Z, Guo Y, Li H (2020) Uptake, translocation and biotransformation of selenium nanoparticles in rice seedlings (*Oryza sativa* L.). J Nanobiotechnol 18(1):103. https://doi.org/10.1186/s12951-020-00659-6

Wang F-H, Qiao K, Shen Y-H, Wang H, Chai T-Y (2021) Characterization of the gene family encoding metal tolerance proteins in Triticum urartu: phylogenetic, transcriptional, and functional analyses. Metallomics 13(7):38. https://doi.org/10.1093/mtomcs/mfab038

Weber M, Harada E, Vess C, Roepenack-Lahaye E, Clemens S (2004) Comparative microarray analysis of Arabidopsis thaliana and Arabidopsis halleri roots identifies nicotianamine synthase, a ZIP transporter and other genes as potential metal hyperaccumulation factors. Plant J 37(2): 269–281. https://doi.org/10.1046/j.1365-313X.2003.01960.x

Xie Q, Ma L, Tan P, Deng W, Huang C, Liu D, Lin W, Su Y (2020) Multiple high-affinity K+ transporters and ABC transporters involved in K+ uptake/transport in the potassium-hyperaccumulator plant Phytolacca acinosa Roxb. Plan Theory 9(4):470. https://doi.org/10.3390/plants9040470

Yamaji N, Ma JF (2021) Metalloid transporters and their regulation in plants. Plant Physiol. https://doi.org/10.1093/plphys/kiab326

Yoshinari A, Takano J (2017) Insights into the mechanisms underlying boron homeostasis in plants. Front Plant Sci 81:951

Yruela I (2005) Copper in plants. Braz J Plant Physiol 17:12. https://doi.org/10.1590/S1677-04202005000100012

Zhang Q, Chen H, He M, Zhao Z, Cai H, Ding G, Shi L, Xu F (2017) The boron transporter BnaC4.BOR1; 1c is critical for inflorescence development and fertility under boron limitation in Brassica napus. Plant Cell Environ 40(9):1819–1833

Heavy Metals: Transport in Plants and Their Physiological and Toxicological Effects

2

Prasann Kumar, E. Lokesh Goud, Priyanka Devi, Shipa Rani Dey, and Padmanabh Dwivedi

Abstract

Heavy metal pollution in the environment is becoming a serious problem and a major concern as it is causes negative effects all over the world. These inorganic pollutants are being discarded in soils, water and the atmosphere as a result of rapidly growing agriculture and metal industries, improper waste disposal, pesticides and fertilisers. This chapter explains how these pollutants enter our environment and cause serious diseases such as cancer by interfering with biological functions and accumulating in various organs. This chapter also describes the toxicological and pharmacokinetic properties of metals and their translocation in the plant, some biochemical properties and physiological effects of such metals in humans as well as in plants.

Keywords

Agriculture · Environment · Food science · Heavy metal · Metal nanoparticles · Soil · Water

Abbreviations

Al	Aluminium
C	Carbon

P. Kumar · E. L. Goud · P. Devi · S. R. Dey
Department of Agronomy, School of Agriculture, Lovely Professional University, Jalandhar, Punjab, India

P. Dwivedi (✉)
Department of Plant Physiology, Institute of Agricultural Sciences, Banaras Hindu University, Varanasi, India

K. Kumar, S. Srivastava (eds.), *Plant Metal and Metalloid Transporters*,
https://doi.org/10.1007/978-981-19-6103-8_2

$Ca(OH)_2$	Calcium hydroxide
CAT	Catalase
Cl	Chloride
CO_2	Carbon dioxide
Cr	Chromium
DNA	Deoxyribonucleic acid
DTPA	Diethylenetriaminepentaacetate acid
GSH	Dlutathione
H_2O_2	Hydrogen peroxide
Hg	Mercury
MDA	Malondialdehyde
MEG3	Maternally expressed gene 3
Mn	Manganese
NaF	Sodium fluoride
O_2	Oxygen
POD	Peroxidase
PSII	Photosystem II
RNA	Ribonucleic acid
S	Sulphur
Si	Silicon
SOD	Superoxide dismutase
TBARS	Thiobarbituric acid relative substances
ZIP	Zinc-iron permease
Zn	Zinc

2.1 Introduction

The environment is defined as the surroundings in which humans, plants, animals and microorganisms live or work. The environment is made up of the Earth's atmosphere, land and water. Four spheres that work in harmony define the Earth's system: the biosphere, atmosphere, lithosphere and hydrosphere. The contaminants, as well as pollutants, are present at higher levels in the environment than in any other section of the environment (Masindi and Muedi 2018). For the last century, the rate of industrialisation has been very fast, resulting in increased demand for exploitation of the Earth's natural resources at an extremely careless rate, which leads to the problem of environmental pollution. The current state of the environment has been severely polluted by various types of pollutants such as organic, inorganic, organo-metallic compounds, gaseous pollutants, radioactive isotopes and nanoparticles (Walker et al. 2005).

Heavy metals are compounds with densities greater than 4.5 g cm^{-3} and atomic weights ranging from 65.4 to 20.59 g/mol. Heavy metals, such as Cd (cadmium), Hg (mercury), Pb (lead), Fe (iron), Cr (chromium) and As (arsenic), have a longer biological life than other elements and can easily persist in the environment for a longer period. All of these heavy metals are thought to be toxic to all living organisms in the environment. They are now classified as contaminants or pollutants because their high concentrations in soil or water cause toxicity and a variety of diseases. Heavy metals are required in trace amounts in some essential micronutrients, and their excess concentration causes toxicity. An increase in the growing industry, particularly the mining industry, is one of the primary causes of heavy metal emissions into the atmosphere, which affects both human health and the plant ecosystem. Both anthropogenic and natural activities emit metals into the atmosphere, which can travel for long distances. Metals released in the air, soil, crop and water reservoirs from natural and anthropogenic activities can be found in areas near urban areas and industrial units. Heavy metal dust particles in the atmosphere have a long-term and toxic effect on human health (Noulas et al. 2018). Because the population is concentrated primarily in urban areas, heavy metal dust particles are inhaled, and consumption of polluted crops may pose health risks. Heavy metal accumulation in crops and vegetables is examined only from the root system accumulation. This is due to metal build-up in the soil system and the plants adsorbed only from the root system. Besides plant roots, aerial plant parts such as fruits, leaves and flowers can also absorb heavy metals. The organ absorbing structure for metal uptake in aerial plant parts is similar to that of the plant's root structure. Following the deposition of atmospheric particles on the surface of the leaf, metal accumulates in plant parts' leaves via foliar transfer (Xiong et al. 2016). Heavy metal concentrations are increased in plants located near mining areas as well as smelting areas. The study was conducted near that industry area to learn more about heavy metal contamination and deposition of these heavy metals. Less research has been done on heavy metal uptake by plant leaves from the atmosphere. The absorption pathway of those metals that are essential for the metabolic activities of plants has been identified and evaluated for the biochemical reaction of plants. Various studies were conducted to learn about foliar absorption of iron, zinc and copper. These metals are absorbed by plants via cuticle penetration and eventually accumulate in plant leaf tissues. The leaves of plants via foliar transfer can absorb non-essential metals such as cadmium, chromium, lead, arsenic and mercury. This review discusses the various mechanisms involved in heavy metal uptake by plants, factors that affect heavy metal uptake by foliar, transportation and compartmentation of foliar heavy metal, toxic and detoxification of heavy metal inside the plant after foliar uptake and a comparison of root and foliar uptake.

Heavy metals are defined as metals with a high atomic mass and a high density. Heavy metals are now defined as metallic chemical elements and metalloids that have toxic effects on the environment and humans. Some metals, such as arsenic, aluminium and selenium, are lighter in weight but still toxic. Some metals are classified as heavy metals, but they are not toxic by nature, unlike gold. Titanium, chromium, manganese, iron, cobalt, nickel, copper, zinc, cadmium, tin, platinum, mercury and lead are some of the heavy metals that have a high density of 5 gm/cm^3 and are more common in our daily lives. The goal of this chapter is to highlight

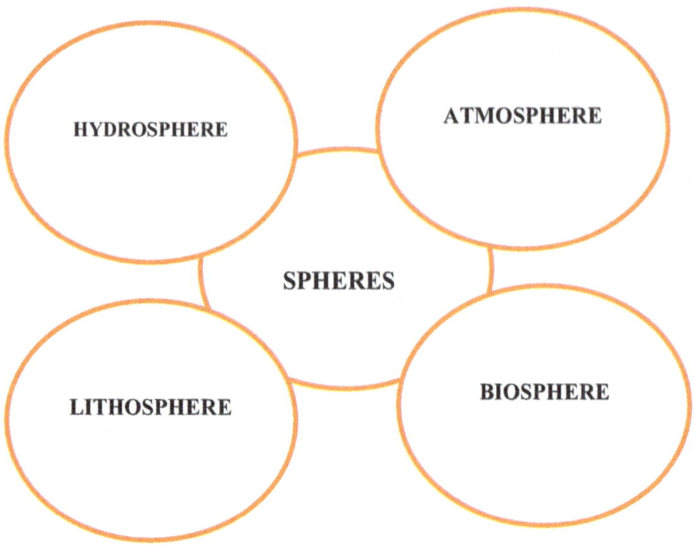

Fig. 2.1 Different relationships of spheres

heavy metal pollution and its entry into our environment, as well as its mechanism and toxicological effects on humans and animals (Fig. 2.1).

2.2 Different Sources of Heavy Metal Pollution

According to the Environmental Protection Agency (EPA), heavy metals were found in drinking water discharged from various industries such as steel, leather and cement and from trivalent chromium naturally deposited by erosion in rocks, soil, plants and volcanic dust found naturally and due to various anthropogenic activities of humans. Heavy metal comes from a variety of sources, including natural processes such as weathering of rocks, and a variety of anthropogenic activities that cause it to enter the environment. Natural processes include the disintegration of rocks, erosion and volcanic eruptions, while human activities include smelting, pesticide and phosphatic fertilizer use, mining, electroplating, industrial effluent and sludge and biosolids in agriculture (Wuana and Okieimen 2011). The main cause of heavy metal concentration in the environment is the increased rate of industrialisation and urbanisation, as well as improper waste disposal. Various anthropogenic activities, such as excessive herbicide use, fertilisation and sludge, are to blame for heavy metal pollution in the atmosphere, but the mining process is the most common source of trace elements (Miransari 2010). When heavy metal concentrations exceed their permissible limits, they have negative effects on microbial activities, plants and the environment. Sharma and Dutta (2017) conducted an experiment in which 240 samples of groundwater were collected from eight districts of the Punjab's Malwa region, and the results revealed that the concentrations of

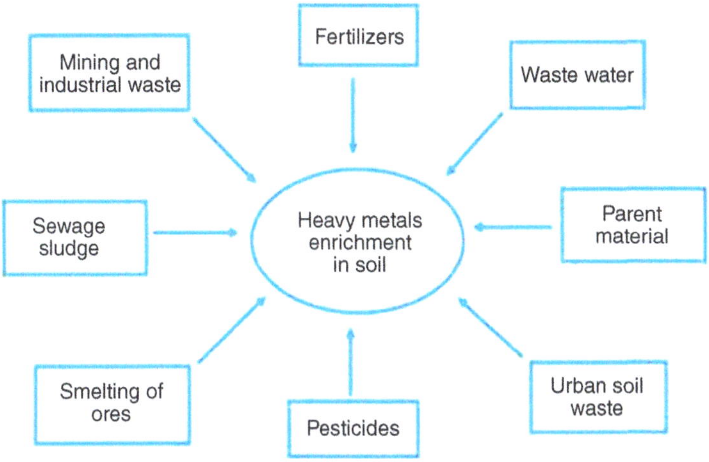

Fig. 2.2 Sources of heavy metals in soil

various heavy metals were higher than their permissible limits, resulting in groundwater contamination. The primary goal was to investigate the distribution of heavy metals in groundwater as well as the effects of heavy metals on human health. They discovered that the water quality was unsafe for drinking as well as other domestic purposes. Different soil samples were collected from a maize field in China to analyse the heavy metal concentrations, and they discovered that the concentration of different heavy metals in the area of wastewater irrigated was higher in the uppermost layer of soil. They stated that the concentration of lead was higher and the concentration of chromium was lower, with the concentration of chromium primarily accumulating in maize roots. The weathering and volcanic eruption of chromium-containing rocks and some human activities such as mining, waste disposal and excessive fertiliser and pesticide use cause chromium accumulation in agriculture (Khan et al. 2016). The holiest place in India, Haridwar, is affected by modern industrialisation. It was discovered that the groundwater around Haridwar was contaminated with five heavy metals, namely, Cr, Co, Fe, Zn and Ni. According to the findings, 216 samples had a percentage of Zn (99%), Co (94%), Fe (99%), Cr (98%) and Ni (90%) that exceeded the permissible limits. Among the five heavy metals, it has been reported that Cr is the most carcinogenic in nature and has the most severe effects on human health as well as in plants (Fig. 2.2).

2.3 Properties of Heavy Metal

Toxicological properties are revealed when a covalent bond is formed as a result of the appearance of metalloids. When metalloids can covalently bind with organic groups, there are two major consequences. As a result, they establish the compounds and ions that produce lipophilic and toxic effects when cellular-type macromolecules bind with non-metallic elements. Because lipophilic behaviour is responsible for metalloid distribution within the biosphere, the response to toxicity is dependent on the simple ionic action formed by similar elements. Methylated and tributyltin oxide are lyophilic compounds with extremely toxic arsenic forms. Heavy metals enter the human body primarily through four routes: inhalation of air from the atmosphere, drinking and ingestion of contaminated water or food and skin contact from pharmaceutical, agricultural, industrial, residential and manufacturing areas (Masindi and Muedi 2018; Walker et al. 2005).

Metal is a non-biodegradable and non-degradable substance. Organisms can detoxify metal ions by hiding those active elements inside proteins and granules of intracellular depositing by them in a non-soluble form that is excreted in organisms' faeces and stored for a long time. When heavy metals are inhaled or swallowed by humans, they bioaccumulate within the human system. As a result, it is classified as hazardous. This bioaccumulation causes physiological and biological complications. Few heavy metals are also called crucial elements that are required for a particular variety of physiological and biochemical functions as well as important for life. Still, the presence of a large number of heavy metals is toxic in nature as well as for humans and animals. Some of them are widely used by various sectors, industries, medicine and agriculture, for the effect it has been spreading into the various environments like soil, water and atmosphere (Tchounwou et al. 2012; Duffus 2002; Wang et al. 2009).

The essential elements are classified into three categories: essential elements for the body, trace elements and macrominerals. For the formation of a block of some living matter, four major important elements are required. According to the atomic number arrangement, hydrogen, nitrogen, oxygen and carbon are examples. The remaining seven essential elements are known as macrominerals because they contain integral-type elements that balance the ions of structural compounds and nucleic and amino acids. Sodium, phosphorous, potassium, sulphur, magnesium, calcium and chlorine were also included in the atomic number sequence. Finally, the trace element group is made up of 13 elements based on their atomic numbers: vanadium, magnesium, cobalt, copper, arsenic, molybdenum, silicon, chromium, iron, iodine, nickel, zinc and selenium. The critical elements play a major role in the structural formation of the skeletal system, the regulation of acid and base equilibrium and the maintenance of the colloidal system. These are also required for enzymes, hormones and structural proteins. For example, iron is required for haemoglobin, and selenium is required for the enzyme glutathione peroxidase (Walker et al. 2005; Tchounwou et al. 2012; Villanueva and Bustamante 2006). Non-essential metals are not required in the human body and play no significant role. However, these metals have a toxic effect and affect the levels of critical elements in the body (Walker et al. 2005).

Toxic metals have been shown to affect cell components and organelles such as nuclei, enzymes, cell membranes, mitochondria and lysosomes. It has been discovered that toxic metal ions interact with nuclear and DNA proteins, causing DNA damage and, as a result, cell cycle modulation, carcinogenesis and apoptosis (Tchounwou et al. 2012). It has been discovered that toxic metal ions interact with

DNA and nuclear proteins, causing damage to a specific site. Toxic ions can cause both direct and indirect harms. Conformational-type changes occur in biomolecules because of direct harm caused by heavy metals. On the other hand, relative oxygen formation and radicals of superoxide or hydroxyl, which are comprised of nitrogen species, endogenous oxidants, nitric oxide and hydrogen peroxide (Valko et al. 2005), cause indirect toxic metal harm. It is caused by lipid peroxidation, sulphydryl homeostasis alteration and DNA damage caused by free radical formations and toxic metals. Few amounts of change have been observed in the calcium homeostasis mediated by metal as a result of membrane damage, which causes a type of calcium-dependent system that is activated by endonuclease participation. Cadmium, nickel, iron, chromium and copper are the most commonly studied metals for free radical formation. Cadmium, chromium and nickel have carcinogenic properties (Valko et al. 2005).

Based on revealing the link between oxidative-type damage and carcinogenesis, free radicals mediated by metals initiated the mutagenicity of the DNA. The formation of free radicals causes various types of base DNA modification, the majority of which are pro-mutagenic. As a result, the essential link between carcinogenicity and oxidative damage caused by toxic metals is demonstrated. The metals arsenic, nickel and cadmium inhibit the DNA repair mechanism. In DNA affected by oxidative stress, there was (1) base modification seen by nickel and chromium; (2) cross-linkage seen by copper, iron and nickel oxidants; (3) strand scission seen by chromium, nickel and cadmium oxidants; and (4) depuration seen by nickel, chromium and copper oxidants (Valko et al. 2005).

Antioxidants, both non-enzymatic and enzymatic, can protect against the attack of free radicals, which is mediated with the help of metals. In general, antioxidants protect iron toxicity by (1) reactions of peroxides and preventing molecular O_2 and also by chelating the ions of ferrous, (2) retaining the state of redox and iron chelation for making them incapable of iron for decreasing the molecular type of oxygen and (3) trapping the radicals produced in this process. Compounds of thiol are the classes of major ones that are classified as highly effective and include glutathione, which traps free radicals, maintains the cell's redox state and decreases peroxide, thus protecting the cells. Damage occurs in the system in vitro due to the presence of metals, which is prevented by non-enzymatic antioxidants such as vitamin E, and an animal taking an overdose of cadmium, copper and iron causes death if it exceeds its permissible limit. The main factors that determined the toxicity effect induced by the metal with carcinogenicity were various reactive-type species and radicles in free form (Valko et al. 2005) (Fig. 2.3; Table 2.1).

2.4 Effects and Transport of Metal Pollutants into the Ecosystem

Pollutants can enter an ecosystem in a variety of ways, including the lithosphere, hydrosphere and atmosphere. Aside from that, there are various natural ways for pollutants to enter the ecosystem, such as rock weathering, anthropogenic activity

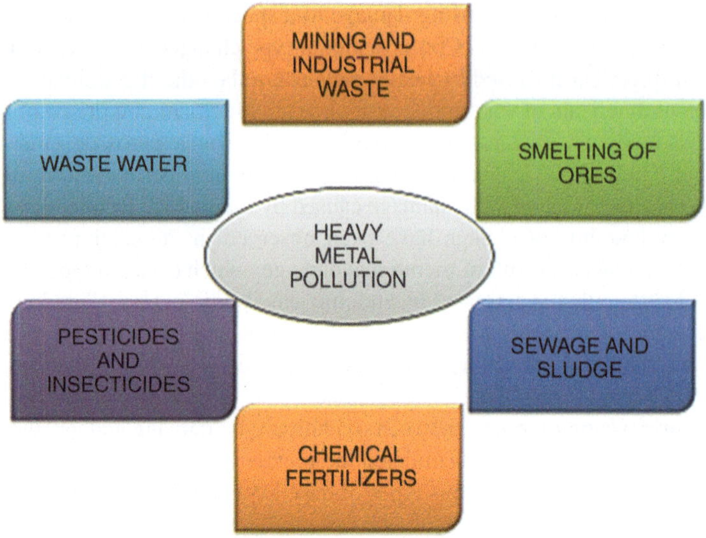

Fig. 2.3 Metal pollution and its sources

and volcanic activity. There is a large amount of wastewater released from various industries, sewage disposal and the overuse of biocides for vector control. Pollutant and heavy metal movement are influenced by a variety of factors such as temperature, air mass circulation, wind speed and water surface direction. Besides that, other factors such as coefficient partition, vapour pressure and molecular stability influence the movement and distribution of pollutants (Walker et al. 2005).

2.4.1 Translocation of Metals in Soil

Wastewater irrigation, pesticides, fertilisers, mines, sewage sludge, coal combustion residues and effluents will cause soil pollution from various industries. Because untreated wastewater and sewage have deposited a large amount of heavy metal in agricultural lands, they will then enter the crop grown on that contaminated field and be consumed by humans. Because the atmosphere only contains oxygen, carbon dioxide and nitrogen, there is widespread concern about rising heavy metal pollution. From the last two to three decades, the increasing industrialisation, as well as urbanisation, leads to the deteriorating of the atmosphere by the emissions of different pollutants. Ozone, sulphur dioxide, carbon dioxide, hydrogen fluoride and nitrogen oxides are examples of common organic and inorganic pollutants. Various heavy metals are emitted into the atmosphere as a result of human anthropogenic activities. Heavy metals are also emitted in large quantities into the atmosphere during the metal processing of ores heating smelters, contaminating the air. The particles suspended in the air are referred to as particulate matter, which can be liquid or solid, whereas aerosols also pose a threat to environmental health, and some

Table 2.1 Threshold limits of various metals by the guidelines of different agencies in air and water

Sl No.	Metal	Threshold limit/exposure limit/permissible limit			
		Air		Water	
		Occupational	Non-occupational	Drinking water	Bottled water
1	Aluminium (Al)	15 mg/m^3 (total dust) and 5 mg/m^3 (respirable fraction)	No data	$\leq$0.2 mg/L	0.2 mg/L
2	Vanadium (V)	0.5 mg/m^3 (V_2O_5 dust) and 0.1 mg/m^3 (V_2O_5 fumes)	1 μg/m^3	No data	No data
3	Chromium (Cr)	Cr(VI), 0.005 mg/m^3; Cr(III), 0.5 mg/m^3; and Cr (0), 1 mg/m^3	1 μg/m^3 for lifetime risk of 4×10^{-2}	0.05 mg/L	0.1 mg/L
4	Manganese (Mn)	5 mg/m^3	0.15 μg/m^3	0.4 mg/L	5 mg/m^3
5	Cobalt (Co)	0.1 mg/m^3	No data	No data	No data
6	Nickel (Ni)	1 mg/m^3	3.8×10^{-4} (μg/m^3)$^{-1}$	0.02 mg/L	No data
7	Copper (Cu)	0.1 mg/m^3(Cu fumes) and 1 mg/m^3 (Cu dust)	No data	2 mg/L EPA: 1.3 mg/L	No data
8	Zinc (Zn)	1 mg/m^3 ($ZnCl_2$ fumes) and 5 mg/m^3 (ZnO fumes and dust)	No data	No data	No data
9	Molybdenum (Mo)	5 mg/m^3 (soluble Mo dust) and 15 mg/m^3 (insoluble Mo dust)	No data NIOSH: 5000 mg/m^3 (insoluble Mo) and 1000 mg/m^3 (soluble Mo)	No data EPA: 0.08 mg/L	No data
10	Arsenic (As)	10 μg/m^3	1.5×10^{-3} μg/m^3 Cancer risk: 1 μg/m^3	0.01 mg/L	10 ppb
11	Cadmium (Cd)	5 μg/m^3	5 ng/m^3	0.003 mg/L	0.005 mg/L
12	Selenium (Se)	0.2 mg/m^3	No data	0.05 mg/L	0.05 mg/L
13	Mercury (Hg)	0.1 mg/m^3 (organic Hg) and 0.05 mg/m^3 (metallic Hg)	No data	0.001 mg/L	No data
14	Lead (Pb)	0.5 mg/m^3	No data	0.01 mg/L	No data

volatile compounds are released from the chimneys of various industries. Different metals are released from various industries, such as chromium, lead, arsenic, nickel, zinc, cadmium and vanadium, which are carcinogenic and have a high toxic potential exposure to living organisms. Because of their ecotoxicity and human bioaccessibility, the potential for oxidative and cytotoxicity metals emitted into the atmosphere and their haziness in the atmosphere. Waste incineration, domestic oil combustion, power generation plants, industrial units, vehicle traffic and various remediation of contaminated sites (Manno et al. 2006) emit heavy metals in the atmosphere. Among all the sources of heavy metal pollution in the atmosphere, industrial and traffic activities are the most significant. Heavy metal emissions into the atmosphere come from industrial processes such as fusion, crushing, reduction, refining and processing. Because heavy metal mobility is high in the atmosphere, pollutants can travel for several kilometres with the wind design.

2.4.2 Translocation of Metals in Water

Water pollution is caused by two major factors: industrialisation and urbanisation. Pollutants containing heavy metals are transported by run-off from industry, villages, cities and towns, where they accumulate in the sediments of bodies of water. This heavy metal is toxic to humans and other ecosystems even at low concentrations. Because of the weathering of Cr-containing rocks, it may enter the water through mechanical operations, soil filtering and other means. With a focus on the oceanic environment, chromium could reduce the oxidation, disintegration and precipitation processes (Kimbrough et al. 1999). The critical level of chromium (III) in water is 8 μg/L, and the critical level of chromium (VI) is 1 μg/L. The level of chromium in effluents from a specific industry or region ranges from 2 to 5 g/L.

2.4.3 Translocation of Metals in Air

Pollutants enter the atmosphere in various forms such as particles, droplets or gaseous forms and are sometimes associated with particles and droplets. Because particles and droplets cannot travel a long distance, they may fall to the ground after a short distance, whereas particles in gaseous form can travel a long distance due to air masses. Heavy metals emitted by industrial smelters are transported several kilometres away from their sources via wet and dry deposition (Dockery 2009). The presence of heavy metals in the targeted organism can pose numerous risks to human health. The World Health Organization establishes health risk guidelines for heavy metals in the air (WHO 2021). Heavy metals can accumulate in plant leaves following the deposition of atmospheric particles on the leaf surfaces via foliar transfer (Schreck et al. 2014). The uptake of metal by the roots of plants has been studied extensively, whereas the uptake of metal by the leaves of plants from the atmosphere has lagged. In the context of pollution research, the foliar transfer of metal is overlooked, which is the major pathway of pollution, primarily when fine ultra-particles interact with the leaves of the plants. The arrangement of the leaves or canopy of plants efficiently filters heavy metals from the atmosphere. Particulate matter is captured on the foliar parts and adsorbs or reduces particulate matter in the atmosphere by the plant canopy. According to various studies, pine forests can retain up to 36.4 tons of airborne particles per hectare per year. The tree's shelterbelt can retain dust particles 38.9% of the time, and it has been reported that plants, as well as growing vegetables, have a high level of foliar heavy metal. The heavy metal concentration in the foliar plant organ easily defined the risk in an environment, which represents the pollution of the environment load. The heavy metal concentration in crop plant tissue was higher than the permissible limits. The vegetables grown in urban areas are enhanced to accumulate heavy metal concentrations of lead and cadmium. Heavy metal uptake is by the foliar surfaces via cuticular cracks, stomata and lenticels and through the ectodermal cell wall, which is the main channel for subsidiary cells and guard cells, or epidermal cell wall. The cuticle is present above the guard cell and is more permeable than the epidermal cell. According to various studies, the waxes of cuticular retain the majority of particulate matter and trichomes on the surfaces of leaves, but some particulate matter enter the plant's leaf tissue. Metal absorption by the foliar uptake is thought to be a surface phenomenon, with the adaxial cuticle playing an important role. According to various studies, copper and nickel transfer rich particles which may enter the leaves of plants via the stomata (Kozlov et al. 2000). Consumption of heavy metal-polluted plants poses a risk to human health, which requires immediate attention, and there is limited data on the health risk in the kitchen garden near the heavy metal contamination area (Manno et al. 2006).

2.5 Heavy Metal Pollution in the Atmosphere: A Need for Great Attention

Because the atmosphere only contains oxygen, carbon dioxide and nitrogen, there is widespread concern about rising heavy metal pollution. Since the last two to three decades, increasing industrialisation and urbanisation have harmed the environment by releasing various pollutants into the atmosphere. Ozone, sulphur dioxide, carbon dioxide, hydrogen fluoride and nitrogen oxides are examples of common organic and inorganic pollutants (Cruz et al. 2015). Various heavy metals are emitted into the atmosphere as a result of human anthropogenic activities. Heavy metals are also emitted in large quantities into the atmosphere during the metal processing of ores heating smelters, contaminating the air. The particles suspended in the air are referred to as particulate matter, which can be liquid or solid, whereas aerosols also pose a threat to environmental health, and some volatile compounds are released from the chimneys of various industries. Different metals are released from various industries, such as chromium, lead, arsenic, nickel, zinc, cadmium and vanadium, which are carcinogenic and have a high toxic potential exposure to living organisms. Waste incineration, domestic oil combustion, power generation plants, industrial units, vehicle traffic and various remediations of contaminated sites (Manno et al.

2006) emit heavy metals in the atmosphere. Industrial and traffic activities are the most significant sources of heavy metal pollution in the atmosphere. Heavy metal emissions into the atmosphere come from various industrial processes such as fusion, crushing, reduction, refining and processing (Zhuang et al. 2009). Because heavy metal mobility is high in the atmosphere, pollutants can travel for several kilometres with the wind design.

2.6 Heavy Metals and Their Translocation in Plants

2.6.1 Chromium

Heavy metals are formed by a variety of processes, including weathering of rocks, which occurs naturally, and a variety of anthropogenic activities, as a result of which they enter the environment. Weathering, erosion and volcanic eruptions are examples of natural processes, whereas anthropogenic activities include smelting, pesticide and phosphatic fertiliser use, mining, electroplating, industrial effluent and sludge and biosolids in agriculture (Wuana and Okieimen 2011). The main cause of heavy metal concentration in the environment is the increased rate of industrialisation and urbanisation, as well as improper waste disposal (Miransari 2010). Various anthropogenic activities, such as excessive herbicide use, fertilisation and sludge, are to blame for heavy metal pollution in the atmosphere, but the mining process is the most common source of trace elements. When heavy metal concentrations exceed their permissible limits, they have negative effects on microbial activities, plants and the environment. Sharma and Dutta (2017) conducted an experiment in which 240 samples of groundwater were collected from eight districts of Punjab's Malwa region, and the results revealed that the concentrations of various heavy metals were higher than their permissible limits, resulting in groundwater contamination. The primary goal was to investigate the distribution of heavy metals in groundwater as well as the effects of heavy metals on human health. They discovered that the water was unfit for drinking and other domestic purposes. Different soil samples were collected from a maize field in China to analyse the heavy metal concentrations, and they discovered that the concentration of different heavy metals in the area of wastewater irrigated was higher in the uppermost layer of soil. They stated that the concentration of lead is higher and the concentration of chromium is lower, with the concentration of chromium primarily accumulating in maize roots. According to Khan et al. (2016), chromium accumulation in agriculture is caused by the weathering and volcanic eruption of chromium-containing rocks, as well as some human activities such as mining, waste disposal and excessive fertiliser and pesticide use. The holiest place in India, Haridwar, is affected by modern industrialisation. It was discovered that the groundwater around Haridwar was contaminated with five heavy metals, namely, Cr, Co, Fe, Zn and Ni. According to the findings, 216 samples had a percentage of Zn (99%), Co (94%), Fe (99%), Cr (98%) and Ni (90%) that exceeded the permissible limits. Among the five heavy metals, it has been reported that Cr is the most carcinogenic in nature and has the most severe effects on human health as well as in plants.

2.6.2 Toxicology Processes

Chromium metal induces oxidative stress, which involves various physiological processes such as lipid peroxidation in plants, resulting in severe damage to the cell membrane, which further leads to the degradation of photosynthetic pigments and a decrease in plant growth. According to Singh and Prasad (2011), the purpose of this study was to investigate the different concentrations of chromium and lead in maize seedlings. The results showed that at 300 micromolar concentrations, the growth, photosynthetic pigments, mineral content and protein content were reduced, whereas metal accumulation increased the content of lipid peroxidation, MDA and hydrogen peroxide content under chromium and lead toxicity. The shoot and root parts were affected by a 200-micromolar concentration of chromium, whereas lead was not as toxic as chromium. Heavy metals such as chromium and cadmium, as well as the biofertiliser mycorrhizal fungus, were tested on various parameters such as growth, proline content, chlorophyll content and metal accumulation in the flax plant. Within 24 days of the experiment, the chromium concentration ranged from 250 to 500 ppm. As a result, it was discovered that there was a 45% decrease in photosynthetic growth and activity. The use of mycorrhizal fungi can increase plant biomass in metal-stressed environments. Under Cd and Cr stress, the content of chlorophyll decreased by 27.5% and 45.5%, respectively, while the content of proline increased by 84% and 71%. Khan et al. (2016) reported the chelating effect on maize-mustard growth grown in cropping system in alluvial soil with artificial chromium stress in the pot. In addition to EDTA, citric acid, DTPA, oxalic acid and humic acid, chromium stress was created at five levels: 0, 5, 10, 20 and 30 ppm. Data were collected at various stages, such as plant height, leaf number, chlorophyll content and dry matter accumulation at 30 and 60 DAS, as well as harvesting. As a result of the increased chromium toxicity in mustard and maize, it was discovered that parameters such as plant height, leaf number, chlorophyll content and dry matter accumulation were reduced. It was discovered that the use of chelating agents significantly increased the growth parameters. The results showed that humic acid at 1 gm/kg produced better results than DTPA at 10 mmol/kg at all chromium levels. Ahmad et al. conducted a pot experiment to assess the effects of Al and Cr on various parameters such as morpho-physiological and biochemical parameters in plant parts. Data were collected at the pre- and post-silking stages when plants were exposed to Al and Cr, with uncontaminated plants serving as the control. As a result, it was demonstrated that both the metal and the photosynthetic reductions reduced maize growth and yield. The toxicity of both metals caused damage to the lipid peroxidation and cell membrane, which activated various antioxidative defences such as superoxide (SOD), peroxidase (POD) and catalase (CAT), which mitigated the glutathione (GSH) content. In metal toxicity, the proline content increased while the protein content decreased. The results also revealed that the toxicity of chromium was greater than that of aluminium when applied individually, but when both metals were applied together, the effect was greater than when applied separately. Overall, the toxic effects of Al + Cr were greater than those of Cr > Al > Ck (control). It was also discovered that the synergistic effects of both metals were more toxic than their individual effects. Islam et al. (2016) investigated the effect of chromium in combination with salicylic acid on plant growth, heavy metal uptake and physiological and biochemical responses under Cr contamination.

2.6.3 Fluoride

The atmosphere contains a trace amount of airborne fluoride concentration. Fluoride in the air is caused by dust particles, coal combustion, phosphate fertiliser production in an industrial area and volcanic activity. Fluoride is present in varying degrees in all natural waters. Drinking water is typically a greater contributor to fluoride consumption regularly. Fluoride concentrations in seawater are typically around 1 mg/L, whereas fluoride concentrations in rivers and lakes are typically less than 0.5 mg/L. Although excess and deficient fluoride concentrations exist in groundwater, they are determined by the fluoride carried by minerals and rocks found in nature. Fluorite is a natural fluoride source in the soil. The fluoride content of the soil ranges between 200 and 300 ppm. Fluoride is a strong soil component that is retained by the soil and does not easily leach down. Fluoride content was found to increase with soil depth, according to scientists (WHO 2021). Water-soluble fluoride accounts for 5–10% of total fluoride in the soil. When the pH of the soil falls below six, fluoride forms complexes with aluminium and iron. Irrigation with fertilised application intensifies the development of Cl^-, SO_4^{2-}, F^- and NO_3^- in underground water. Fluoride concentrations in irrigated lands with groundwater were high due to alkalisation. Fluoride is normally absorbed by the soil surface, with only a trace amount absorbed by the plant. The F accumulation is a direct risk in the food chain caused by fluoride bioaccumulation in grains of various crops, primarily legumes, cereals and edible parts of various vegetables (Mombo et al. 2016). A large amount of F accumulates in various popular plants such as *Cichorium* sp., *Spinacia* sp., *Brassica* sp. and *Apium graveolens*. The primary cause of accumulation is efficient translocation from contaminated soil, which is rooted in shoots, and direct absorption from the atmosphere. An experiment was conducted and reported that in rice seedlings that were irrigated with contaminated water with sodium fluoride concentrations of 20 and 30 mg/dm^{-3} for 15 days, there was an accumulation of Fluoride which was less than 75 mg/kg dry mass. Plants with a high fluoride concentration are not suitable for animal feeding, and they served as an indicator for the accumulation of F in grains. When the rice plant was subjected to metal stress by administering sodium fluoride at a concentration of 5 mg/dm^{-3}, the accumulation of F translocated in the part of the shoot was greatest (Mombo et al. 2016). In Tamil Nadu, rice seedlings and black gram were irrigated with groundwater, resulting in F accumulation that was higher than the WHO safe limit of 1.5 mg/kg^{-1}. When *Camellia* sp. was grown in fluoride-contaminated soil, the leaves contained highly accumulated F, that is, 214 mg/kg^{-1}.

2.6.4 Toxicological Processes

Stressed seedlings showed significant fluoride bioaccumulation, cell disruption and growth inhibition. Calcium compounds improve plant efficiency by increasing seed germination percentage, seedling biomass and shoot and root length; preventing chlorophyll deterioration, electrolyte leakage and H_2O_2; and preventing malondialdehyde levels from decreasing. Calcium priming reduces the stress between total phenolic and carotenoids and flavonoids that are increased by stress. Overall, the results show that a $Ca(OH)_2$ concentration of 0.3mM is the most beneficial for all three primings. Compounds of calcium are used as a priming agent for revoking the toxicity of fluoride in rice. Ghassemi-Golezani and Farhangi-Abriz (2019) reported that they were able to reduce the toxicity of fluoride at various concentrations (free of toxicity, 800, 400, 200 and 100 mg NaF) per kg in safflower seedlings by inducing the application of biochar at 50, 25 and 0 g/kg in soil. When fluoride toxicity exceeds 200mg NaF kg^{-1}, safflower growth did not slow. Plant growth is reduced when the fluoride concentration in the soil ranges between 800 and 400 mg NaF per kg. Scarcity in plant growth as a result of increased fluoride content in plant tissues, formation of superoxide radicals, H_2O_2, increase in lipid peroxidation and reduction in photosynthetic pigment synthesis containing carotenoid, chlorophyll and anthocyanin is noticed. A suitable amount of biochar application reduces fluoride solubility, fluoride content in plant tissues and oxidative stress; increases pH, calcium and potassium uptake; and improves photosynthetic pigment synthesis and plant growth. The results show that a 1 kg soil application of 50 g biochar was the best concentration for mitigating fluoride toxicity in the safflower plant. Manoharan et al. (2007) investigated whether the amount of fluoride added in acidic soil with phosphatic fertiliser was harmful to the root growth of the barley crop. The increase in fluoride concentration in soil naturally increases the concentration of aluminium in the soil solution with a pH of 4.2–5.8. Plant root growth in acidic soils has been observed to be halted due to the addition of an excessive amount of fluoride. According to calculations, increasing fluoride concentration improves the concentration of Al-F complexes in soil. The data showed that the constant application of fluoride into soils resulted in increased acidification of the soil, which could pose a problem for fluoride in the future. Singh and Prasad (2011) reported fluoride pollution mitigation strategies in plant water and soil. Fluoride is regarded as the 14th essential nutrient physiologically required for human growth and development. FL-1 becomes toxic to humans when the fluoride concentration in drinking water exceeds 1.5 mg. Primarily geogenic and anthropogenic processes cause fluoride pollution in the ecosystem. Fluoride is widely distributed in all environmental resources, with 0.11–0.61 mg L^{-1} in the air, 151–401 mg kg^{-1} in soil and 151–401 mg kg^{-1} in water (1.01–38.2 mg L^{-1}). Humans and animals were exposed to fluoride when the concentrations of fluoride in water and plants ranged between 0.21–42.1 mg L^{-1} and 0.78–29.6 mg g^{-1}. The primary goal of this study was to focus on biological and physiochemical methods for reducing fluoride's impact on the environment.

2.6.5 Manganese

Manganese is abundant in both the aquatic and terrestrial environments, accounting for 0.1% of the Earth's crust. Manganese is always found in compounds with other elements such as Cl, C, S, Si and O, all of which have a solid physical state and some of which are water-soluble or small particles suspended in the air. Despite the fact that Mn can exist in a variety of oxidation states ranging from –3 to +7, the majority of compounds are in +4 and +2 oxidation states. This element has a concentration range of 40–900 mg/kg in soil (Millaleo et al. 2010). The concentration of dissolved oxygen in major waterbodies on Earth ranges between 10 and 10,000 g/L. They tend to settle at the bottom of water resources. Volcanic activity, weathering of rocks in the Earth's crust, forest fires, ocean spray and vegetation are all natural sources of Mn. The Mn's common oxidation states are interconvertible via redox reactions caused by either biotic or abiotic processes. Mn^{2+} compounds are soluble, whereas Mn^{4+} compounds are insoluble oxides. The Mn chemical reactions in an oceanic environment are influenced by O_2 concentration, pH and redox conditions. Manganese oxides have been discovered to settle in oceanic sediments. However, the low oxygen condition, combined with the high temperature, releases free manganese ions into the overlying water, resulting in a 1000-fold increase in Mn concentration in water. Hypoxic (low O_2 concentration) conditions in the marine ecosystem can be caused by either anthropogenic or natural factors.

Despite its widespread distribution in nature, Mn is only required in trace amounts by living organisms to regulate enzyme activation, metabolic reactions, connective tissue formation, cellular defence, reproductive hormone regulation, bone mineralisation, nervous system activities, blood clotting, glycosaminoglycan formation and protein, carbohydrate, lipid and amino acid metabolism. Mn-containing enzymes include manganese superoxide dismutase, pyruvate carboxylase and arginase, while manganese-activated enzymes include decarboxylases, hydrolases, DNA and RNA polymerases, kinases and transferases.

The Mn can be released into the environment through industrial processes such as welding, mining, alloy production, agrochemical production, Fe-Mn operations and goods processing. Because the metal is bioavailable, there is a greater chance of uptake by living organisms through the water. The gastrointestinal tract is the primary absorption route, followed by the inhalation pathway, and the use of narcotics intravenously is a recently reported source. Occupational exposure occurs frequently during welding and smelting through inhalation, from which it enters the bloodstream and eventually the brain (Lucchini et al. 2012). Toxic effects can occur as a result of contaminated water and food intake. For example, studies show that water in Bangladesh is contaminated with 2 mg L^{-1}, a value four times higher than WHO limits (400 g L^{-1}), and the mean daily uptake of Mn from Western diets was 2.3–8.8 mg (Frisbie et al. 2002; Khan et al. 2012). Excess Mn in soils near a ferroalloy plant in Italy caused serious health problems in schoolchildren, such as impaired odour identification, hand dexterity and motor coordination (Lucchini et al. 2012).

To absorb manganese, the gut can use a simple diffusion or active transport mechanism. Because of the large metal particle size of Mn, which is swallowed after coughing up by mucociliary action, the diffusion process through the alveoli is difficult. The presence of calcium, iron and phosphorus in the diet can reduce Mn intake and uptake (Briffa et al. 2020). The entry of Mn into the brain can be through three routes:

1. Nasal mucosa → olfactory neuron → brain olfactory bulb
2. Blood → capillary endothelial cells → blood-brain barrier cells
3. Blood → choroid plexus → cerebral spinal fluid

The Mn accumulated in the mitochondria of cells such as neurons, astrocytes and oligodendrocytes. It can inhibit NADH dehydrogenase and F1/F0 ATP synthase, causing ATP synthesis to be disrupted. This results in the release of free radicals and, as a result, oxidative stress. Manganism is a neurological disorder caused by Mn exposure that is similar to Parkinson's disease (Harischandra et al. 2019).

Manganese is one of the essential micronutrients required by plants, with concentrations ranging from 30 to 500 mg/kg dry weight of different plant species (Clarkson 1988). Plants become toxic when this level is exceeded. A decrease in growth rate, a decrease in root and shoot length, marginal and interveinal chlorosis, necrotic spots on the leaf, dark inclusions, leaf crinkling, a reduction in photosynthesis, a decrease in chlorophyll content, a decrease in carotenoids and interference in thylakoid stacking due to Mn accumulation in thylakoids are some of the effects of excess Mn concentration in plants (Millaleo et al. 2010).

Excess manganese exhibits an antagonistic reaction with similar ions, interfering with their normal functional roles. The proposed theories about Mn phytotoxicity are based on experimental findings. Leaf black spots and callose are formed as a result of apoplastic-mediated phytotoxicity due to the accumulation of oxidised phenols and manganese oxides (Wissemeier and Horst 1992). Solar radiation causes Mn toxicity by causing ROS (reactive oxygen species) production in the plant. Because the antioxidative enzymatic system lacks the cofactor, the antagonistic activity of Mn has been found to cause oxidative stress.

Chlorosis is a notable symptom of manganese toxicity caused by photobleaching or the destruction of chlorophyll in the presence of light. As a result, sunlight causes Mn accumulation in leaves. Plants will face competition for elements such as K, Fe, Ca and Mg, resulting in nutrient stress in plants grown in excess manganese concentrations (Fernando and Lynch 2015). Excess Mn in soil resulted in low shoot biomass in the soybean plant due to low stomatal conductance and carbon dioxide assimilation rate but increased the activity of antioxidant enzymes and hypertrophying symptoms in the adaxial epidermis of the leaves, forming necrotic regions with purple veins. The Mn toxicity causes cell death in plants by interfering with metabolic activities and causing the degradation of nucleic acids, lipids, proteins and carbohydrates. Even the photosynthetic apparatus is degraded, resulting in a decrease in photosynthesis rate (Santos et al. 2017). However, some plants, such as *Phytolacca acinosa*, *Polygonum lapathifolium*, *Polygonum hydropiper*,

Phytolacca americana, Polygonum pubescent, Celosia argentea, Schima superba and others, can withstand the toxic effects of Mn in high concentrations (Liu et al. 2016).

2.6.6 Cobalt

Cobalt, an element with the atomic number 27, is required in trace amounts by living organisms and exists as inorganic compounds, most commonly in the +2 and +3 oxidation states. Root nodule development in a leguminous crop can be influenced by the Co. Other roles of cobalt include enzymatic activation, being a component of vitamin B12, being a coenzyme in DNA synthesis, being a coenzyme in fatty acid oxidation and so on. Diseases such as anaemia, apatite loss, bone fragility and scaly skin can occur in ruminant animals fed with co-deficient fodders/feed additives (Smart et al. 1981). However, excessive concentrations can have toxic effects on plants such as low shoot biomass, premature leaf fall, vein discolouration, premature leaf closure and inhibition of greening. It also causes the production of ROS such as hydrogen peroxide, hydroxyl radicals, proline and malondialdehyde, which disrupts the electron transport chain, reduces photosynthetic pigments and impedes N metabolism within the plant body. Grains and grasses have a lower distribution of Co than leguminous crops, indicating that it is species dependent. Leaf necrosis, interveinal chlorosis, chromosome damage, cellular mitosis, inhibition of hypocotyl elongation and root and seed germination inhibit Fe and Ca uptake (Mahey et al. 2020).

Cobalt occurs naturally as carbonates or in association with other metals such as Ag, Pb, Ni, Fe, Cu, Mn and Cu, and it is classified as a chalcophile, lithophile and siderophile element. Mineral forms of Co such as cobaltite, erythrite, heterogeneity, skutterudite and spherocobaltite are abundant in igneous and sedimentary rocks, but their abundance in the Earth's crust is very low in comparison to other heavy metals. Cobalt concentrations of around 100 mg kg^{-1} are found in ultramafic rocks in greater abundance. Hard metal manufacturing units, cement industries, e-waste processing, diamond polishing, paint and pigment industries, incinerators, motor fuel combustion, mining, coal combustion, extraction of metals from Co containing ores, fertiliser industry working with Co salts, cosmetic products, pharmaceuticals and so on are examples of anthropogenic Co release activities. Co dust is a product that is released and suspended in the air, and certain industrial activities cause leaching into groundwater and soil (Mahey et al. 2020).

Some plant parts, such as grains and fruits, can deposit Co in their tissues, fulfilling Co's essential role in a living body. However, in mining areas, the plant accumulates Co in the range of 111–245 mg kg^{-1}, which is significantly higher than normal values in the range of 0.05–5 mg kg^{-1} dry weight (Yaman 2014). To ensure secure and safe food production, proper methods for dealing with the remediation of metal-polluted sites must be used, as they do not biodegrade. The uptake of Co and transport across the cell membrane is governed by high-affinity binding sites within cells and transport proteins. Toxic metal ion homeostasis is the function of a wide range of proteins like the following:

1. CPx-type ATPases that is a class of exporter or transporter protein bound to the cell membrane and facilitates ion export from cells.
2. NRAMP family, elaborated as natural resistance—associated macrophage protein—is found in a broad spectrum of organisms from bacteria to human as an integral membrane-bound protein and evolutionarily conserved that transports metal ions.
3. CDF protein family, known as cation diffusion facilitator, is involved in the efflux of divalent heavy metal ions and found in prokaryotes and eukaryotes, while in plants, they named to be metal tolerance proteins.
4. ZIP protein family (zinc-iron permease) can transport heavy metals primarily cadmium, cobalt and zinc with a structure having six TM domains, C-terminal cation binding domain and N-terminal signature sequence (Hall and Williams 2003; Kolaj-Robin et al. 2015).

In the tonoplast of *Arabidopsis thaliana*, heavy metal-associated 3 (AtHMA3) is located which is a protein in the family of P-type ATPase and assists in the transport of metals across the membrane. Passive transport is adopted for the translocation of cobalt within the higher plants after the entry via roots, and IRT1 transporters assist in their translocation (Sarma 2011).

Iron deficiency is common in young leaves of plants subjected to excessive Co stress. Catalase activity, chlorophyll concentration and shoot biomass were reduced, while leaf P and carbohydrate fractions, as well as acid phosphatase, peroxidase and ribonuclease activities, were increased. Co affects the rate of transpiration and water potential in cauliflower, as well as the translocation of elements such as sulphur, copper and phosphorus (Nagajyoti et al. 2010). At higher Co concentrations, radical growth is hampered in the early stages of the plant, with a reduced rate of germination and the appearance of chlorosis on young leaves. In some cases, the leaf margin is purple. Other visible signs of Co toxicity include leaves with white and dead edges, marginal scorching, rudimentary leaflets with a hooklike appearance at the plant's top, black patches on tomatoes and diffused chlorosis. The reactive oxygen species produced by metal stress in the bean plant is effectively mitigated by de novo synthesis of phenols and anti-stress protein. As a result, the bean plant is thought to be a metal-resistant species. Cobalt tolerance mechanisms in plants include gene upregulation, reduced root-to-shoot Co translocation, activation of antioxidant enzymes, amino acid deposition, phytochelatin synthesis, osmoprotectant accumulation, chelation and organic acid deposition. Exogenous application of epibrassinolide in *Brassica juncea* L., calcium supplementation in a growing medium and salicylic acid application can all help to mitigate the effects of excess Co stress in plants. Many plants, including *Alyssum murale*, *Lycopersicon esculentum* and *Thlaspi caerulescens*, are cobalt hyperaccumulator species. *Epipremnum* and *Pistia stratiotes* are effective for rhizofiltering Co from an aqueous medium (Akeel and Jahan 2020).

Because of Co's strong affinity for the sulphydryl group, enzymes with critical functions are harmed. This improves mitochondrial respiration, and Co^{2+} creates an antagonistic and inhibitory pathway to Ca^{2+} in binding with the appropriate sites, as well as metal-activated enzymes. The Fenton reaction generates ROS, which causes oxidative stress in macromolecules such as DNA, lipids and proteins. Co can directly cause DNA damage through the harmful effects of DNA—protein crosslinking and sister chromatid exchange. The carcinogenic impact due to the cobalt is attributed to the free radicals released in reactions mediated by cobalt. The presence of Co^{2+} in the Fenton reaction increases DNA cleavage at all bases and causes the peptides to act as pro-oxidants, causing cell damage. The lysosome is where Co^{2+}-induced ROS are produced (Valko et al. 2005).

Cobalt can cause cancer in animals, but according to IARC, it is not harmful enough to cause cancer in humans. Cardiomyopathy is a medical condition caused by Co poisoning in humans. Furthermore, the hypoxic condition caused by the Co ions prevents the kidney from receiving oxygen and producing erythropoietin. Furthermore, Co and its ions are genotoxic and induce apoptosis, necrosis and inflammatory responses (Simonsen et al. 2012). The biological solubility of the metal influences inhalation of Co dust and subsequent absorption into tissues. Soluble metal ions enter the bloodstream via the alveolar and bronchial walls. The oral pathway for metal ingestion is also affected by Co solubility, fasting, Fe deficiency and other factors that make Fe and Co compete for absorptive sites in the intestine. In terms of dermal absorption, the broken skin is more susceptible to Co uptake than the intact skin. The liver absorbs the most Co, followed by the kidneys and lungs, and it is excreted more through urine than faeces and bile.

2.6.7 Nickel

Nickel, a group 10 element in the periodic table with atomic number 28 and the 24th most abundant element in the Earth's crust, accounts for 3% of the Earth's composition but ranks 5th in terms of abundance in terms of weight. It is a ferromagnetic element that can be found naturally as sulphides and oxides or combined with As and Sb. This transition metal is highly resistant to corrosive agents such as water, air and alkali, but dilute oxidising acids can dissolve Ni. Organic salts and strong acid salts of Ni have high water solubility, whereas weak acid salts and inorganic salts of Ni have low water solubility. Nickel prefers to exist in +2 oxidation states, though a range of −1 to +4 is possible. A large portion of Ni is inaccessible on Earth due to its confinement in the outer core as Fe-Ni alloy, of which 10% is Ni (Cempel and Nikel 2006).

Nickel is used in armaments, electrical appliances, utensils, stainless steel, clothing fasteners, inexpensive jewellery, coins, keys, alloy, paper clips, catalyst, metallurgical process, pigments, electroplating, Ni-Cd batteries, orthodontic materials, electroforming, medical prostheses and so on. This trace metal is widely distributed in soil, air, water and biological material due to natural and human sources. Natural sources of Ni include volcanic eruptions, forest fires, vegetation, direct leaching of sediments and rocks into water resources and weathering of rocks and soils. The wind can transport Ni dust from one location to another. Anthropogenic Ni release into the environment is caused by the combustion of fossil fuels, tobacco smoking, food processing in stainless steel utensils, kitchen kettles, sewage, waste incineration, e-waste processing and various applications of Ni nanoparticles. Nickel carbonyl, a hazardous chemical form, is found in tobacco smoke. The presence of high levels of Ni in nuts, cocoa, chocolate, broccoli, asparagus, tomato, carrots, spinach and green beans is a serious concern (Genchi et al. 2020).

In industrialised areas, the mean concentration of Ni in the air is higher. Ni was found to have an average concentration in the air of 0.00001–0.003 g m^{-3} in remote areas while 0.07–0.77 g m^{-3} in urban areas with metallurgical or Ni processing industries. This is critical in occupational exposure to Ni dust and fumes via inhalation and dermal contact. Around 0.2% of the total workforce is at risk of being exposed to Ni fumes or dust. Ni inhalation pathways include insoluble Ni compounds (dust), nickel solutions (aerosols) and nickel carbonyl (gas). Drinking water samples from areas near the ore mining have been reported to contain as much as 200 g nickel L^{-1} of water. Metal smelters and domestic effluent release are to blame for trace metal contamination in the aquatic ecosystem. Ni deposits phytoplankton, which is known as sensitive bioindicator of aquatic pollution. Nickel concentrations in soil range between 3 and 1000 mg kg^{-1}. Water-soluble Ni, Ni adsorbed on inorganic cation surfaces, inorganic crystalline minerals, Ni-organic cation exchange complex, chelated metal complexes and free ion in the soil solution are the various forms of Ni in soil. Various foods were tested for Ni levels ranging from 0.1 to 0.5 mg kg^{-1}. Processing and manufacturing can add up to some amounts, either through catalytic hydrogenation of fats/oils by Ni-containing catalysts like Raney nickel or flour milling or leaching of Ni from stainless steel food processing equipment. With increased consumption of oatmeal, soy products, nuts, dried beans, dark chocolate and peas, human Ni uptake can reach nearly 900 g per person per day.

The Ni absorption in the gastrointestinal tract is attributed to the metal ion's lipophilicity. Animals absorb trace amounts of Ni via facilitated diffusion and active transport. However, due to carrier saturation, passive diffusion is used for a large amount of Ni. Various ligands and ions can influence these processes. Nickel travels through the bloodstream due to its interactions with albumin and ultrafilterable ligands. Ni competes with Cu to bind to albumin. Ni enters the liver through hepatocytes' Ca channels. Nickel removal or excretion from the body is via the urine, sweat, milk, hair and skin. Because of its solubility in fat, nickel carbonyl passes through the cell membrane, and its reaction in blocking the Ca channel releases free Ca into intracellular compartments, resulting in a negative impact on cell growth, differentiation and apoptosis. Nickel activates the transcription factor called ATF-1, which subsequently lessened the activity of the TSP1 regulator, which also stimulated angiogenesis that led to the growth of the tumour. Also, inflammatory response and apoptosis in the body are due to the Ni-mediated activation of another transcriptional factor, NF-κB. The carcinogenic mechanism of nickel is the combination of free radical production, transcription factor stimulation and gene expression at a controlled level. MEG3 (maternally expressed gene 3) is downregulated as a result of the Ni-induced methylation of the promoter which consequently upregulates two proteins, HIF-1α, and results in carcinogenesis (Engwa et al. 2019).

A decrease in biomass yield and growth parameters caused by Ni toxicity has a negative impact on a plant's vegetative growth. The dimensions of the vascular bundle, mesophyll cells and epidermal cells on the leaf and the ultrastructure of the leaf all had negative effects. Cell wall plasticity is compromised, as is the volume of intercellular spaces. Under Ni stress, certain plants may experience disrupted osmosis and diffusion processes, instability in the cytoplasmic ion balance, disruptive membrane functions, decreased rate of transpiration, low rate of stomatal conductance, low water potential in leaf and reduced leaf area. Other mineral nutrients with similar radii to Ni^{2+} were found to be competitive in plant uptake and binding sites. This will result in visual deficiency symptoms caused by those specific metal ions. The lipid peroxidation caused by ROS generation in plants exposed to Ni toxicity causes genetic instability, ionic leakage, membrane instability, loss of turgor pressure and disruption of membrane-bound enzymes. This is demonstrated by the high level of malondialdehyde (MDA). Any plant species with an active enzymatic or non-enzymatic antioxidative system can recover from stress-induced injuries. Competitive uptake of Ni causes a deficiency of iron (Fe^{2+}), which is an essential component of the antioxidant enzyme cofactor; as a result, enzyme activity is reduced and membrane permeability is increased. Plant photosynthetic activity is hampered by Ni toxicity, as evidenced by structural deformation of chlorophyll, a hindrance to chlorophyll biosynthesis, decreased activity of Calvin cycle enzymes (aldolase, RuBisCO, NADP-dependent phosphoglyceraldehyde dehydrogenase, fructose 1,6-bisphosphate, 3-phosphoglycerate kinase) and electron transport reactions, and the number of grana and thylakoids decreased as the lipid membrane's composition changed. The displacement of certain essential metal ions by Ni^{2+}, such as Mg^{2+}, alters the structure of chlorophyll and challenges the activity of RuBisCO. The low content of cytochrome b559, cyt. b6f, plastocyanin and ferredoxin causes a decrease in the efficiency of the electron transport chain in the photosynthetic process (Shahzad et al. 2018).

Some plants in the Brassicaceae family, as well as arbuscular mycorrhizal fungi, can accumulate Ni in greater quantities. This is a promising technique for the remediation of Ni-contaminated soil that is less expensive and more environmentally friendly than other costly and non-environmentally friendly techniques. They can also compartmentalise the metal absorbed into vacuoles, which is facilitated by chelating with organic acids. *Vigna radiata*, *Berkheya coddii*, *Sebertia acuminate*, *Pistia stratiotes*, *Isatispinnatiloba*, *Alyssum murale*, *Alyssum heldreichii*, *Melastoma malabathricum* and others are examples.

2.6.8 Copper

Copper (Cu), with atomic number 29, is an essential micronutrient with a high thermal conductivity as well as electrical conductivity, as well as ductile and malleable properties. The most common oxidation states are +1 and +2, which are pronounced cuprous and cupric. They can be involved in a variety of physiological processes such as photosynthetic electron transport, mitochondrial respiration, symbiotic nitrogen fixation, ethylene sensing, photosynthesis, oxidation-reduction reactions, carbohydrate and protein metabolic pathways and so on. Cu ranks 25th in abundance among all the elements in the Earth's crust. It is the third most commonly used metal in the world. Copper is a structural component of many proteins involved in the regulatory function. Plastocyanins are electron carriers in Cu-based plastids (WHO 2021).

Fertiliser, metallurgical, refining fungicides, paints and pigments, printed circuit board production, chemical manufacturing, municipal waste, sewage, agricultural waste, storm water runoff, mining, traffic emissions and so on are the major anthropogenic causes of Cu release into the environment. Copper alloys are widely used in pipelines, wire production, power plants, cooling towers, heat exchangers, marine industries and electronic industries. Copper resists corrosive agents to a large extent by forming a non-conductive layer on its surface. Copper compounds such as Bordeaux mixture, copper hydroxides, copper oxychlorides and others are used to protect plants from plant diseases, which increases the release of Cu at toxic levels. Cu's antimicrobial properties allow it to be used in clinical or medical settings. The annual increase in copper consumption necessitates more mining. Air, soil, rock and water are examples of natural pathways. Copper minerals include covellite, malachite, digenite, azurite, tetrahedrite, chalcopyrite, bornite and chrysocolla.

Copper-based agro-inputs or agrochemicals, such as fungicides, herbicides, miticides, fertilisers, insecticides and nematicides, are the most common causes of soil contamination. This could amount to more than 5000 tons/year. When copper-rich animal feeds are used in crop production, they cause copper deposition in the soil. Cu is also abundant in some phosphatic fertilisers. Cu has an average abundance of 60 mg/kg in the Earth's crust. It is found attached to the organic matter in the upper soil layers. The majority of copper in the soil cannot be degraded, posing a toxic effect on microbes and disrupting the nutrient cycle. As the degree of contamination in soil increases, so does NH_3 oxidation, so quantification of microbes involved in this process, such as archaea and bacteria, can determine the extent of contamination. Certain micronutrients such as Mn, Zn, Fe and macroelement phosphorus were found to decrease as Cu increased in soils. Cu can be found in water resources and drinking water as particulate matter or complexes. Copper levels in surface water in the United States range from 0.0005 to 1 mg L^{-1}. Aquatic organisms pose a threat to dissolved copper. Particulates released from mine tailings, processing of Cu-containing materials, soil suspension in the air and combustion sources are all sources of airborne copper (Rehman et al. 2019).

When exposed to copper toxicity, plant growth is slowed, with a bluish appearance that eventually turns brown/yellow. The binding of copper with clay minerals and organic matter creates an imbalance in nutrient availability, reducing productivity. Cu toxicity symptoms include leaf chlorosis with white or cream spots and lesions, necrotic regions on the leaf's edges and tips and wilting. Cu phytoavailability and medium pH have an inverse relationship. Organic matter content; waterlogged soil; N, P and Zn levels; and root growth are other governing factors. The root has a greater proclivity to accumulate Cu and little translocation to upper parts, resulting in a more toxic effect on roots such as structural deformation, less proliferated root hairs and a disrupted root cuticle. Free Cu ions can cause ROS to be produced and the signal transduction pathway for cellular damage to be activated, and this reaction is known as the Fenton reaction or the Haber-Weiss reaction. By the Haber-Weiss reaction, Cu^{2+} is reduced to Cu^{+} in the presence of any reducing agent, such as superoxide, which then generates hydroxyl radical from H_2O_2. Unsaturated fatty acids react with OH^{-} in the cell membrane to form lipid radicals and MDA, which is a cytotoxic indicator of tissue damage. Along with the disruption of protein and pigment composition, the inefficiency of PS II reduced CO_2 diffusion and a negative effect on gas exchange parameters such as net photosynthetic rate, light saturation point, rate of dark respiration and light compensation point; it also impedes electron flow in the electron transport chain by inactivating quinone acceptors (Shabbir et al. 2020).

ROS production and activity, which is stimulated by Cu, are lethal to almost all cell organelles, including DNA. The Cu^{2+} ions have the ability to inhibit enzymes involved in DNA repair, base modification, structural changes of the cell wall, cleaving of the double and single strands of genetic material, cross-linking of DNA and protein, lowering of the mitotic index, chromosomal aberrations and cell division. The formation of micronuclei in plants is the endpoint of genotoxicity caused by any metal. Copper homeostasis in plants is maintained by limiting the movement of Cu ions to maintain an optimal level in the plant body, which is accomplished through the combined efforts of transporter proteins and a number of genes. Some non-enzymatic antioxidants, such as phytochelatins, salicylic acid, proline, metallothioneins, ascorbic acid and glutathione, are ready for an effective defence response against Cu-induced damage.

Copper becomes toxic to the living body at a concentration of 20 g/g, and 0.9 mg day^{-1} is recommended for adults to avoid deficiency symptoms. At toxic levels of Cu in the human body, the kidney and liver produce metallothionein (a low molecular weight protein), which binds with copper and produces a water-soluble complex that is easily excreted. The hCtrl and hCtr2 are two copper transporters that regulate copper intake and absorption in living organisms. Another Cu-transporting protein is MNK (P-type ATPase) which allows the passing of Cu via the basolateral membrane of the intestine and delivers the metal to specific enzymes and copper efflux from the cell. In the bloodstream, translocation of Cu to the liver is possible after binding with histidine and albumin through the portal circulation. In hepatocytes, copper chaperons and copper form complexes that enable the intracellular distribution of this ion. In the liver, another protein named WND protein acts during the high Cu concentration by releasing copper ions at the canalicular membrane which is ultimately excreted through bile. So, a large part of Cu entered in the living body is released through bile and small portions via sweat, faeces and urine.

2.6.9 Zinc

The zinc enters the aquatic system via artificial pathways such as municipal waste, industrial waste, waste incineration, urban runoff, mine drainage, soil erosion, galvanised materials and coal-burning power plants. Leachable sources of zinc to groundwater include the use of old and galvanised metal pipes, mineral fertilisers and well cribbings with Zn coating. Zinc at higher levels in water imparts a metallic and bitter taste, with a 4 mg L^{-1} threshold limit for the particular taste. Above the 5 mg Zn L^{-1}, water has a milky appearance. Boiling drinking water with a Zn content greater than 3 mg/L produces an astringent taste and a greasy layer on top. Because most agricultural fields have zinc-deficient soils, Zn levels below 2 mg L^{-1} are acceptable for irrigation water, but levels above this level are toxic to plants (Noulas et al. 2018).

Excess Zn has a negative impact on plants, causing symptoms such as stunted shoot growth, chlorosis, leaf tip death and young leaves curling and rolling. Some of the reported effects on plants include root blunting with impaired cell elongation and division, inhibited root growth, damage to cortical cells of roots, late reproductive development and sometimes no flowers. Any metal's bioavailability is determined by the soil's adsorptive capacity and the release of metal chelators via root exudation. Zinc toxicity affects iron translocation, membrane disruption, cell organelle disintegration, increased nucleus number, distorted mitochondria and chromatid condensation (Rout and Das 2011).

More than 300 enzymes, including carboxypeptidase, carbonic anhydrase, alkaline phosphatase, RNA transcriptase, alcohol dehydrogenase, DNA polymerases and Cu, Zn superoxide dismutase, rely on Zn to regulate their functioning, catalytic properties and structural stability. Zn's primary functions are cell proliferation and DNA and RNA synthesis. A diffusible intracellular zinc carrier, CRIP, binds Zn to the intestine's mucosal layer in a saturable process. This carrier protein competes with metallothionein, preventing absorption and allowing transport through the intestinal lumen. When the concentration of zinc increases, metal binding sites in CRIP become rapidly saturated. Zn exhibits non-specific binding on various proteins and ligands, causing brush border cells in the intestinal lumen to be damaged. Citric acid, cysteine, methionine, prostaglandin E2, reduced glutathione, histidine, copper, cadmium, iron and calcium are all factors that influence Zn uptake in the intestine. Metal complexes with phytate or dietary fibres reduce zinc availability. Carriers in the bloodstream include albumin, 2-macroglobulin and amino acids. Zinc is stored in various body parts after ingestion, including the liver, kidney, pancreas, bone and muscles. A transmembrane protein, ZIP (zinc-iron-related transporter protein), present on the biological membrane, regulates zinc concentration in eukaryotic cells. Zinc's activities include transcription factor activation, dephosphorylation, second messenger metabolism, protein phosphorylation and extracellular signal recognition. Zinc deficiency can cause apoptosis because zinc is an essential component of cell growth and proliferation. Furthermore, higher and excess zinc concentrations destabilise zinc homeostasis, resulting in cytotoxicity and apoptosis. When Zn levels are high, cells exhibit necrotic symptoms (Briffa et al. 2020).

2.6.10 Mercury

Mercury is a global pollutant that has been identified as a bioaccumulative toxin and ranks third on the list of priority-based hazardous substances (Hg). They can survive in the atmosphere for up to 2 years. Mercury exists in three different forms in the environment: (a) elemental mercury, denoted as Hg0; (b) inorganic mercury, with oxidation states +1 and +2; and (c) organic mercury. Organic mercury is the most toxic of the three, existing as monomethyl or dimethyl, Hg and methylation of inorganic mercury in the absence of oxygen, and the aquatic system can obtain organic mercury (Kumari et al. 2020). Mercury is used as a vaccine preservative, cosmetic industry, medicinal uses, disinfectant against microbes, switches, metal amalgamation, electrolysis chlorine production, barometer, thermometer, manometer and industries related to dyes, industries, paints, plastic and pharmaceuticals. Mercury's natural sources include forest fires, volcanoes, fossil fuels, cinnabar ore and so on. Both human and natural sources are estimated to emit 5000–8000 metric tons of mercury per year. Gold mining, coal combustion, non-Fe metal manufacturing and cement manufacturing all contribute 38%, 21%, 15% and 11% of anthropogenic mercury release, respectively. Effluent dumping from mining, paper manufacturing, hydroelectric power projects and pulp production causes an alarming rate of mercury release. Other human-caused sources include landfills, municipal waste incineration, medical waste incineration, laboratory usages, dental amalgamation, alkali manufacturing, power generation via coal combustion, agricultural activities and so on.

The global mean concentration of mercury in the soil is estimated to be 1.1 mg kg^{-1}. Contamination, on the other hand, adds a several-fold increase to this value, sometimes near industrial sites 2456 mg kg^{-1}. The determinant of mercury toxicity is chemical speciation, which is determined by soil chemical properties such as pH, cation exchange capacity, electrical conductivity and organic matter. The elemental form is gaseous because of its high vapour pressure, which allows it to vaporise easily. The Hg reductase enzyme converts Hg^{2+} to Hg^0 in soil microbes and is found in bacteria with the Mer gene, including *Cryptococcus* sp., *Thiobacillus ferrooxidans*, *Pseudomonas* sp., *Streptomyces* sp. and *Staphylococcus aureus*. The increase in temperature boosts microbial activity in the soil, which in turn boosts Hg^{2+} methylation.

Mercury is primarily absorbed through inhalation, followed by oral ingestion. Incineration of waste containing mercury-related compounds or substances can produce Hg vapours that are inhaled and have a half-life of 60 days. With a half-life of 20 days, these vapours are lipid-soluble and encourage bioaccumulation in body parts. Hg exposure can have a negative impact on the cell, ranging from changes in cell membrane permeability to DNA damage. Because of its affinity for the sulphydryl group, mercury forms bonds with S-containing amino acids such as cysteine and methionine, which disrupts the cell's normal functioning and metabolism. Hg binds to RBC, allowing it to pass easily through the placenta and blood-brain barrier. The pancreas, lungs, breast, thyroid, adrenals, liver, myocardium, salivary glands, testes, skin, kidney, sweat glands, prostate, enterocytes and breast milk are all sites of Hg deposition. Elemental mercury is excreted from the body as mercuric Hg. Mercurous chloride, also known as calomel, has a low water solubility and intestinal absorption. However, a small amount is oxidised and absorbed in the body, which can result in acrodynia or pink disease. Mercuric Hg is excreted primarily through urine and faeces, whereas organic mercury is excreted primarily through faeces and breast milk. Cardiomyopathies, pulmonary fibrosis, bronchitis and Young's syndrome are all heart-related impairments and disorders. Chronic renal disease, nephrotic syndrome, glomerulonephritis, renal cancer and acute tubular necrosis are kidney disorders caused by Hg exposure. Because the nervous system is the primary target of mercury toxicity, a chronic malfunction state has resulted from the blockage of the enzymatic process, P-450, which inhibits energy generation in neurons.

Transmembrane proteins in plants can mediate mercury in-and-out translocation. In higher eukaryotes, ABC transporters (ATP binding cassette) are in charge of the influx and efflux of Hg ions. Mercury, whether present as Hg^{2+} in organic or inorganic form, is primarily toxic. Low productivity, genotoxicity, lipid peroxidation, nutrient imbalance and impaired photosynthetic process all impede plant growth. Mercury, a genotoxin, can cause genetic mutations by interfering with the proteins that control DNA repair, and they can break the DNA strands. The cell cycle is disrupted by low mitotic and cell proliferation indices. Excessive mercury reduces the total protein content of the plant and also causes the proteins to precipitate. The normal function of some components in the photosynthetic apparatus in the chloroplast is altered in the presence of Hg, such as the protein involved in oxygen evolution, the donor site of PS II, the subunit of ATP synthase, substitutive inhibition of Hg ions in the place of other essential metallic ions, inhibition of chlorophyll biosynthesis enzymatic activity and chlorophyll degradation, and lipid bilayer of the cell membrane is damaged by the hydrogen peroxides or other free radicals by a process called lipid peroxidation, which mostly yield two cytotoxic substances, namely, thiobarbituric acid relative substances (TBARS) and malondialdehyde. These compounds are the indicators of cells under stress. In *Cucumis sativus*, $HgCl_2$ applied exogenously was observed to have an increase in TBARS by 250%. Consumption of Hg-contaminated food items and its presence in edible parts pose potential health risks to the public. Around 22–42% of mercury exposure is through vegetable consumption. A severe mercury poisoning, first reported to be in Japan, was named Minamata disease which is a neurological syndrome, suffered by the people who consumed the fish and shellfishes with methyl mercuric contamination discharged from a nearby chemical plant.

2.6.11 Lead

Lead, the second most toxic metal and defined as a non-disintegrative and noxious heavy metal, accounts for 0.002% of the Earth's crust. Alloys can be formed due to their ductility and malleability. Inorganic lead can be found in old paint, soil and dust, whereas organic lead, tetraethyl lead, can be found in leaded gasoline. Both types of lead are toxic, but organic lead complexes have a greater toxic impact on the biological system. Physiological body functions require 29 ng per gram of dietary intake for reproductive functions, cell development and haematopoiesis to occur. Lead enters the body via contaminated water and food, the canning industry, atmospheric dust, paint and automobile exhaust, among other routes. The human body can absorb a large portion of ingested lead, approximately 70%. Natural causes include volcanic activity; weathering; remobilisation of water, soil or sediment from mining areas; and sea spray emissions, while anthropogenic causes include the use of paint, stained glass, batteries, cosmetics, agrochemicals, toys, jewellery, pottery, water from old pipes, smoking, radiators of cars and trucks, door frames, smelting, peeling windows, oil processing and plumbing. Lead can be found in soils as a free metal ion, in association with sulphates, carbonates, chlorides and bicarbonates or as organic ligands such as humic acid, fulvic acid and amino acid. Because plants have a high affinity for organic matter, their uptake of Pb is more pronounced (Kumar et al. 2020).

The devastation caused by Pb in the human body is determined by age, duration of exposure, amount of intake, route of intake and other factors. The Centers for Disease Control and Prevention has established a standard level of lead in the blood of adults as 10 g/dL and children as 5 g/dL. From the total absorbed Pb, only a small fraction (5%) is moved to the upper ground portion of the plant. Lead from the soil solution was found to be bound to polysaccharide or uronic acid in the rhizome of some plant species, including *Festuca rubra*, *Lactuca sativa*, *Brassica juncea*, *Vigna unguiculata* and *Funaria hygrometrica*, after adsorption on the roots. Passively entered Pb uses the apoplastic water stream for translocation up to the endodermis, but due to the impermeable Casparian strips, symplastic movement of Pb is observed after that. Pectins immobilise Pb in the root cell wall, while insoluble salts of Pb precipitate in the intercellular spaces. Pb sequestration in the vacuole or accumulation in the cell membrane has also been reported in root cells. A large portion of the Pb is excreted from the endodermis region as part of the plant's detoxification mechanisms. These may or may not be cited as reasons for the low Pb translocation to the upper ground portion.

Pb toxicity causes a decrease in growth, disrupted cell division, shortened roots, impaired chlorophyll structure and chloroplast lamella, reduced seed germination and seedling development in plants. Some plants' internal detoxification mechanisms include compartmentalisation (vacuolar metal sequestration), maintaining selectivity in heavy metal uptake, allowing specific ligands to bind with the metal for complexation, excretion and so on. Plants have a critical limit of 2 mg Pb kg^{-1}, above which there may be biochemical, physiological and morphological effects. Excess lead inhibits radical emergence from seeds by increasing protein and carbohydrates via negative effects on the enzymes peroxidase, acid phosphatases (amylase), acid invertases (amylase) and polyphenol oxidase. Roots lose their ability to oxidise. The Pb toxicity reduced the turgidity of a cell, decreased the plasticity of the cell wall, decreased the rate of transpiration, disrupted stomatal movements, reduced water use efficiency in plants and relative water content and caused the development of a waxy layer on a leaf, which reduced the respiration rate and transpiration ratio. It competes with the plants for the uptake of divalent cations as well as monovalent K^+.

It can be inhaled as Pb-laden dust, ingested in soils or ingested orally through polluted water and food, resulting in adverse health conditions. People who live near industrially polluted areas have varying lead levels in their blood during different seasons, such as summer and early fall. Lead primarily binds to haemoglobin and is transported throughout the body in the bloodstream. Large Pb particles are prevented from entering the respiratory tract by mucociliary cells, but they are swallowed after passing through the oropharynx. The distribution of lead in the body is independent of the route of exposure and is primarily confined to the bones. Body conditions such as lactation, pregnancy, osteoporosis and menopause increase bone resorption. This raises the blood lead level once more. Breastfeeding can be a way for the newborn to be exposed to the mother. Pb forms complexes with ligands such as albumin, ALAD, sulphydryls and others. Human excretion pathways include faeces, urine, saliva, breast milk, hair, sweat, nails and seminal fluids. Some clinical symptoms of lead poisoning in humans include hearing loss, hallucinations, a blue line in the gums, jaundice, pallor, depression, decreased libido, memory loss, coma, encephalopathy, insomnia, nausea, diarrhoea, reproductive impairments, anaemia, fatigue, seizers, convulsions, malaise and decreased bone density.

2.7 Conclusion

The pollution of heavy metal is of great concerning issue among all. It is increasing day by day as the industrialisation and other activities of day to day are increasing. With the increasing population today, it has become difficult to produce and feed quality products to the population. The pollution of heavy metal causes toxicity to the land and water and may lead to serious health hazards to the population. It is very important to deal with this problem so that the quality of the food products can be maintained. Different chemical and physical processes are used to remove the heavy metal pollutants, but these are very difficult to remediate. These methods are very expensive and effective results also not are there. So the process called phytoremediation is used to remediate the heavy metal pollutants from the soil as well as wastewater bodies. It is a safe and efficient method to remove the pollutants of heavy metals and degrade them easily. It is an ecological and cost-effective approach. It is a new study and requires more research in this field to get better and more effective results. The process of phytoremediation is associated with other studies also like soil microbiology, biotechnology, soil health and quality, plant metabolic and catabolic processes and plant physiology. There is a need to multiply the diversity of hyperaccumulator plants so that the pollution due to heavy metals can be reduced or the quality of the plant products can be increased for the betterment of the population. Breeding and transgenic techniques can be followed for the production of hyperaccumulator plants. More research in this field is required to get better results. More modifications in this research can help to understand the efficiency and to increase the mechanism of phytoremediation by hyperaccumulator plants. The uptake of heavy metal pollutants is only understood by the knowledge of soil physical and chemical processes. The knowledge of plant physiology is also very important to understand the translocation of the heavy metal pollutants inside the plant system. It is an effective technique and requires much attention so that it can become more beneficial further. In the future, this technique is expected to be more efficient and commercially viable.

References

Akeel A, Jahan A (2020) Role of cobalt in plants: its stress and alleviation. Contam Agric 25:339–357

Briffa J, Sinagra E, Blundell R (2020) Heavy metal pollution in the environment and their toxicological effects on humans. Heliyon 6(9):e04691

Cempel M, Nikel GJ (2006) Nickel: a review of its sources and environmental toxicology. Pol J Environ Stud 15(3):375–382

Clarkson DT (1988) The uptake and translocation of manganese by plant roots. In: Manganese in soils and plants. Springer, Dordrecht, pp 101–111

Cruz AM, Sarmento S, Almeida SM, Silva AV, Alves C, Freitas MC, Wolterbeek H (2015) Association between atmospheric pollutants and hospital admissions in Lisbon. Environ Sci Pollut Res 22(7):5500–5510

Dockery DW (2009) Health effects of particulate air pollution. Ann Epidemiol 19(4):257–263

Duffus JH (2002) "Heavy metals" a meaningless term? Pure Appl Chem 74(5):793–807

Engwa GA, Ferdinand PU, Nwalo FN, Unachukwu MN (2019) Mechanism and health effects of heavy metal toxicity in humans. In: Poisoning in the modern world-new tricks for an old dog. IntechOpen, London, p 10

Fernando DR, Lynch JP (2015) Manganese phytotoxicity: new light on an old problem. Ann Bot 116(3):313–319

Frisbie SH, Ortega R, Maynard DM, Sarkar B (2002) The concentrations of arsenic and other toxic elements in Bangladesh's drinking water. Environ Health Perspect 110(11):1147–1153

Genchi G, Carocci A, Lauria G, Sinicropi MS, Catalano A (2020) Nickel: human health and environmental toxicology. Int J Environ Res Public Health 17(3):679

Ghassemi-Golezani K, Farhangi-Abriz S (2019) Biochar alleviates fluoride toxicity and oxidative stress in safflower (Carthamus tinctorius L.) seedlings. Chemosphere 223:406–415

Hall JÁ, Williams LE (2003) Transition metal transporters in plants. J Exp Bot 54(393):2601–2613

Harischandra DS, Ghaisas S, Zenitsky G, Jin H, Kanthasamy A, Anantharam V, Kanthasamy AG (2019) Manganese-induced neurotoxicity: new insights into the triad of protein misfolding, mitochondrial impairment, and neuroinflammation. Front Neurosci 13:654

Islam F, Yasmeen T, Arif MS, Riaz M, Shahzad SM, Imran Q, Ali I (2016) Combined ability of chromium (Cr) tolerant plant growth promoting bacteria (PGPB) and salicylic acid (SA) in attenuation of chromium stress in maize plants. Plant Physiol Biochem 108:456–467

Khan K, Wasserman GA, Liu X, Ahmed E, Parvez F, Slavkovich V, Levy D, Mey J, van Geen A, Graziano JH, Factor-Litvak P (2012) Manganese exposure from drinking water and children's academic achievement. Neurotoxicology 33(1):91–97

Khan MU, Shahbaz N, Waheed S, Mahmood A, Shinwari ZK, Malik RN (2016) Comparative health risk surveillance of heavy metals via dietary foodstuff consumption in different land-use types of Pakistan. Hum Ecol Risk Assess 22(1):168–186

Kimbrough DE, Cohen Y, Winer AM, Creelman L, Mabuni C (1999) A critical assessment of chromium in the environment. Crit Rev Environ Sci Technol 29(1):1–46

Kolaj-Robin O, Russell D, Hayes KA, Pembroke JT, Soulimane T (2015) Cation diffusion facilitator family: structure and function. FEBS Lett 589(12):1283–1295

Kozlov MV, Haukioja E, Bakhtiarov AV, Stroganov DN, Zimina SN (2000) Root versus canopy uptake of heavy metals by birch in an industrially polluted area: contrasting behaviour of nickel and copper. Environ Pollut 107(3):413–420

Kumar A, Chaturvedi AK, Shabnam AA, Subrahmanyam G, Mondal R, Gupta DK, Malyan SK, Kumar SS, Khan S, Yadav KK (2020) Lead toxicity: health hazards, influence on food chain, and sustainable remediation approaches. Int J Environ Res Public Health 17(7):2179

Kumari S, Jamwal R, Mishra N, Singh DK (2020) Recent developments in environmental mercury bioremediation and its toxicity: a review. Environ Nanotechnol 13:100283

Liu K, Yu F, Chen M, Zhou Z, Chen C, Li MS, Zhu J (2016) A newly found manganese hyperaccumulator—Polygonum lapathifolium Linn. Int J Phytoremediation 18(4):348–353

Lucchini RG, Guazzetti S, Zoni S, Donna F, Peter S, Zacco A, Salmistraro M, Bontempi E, Zimmerman NJ, Smith DR (2012) Tremor, olfactory and motor changes in Italian adolescents exposed to historical ferro-manganese emission. Neurotoxicology 33(4):687–696

Mahey S, Kumar R, Sharma M, Kumar V, Bhardwaj R (2020) A critical review on toxicity of cobalt and its bioremediation strategies. Appl Sci 2(7):1–2

Manno E, Varrica D, Dongarra G (2006) Metal distribution in road dust samples collected in an urban area close to a petrochemical plant at Gela, Sicily. Atmos Environ 40(30):5929–5941

Manoharan V, Loganathan P, Tillman RW, Parfitt RL (2007) Interactive effects of soil acidity and fluoride on soil solution aluminium chemistry and barley (Hordeum vulgare L.) root growth. Environ Pollut 145(3):778–786

Masindi V, Muedi KL (2018) Environmental contamination by heavy metals. Heavy Metals 10: 115–132

Millaleo R, Reyes-Díaz M, Ivanov AG, Mora ML, Alberdi M (2010) Manganese as essential and toxic element for plants: transport, accumulation and resistance mechanisms. J Soil Sci Plant Nutr 10(4):470–481

Miransari M (2010) Contribution of arbuscular mycorrhizal symbiosis to plant growth under different types of soil stress. Plant Biol 12(4):563–569

Mombo S, Foucault Y, Deola F, Gaillard I, Goix S, Shahid M, Schreck E, Pierart A, Dumat C (2016) Management of human health risk in the context of kitchen gardens polluted by lead and cadmium near a lead recycling company. J Soils Sediments 16(4):1214–1224

Nagajyoti PC, Lee KD, Sreekanth TV (2010) Heavy metals, occurrence and toxicity for plants: a review. Environ Chem Lett 8(3):199–216

Noulas C, Tziouvalekas M, Karyotis T (2018) Zinc in soils, water and food crops. J Trace Elem Med Biol 49:252–260

Rehman M, Liu L, Wang Q, Saleem MH, Bashir S, Ullah S, Peng D (2019) Copper environmental toxicology, recent advances, and future outlook: a review. Environ Sci Pollut Res 26(18): 18003–18016

Rout G, Das P (2011) Effect of metal toxicity on plant growth and metabolism, review article Effect of metal toxicity on plant growth and metabolism. Agronomie 144:53

Santos EF, Santini JM, Paixão AP, Júnior EF, Lavres J, Campos M, Dos Reis AR (2017) Physiological highlights of manganese toxicity symptoms in soybean plants: Mn toxicity responses. Plant Physiol Biochem 113:6–19

Sarma H (2011) Metal hyperaccumulation in plants: a review focusing on phytoremediation technology. J Environ Sci Technol 4(2):118–138

Schreck E, Dappe V, Sarret G, Sobanska S, Nowak D, Nowak J, Stefaniak EA, Magnin V, Ranieri V, Dumat C (2014) Foliar or root exposures to smelter particles: consequences for lead compartmentalization and speciation in plant leaves. Sci Total Environ 476:667–676

Shabbir Z, Sardar A, Shabbir A, Abbas G, Shamshad S, Khalid S, Murtaza G, Dumat C, Shahid M (2020) Copper uptake, essentiality, toxicity, detoxification and risk assessment in soil-plant environment. Chemosphere 259:127436

Shahzad B, Tanveer M, Rehman A, Cheema SA, Fahad S, Rehman S, Sharma A (2018) Nickel; whether toxic or essential for plants and environment: a review. Plant Physiol Biochem 132: 641–651

Sharma R, Dutta A (2017) A study of heavy metal pollution in groundwater of Malwa Region of Punjab, India: current status, pollution and its potential health risk. Int J Eng Res Appl 7(3): 81–91

Simonsen LO, Harbak H, Bennekou P (2012) Cobalt metabolism and toxicology—a brief update. Sci Total Environ 432:210–215

Singh A, Prasad SM (2011) Reduction of heavy metal load in food chain: technology assessment. Rev Environ Sci 10(3):199–214

Smart ME, Gudmundson J, Christensen DA (1981) Trace mineral deficiencies in cattle: a review. Can Vet J 22(12):372

Tchounwou PB, Yedjou CG, Patlolla AK, Sutton DJ (2012) Heavy metal toxicity and the environment. Mol Clin Environ Toxicol 101:133–164

Valko MM, Morris H, Cronin MT (2005) Metals, toxicity and oxidative stress. Curr Med Chem 12(10):1161–1208

Villanueva R, Bustamante P (2006) Composition in essential and non-essential elements of early stages of cephalopods and dietary effects on the elemental profiles of Octopus vulgaris paralarvae. Aquaculture 261(1):225–240

Walker CH, Sibly RM, Peakall DB (2005) Principles of ecotoxicology. CRC Press, Boca Raton

Wang LK, Chen JP, Hung YT, Shammas NK (2009) Heavy metals in the environment. CRC Press, Boca Raton

WHO (2021) Manganese in drinking water: background document for development of WHO Guidelines for drinking-water quality. World Health Organization, Geneva

Wissemeier AH, Horst WJ (1992) Effect of light intensity on manganese toxicity symptoms and callose formation in cowpea (Vigna unguiculata (L.)Walp.). Plant Soil 143(2):299–309

Wuana RA, Okieimen FE (2011) Heavy metals in contaminated soils: a review of sources, chemistry, risks and best available strategies for remediation. Int Sch Res Notices 2011:402647

Xiong T, Dumat C, Pierart A, Shahid M, Kang Y, Li N, Bertoni G, Laplanche C (2016) Measurement of metal bioaccessibility in vegetables to improve human exposure assessments: field study of soil–plant–atmosphere transfers in urban areas, South China. Environ Geochem Health 38(6):1283–1301

Yaman M (2014) Teucrium as a novel discovered hyperaccumulator for the phytoextraction of Ni-contaminated soils. Ekoloji Dergisi 23:90

Zhuang P, McBride MB, Xia H, Li N, Li Z (2009) Health risk from heavy metals via consumption of food crops in the vicinity of Dabaoshan mine, South China. Sci Total Environ 407(5): 1551–1561

The Role of ABC Transporters in Metal Transport in Plants

3

Siddhi Kashinath Jalmi

Abstract

The metal ions are very essential for the survival, being an important part of many biological processes; however excess of metals causes toxicity. Correct balance of metal ions is necessary in the acquisition of metals depending on physiological needs. The uptake and transport of the essential metals occur with the help of metal transporters, which are also involved in homeostasis of metals. The metal transporters are found to be localized in different cellular compartments in different tissues depending upon their uptake in root cells and transport from root to xylem and to shoot and sequestration in vacuole to maintain the homeostasis. Different classes of transporters which are involved in uptake, transport, and sequestration of metals are discussed of which ATP binding cassette (ABC) family of transporters constitutes the largest family of transporters. Originally, ABC transporters in plants were identified in detoxification process where they exhibited important role in heavy metal detoxification and in metal homeostasis in cell. This chapter discusses the different subfamilies of ABC transporters and their role in metal transport and sequestration.

Keywords

ABC transporters · Heavy metal toxicity · Metal uptake · Metal sequestration · Phytoremediation

S. K. Jalmi (✉)
Department of Botany, School of Biological Sciences and Biotechnology, Goa University, Taleigao, Goa, India
e-mail: siddhi@unigoa.ac.in

K. Kumar, S. Srivastava (eds.), *Plant Metal and Metalloid Transporters*,
https://doi.org/10.1007/978-981-19-6103-8_3

3.1 Introduction

The acquisition of metal ions is very essential for the survival of all the organisms as these metals are an important part of many biological processes and serves as cofactors as a component of metalloproteins. Deficiency of the metals causes growth retardation, and developmental defects in organisms and uptake of excess of metals cause toxicity. Hence, proper balance is necessary in the acquisition of metals depending on physiological needs. Like other organisms, plants show mechanisms to tightly regulate metal uptake, allocation, and storage. The essential metals like iron (Fe), zinc (Zn), manganese (Mn), and copper (Cu) are trace metals that are required only in traces usually less than one parts per million (ppm), and their accumulation leads to toxicity. Apart from these, nature does contain nonessential heavy metals that are toxic and interfere with transporters required for uptake and transport of essential metals. These heavy metals possess deleterious effects on environment and ecosystem and are accumulated in environment naturally due to volcanic eruption and anthropogenically due to mining, incinerations, the use of pesticides and fertilizers, etc. Heavy metals include arsenic (As), cadmium (Cd), chromium (Cr), iron (Fe), lead (Pb), mercury (Hg), silver (Ag), and zinc (Zn) (Phillips 1981).

The uptake and transport of the essential metals occur with the help of metal transporters, which are also involved in homeostasis of metals. Many of the heavy metals are known to hijack metal transporters for their uptake and transport. Plants have acquired specialized genetic and biochemical mechanism to sense and transport essential metals at proper physiological concentration and remove nonessential metals from the cellular environment (Verbruggen et al. 2009). The metal transporters are found to be localized in different cellular compartments in different tissues depending upon their uptake in root cells and transport from root to xylem and to shoot and sequestration in vacuole to maintain the homeostasis. Different classes of transporters which are involved in uptake, transport, and sequestration of metals are discussed below.

One of the major metal transporter families involved in metal uptake is ZIP family. The members of this family are zinc-regulated transporter (ZRT-like proteins) and iron-regulated transporter (IRT), and this is where the name of family arises (Claus and Chavarría-Krauser 2012). These transporters translocate divalent cations across the membrane such as Fe, Zn, Cd, Mn by IRT1, and Zn by ZRT1 and ZRT2 (Grotz et al. 1998; Vert et al. 2002). These cationic or divalent metal transporters also take part in transport of heavy metal Cd (Thomine et al. 2000). Another family of transporter associated with Cu transport and homeostasis is copper transporters (COPT), having transmembrane domains in protein. These are also involved in transport of Zn and Fe (Morel et al. 2009; Yuan et al. 2011) (Fig. 3.1).

Natural resistance-associated macrophage protein (NRAMP) transporters play an important role in divalent metal transport including heavy metals like Fe, Mn, Cu, Zn, Ni, Co, and Cd (Sasaki et al. 2012). In *Arabidopsis*, Fe, Mn, and Cd transport is carried out by NRAMP1, NRAMP3, and NRAMP4, wherein NRAMP1 is localized

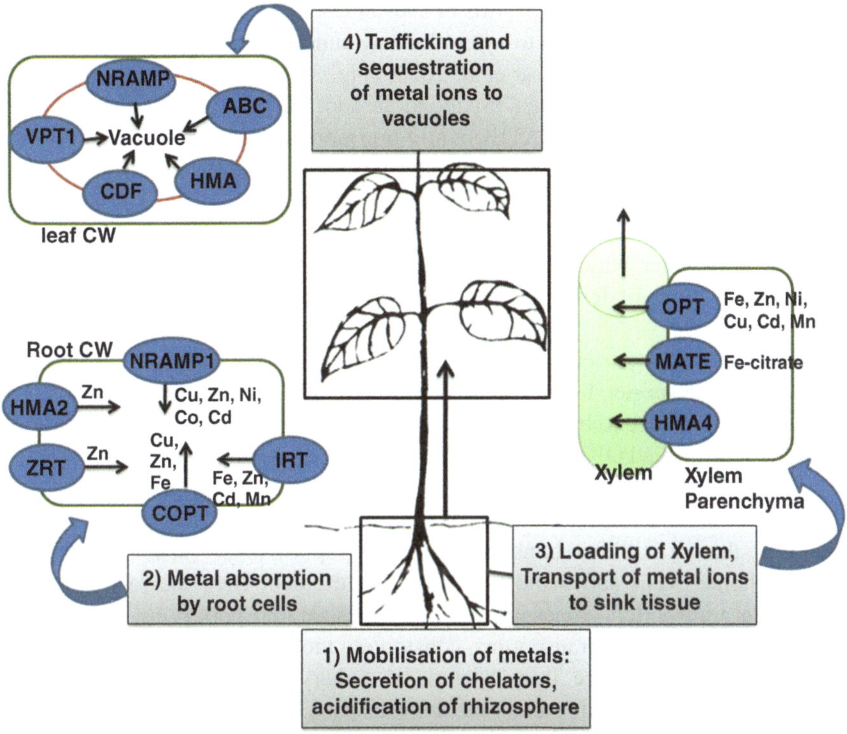

Fig. 3.1 Overview of metal transporters in various stages of metal transport and sequestration. The uptake of metal ions and their transport and sequestration occur in various stages. Firstly, the metals are mobilized by the roots by secretion of chelators and acidification of the soil surrounding the roots. After mobilization, metal ions are imported into the root cells by various transporters present on the cell membrane of root cells (as described in the figure). Following the metal absorption, metal ions are transported to the xylem parenchyma for the loading of xylem wherein a set of metal transporters are involved. Through the xylem, metals are transported to sink tissues wherein their sequestration in vacuoles takes place depending on the concentration and homeostasis of metal ions

in cell membrane of root cells and NRAMP3 and NRAMP4 are localized in tonoplast. NRAMP2 is studied to be involved in Fe transport, whereas Cd is transported by NRAMP6 (Sasaki et al. 2012; Bozzi and Gaudet 2021) (Fig. 3.1).

Apart from contribution of metal transporters in the uptake of metals by root cells, they also contribute in their transport from root to shoot. For this, the loading of the xylem from the adjacent parenchymal cells is necessary, which is performed by the transporters discussed below. One of the transporters contributing in root-to-shoot transport is heavy metal transporting P-type ATPase (HMAs). This transports metals within the cellular compartments, that is, from cytoplasm to xylem or from plasma membrane to organelles. Of these P-type ATPase, HMA2 transports Zn in developing tissues, and HMA3 transports Cd and is present in tonoplast; however, HMA4 plays a role in loading metal to xylem tissue (Hussain et al. 2004; Yamaji et al. 2013). P-type ATPase are the ATP-dependent ion transporters which are the largest and most diverse transporters in terms of substrate specificity (Palmgren and Nissen 2011). These are also reported to xylem loading and shoot transport of heavy metal like Cd in addition to Zn (Ceasar et al. 2020) (Fig. 3.1).

Another family of transporters involving the root-to-shoot transport is multidrug toxic compound extrusion (MATE) family transporters. These are basically efflux proteins studied to detoxicate the cellular environment by extruding the toxic compounds out of the cell. These transporters mainly transport Fe from root to shoot and extrude Al out of cells. Example of transporter belonging to MATE class is citrate transporter which participates in loading of Fe and citrate in vascular tissues, in the form of ferric citrate complex (Durrett et al. 2007; Delhaize et al. 2012) (Fig. 3.1).

A family of transporter capable of transporting diverse range of substrates is oligopeptide transporter family (OPT), of which yellow strip protein 1 (YS1) is characterized in maize to be involved in transporting Fe, Zn, Ni Cu, Cd, and Mn. In rice, several identified OPTs are known to translocate chelated Fe (Lubkowitz 2011; Liu et al. 2012) (Fig. 3.1).

Whenever a metal ion is not utilized in any metabolic activity or if its concentration rises, then it is either transported to the apoplast or to vacuole in order to maintain homeostasis and avoid toxification of the cell. Here the role of several intracellular transporters comes into play. One of such transporters are ATP binding cassette (ABC) family of transporters, constituting the largest family of transporters, ranging from its presence from bacteria to eukaryotes. These are ATP-driven transporters containing transmembrane domains used in the transport of xenobiotics and chelated metals into vacuole (Henikoff et al. 1997; Wanke and Üner Kolukisaoglu 2010). Apart from ABC transporters, cation diffusion facilitator (CDF) family and HMA and NRAMP transporter family are also involved in sequestration of metals such as Pb, Zn, Co, Cd, and Fe into the vacuoles and vacuolar phosphate transporters (VPT1) involved in the intake of heavy metal As (Kawachi et al. 2008; Sasaki et al. 2012) (Fig. 3.1).

ABC transporters being the largest family of transporters, and despite of the availability of genome sequence of model plant *Arabidopsis*, the functional knowledge about this large ABC transporter family is still scarce. This chapter focuses on the role of ABC transporters in the metal transport and their sequestration in plants, giving information about the structure of ABC transporters and their localization in plants and their role in metal transport.

3.2 ABC Transporter Family

As discussed above, ABC transporter constitutes largest family ranging from its presence in bacteria to humans, primarily serving as membrane-intrinsic primary activated pumps. ABC transporters are known to be transporters of diverse substances like lipids, carbohydrates, phytohormones, chlorophyll catabolites, xenobiotic conjugates, and heavy metals across the cellular membrane. Observing their diverse role, these transporters are localized on diverse membrane like plasma membrane, tonoplast, and mitochondrial, plastidial, and peroxisomal membrane, implicating their function in cellular secretion and vacuolar sequestration. The first ABC transporter to be studied was involved in nutrient uptake in bacteria and was called as prokaryotic periplasmic permease (Ames 1986). Much of the studies have been performed on other members of ABC transporter superfamily and are identified as transporters from human, yeast, and bacteria that mediate multiple drug resistance (MDR), also including cystic fibrosis transmembrane conductance regulator (CFTR) and sulfonylurea receptor (SUR) (Rommens et al. 1989; Aguilar-Bryan et al. 1995; Prasad et al. 1996). In addition to their importance in animals and other organisms, ABC transporters also have emerged as fascinating transporters in plants. In the review by Higgins in 1992, only one plant ABC transporter was reported. Two pioneering studies marking the start of research on plant ABC transporters describe the isolation of MDR homolog in *Arabidopsis thaliana* and transport of glutathione conjugate into the vacuole by MRP-like transporter in barley (Dudler and Hertig 1992; Martinoia et al. 1993). Currently, in model plant *Arabidopsis* and crop plant rice, there are around 120 members of ABCs identified. In *Arabidopsis*, there are nine subclasses of these transporters, namely, ABCA, ABCB, ABCC, ABCD, ABCE, ABCF, ABCG, ABCH, and ABCI, having different locations in cell, with involvement in different metal transports (Verrier et al. 2008). However, based on the size of the protein, orientation, and presence of transmembrane domain, plant ABC transporters can be classified into 13 subfamilies and are categorized into full-molecule transporters, half-molecule transporters, and quarter molecule transporters (Rea 2007). ABC transporters being an important part of detoxification machinery are also recognized to participate in a range of physiological processes like polar auxin transport, hormonal regulation of abscisic acid (ABA), lipid metabolism, stomatal regulation, disease resistance, etc., allowing plant to cop up with environmental biotic and abiotic stresses (Geisler and Murphy 2006; Stein et al. 2006; Kang et al. 2010; Kuromori et al. 2010). ABC transporters exhibit peculiar features as compared to other families of transporters. Firstly, ABC transporters are ATP driven, but the form of ATP that is utilized is Mg-ATP and not the free ATP. However, other nucleoside triphosphates like UTP or GTP can partially be substituent for ATP in the transport process. Also, vanadate, a metastable analog of orthophosphate, can substitute for ATP thus affecting the transporters. Secondly, the transmembrane electrochemical potential difference does not affect the transport process by ABCs (Rea et al. 1998).

3.3 Molecular Structure of ABC Transporters

Modular structure of ABC transporters includes four core structural domains consisting of two transmembrane domains (TMD) and two nucleotide folds (NBF). The TMD consist of multiple membrane spanning alpha helices, and NBF is involved in binding to ATP and its hydrolysis. In bacteria, these domains are encoded on different polypeptides by different genes and are thus called as half-size ABC transporters. While in eukaryotes, these transporters are encoded as one full polypeptide, so-called as full-size transporters. Full-size transporters can be further divided based on their orientation as multidrug-resistant proteins (MDRPs), multidrug-resistant-like proteins (MRPs) having the topology of TMD-NBF-TMD-NBF, while the pleiotropic drug resistance (PDRs) and ABC-like proteins exhibit mirrored topology of NBF-TMD-NBF-TMD (Bungert et al. 2001) (Table 3.1). The NBF domain is made up of two conserved sequence motifs: one is ATP binding site consisting of two boxes Walker A and Walker B, separated by 120 amino acids of specific consensus sequence of ABC transporters. The NBF domain sequence exhibits 30–40% identity over a range of 200–400 amino acids among different ABC transporters (Martinoia et al. 2002; Linton 2007).

Sequencing of *Arabidopsis* genome has revealed the presence of 53 genes encoding full-size ABC transporters which seemed to be relatively more than ABC transporter genes found in yeast and mammals. This large number of ABC genes and higher sequence homology suggest functional redundancy (Martinoia et al. 2002; Rea 2007).

3.4 Primary Superfamilies of Plant ABC Transporters

3.4.1 MDR Superfamily

The first homolog of MDR to be cloned was AtPGP1 from *Arabidopsis*. AtPGP1 exhibit similar intron and exon structure and similar structural domain organization as that of mammalian homolog (Dudler and Hertig 1992). Another homolog AtPGP2 sharing 44% amino acid identity with AtPGP1 was cloned, and this showed 45% identity to mammalian MDR1 (Dudler and Sidler 1998). The function of AtPGP1 was demonstrated in hypocotyl cell elongation and export of peptide hormone from shoot apex. Homologs of mammalian MDR1 have also been reported in other plant species like PMDR1 in potato (showing 86% identical to AtPGP1/AtPGP2) and HvMDR1 and HvMDR2 in barley (showing 43% identical to AtPGP1/AtPGP2) (Wang et al. 1996; Davies et al. 1997). Much of the information about primary structure, topology, and domain organization can be revealed by comparing these homologs. Plant MDR proteins have significantly smaller molecular weight (between 134 and 144 kDa) than mammalian counterparts (~180 kDa) (Higgins 1992).

3.4.2 MRP Superfamily

Investigations on the function of ABC transporters in xenobiotic detoxification led to the identification and characterization of MRP superfamily transporters (Rea et al. 1998). Based on the MRP members in mammals and yeast cadmium factor YCF1 (MRP1 homolog in yeast), search for the homologs was carried out in model plant *Arabidopsis* (Li et al. 1997). AtMRP1, AtMRP2, AtMRP3, AtMRP4, and AtMRP5 were cloned and identified to be the homologs, encoding glutathione-conjugated

Table 3.1 Subfamilies of ABC transporters in plants

	ABC subfamily	Protein members	Domain orientation	Transmembrane domain	Localization
1	ABCA ATH (half size) and AOH (full size)	ABCA1 (full size) and ABCA2–ABCA12 (half size)	Forward orientation TMD-NBF-TMD-NBF	TM domain present	Not known
2	ABCB TAP/HMT (half size) MDR/DGP (full size)	ABCB1–ABCB7, ABCB9– ABCB22 (full size), and ABCB23–ABCB29-(half size)	Forward orientation TMD-NBF-TMD-NBF	TM domain present	Full-size proteins are localized to plasma membrane, and half-size proteins are localized to mitochondria; ABCB27 is localized in vacuolar membrane
3	ABCC MRP (full size)	ABCC1– ABCC15 (full size)	Forward orientation TMD-NBF-TMD-NBF	Additional TM domain (TMD0) present	Localized in vacuolar membrane
4	ABCD (PMP)	ABCD1 (full size) and ABCD2 (half size)	Forward orientation TMD-NBF-TMD-NBF	TM domain present	Localized in peroxisomes
5	ABCE/F RLI/GCN	ABCE1, ABCE2, and ABCE3 and ABCF1–ABCF5	–	Lack TM domain	Soluble cytosolic protein
6	ABCG PDR (full size) WBC (half size)	ABCG1– ABCG18 (half size), ABCG29–ABCG39, and ABCG43 (full size)	Reverse orientation NBF-TMD-NBF-TMD	TM domain present	Localized to plasma membrane except ABCG19
7	ABCH	Not identified in plants			
8	ABCI (bacterial type)	Seven members having only one TMD, ten members having only one NBF, four members having other domains	–	–	ABCI with NBF are localized to cytosol; rest ABCI are localized to mitochondria and chloroplast

transporters, located to be in tonoplast vesicles and isolated vacuoles (Lu et al. 1997, 1998; Marin et al. 1998; Sánchez-Fernández et al. 1998; Tommasini et al. 1998). AtMRP1 showed the highest transport capacity and is able to reduce the Cd sensitivity (Tommasini et al. 1998).

3.5 Classes of Plant ABC Transporters

ABCA subfamily of *Arabidopsis* includes 12 proteins of which ABCA1 is orthologous to mammalian ABC1 and is a full-size protein having the large linker domain. The rest of 11 proteins in ABCA subfamily are half-size proteins missing the linker domain, and these are only found to be present in plants (Kovalchuk and Driessen 2010). ABCB subfamily contains 21 full-size and half-size proteins in *Arabidopsis*, which are much more than what is present in mammals (Lee et al. 2008). ABCC subfamily contains only full-size proteins, having N-terminal transmembrane domain, the role of which is not studied in plants; however, it has role in protein targeting in yeast and mammals (Klein et al. 2006). *Arabidopsis* ABCD subfamily contains one full-size ABCD1 and one half-size transporter (Hayashi et al. 2002). Members of ABCE subfamily include three proteins and that of ABCF include five proteins, and they have reported to be functioning as soluble proteins performing other functions like ribosomal recycling and not as transporters, since they lack transmembrane domain (Pisarev et al. 2010; Dong et al. 2017). ABCG subfamily is the largest subfamily having reverse orientation of NBF-MSD domain organization. This subfamily includes 12 full-size and 28 half-size proteins, and they have been identified to be present in plants, fungi, oomycetes, and even slime molds (Stein et al. 2006; Choi et al. 2011). ABCH subfamily also contains reverse orientation of domains; however, phylogenetically it is not related to ABCG and is not characterized in plants. ABCI subfamily have 21 members and are encoded as individual NBF as cytosolic protein lacking transmembrane domain (Zeng et al. 2017) (Table 3.1).

3.6 Role of ABC Transporters

ABC transporters in bacteria were studied to be involved in transport of primary metabolites and export of lipids, enzymes like protease, lipases, antibiotics, cytotoxin, etc. In eukaryotes, ABC transporters were identified to play a role in cellular detoxification that helped in extrusion of toxic compounds from the cytosol. In plants, the first report stating ABC-mediated transport was observed in investigation of xenobiotic detoxification. In this study, it was noted that toxic compounds in plants get conjugated with glutathione tripeptide through thiol group, which are then transported to vacuole independently by vacuolar proton pumps. Xenobiotic detoxification is a multistep process, composed of stage 1 of activation, stage 2 of conjugation, and stage 3 of compartmentalization. In stage 1, the activation is caused by either the hydrolysis reaction by esterases and amidase or by oxidation reaction by cytochrome P-450. This activation stage generates products with higher toxicity; however, it is an important step in detoxification since in this the functional groups are generated which are required in the next conjugation stage. In conjugation stage, the activated products of toxic compounds covalently linked to hydrophilic substances like glucose/glutathione/malonate in plants and glutathione/ glucuronate/sulfate in animals. In stage 3 of compartmentalization, the inactive water-soluble conjugates are transported from cytoplasm to compartments like vacuoles (Kreuz et al. 1996; Coleman et al. 1997). In animals, these conjugated compounds are secreted out of the cell through the plasma membrane, with the help of specific ATPase. However, in plants, these compounds are transported into the vacuoles, wherein they can be further metabolized. The transport to the vacuoles in plants is mediated by Mg-ATP-dependent transporters and not by nonhydrolyzable ATP or by inorganic phosphate (PPi). These pumps later were identified to be vacuolar glutathione-conjugate ABC transporters ABCC1 and ABCC2, performing transport of conjugated toxin to the vacuole (Martinoia et al. 1993; Kreuz et al. 1996). Both these ABCC transporters are involved in the transport of a broad range of glutathione conjugates. *Arabidopsis* ABCC1 is also studied to be an efficient transporter of folate which is known to be stored in vacuole and is necessary for methylation processes (Raichaudhuri et al. 2009). Initial stages of detoxification are similar in plants and animals, wherein cytochrome P450, glutathione-S-transferase, and glycosyltransferase are involved in conjugation. But in later processes in animals, conjugated toxic compounds are excreted out, whereas in plants, they are internalized in large central vacuole (Lu et al. 1998).

In eukaryotes, the ABC transporters are studied to perform transport in both directions irrespective of the side of the presence of NBF domain and the substrate (Shitan et al. 2003). Originally ABC transporters in plants were identified in detoxification process where they exhibited important role in heavy metal detoxification and in metal homeostasis in cell. However, apart from metal detoxification, ABC transporters are involved in diverse processes like transport of phytohormones, phytate accumulation in seed, and surface lipid deposition; hence, they play an important role in plant growth and development and stress management (Martinoia et al. 2002).

3.6.1 Role in Growth and Development: Transport of Hormones, Fatty Acids, and Phytate

Other than toxic compound, ABC transporters have role in transport of important compounds like fatty acids, hormones, phytates, etc. Phytohormones serve to be crucial components of plant growth and development; hence, their biosynthesis and transport become vital for their proper function. For example, indole-3-acetic acid (auxin) is a central phytohormone performing many developmental processes like shoot development, lateral root formation, floral bud development, phototropism, and gravitropism. This is produced in shoot apical meristem and transported throughout the plant via xylem parenchyma cell-cell transport (Vanneste and Friml

2009). The cellular import and polar export of auxin are performed by two vital transporters, AUX1/LAX family and PIN family, respectively (Kramer 2004). In addition to these transporters, two ABC plasma membrane intrinsic ABC transporters ABCB1 and ABCB19 also contribute to intercellular auxin transport. It has been studied that ABC transporters and PIN proteins interact and modulate overall auxin transport activity (Geisler et al. 2005; Blakeslee et al. 2007). Besides auxins, ABC transporters transport stress hormone ABA from the site of its production, that is, from vascular tissues of root and shoot to foliar tissues. ABCG half-size transporter ABCG25, which is expressed in vascular parenchyma of root and shoot, exports ABA and shows high affinity to ABA (Kuromori et al. 2010). ABCG40 is involved in import of ABA across the stomatal plasma membrane which is also implicated in heavy metal resistance. Owing to the role of ABCG40, it is localized in stomatal plasma membrane along with another transporter ABCB14, regulating the import of apoplastic malate (Lee et al. 2008).

As noted earlier, ABC transporters are also involved in fatty acid transport from cytosol to peroxisomes, where fatty acids are involved in production of acetyl CoA. A study in *Arabidopsis* reported the role of ABCD1 in fatty acid import in peroxisomes (Zolman et al. 2001). The fatty acids are also important components of cuticle in plants, which is a protective layer covering the epidermis of aerial organs in plants. Cuticle formation is also necessary for proper organ development and in maintaining the morphology of organs. Two half-size ABC transporters, ABCG11/ABCG12, play a crucial role in cuticle formation (Pighin et al. 2004).

3.6.2 Role in Pathogen Defense

Plant's secondary metabolites are considered to be important in pathogen defense, giving first line of defense against host and nonhost pathogens. Plants produce these metabolites and transport them to different plant parts like aerial surfaces, rhizospheric region, and apoplast and near the site of infection. This transport is studied to be mediated by a number of full-size ABC transporters of ABCG subfamily. ABCG30 mutation resulted in change in microflora in the plant's rhizosphere, which suggested the change in root exudate composition due to nonfunctionality of ABCG transporter. Apart from this, six other full-size ABC transporters and a half-size transporter were involved in transport and secretion of root exudates, mutation of which resulted in the alteration of plant's secondary metabolites in root exudate (Badri et al. 2008, 2009). The putative role of ABC transporter in pathogen defense was suggested in expression analysis of ABC transporters in response to biotic stress hormones SA and JA (Moons 2008). A full-size ABCG transporter PDR1 was reported to be the first transporter involved in active transport of terpenoids in plants, thus imparting defense against pathogen (Jasiński et al. 2001). This transporter is found to be overexpressed in the leaf epidermis, trichomes, petals, and roots, and its transcripts get induced upon the treatment of terpenoids, SA and JA. Mutation or downregulation of PDR1 causes plant susceptible to nonhost necrotrophic pathogens (Stukkens et al. 2005). Another study suggested the role of ABCG38/PDR8 in nonhost resistance, wherein the mutation of this transporter increased the susceptibility toward nonhost pathogens; however, it provided hypersensitivity against compatible host pathogen (Stein et al. 2006).

Besides acting as pathogen repellent, plant's secondary metabolites like isoflavonoids also act as attractant for beneficial microbes and plant-derived signaling molecules for establishing symbiosis with rhizobia, mycorrhizae, etc., thus helping in promoting the plant growth (Sugiyama et al. 2008). Considering the role of ABC transporters in transporting plant's secondary metabolites, it can be also assumed that these transporters play a role in defense against herbivores.

3.7 ABC Transporters in Metal Transport and Sequestration

Although the ABC transporters have extended role other than transporting metals and heavy metals, however, involvement of ABC transporters in detoxification mechanism has been always a focal point, taking into consideration the metal contamination of agricultural soils becoming a serious problem. In this section, we will focus on function of metal, the role of ABC transporters in *Arabidopsis*, and some other plant species. ABC transporters present on plasma membrane are involved in metal uptake and transport, while those present on vacuolar membrane are involved in detoxification. Most of ABC transporters found in the vacuolar membrane belong to ABCC subfamily.

The detoxification of metal by ABC transporters on vacuolar membrane in plants is largely dependent on formation of phytochelatin (PC), that is, the peptide-type chelators, which are synthesized by heavy metal-activated phytochelatin synthase from glutathione (Cobbett 2000). These PCs are produced in response to increased level of metals and heavy metals in plants. The first metal phytochelatin transporter that was identified in tonoplast membrane of yeast was HMT1, of which the homologs have been studied in animals while its homolog has not been identified in vacuolar membrane of plants (Ortiz et al. 1995; Schwartz et al. 2010). However, a mitochondrial ABC transporter ABCB25 that is involved in biogenesis of iron-sulfur (Fe-S) cluster and biosynthesis of molybdenum cofactor in plants is studied to be close homolog of HMT of yeast. ABCB25 is also involved in transport of glutathione-Cd, Cd-S, and Fe-S from mitochondria to cytosol (Kim et al. 2007) (Fig. 3.2).

After a long time of identification of vacuolar phytochelatin transporter in yeast, Song and his team in 2010 succeeded in identifying the vacuolar phytochelatin transporters in plants. These were two ABCC proteins ABCC1 and ABCC2, having redundant function identified in *Arabidopsis*. These were studied to be main important transporters of arsenic-PC complex. The ABCC1/2 showed marginal effect in arsenic stress tolerance in *Arabidopsis* (Song et al. 2010). The study suggested that to obtain tolerance to arsenic overexpression of both ABCC transporter and PC synthase was mandatory and not of just the ABCC transporters (Song et al. 2014a, b). Subsequently, the role of these ABCC1 and ABCC2 was identified in

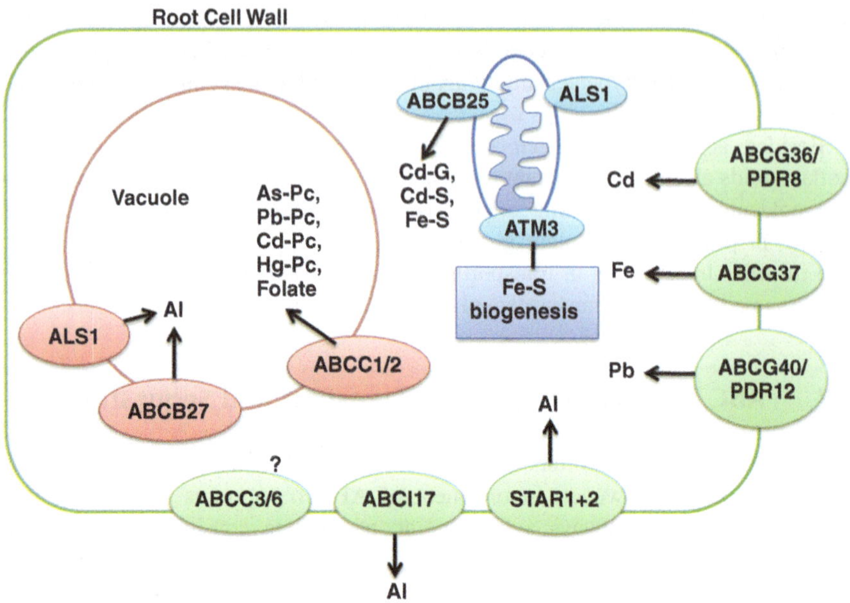

Fig. 3.2 Metal transport and sequestration by ABC transporters. The members of ABC transporter family playing an important role in metal transport and sequestration are described in this figure. Several ABC transporters are known to be localized on cell membrane (mainly ABCG and ABCI subfamily) importing metals inside the root cell from the soil. Some ABC transporters ABCB and ABCC subfamily present on vacuolar membrane are involved in sequestration of metals. Also, ABC transporters are found to be present on mitochondrial membrane transporting metals in and out of mitochondria

vacuolar sequestration of cadmium and mercury, thus providing tolerance to Cd and Hg stress (Park et al. 2012) (Fig. 3.2). The homolog of ABCC1 in yeast is yeast cadmium factor (YCF1) which was the first transporter identified to be involved in heavy metal resistance. YCF1 mediate the transport of glutathione metal complexes for As and Cd, similar to its function in plants (Ghosh et al. 1999). The transport of Cd is also exhibited by other two proteins of ABCC family ABCC3 and ABCC6; however, the mechanism of transport is not examined in plants (Wanke and Üner Kolukisaoglu 2010). Apart from these, ABCB family member, ALS1, found to be localized on vacuolar membrane performs the transport of aluminum in rice (Huang et al. 2012) (Fig. 3.2).

In addition to their prominent role in exporting metals to vacuole for sequestration, some of ABC transporters have shown their role in metal uptake from soil. These transporters belong to ABCG subfamily and bacterial-type ABC subfamily having their localization on plasma membrane vesicle of root cells. The bacterial-type ABC proteins are STAR1 and STAR2 that play a crucial role in imparting tolerance to aluminum stress in rice. STAR1 corresponds to nucleotide binding domain whereas STAR2 corresponds to transmembrane domain, where both these domains make single functional transporter protein (Huang et al. 2009). ABCG subfamily transporter, ABCG36, also called as PDR8/PEN3 is also reported to be localized at plasma membrane which was upregulated in response to Cd metal, and its overexpression imparted tolerance to Cd metal stress (Stein et al. 2006) (Fig. 3.2). Apart from metal transport, ABCG36 also exhibits an important role in defense against pathogen (Stein et al. 2006).

Another member of ABCG family, ABCG40/PDR12, was studied to be involved in lead detoxification; this fact was proven by using the overexpression and mutant lines of ABCG40, wherein overexpression lines were more tolerant and accumulated less Pb while mutant lines were more sensitive and accumulated more Pb inside the plants (Lee et al. 2005) (Fig. 3.2). This Pb tolerance driven by ABCG40 may be due to direct transport of Pb causing its detoxification or may be due to uptake of stress hormone ABA (Kang et al. 2010).

AtATM3 is an ATP binding cassette transporter from *Arabidopsis* and was studied to be involved in biogenesis of iron-sulfur cluster and iron homeostasis in plants. AtATM3 is a mitochondrial protein and was seen to be upregulated in root cells of plants treated with lead and cadmium and imparted tolerance to Pb and Cd heavy metals (Kim et al. 2006) (Fig. 3.2).

3.8 Future Prospects

Heavy metal contamination and its toxicity have posed serious effects on the agriculture and the health of living organisms. For past many years, efforts have been made to develop plants that are able to accumulate large amounts of heavy metals and at same time produce high biomass. Observing the roles of these ABC transporters in metal uptake and sequestration, researchers are trying to make use of these in phytoremediation of metal-contaminated sites. Approaches being used are either overexpressing these metal transporters or producing higher amount of glutathione or phytochelatin in plants. Looking at the promising role of yeast ABC transporter YCF1 in heavy metal sequestration in various plants, transgenic popular plants overexpressing YCF1 are produced. These plants are capable of accumulating much of Cd from contaminated soils and were found to be much more tolerant and produced higher biomass (Shim et al. 2013). These transgenic popular trees due to their larger root system are efficient for long-term stabilization of metal-polluted soils. The use of the metal transporters particularly ABC transporters will provide useful tools for genetic engineering of plant with enhanced metal accumulation and tolerance, with their efficient use in phytoremediation.

References

Aguilar-Bryan L, Nichols CG, Wechsler SW, Clement J, Boyd A, Gonzalez G, Nelson DA (1995) Cloning of the beta cell high-affinity sulfonylurea receptor: a regulator of insulin secretion. Science 268(5209):423–426

Ames GFL (1986) Bacterial periplasmic transport systems: structure, mechanism, and evolution. Annu Rev Biochem 55(1):397–425

Badri DV, Loyola-Vargas VM, Broeckling CD, De-la-Peña C, Jasinski M, Santelia D, Vivanco JM (2008) Altered profile of secondary metabolites in the root exudates of Arabidopsis ATP-binding cassette transporter mutants. Plant Physiol 146(2):323–324

Badri DV, Quintana N, El Kassis EG, Kim HK, Choi YH, Sugiyama A, Vivanco JM (2009) An ABC transporter mutation alters root exudation of phytochemicals that provoke an overhaul of natural soil microbiota. Plant Physiol 151(4):2006–2017

Blakeslee JJ, Bandyopadhyay A, Lee OR, Mravec J, Titapiwatanakun B, Sauer M, Murphy AS (2007) Interactions among PIN-FORMED and P-glycoprotein auxin transporters in Arabidopsis. Plant Cell 19(1):131–147

Bozzi AT, Gaudet R (2021) Molecular mechanism of Nramp-family transition metal transport. J Mol Biol 433:166991

Bungert S, Molday LL, Molday RS (2001) Membrane topology of the ATP binding cassette transporter ABCR and its relationship to ABC1 and related ABCA transporters: identification of N-linked glycosylation sites. J Biol Chem 276(26):23539–23546

Ceasar SA, Lekeux G, Motte P, Xiao Z, Galleni M, Hanikenne M (2020) di-Cysteine residues of the Arabidopsis thaliana HMA4 C-terminus are only partially required for cadmium transport. Front Plant Sci 11:560

Choi H, Jin JY, Choi S, Hwang JU, Kim YY, Suh MC, Lee Y (2011) An ABCG/WBC-type ABC transporter is essential for transport of sporopollenin precursors for exine formation in developing pollen. Plant J 65(2):181–193

Claus J, Chavarría-Krauser A (2012) Modeling regulation of zinc uptake via ZIP transporters in yeast and plant roots. PLoS One 7(6):e37193

Cobbett CS (2000) Phytochelatins and their roles in heavy metal detoxification. Plant Physiol 123(3):825–832

Coleman J, Blake-Kalff M, Davies E (1997) Detoxification of xenobiotics by plants: chemical modification and vacuolar compartmentation. Trends Plant Sci 2(4):144–151

Davies TE, Theodoulou FL, Hallahan DL, Forde BG (1997) Cloning and characterisation of a novel P-glycoprotein homologue from barley. Gene 199(1-2):195–202

Delhaize E, Ma JF, Ryan PR (2012) Transcriptional regulation of aluminium tolerance genes. Trends Plant Sci 17(6):341–348

Dong J, Piñeros MA, Li X, Yang H, Liu Y, Murphy AS, Liu D (2017) An Arabidopsis ABC transporter mediates phosphate deficiency-induced remodeling of root architecture by modulating iron homeostasis in roots. Mol Plant 10(2):244–259

Dudler R, Hertig C (1992) Structure of an MDR-like gene from Arabidopsis thaliana. Evolutionary implications. J Biol Chem 267(9):5882–5888

Dudler R, Sidler M (1998) Arabidopsis MDR genes: molecular cloning and protein chemical aspects. Methods Enzymol 292:162–173

Durrett TP, Gassmann W, Rogers EE (2007) The FRD3-mediated efflux of citrate into the root vasculature is necessary for efficient iron translocation. Plant Physiol 144(1):197–205

Geisler M, Murphy AS (2006) The ABC of auxin transport: the role of p-glycoproteins in plant development. FEBS Lett 580(4):1094–1102

Geisler M, Blakeslee JJ, Bouchard R, Lee OR, Vincenzetti V, Bandyopadhyay A, Titapiwatanakun B, Peer WA, Bailly A, Richards EL et al (2005) Cellular efflux of auxin catalyzed by the Arabidopsis MDR/PGP transporter AtPGP1. Plant J 44(2):179–194

Ghosh M, Shen J, Rosen BP (1999) Pathways of As (III) detoxification in Saccharomyces cerevisiae. Proc Natl Acad Sci 96(9):5001–5006

Grotz N, Fox T, Connolly E, Park W, Guerinot ML, Eide D (1998) Identification of a family of zinc transporter genes from Arabidopsis that respond to zinc deficiency. Proc Natl Acad Sci 95(12): 7220–7224

Hayashi M, Nito K, Takei-Hoshi R, Yagi M, Kondo M, Suenaga A, Nishimura M (2002) Ped3p is a peroxisomal ATP-binding cassette transporter that might supply substrates for fatty acid β-oxidation. Plant Cell Physiol 43(1):1–11

Henikoff S, Greene EA, Pietrokovski S, Bork P, Attwood TK, Hood L (1997) Gene families: the taxonomy of protein paralogs and chimeras. Science 278(5338):609–614

Higgins CF (1992) ABC transporters: from microorganisms to man. Annu Rev Cell Biol 8:67–113

Huang CF, Yamaji N, Mitani N, Yano M, Nagamura Y, Ma JF (2009) A bacterial-type ABC transporter is involved in aluminum tolerance in rice. Plant Cell 21(2):655–667

Huang CF, Yamaji N, Chen Z, Ma JF (2012) A tonoplast-localized half-size ABC transporter is required for internal detoxification of aluminum in rice. Plant J 69(5):857–867

Hussain D, Haydon MJ, Wang Y, Wong E, Sherson SM, Young J, Cobbett CS (2004) P-type ATPase heavy metal transporters with roles in essential zinc homeostasis in Arabidopsis. Plant Cell 16(5):1327–1339

Jasiński M, Stukkens Y, Degand H, Purnelle B, Marchand-Brynaert J, Boutry M (2001) A plant plasma membrane ATP binding cassette–type transporter is involved in antifungal terpenoid secretion. Plant Cell 13(5):1095–1107

Kang J, Hwang JU, Lee M, Kim YY, Assmann SM, Martinoia E, Lee Y (2010) PDR-type ABC transporter mediates cellular uptake of the phytohormone abscisic acid. Proc Natl Acad Sci 107(5):2355–2360

Kawachi M, Kobae Y, Mimura T, Maeshima M (2008) Deletion of a histidine-rich loop of AtMTP1, a vacuolar Zn2+/H+ antiporter of Arabidopsis thaliana, stimulates the transport activity. J Biol Chem 283(13):8374–8383

Kim DY, Bovet L, Kushnir S, Noh EW, Martinoia E, Lee Y (2006) AtATM3 is involved in heavy metal resistance in Arabidopsis. Plant Physiol 140(3):922–932

Kim DY, Bovet L, Maeshima M, Martinoia E, Lee Y (2007) The ABC transporter AtPDR8 is a cadmium extrusion pump conferring heavy metal resistance. Plant J 50(2):207–218

Klein M, Burla B, Martinoia E (2006) The multidrug resistance-associated protein (MRP/ABCC) subfamily of ATP-binding cassette transporters in plants. FEBS Lett 580(4):1112–1122

Kovalchuk A, Driessen AJ (2010) Phylogenetic analysis of fungal ABC transporters. BMC Genomics 11(1):1–21

Kramer EM (2004) PIN and AUX/LAX proteins: their role in auxin accumulation. Trends Plant Sci 9(12):578–582

Kreuz K, Tommasini R, Martinoia E (1996) Old enzymes for a new job (herbicide detoxification in plants). Plant Physiol 111(2):349

Kuromori T, Miyaji T, Yabuuchi H, Shimizu H, Sugimoto E, Kamiya A, Shinozaki K (2010) ABC transporter AtABCG25 is involved in abscisic acid transport and responses. Proc Natl Acad Sci 107(5):2361–2366

Lee M, Lee K, Lee J, Noh EW, Lee Y (2005) AtPDR12 contributes to lead resistance in Arabidopsis. Plant Physiol 138(2):827–836

Lee M, Choi Y, Burla B, Kim YY, Jeon B, Maeshima M, Lee Y (2008) The ABC transporter AtABCB14 is a malate importer and modulates stomatal response to CO2. Nat Cell Biol 10(10): 1217–1223

Li ZS, Lu YP, Zhen RG, Szczypka M, Thiele DJ, Rea PA (1997) A new pathway for vacuolar cadmium sequestration in Saccharomyces cerevisiae: YCF1-catalyzed transport of bis (glutathionato) cadmium. Proc Natl Acad Sci 94(1):42–47

Linton KJ (2007) Structure and function of ABC transporters. Physiology 22(2):122–130

Liu T, Zeng J, Xia K, Fan T, Li Y, Wang Y, Zhang M (2012) Evolutionary expansion and functional diversification of oligopeptide transporter gene family in rice. Rice 5(1):1–14

Lu YP, Li ZS, Rea PA (1997) AtMRP1 gene of Arabidopsis encodes a glutathione S-conjugate pump: isolation and functional definition of a plant ATP-binding cassette transporter gene. Proc Natl Acad Sci 94(15):8243–8248

Lu YP, Li ZS, Drozdowicz YM, Hörtensteiner S, Martinoia E, Rea PA (1998) AtMRP2, an Arabidopsis ATP binding cassette transporter able to transport glutathione S-conjugates and chlorophyll catabolites: functional comparisons with AtMRP1. Plant Cell 10(2):267–282
Lubkowitz M (2011) The oligopeptide transporters: a small gene family with a diverse group of substrates and functions? Mol Plant 4(3):407–415
Marin E, Leonhardt N, Vavasseur A, Forestier C (1998) Cloning of AtMRP1, an Arabidopsis thaliana cDNA encoding a homologue of the mammalian multidrug resistance-associated protein. Biochim Biophys Acta 1369(1):7–13
Martinoia E, Grill E, Tommasini R, Kreuz K, Amrhein N (1993) ATP-dependent glutathione S-conjugate 'export' pump in the vacuolar membrane of plants. Nature 364(6434):247–249
Martinoia E, Klein M, Geisler M, Bovet L, Forestier C, Kolukisaoglu U, Schulz B (2002) Multifunctionality of plant ABC transporters–more than just detoxifiers. Planta 214(3):345–355
Moons A (2008) Transcriptional profiling of the PDR gene family in rice roots in response to plant growth regulators, redox perturbations and weak organic acid stresses. Planta 229(1):53–71
Morel M, Crouzet J, Gravot A, Auroy P, Leonhardt N, Vavasseur A, Richaud P (2009) AtHMA3, a P1B-ATPase allowing Cd/Zn/co/Pb vacuolar storage in Arabidopsis. Plant Physiol 149(2): 894–904
Ortiz DF, Ruscitti T, McCue KF, Ow DW (1995) Transport of metal-binding peptides by HMT1, a fission yeast ABC-type vacuolar membrane protein (∗). J Biol Chem 270(9):4721–4728
Palmgren MG, Nissen P (2011) P-type ATPases. Annu Rev Biophys 40:243–266
Park J, Song WY, Ko D, Eom Y, Hansen TH, Schiller M, Lee Y (2012) The phytochelatin transporters AtABCC1 and AtABCC2 mediate tolerance to cadmium and mercury. Plant J 69(2):278–288
Phillips DA (1981) Chemistry and biochemistry of trace metals in biological systems. In: Lepp NW (ed) Effect of heavy metal pollution on plants, vol 1. Applied Science Publishers, London
Pighin JA, Zheng H, Balakshin LJ, Goodman IP, Western TL, Jetter R, Samuels AL (2004) Plant cuticular lipid export requires an ABC transporter. Science 306(5696):702–704
Pisarev AV, Skabkin MA, Pisareva VP, Skabkina OV, Rakotondrafara AM, Hentze MW, Pestova TV (2010) The role of ABCE1 in eukaryotic posttermination ribosomal recycling. Mol Cell 37(2):196–210
Prasad R, Murthy SK, Prasad R, Gupta V, Lata S (1996) Multidrug resistance: an emerging threat. Curr Sci 71:205–213
Raichaudhuri A, Peng M, Naponelli V, Chen S, Sánchez-Fernández R, Gu H, Rea PA (2009) Plant vacuolar ATP-binding cassette transporters that translocate folates and antifolates in vitro and contribute to antifolate tolerance in vivo. J Biol Chem 284(13):8449–8460
Rea PA (2007) Plant ATP-binding cassette transporters. Annu Rev Plant Biol 58:347–375
Rea PA, Li ZS, Lu YP, Drozdowicz YM, Martinoia E (1998) From vacuolar GS-X pumps to multispecific ABC transporters. Annu Rev Plant Biol 49(1):727–760
Rommens JM, Iannuzzi MC, Kerem BS, Drumm ML, Melmer G, Dean M, Hidaka N (1989) Identification of the cystic fibrosis gene: chromosome walking and jumping. Science 245(4922): 1059–1065
Sánchez-Fernández R, Ardiles-Díaz W, Van Montagu M, Inzé D, May MJ (1998) Cloning and expression analyses of AtMRP4, a novel MRP-like gene from Arabidopsis thaliana. Mol Gen Genet 258(6):655–662
Sasaki A, Yamaji N, Yokosho K, Ma JF (2012) Nramp5 is a major transporter responsible for manganese and cadmium uptake in rice. Plant Cell 24(5):2155–2167
Schwartz MS, Benci JL, Selote DS, Sharma AK, Chen AG, Dang H, Vatamaniuk OK (2010) Detoxification of multiple heavy metals by a half-molecule ABC transporter, HMT-1, and coelomocytes of Caenorhabditis elegans. PLoS One 5(3):e9564
Shim D, Kim S, Choi YI, Song WY, Park J, Youk ES, Lee Y (2013) Transgenic poplar trees expressing yeast cadmium factor 1 exhibit the characteristics necessary for the phytoremediation of mine tailing soil. Chemosphere 90(4):1478–1486

Shitan N, Bazin I, Dan K, Obata K, Kigawa K, Ueda K, Yazaki K (2003) Involvement of CjMDR1, a plant multidrug-resistance-type ATP-binding cassette protein, in alkaloid transport in Coptis japonica. Proc Natl Acad Sci 100(2):751–756

Song WY, Park J, Mendoza-Cózatl DG, Suter-Grotemeyer M, Shim D, Hörtensteiner S, Martinoia E (2010) Arsenic tolerance in Arabidopsis is mediated by two ABCC-type phytochelatin transporters. Proc Natl Acad Sci 107(49):21187–21192

Song WY, Park J, Eisenach C, Maeshima M, Lee Y, Martinoia E (2014a) ABC transporters and heavy metals. In: Plant ABC transporters. Springer, Cham, pp 1–17

Song WY, Yamaki T, Yamaji N, Ko D, Jung KH, Fujii-Kashino M, Ma JF (2014b) A rice ABC transporter, OsABCC1, reduces arsenic accumulation in the grain. Proc Natl Acad Sci 111(44): 15699–15704

Stein M, Dittgen J, Sánchez-Rodríguez C, Hou BH, Molina A, Schulze-Lefert P, Somerville S (2006) Arabidopsis PEN3/PDR8, an ATP binding cassette transporter, contributes to nonhost resistance to inappropriate pathogens that enter by direct penetration. Plant Cell 18(3):731–746

Stukkens Y, Bultreys A, Grec S, Trombik T, Vanham D, Boutry M (2005) NpPDR1, a pleiotropic drug resistance-type ATP-binding cassette transporter from Nicotiana plumbaginifolia, plays a major role in plant pathogen defense. Plant Physiol 139(1):341–352

Sugiyama A, Shitan N, Yazaki K (2008) Signaling from soybean roots to rhizobium: an ATP-binding cassette-type transporter mediates genistein secretion. Plant Signal Behav 3(1): 38–40

Thomine S, Wang R, Ward JM, Crawford NM, Schroeder JI (2000) Cadmium and iron transport by members of a plant metal transporter family in Arabidopsis with homology to Nramp genes. Proc Natl Acad Sci 97(9):4991–4996

Tommasini R, Vogt E, Fromenteau M, Hörtensteiner S, Matile P, Amrhein N, Martinoia E (1998) An ABC-transporter of Arabidopsis thaliana has both glutathione-conjugate and chlorophyll catabolite transport activity. Plant J 13(6):773–780

Vanneste S, Friml J (2009) Auxin: a trigger for change in plant development. Cell 136(6): 1005–1016

Verbruggen N, Hermans C, Schat H (2009) Mechanisms to cope with arsenic or cadmium excess in plants. Curr Opin Plant Biol 12(3):364–372

Verrier PJ, Bird D, Burla B, Dassa E, Forestier C, Geisler M, Theodoulou FL (2008) Plant ABC proteins–a unified nomenclature and updated inventory. Trends Plant Sci 13(4):151–159

Vert G, Grotz N, Dédaldéchamp F, Gaymard F, Guerinot ML, Briat JF, Curie C (2002) IRT1, an Arabidopsis transporter essential for iron uptake from the soil and for plant growth. Plant Cell 14(6):1223–1233

Wang W, Takezawa D, Poovaiah BW (1996) A potato cDNA encoding a homologue of mammalian multidrug resistant P-glycoprotein. Plant Mol Biol 31(3):683–687

Wanke D, Üner Kolukisaoglu H (2010) An update on the ABCC transporter family in plants: many genes, many proteins, but how many functions? Plant Biol 12:15–25

Yamaji N, Xia J, Mitani-Ueno N, Yokosho K, Feng Ma J (2013) Preferential delivery of zinc to developing tissues in rice is mediated by P-type heavy metal ATPase OsHMA2. Plant Physiol 162(2):927–939

Yuan M, Li X, Xiao J, Wang S (2011) Molecular and functional analyses of COPT/Ctr-type copper transporter-like gene family in rice. BMC Plant Biol 11(1):1–12

Zeng X, Tang R, Guo H, Ke S, Teng B, Hung YH, Zhang XQ (2017) A naturally occurring conditional albino mutant in rice caused by defects in the plastid-localized OsABCI8 transporter. Plant Mol Biol 94(1):137–148

Zolman BK, Silva ID, Bartel B (2001) The Arabidopsis pxa1 mutant is defective in an ATP-binding cassette transporter-like protein required for peroxisomal fatty acid β-oxidation. Plant Physiol 127(3):1266–1278

4 Cadmium, a Nonessential Heavy Metal: Uptake, Translocation, Signaling, Detoxification, and Impact on Amino Acid Metabolism

Kaiser Iqbal Wani, Andleeb Zehra, Sadaf Choudhary, M. Naeem, and Tariq Aftab

Abstract

Cadmium (Cd) is a toxic nonessential heavy metal for higher plants, which has a long biological half-life. Based on the fact that Cd is harmful to the environment and plants, and due to the sessile nature of the plants, they need to elevate their protective mechanisms to cope with Cd stress. Due to its high solubility and being easily taken up by the plants using the transporters of various essential elements like iron, zinc, and manganese, it not only interferes with the uptake of these essential elements but also induces various structural, physio-biochemical, and morphological changes in plants at high concentration and results in nutrient imbalance and deficiency of these essential elements when they are low in the soil. In order to cope with Cd toxicity, a number of signaling pathways are activated in plants like the MAPK pathway. These signaling pathways help in Cd detoxification via ion transport, regulating metabolism, ROS homeostasis, and activation of different transcription factors which in turn activate different stress-responsive genes. Cd detoxification through peptides such as phytochelatins and metallothioneins which are produced in plants in response to heavy metal stress constitutes an important part of plants' defense against Cd toxicity. Moreover, the metabolism of the nitrogenous compounds like amino acids also plays an important role in plants in alleviating the negative impacts of Cd stress. The production of a number of these amino acids like proline, serine, arginine, asparagine, and many others is escalated under Cd stress, which then act as signaling molecules, osmolytes, and free radical scavengers and improve the growth parameters of plants. In the present chapter, we have tried to describe a holistic view of Cd uptake, translocation, signaling, and detoxification via phytochelatins and metallothioneins and its impact on amino acid metabolism.

K. I. Wani · A. Zehra · S. Choudhary · M. Naeem · T. Aftab (✉)
Department of Botany, Aligarh Muslim University, Aligarh, India

K. Kumar, S. Srivastava (eds.), *Plant Metal and Metalloid Transporters*,
https://doi.org/10.1007/978-981-19-6103-8_4

Keywords

Nonessential metal · MAPK pathway · Phytochelatins · Metallothioneins

4.1 Introduction

Heavy metal (density >5 g cm^{-3}) contamination is one of the most serious global environmental problems adversely affecting the growth and development of plants. Cadmium (Cd) is a natural element in the Earth's crust and is usually found as a mineral in combination with other elements such as sulfur, oxygen, or chlorine. Cd (density = 8.6 g cm^{-3}) is a nonessential element for plants that occurs naturally in the soil at low concentration but becomes toxic when it breaches a certain concentration limit (Zafarzadeh et al. 2018). There has been a significant rise in Cd emission into the environment during the last few decades due to the widespread use of Cd in industries resulting in the release of Cd at higher concentrations than originating from natural sources (Nriagu 1988; Liu et al. 2007). It poses a serious threat to both animal and plant lives due to its high solubility, high toxicity, and being easily taken up by plant roots from the soil (Pinto et al. 2004). Just like other abiotic stressors which induce many structural, physio-morphological, and biochemical changes in plants, with some stressors reported to boost secondary metabolite production like artemisinin in *Artemisia annua* (Wani et al. 2021a), Cd has also been found to induce some of these changes in plants depending upon its concentration and exposure period, type of tissue/organ, and species (Benavides et al. 2005), with its accumulation resulting in excessive reactive oxygen species (ROS) production in plants, contributing to cytotoxicity, damaging proteins, nucleic acids, and lipids (Garnier et al. 2006; Feng et al. 2016). Cd accumulation also results in decreased crop productivity due to interference with mineral uptake, inhibition of enzyme activity, oxidative stress, impaired amino acid biosynthesis, and disturbed plant water status (Perfus-Barbeoch et al. 2002; Liu et al. 2015; Zhu et al. 2018). The symptoms of Cd toxicity in plants include browning of roots, leaf rolling, chlorosis and necrosis of leaves, and cell apoptosis (Zemanová et al. 2015). Cd toxicity is also reported to be associated with chloroplast damage and reduction in chlorophyll and carotenoid content, resulting in lower photosynthetic quantum yield (Perfus-Barbeoch et al. 2002).

Many plant species have evolved a range of intricate mechanisms to deal with the problem of Cd toxicity. These mechanisms include sequestration, chelation, Cd efflux, detoxification, production of antioxidant enzymes, metabolic utilization, accumulation of osmolytes, and production of thiol-related compounds (Cobbett and Goldsbrough 2002; Clemens et al. 2013; Rahman et al. 2017; Sakouhi et al. 2018; Rizwan et al. 2018). At the molecular level, plants produce different stress-related proteins and signaling molecules in response to Cd stress. Nicotianamine, phytosiderophores, and organic acids are few chelating compounds released by the plant roots which might influence heavy metal uptake, including that of Cd (Dalcorso et al. 2010). The potential alleviation approaches for Cd toxicity in plants include the use of plant growth regulators, mineral nutrients, biochar, and compost and the use of different bioremediation and chemical methods. However, the knowledge about the mechanism and the effects of these diverse alleviation approaches is still limited and needs further analysis.

4.2 Cadmium Transporters: Uptake and Translocation

Plants take up Cd from the soil and water with the help of root cells, and its availability to plants is governed by a number of factors like soil pH, its concentration in the soil, and the presence of organic acids in the rhizosphere (Benavides et al. 2005). The buildup of Cd in shoots depends upon its entry through the roots, sequestration within root vacuoles, and translocation through xylem and phloem tissues. The uptake of Cd occurs via the transporters which are involved in the uptake of iron (Fe), calcium (Ca), zinc (Zn), magnesium (Mg), manganese (Mn), and copper (Cu) (Clemens 2006). A list of transporters is given in Table 4.1.

4.3 NRAMP Transporters

The natural resistance-associated macrophage proteins (NRAMP) are a large family of proteins associated with metal ion transport that is evolutionarily conserved throughout the organisms including plants, animals, algae, bacteria, and yeast (Nevo and Nelson 2006). *Nramp1*, the first *NRAMP* gene identified in mammals, encodes an integral membrane protein recruited in infected macrophages (Vidal et al. 1993). Further studies have shown that NRAMP proteins act as metal transporters and are involved in translocation, detoxification, and intracellular transport of metals like Fe^{2+}, Cd^{2+}, Zn^{2+}, Cu^{2+}, and many other metal ions (Nevo and Nelson 2006; Sasaki et al. 2012). In yeast, *Arabidopsis thaliana*, and *Oryza sativa* genomes, there are three, six, and seven *NRAMP* homologs, respectively, with high amino acid sequence conservation (Thomine and Schroeder 2013). Based on northern blot and promoter-reporter gene fusion studies, it has been found that *NRAMP* genes are expressed in both shoots and roots, unlike the metal transporter genes from the ZIP family, *IRT1* and *IRT2*, which are strictly root-specific (Belouchi et al. 1997; Curie et al. 2000; Thomine et al. 2000; Bereczky et al. 2003).

As a nonessential and toxic metal ion, Cd is taken up by the plants from soil through the passage of essential elements (Fe^{2+}, Zn^{2+}, Mn^{2+}) as shown in Fig. 4.1 (Uraguchi et al. 2011; Clemens et al. 2013). NRAMP1, NRAMP3, and NRAMP4 from *A. thaliana* have shown transport activity for Fe^{2+}, Mn^{2+}, and Cd^{2+} in yeast (Thomine et al. 2000; Curie et al. 2000; Lanquar et al. 2005). *NRAMP1* has shown activity for Cd and Fe uptake in rice, and its expression is higher in high Cd-accumulating indica cultivars of rice (Takahashi et al. 2011). *NRAMP1* (localized in root cells) knockout lines have shown reduced growth at low Mn concentration, depicting its role in manganese uptake (Cailliatte et al. 2010). NRAMP5 located in the plasma membrane of root cells is a major transporter of Cd

Table 4.1 List of transporters along with their location within the plants and function

Transporter	Location	Function	Reference
NRAMP1	Plasma membrane of root and shoot cells	Cadmium, iron, and manganese transport	Cailliatte et al. (2010), Takahashi et al. (2011)
NRAMP3	Root stele and leaf vascular bundles (tonoplast)	Iron, manganese, and cadmium transport	Thomine et al. (2000), Oomen et al. (2009)
NRAMP4	Root, vascular bundles, and plastids	Cadmium, manganese, and iron transport Exports manganese from vacuoles of mesophyll cells for photosystem II	Thomine et al. (2000), Oomen et al. (2009)
NRAMP5	Root plasma membrane	Cadmium and manganese uptake	Tang et al. (2017)
NRAMP6	Roots and leaves	Cadmium and manganese transmembrane transporter activity	Cailliatte et al. (2009)
IRT1	Plasma membrane of roots	Cadmium, iron, zinc, and manganese uptake	Connolly et al. (2002), Nakanishi et al. (2006)
IRT2	Plasma membrane of roots	Cadmium and iron uptake	Nakanishi et al. (2006)
OsZIP1	Roots	Zinc uptake and cadmium uptake in yeast cells expressing *OsZIP1*	Ramesh et al. (2003)
YSL2	Vascular tissues and roots	Phloem transport of nicotianamine complexes with iron, manganese, and cadmium	Koike et al. (2004)
YSL3	Vascular tissues and epidermal cells of roots and stem	Transports nicotianamine complexes containing cadmium, zinc, copper, and iron	Feng et al. (2017)
HMA2 and HMA4	Roots and vascular tissues	Cadmium and zinc transport into xylem for upward translocation	Hussain et al. (2004), Takahashi et al. (2012)
HMA3	Tonoplast	Vacuolar sequestration of cadmium	Yan et al. (2016)
LCT1	Phloem parenchyma and leaves	Efflux of cadmium, potassium, calcium, manganese, and magnesium	Uraguchi et al. (2011)
PDR8	Root hairs and epidermal cells	Cadmium and lead efflux	Kim et al. (2007)

and Mn uptake in plants, and NRAMP5 knockout lines show reduced Cd uptake (Yang et al. 2014; Sui et al. 2018). The shared transporters between Cd and Mn explain the reduction seen in Cd uptake in *O. sativa* upon increasing Mn concentration (Yang et al. 2014). In *Arabidopsis*, NRAMP3 and NRAMP4 have shown transport activity for Cd and iron (Thomine et al. 2000; Thomine and Schroeder 2013). The orthologs of *Arabidopsis NRAMP3* and *NRAMP4* genes show increased

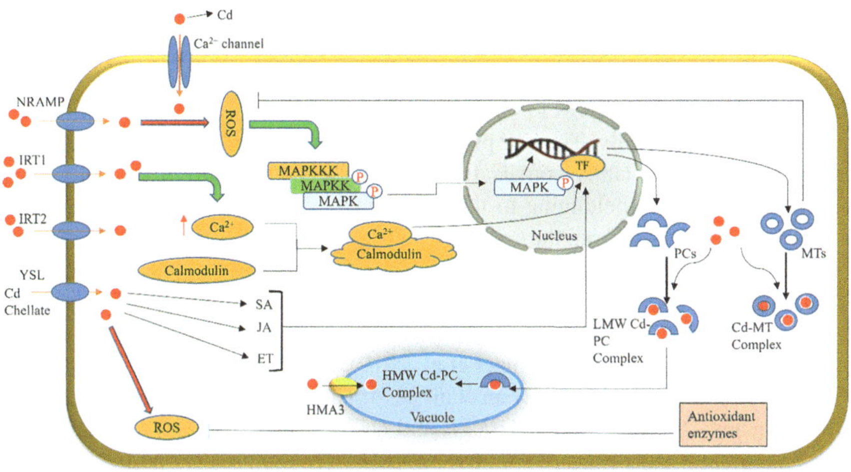

Fig. 4.1 Schematic representation of cadmium influx into cells and the activation of different signaling pathways like MAPK kinase activated due to ROS production. Furthermore, Ca-calmodulin system and stress-related hormones activate various stress-responsive genes via the regulation of different transcription factors and help in cadmium detoxification and compartmentalization into vacuoles. *Cd* cadmium, *MAPK* mitogen-activated protein kinase, *PC* phytochelatin, *MT* metallothionein, *ROS* reactive oxygen species, *TF* transcription factor, *LMW* low molecular weight, *HMW* high molecular weight, *SA* salicylic acid, *JA* jasmonic acid, *ET* ethylene, *NRAMP* natural resistance-associated macrophage proteins, *YSL* yellow stripe-like

expression in metal hyperaccumulator *Thlaspi caerulescens* (Oomen et al. 2009). Certain other transporters from the ZIP family such as IRT1 and IRT2 may also have a role in Cd uptake in *O. sativa*, but their contribution seems meager (Sui et al. 2018). The reduction of Cd uptake upon increasing Zn concentration suggests the involvement of Zn transporters in Cd uptake (Hart et al. 2005). In contrast to wheat and maize, rice has higher Cd uptake ability due to higher expression of *Nramp5* in rice in comparison to its orthologs in wheat and maize. Higher Cd uptake ability in rice is also confirmed by the presence of excess Cd in rice grains as compared to wheat and maize (Sui et al. 2018). Therefore *Nramp5* has been found to be a primary target for decreasing the Cd concentration in rice grains, and this has been achieved by knocking out *Nramp5* via gene editing or mutation (Tang et al. 2017). However, mutating *Nramp5* may also result in Mn deficiency in plants growing in low Mn soil, due to its role as a Mn transporter. NRAMP6 does not show any transport activity for Fe, but it shows transport activity for Cd and has been associated with Cd toxicity in plants (Cailliatte et al. 2009). NRAMP6 functions as an intracellular metal ion transporter, with *NRAMP6* overexpression lines showing Cd toxicity symptoms, without any change in Cd concentration (Cailliatte et al. 2009). These observations suggest the involvement of this gene in intracellular Cd distribution (Fig. 4.1).

4.4 ZIP Transporters

The ZIP (ZRT- and IRT-like proteins) transport proteins are named due to their sequence similarity with ZRT1 and IRT1 transporters from *Saccharomyces* and *Arabidopsis*, respectively. The ZIP proteins have been found in a variety of different organisms including plants, animals, fungi, and bacteria (Connolly et al. 2002). IRT1 is a metal transporter belonging to the ZIP family which was initially found to be involved in iron uptake from the soil. Later on, it was also found to be involved in the transport of certain other divalent metal ions including Cd^{2+}, Cu^{2+}, Zn^{2+}, and Fe^{2+} (Grotz and Guerinot 2006). The overexpression of *IRT1* in transgenic *A. thaliana* results in Cd toxicity in such plants under iron starvation, confirming its role in Cd uptake (Connolly et al. 2002; Ismael et al. 2019). Furthermore, *irt1-1* mutants also show reduced Cd accumulation (Fan et al. 2014; Ismael et al. 2019). The expression of three genes, namely *OsZIP1*, *AtIRT*1, and *OsZIP3*, in yeast resulted in increased Cd accumulation and sensitivity under Cd stress, depicting their role in Cd uptake (Zheng et al. 2018). In *O. sativa*, the homolog of *Arabidopsis* IRT1 shows predominant expression in roots and is upregulated upon Fe deficiency followed by Cd accumulation which shows its role in Cd transport (Bughio et al. 2002). In *A. Thaliana*, the overexpression of another ZIP protein IRT2 results in increased Cd absorption; however, it has not been found to alter Cd sensitivity in yeast cells and has been found to be involved in the uptake of Zn and Fe and not Cd (Ismael et al. 2019). The expression of another ZIP IRT3 is higher in Zn and Cd hyperaccumulating *Arabidopsis halleri* as compared to the expression of its ortholog in *A. thaliana* (Ismael et al. 2019). The IRT3 overexpressing *A. thaliana* plants show increased Zn and Fe accumulation, with no impact on Cd uptake (Lin et al. 2009) signifying its role as Zn and Fe transporter. OsZIP1 has been found to transport Cd in addition to Zn (Ramesh et al. 2003). Yeast cells expressing *OsZIP1* (from rice plants) are unable to grow in presence of Cd which confirms the involvement of OsZIP1 in Cd uptake in rice plants (Nakanishi et al. 2006).

4.5 YSL Transporters

Certain yellow stripe-like (YSL) transporter proteins have also been reported to be involved in Cd transport, with Cd first forming metal-ligand complex with low molecular weight ligands, such as mugineic acid secreted by the plants. These metal-ligand complexes are then transported via YSL transporters. In plants, the first YSL transporter gene (*zmYSL*) was reported in *Zea mays*, and it has been found to transport Cd in the form of Cd complexes formed with phytosiderophore or nicotianamine (Schaaf et al. 2004). *Solanum nigrum* shows constitutive expression of *snYSL3* which is enhanced under increased Cd stress, with *snYSL3* expression in yeast showing its role in the transport of Cd, Zn, Cu, and Fe (Wang et al. 2013). It transports these metals in the form of metal-nicotianamine complexes. The calcium ion channels in the guard cells of *Arabidopsis* have also been found to be permeable for Cd (Vavasseur and Forestier 2002).

4.6 Transporters Involved in Shoot Uptake of Cadmium

The accumulation of Cd in the aboveground parts depends upon two factors: root uptake of Cd from the soil and its translocation from root to shoot. The root-to-shoot translocation of Cd is an important step in controlling Cd accumulation in shoots. Despite having higher root Cd uptake ability, rice has low root-to-shoot Cd translocation, and wheat shows high root-to-shoot translocation of Cd despite low Cd uptake via roots (Sui et al. 2018). Inside root cells, Cd may be stored in root vacuoles, transferred into vascular tissues via P_{1B}-type heavy metal-transporting ATPases (HMAs) HMA2 and HMA4, or expelled out of the cells (Kim et al. 2007; Ismael et al. 2019). In *Arabidopsis*, AtHMA2 and AtHMA4 are essential for pumping Cd ions into the xylem wherefrom they are transported to the shoot (Wong and Cobbett 2009). Moreover, these transporters are also predominantly expressed in neighboring tissues of vessels (Verret et al. 2004), making shoot uptake of Cd uptake easier. In rice, OsHMA2 is involved in the loading of Cd into xylem (Takahashi et al. 2012), whereas OsHMA3 is a tonoplast Cd transporter involved in vacuolar sequestration of Cd, and the loss-of-function *OsHMA3* rice cultivars show decreased sequestration of Cd into root vacuoles, resulting in increased Cd translocation into shoot portion and grains (Miyadate et al. 2011; Yan et al. 2016). In *Arabidopsis*, AtPDR8 is an ABC transporter localized on the plasma membrane of epidermal cells and root hairs involved in Cd export. Its overexpression has been found to be associated with Cd tolerance whereas its loss of function mutants show increased Cd accumulation (Kim et al. 2007). Certain genes that control Cd translocation in rice include *OsCAL1* and *OsHMA2* (Luo et al. 2018). OsLCT1 is an important low-affinity cation transporter involved in the transport of Cd from vegetative tissue into the grains (Uraguchi et al. 2011).

4.7 Cadmium Stress Signaling

In plants, roots are the primary organs through which heavy metals gain access. The Cd stress in plants activates different types of signaling pathways like the calcium-calmodulin pathway, ROS-mediated signaling, and phytohormone synthesis (Dobrikova and Apostolova 2019; El Rasafi et al. 2020). Electron microscopic analysis of Cd and copper localization in plants has revealed that the majority of the heavy metals are concentrated in root cell wall than root cytoplasm (Arduini et al. 1996). The higher heavy metal retention capacity of cell walls might be due to their negative charge, giving them higher heavy metal binding and retention capacity (Polle and Schuetzenduebel 2003). A number of signaling molecules are generated on cell walls, in response to extracellular stimuli like heavy metal stress. Some of these signaling molecules result in the activation of genes responsible for counteracting heavy metal toxicity. In plants, one of the important signaling cascades in heavy metal stress alleviation including that of Cd is MAPK (mitogen-activated protein kinase) pathway, consisting of three protein kinases (MAPKKK, MAPKK, MAPK) which are sequentially activated via phosphorylation, which then activate various kinase enzymes (Jonak et al. 2004; DalCorso et al. 2010). Therefore, the phosphorylation cascade is thought to be involved in Cd signaling perception by the nucleus. Two important second messengers, that is, calcium (Ca) and calmodulin, have been found to increase under Cd stress, with Ca ions increasing first which in turn stimulate calmodulin-like proteins (Dalcorso et al. 2008). The increased cytosolic Ca ion concentration in response to Cd stress occurs due to the perturbation of phospholipase C activity and certain channels such as ADP ribose-gated channels and Ca ion channels stimulated by IP_3 (inositol-3-phosphate) (Chmielowska-Bak et al. 2014). The interaction of Ca ions with calmodulin-like proteins changes the conformation of these proteins which in turn regulate plant response to Cd stress via gene regulation, ion transport, and regulating metabolism (DalCorso et al. 2010). Cd stress has also been associated with enhanced expression of *MAPKK2* in the roots of *Glycine max* seedlings (Chmielowska-Bak et al. 2013). Furthermore, MAPKs have also been found to be associated with altering responses of various genes such as *WRKY33*, which is involved in H_2O_2 regulation (Sun et al. 2020).

The reduction in GSH/GSSG ratio under Cd stress is believed to be involved in Cd stress alleviation by activation of response genes (Romero-Puertas et al. 2007). Under heavy metal stress, glutathione (GSH) has been found to control the expression of many antioxidant enzymes like glutathione reductase, superoxide dismutase, chalcone synthase, and phenylalanine ammonia lyase (Romero-Puertas et al. 2007). The exposure of plants to Cd and certain other heavy metals has also been found to be associated with enhanced biosynthesis of certain phytohormones like jasmonic acid, ethylene, and salicylic acid. Plant hormones control the growth and development of plants under normal as well as stressful conditions (Aftab et al. 2011; Wani et al. 2020). Maksymiec (2007) found that the increased production of ethylene under Cd stress is due to the increased activity of ACC synthase enzyme. Furthermore, they also reported that the treatment of barley seedlings with salicylic acid before Cd treatment conferred protection against Cd stress. The exogenous application of IAA on tomato seedlings under Cd stress has been found to inhibit H_2O_2 and superoxide accumulation and thus increasing Cd tolerance (Khan et al. 2019).

The increased accumulation of Cd in plants is also related to a surge in reactive oxygen species (ROS), which act as signaling molecules and have a dual role in plants, acting as both beneficial and harmful, depending on their concentration (Jalmi et al. 2018; Wani et al. 2021b). The increased ROS production under Cd stress has been reported in many plants such as tomato (Khan et al. 2019), rice (Rahman et al. 2016), maize (Liu et al. 2019), and fenugreek (Zayneb et al. 2015). The role of ROS in signaling pathways under Cd stress has also been reported in several research studies (Jain et al. 2018; Mei et al. 2018). The increased ROS production due to Cd stress in *Arabidopsis* causes the activation of *MPK3* and *MPK6* genes involved in signal transduction (Liu et al. 2010). Ultimately, the various signaling pathways induced by Cd converge regulating transcription factors which in turn activate different stress-responsive genes (Fig. 4.1).

4.8 Phytochelatins and Metallothioneins: Role in Cd Detoxification

Phytochelatins (PCs) are the heavy metal binding peptides whose production from GSH is catalyzed by PC synthase (PCS), a specific g-glutamylcysteine dipeptidyl transpeptidase. PCs are important in the detoxification of certain metals like Cu, Cd, and arsenic (As), while nickel (Ni) and Zn are poor inducers of PC biosynthesis and have a low binding affinity. The PCS (EC 2.3.2.15) is constitutively expressed in cells and is activated posttranslationally upon heavy metal exposure. It has been found that PCS is regulated by Cd-dependent phosphorylation on a threonine residue adjacent to the catalytic site, and it could act as a Cd sensor (Wang et al. 2009). *A. thaliana* Cd-sensitive mutants like *cad1* have been quite useful in elucidating the role of PCs in Cd detoxification (Howden and Cobbett 1992). In cells, Cd is chelated by GSH and PCs which contain thiol groups, and the complexes formed are transported into the vacuole or apoplast via ATP-driven membrane pumps (Wolf et al. 1996). Inside the vacuoles, the Cd-PC complexes form high molecular weight complexes, with a molecular mass of about 2500–3600 Da (Clemens 2006). Moreover, it has also been found that the overexpression of PCS in transgenic plants results in increased Cd tolerance capacity in such plants (Heiss et al. 2003), with prolonged Cd exposure also enhancing the expression of PCS. The PCS-encoding genes (*PCS1*, *PCS2*) have been reported and cloned in a number of plants like *Arabidopsis*, *Brassica juncea*, *Triticum aestivum*, and *Oryza sativa* (Ha et al. 1999; Heiss et al. 2003; DalCorso et al. 2010). PCS1 and PCS2 in *Arabidopsis* form important constituents of metal-dependent transpeptidation and enhance plant tolerance against Cd stress (Brunetti et al. 2011; Kuhnlenz et al. 2014). The chelation of Cd with thiol groups of cysteine residues prevents its circulation as Cd^{+2} in cytosol. Certain Cd-inducible genes like *VTC2–1*, *PAD2–1* (Koffler et al. 2014), *HsfA4a* (Shim et al. 2009), *MAN3* (Chen et al. 2015), *ZAT6* (Chen et al. 2016), and *HMT1* (Huang et al. 2012) have been reported to be involved in PC-dependent pathway. The *MAN3* gene increases mannanase activity and mannose content in the cell wall, with its overexpression also increasing Cd tolerance through GSH-dependent PC synthetic pathway (Chen et al. 2015). *GSH* overexpression in poplar and mustard has shown increased Cd tolerance due to increased production of GSH and PCs (Hernández et al. 2015; Song et al. 2017). Furthermore, the co-expression of *GSH1* and *PCS1* showed coordinated regulation of PC production and Cd tolerance (Guo et al. 2008). Ferrochelatase-1 (EC4.99.1.1), a terminal enzyme of heme biosynthesis, has been found to be transcriptionally upregulated in *Arabidopsis* upon Cd exposure, with *Ferrochelatase-1* overexpression lines showing enhanced production of GSH and PCs, thus increasing tolerance against Cd toxicity (Song et al. 2017.

Metallothioneins (MTs) are low molecular weight cysteine-rich peptides that help in the detoxification of cells from Cd toxicity. They are produced in response to heavy metal stress and protect the plants against the negative impacts of heavy metal accumulation. In plant MTs, the cysteine residues are present as Cys-Cys, Cys-x-Cys, or Cys-x-x-Cys clusters (x is any amino acid other than cysteine). There are four types of MTs in plants, and in *Arabidopsis*, six MTs belonging to these four types have been reported. These include MT1a, MT2a, MT2b, MT3, MT4a, and MT4b. Among these, MT1, MT2, and MT3 types have been found to enhance Cd tolerance (Guo et al. 2008). The expression of *Arabidopsis* MTs (MT2a and MT3) in the guard cells of *Vicia faba* increases its tolerance against Cd toxicity (Lee et al. 2004). Similarly, MT2 from mustard also results in increased Cd tolerance in transgenic *A. thaliana* and also in *E. coli* (Zhigang et al. 2006). Besides acting as metal chelators, MTs also work as ROS scavengers, with cysteine residue working as a redox sensor and help in alleviating oxidative stress (Hassinen et al. 2011). In *Tamarix hispida*, MT3 acts as a ROS scavenger contributing to Cd tolerance; and it also increases the activity of catalase, superoxide dismutase, and glutathione reductase.

4.9 Cadmium Toxicity and Amino Acid Metabolism

Metabolism of nitrogenous compounds is vital in the plant response to heavy metals, including Cd (Chaffei et al. 2004). Plants produce a range of low molecular weight substances, mainly specific free amino acids known as compatible solutes, upon heavy metal exposure. These substances act as signaling molecules and also play a variety of roles in plants like acting as radical scavengers, osmolytes, modulating stomatal aperture, and heavy metal detoxification (Pavlíková et al. 2014; Zhu et al. 2018). Amino acid metabolism plays a central role in pH regulation, especially γ-aminobutyrate (GaBa) and alanine (Miyashita et al. 2007; Seher et al. 2013). The buildup of amino acids and amino acid-derived molecules plays a key role in regulating the adaptive response of plants against Cd toxicity. The expression of some genes, synthesis and activity of enzymes, and redox homeostasis are also regulated by certain amino acids.

The fate of Cd ions inside cells depends strongly on thiol-containing molecules, including sulfhydryl-containing amino acids. Wu et al. (2004) found that Cd stress greatly alters the concentration and composition of free amino acids in different plant parts, with a significant increase in the concentration of proline, serine, alanine, and histidine in roots. In a study conducted on tomato by Chaffei et al. (2004), it was found that Cd stress resulted in increased total amino acid content in roots; with proline and asparagine showing 14- and 8-fold increase respectively, glutamate concentration decreased, whereas aspartate and glutamine remained unaffected. They further reported that in phloem sap, there was a drastic decrease in the concentration of GaBa, with arginine and histidine showing a significant increase. Cd toxicity has been found to result in increased accumulation of proline in barley and mung bean seedlings (Zhang et al. 2000; Vassilev and Lidon 2011). In two *Noccaea* metallophyte species, Cd stress is associated with increased tryptophan, ornithine, threonine, and phenylalanine and decreased glycine and alanine (Zemanová et al. 2017). The exogenous application of glycinebetaine and proline on tobacco bright yellow-2 cells under Cd stress results in increased intracellular proline and/or glycinebetaine production (Islam et al. 2009). Moreover, it also resulted in increased ascorbate-GSH cycle enzyme activities, restored membrane integrity, and increased activity of antioxidant enzymes like catalase and superoxide dismutase. Proline has been demonstrated to protect nitrate reductase and glucose-6-phosphate dehydrogenase in vitro under Cd toxicity by the formation of a metal-proline complex (Sharma et al. 1998; Sharma and Dietz 2006). Cd accumulation and Cd tolerance in *Crassocephalum crepidioides* which is a Cd hyperaccumulator are associated with an increased buildup of free amino acids, particularly asparagine and glutamine (Zhu et al. 2018). Based on the available literature, amino acid metabolism in plants is affected differently by heavy metal stress depending upon the type of heavy metal, plant species, genotype, and organ affected.

4.10 Conclusion

The impact of heavy metal stress on plant growth and crop productivity has become a major concern during the past few decades due to the alarming rise in the release of heavy metals into soil. Making the matter worse, the global demand for food production is also increasing due to the population explosion, particularly in developing countries. The effect of different biotic and abiotic stressors including heavy metal pollution on plant health is a major constraint on food crop production. Plants have developed various mechanisms to cope with heavy metal stress including Cd stress, and these responses operate at morphological, physiological, biochemical, molecular, and anatomical levels. Under heavy metal stress, some plants show increased metal sequestration, metal exclusion, and reduced metal accumulation in cells in order to cope with these conditions. The development of heavy metal-tolerant transgenic plants with enhanced expression of stress-responsive genes and metabolite production and the selection of heavy metal-tolerant varieties of crop plants are viable options to deal with this problem. However, the application and success of these procedures in the actual field conditions require a comprehensive understanding of the various activities that are stimulated in plants in response to heavy metal toxicity and thus require further comprehensive research.

References

Aftab T, Khan MMA, da Silva JAT, Idrees M, Naeem M (2011) Role of salicylic acid in promoting salt stress tolerance and enhanced artemisinin production in Artemisia annua L. J Plant Growth Regul 30(4):425–435

Arduini I, Godbold DL, Onnis A (1996) Cadmium and copper uptake and distribution in Mediterranean tree seedlings. Physiol Plant 97:111–117

Belouchi A, Kwan T, Gros P (1997) Cloning and characterization of the OsNramp family from Oryza sativa, a new family of membrane proteins possibly implicated in the transport of metal ions. Plant Mol Biol 33:1085–1092

Benavides MP, Gallego SM, Tomaro ML (2005) Cadmium toxicity in plants. Braz J Plant Physiol 17(1):21–34. https://doi.org/10.1590/S1677-04202005000100003

Bereczky Z, Wang H-Y, Schubert V et al (2003) Differential regulation of nramp and irt metal transporter genes in wild type and iron uptake mutants of tomato. J Biol Chem 278:24697–24704. [PubMed: 12709425]

Brunetti P, Zanella L, Proia A, De Paolis A, Falasca G, Altamura MM, Sanita di Toppi L, Costantino P, Cardarelli M (2011) Cadmium tolerance and phytochelatin content of Arabidopsis seedlings overexpressing the phytochelatin synthase gene AtPCS1. J Exp Bot 62:5509–5519

Bughio N, Yamaguchi H, Nishizawa NK, Nakanishi H, Mori S (2002) Cloning an iron-regulated metal transporter from rice. J Exp Bot 53(374):1677–1682

Cailliatte R, Lapeyre B, Briat JF, Mari S, Curie C (2009) The NRAMP6 metal transporter contributes to cadmium toxicity. Biochem J 422:217–228

Cailliatte R, Schikora A, Briat JF, Mari S, Curie C (2010) High-affinity manganese uptake by the metal transporter NRAMP1 is essential for Arabidopsis growth in low manganese conditions. Plant Cell 22(3):904–917

Chaffei C, Pageau K, Suzuki A, Gouia H, Ghorbel MH, Masclaux-Daubresse C (2004) Cadmium toxicity induced changes in nitrogen management in Lycopersicon esculentum leading to a metabolic safeguard through an amino acid storage strategy. Plant Cell Physiol 45(11):1681–1693

Chen J, Yang L, Gu J, Bai X, Ren Y, Fan T, Han Y, Jiang L, Xiao F, Liu Y, Cao S (2015) MAN3 gene regulates cadmium tolerance through the glutathione-dependent pathway in Arabidopsis thaliana. New Phytol 205:570–582

Chen J, Yang L, Yan X, Liu Y, Wang R, Fan T, Ren Y, Tang X, Xiao F, Liu YS, Cao S (2016) Zinc-finger transcription factor ZAT6 positively regulates cadmium tolerance through the glutathione-dependent pathway in Arabidopsis. Plant Physiol 171:707–719

Chmielowska-Bak J, Lefevre I, Lutts S, Deckert J (2013) Short term signaling responses in roots of young soybean seedlings exposed to cadmium stress. J Plant Physiol 170(18):1585–1594. https://doi.org/10.1016/j.jplph.2013.06.019

Chmielowska-Bak J, Gzyl J, Rucinska-Sobkowiak R, Arasimowicz-Jelonek M, Deckert J (2014) The new insights into cadmium sensing. Front Plant Sci 5:245–213. https://doi.org/10.3389/fpls.2014.00245

Clemens S (2006) Evolution and function of phytochelatin synthases. J Plant Physiol 163:319–332

Clemens S, Aarts MGM, Thomine S, Verbruggen N (2013) Plant science: the key to preventing slow cadmium poisoning. Trends Plant Sci 18:92–99

Cobbett C, Goldsbrough P (2002) Phytochelatins and metallothioneins: roles in heavy metal detoxification and homeostasis. Plant Biol 53:159–182

Connolly EL, Fett JP, Guerinot M (2002) Expression of the IRT1 metal transporter is controlled by metals at the levels of transcript and protein accumulation. Plant Cell 14:1347–1357

Curie C, Alonso JM, Le Jean M et al (2000) Involvement of NRAMP1 from Arabidopsis thaliana in iron transport. Biochem J 347:749–755

DalCorso G, Farinati S, Maistri S, Furini A (2008) How plants cope with cadmium: staking all on metabolism and gene expression. J Integr Plant Biol 50:1268–1280

DalCorso G, Farinati S, Furini A (2010) Regulatory networks of cadmium stress in plants. Plant Signal Behav 5(6):663–667. https://doi.org/10.4161/psb.5.6.11425

Dobrikova AG, Apostolova EL (2019) Damage and protection of the photosynthetic apparatus under cadmium stress. In: Hasanuzzaman M, Prasad MNV, Fujita M (eds) Cadmium toxicity and tolerance in plants. Elsevier, Amsterdam, pp 275–298

El Rasafi T, Oukarroum A, Haddioui A, Song H, Kwon EE, Bolan N, Tack FMG, Abin Sebastian MNV, Rinklebe J (2020) Cadmium stress in plants: a critical review of the effects, mechanisms, and tolerance strategies. Crit Rev Environ Sci Technol. https://doi.org/10.1080/10643389.2020.1835435

Fan SK, Fang XZ, Guan MY, Ye QY, Lin XY, Du ST, Jin C (2014) Exogenous abscisic acid application decreases cadmium accumulation in Arabidopsis plants, which is associated with the inhibition of IRT1-mediated cadmium uptake. Front Plant Sci 5:721

Feng SJ, Liu XS, Tao H, Tan SK, Chu SS, Oono Y, Zhang XD, Chen J, Yang ZM (2016) Variation of DNA methylation patterns associated with gene expression in rice (Oryza sativa) exposed to cadmium. Plant Cell Environ 39:2629–2649

Feng S, Tan J, Zhang Y, Liang S, Xiang S, Wang H, Chai T (2017) Isolation and characterization of a novel cadmium-regulated yellow stripe-like transporter (SnYSL3) in Solanum nigrum. Plant Cell Rep 36(2):281–296

Garnier L, Simon-Plas F, Thuleau P, Agnel JP, Blein JP, Ranjeva RJ, Montillet L (2006) Cadmium affects tobacco cells by a series of three waves of reactive oxygen species that contribute to cytotoxicity. Plant Cell Environ 29:1956–1969

Grotz N, Guerinot ML (2006) Molecular aspects of Cu, Fe and Zn homeostasis in plants. Biochim Biophys Acta 1763(7):595–608. https://doi.org/10.1016/j.bbamcr.2006.05.014

Guo WJ, Meetam M, Goldsbrough PB (2008) Examining the specific contributions of individual Arabidopsis metallothioneins to copper distribution and metal tolerance. Plant Physiol 146(4): 1697–1706

Ha SB, Smith AP, Howden R, Dietrich WM, Bugg S, O'Connell MJ et al (1999) Phytochelatin synthase genes from Arabidopsis and the yeast Schizosaccharomyces pombe. Plant Cell 11: 1153–1164

Hart JJ, Welch RM, Norvell WA, Clarke JM, Kochian LV (2005) Zinc effects on cadmium accumulation and partitioning in near-isogenic lines of durum wheat that differ in grain cadmium concentration. New Phytol 167:391–401

Hassinen VH, Tervahauta AI, Schat H, Karenlampi SO (2011) Plant metallothioneins—metal chelators with ROS scavenging activity? Plant Biol 13(2):225–232

Heiss S, Wachter A, Bogs J, Cobbett C, Rausch T (2003) Phytochelatin synthase (PCS) protein is induced in Brassica juncea leaves after prolonged Cd exposure. J Exp Bot 54:1833–1839

Hernández LE, Sobrino-Plata J, Montero-Palmero MB, Carrasco-Gil S, FloresCáceres ML, Ortega-Villasante C, Escobar C (2015) Contribution of glutathione to the control of cellular redox homeostasis under toxic metal and metalloid stress. J Exp Bot 66:2901–2911

Howden R, Cobbett CS (1992) Cadmium-sensitive mutants of Arabidopsis thaliana. Plant Physiol 99:100–107

Huang J, Zhang Y, Peng JS, Zhong C, Yi HY, Ow DW, Gong JM (2012) Fission yeast HMT1 lowers seed cadmium through phytochelatin-dependent vacuolar sequestration in Arabidopsis. Plant Physol 158:1779–1788

Hussain D, Haydon MJ, Wang Y, Wong E, Sherson SM, Young J, Cobbett CS (2004) P-type ATPase heavy metal transporters with roles in essential zinc homeostasis in Arabidopsis. Plant Cell 16(5):1327–1339

Islam MM, Hoque MA, Okuma E, Jannat R, Banu MNA, Jahan MS, Murata Y (2009) Proline and glycinebetaine confer cadmium tolerance on tobacco bright yellow-2 cells by increasing ascorbate-glutathione cycle enzyme activities. Biosci Biotechnol Biochem 73:2320–2323

Ismael MA, Elyamine AM, Moussa MG, Cai M, Zhao X, Hu C (2019) Cadmium in plants: uptake, toxicity, and its interactions with selenium fertilizers. Metallomics 11(2):255–277

Jain S, Muneer S, Guerriero G, Liu S, Vishwakarma K, Chauhan DK, Dubey NK, Tripathi DK, Sharma S (2018) Tracing the role of plant proteins in the response to metal toxicity: a comprehensive review. Plant Signal Behav 13(9):1–11. https://doi.org/10.1080/15592324.2018.1507401

Jalmi SK, Bhagat PK, Verma D, Noryang S, Tayyeba S, Singh K, Sharma D, Sinha AK (2018) Traversing the links between heavy metal stress and plant signaling. Front Plant Sci 9:12–21. https://doi.org/10.3389/fpls.2018.00012

Jonak C, Nakagami H, Hirt H (2004) Heavy metal stress. Activation of distinct mitogen-activated protein kinase pathways by copper and cadmium. Plant Physiol 136(2):3276–3283

Khan MY, Prakash V, Yadav V, Chauhan DK, Prasad SM, Ramawat N, Singh VP, Tripathi DK, Sharma S (2019) Regulation of cadmium toxicity in roots of tomato by indole acetic acid with special emphasis on reactive oxygen species production and their scavenging. Plant Physiol Biochem 142:193–201. https://doi.org/10.1016/j.plaphy.2019.05.006

Kim DY, Bovet L, Maeshima M, Martinoia E, Lee Y (2007) The ABC transporter AtPDR8 is a cadmium extrusion pump conferring heavy metal resistance. Plant J 50(2):207–218

Koffler BE, Polanschütz L, Zechmann B (2014) Higher sensitivity of pad2-1 and vtc2-1 mutants to cadmium is related to lower subcellular glutathione rather than ascorbate contents. Protoplasma 251:755–769

Koike S, Inoue H, Mizuno D, Takahashi M, Nakanishi H, Mori S, Nishizawa NK (2004) OsYSL2 is a rice metal-nicotianamine transporter that is regulated by iron and expressed in the phloem. Plant J 39:415–424

Kuhnlenz T, Schmidt H, Uraguchi S, Clemens S (2014) Arabidopsis thaliana phytochelatin synthase 2 is constitutively active in vivo and can rescue the growth defect of the PCS1-deficient cad1-3 mutant on Cd-contaminated soil. J Exp Bot 65:4241–4253

Lanquar V, Lelièvre F, Bolte S, Hamès C, Alcon C, Neumann D, Vansuyt G, Curie C, Schröder A, Krämer U, Barbier Brygoo H, Thomine S (2005) Mobilization of vacuolar iron by AtNRAMP3 and AtNRAMP4 is essential for seed germination on low iron. EMBO J 24:4041–4051

Lee J, Shim D, Song WY, Hwang I, Lee Y (2004) Arabidopsis metallothioneins 2a and 3 enhance resistance to cadmium when expressed in Vicia faba guard cells. Plant Mol Biol 54:805–815

Lin YF, Liang HM, Yang SY, Boch A, Clemens S, Chen CC, Yeh KC (2009) Arabidopsis IRT3 is a zinc-regulated and plasma membrane localized zinc/iron transporter. New Phytol 182(2): 392–404

Liu Y, Wang X, Zeng G, Qu D, Gu J, Zhou M, Chai L (2007) Cadmium-induced oxidative stress and response of the ascorbate-glutathione cycle in Boehmeria nivea (L.) Gaud. Chemosphere 69(1):99–107. https://doi.org/10.1016/j.chemosphere.2007.04.040

Liu Z, Chen W, He X (2015) Influence of Cd^{2+} on growth and chlorophyll fluorescence in a hyperaccumulator: Lonicera japonica Thunb. J Plant Growth Regul 34:672–676

Liu Y, Liu L, Qi J, Dang P, Xia T (2019) Cadmium activates ZmMPK3-1 and ZmMPK6-1 via induction of reactive oxygen species in maize roots. Biochem Biophys Res Commun 516(3): 747–752

Luo J-S, Huang J, Zeng D-L, Peng J-S, Zhang G-B, Ma H-L, Guan Y, Yi H-Y, Fu Y-L, Han B, Lin H-X, Qian Q, Gong J-M (2018) A defensin-like protein drives cadmium efflux and allocation in rice. Nat Commun 9:645

Maksymiec W (2007) Signaling responses in plants to heavy metal stress. Acta Physiol Plant 29: 177–187

Mei H, He CQ, Ding NZ (2018) Abiotic stresses: general defenses of land plants and chances for engineering multistress tolerance. Front Plant Sci 871:1–18. https://doi.org/10.3389/fpls.2018.01771

Miyadate H, Adachi S, Hiraizumi A, Tezuka K, Nakazawa N, Kawamoto T, Katou K, Kodama I, Sakurai K, Takahashi H, Satoh-Nagasawa N, Watanabe A, Fujimura T, Akagi H (2011) OsHMA3, a P-1B-type of ATPase affects root-to shoot cadmium translocation in rice by mediating efflux into vacuoles. New Phytol 189:190–199

Miyashita Y, Dolferus R, Ismond KP, Good AG (2007) Alanine aminotransferase catalyses the breakdown of alanine after hypoxia in Arabidopsis thaliana. Plant J 49(6):1108–1121. https://doi.org/10.1111/j.1365-313X.2006.03023.x

Nakanishi H, Ogawa I, Ishimaru Y, Mori S, Nishizawa NK (2006) Iron deficiency enhances cadmium uptake and translocation mediated by the Fe2+ transporters OsIRT1 and OsIRT2 in rice. Soil Sci Plant Nutr 52(4):464–469

Nevo Y, Nelson N (2006) The NRAMP family of metal-ion transporters. Biochem Biophys Acta 1763:609–620

Nriagu J (1988) A silent killer of environmental metal poisoning. Environ Pollut 50(1–2):139–161

Oomen RJ, Wu J, Lelièvre F, Blanchet S, Richaud P, Barbier-Brygoo H, Thomine S (2009) Functional characterization of NRAMP3 and NRAMP4 from the metal hyperaccumulator Thlaspi caerulescens. New Phytol 181(3):637–650

Pavlíková D, Zemanová V, Procházková D, Pavlík M, Száková J, Wilhelmová N (2014) The long-term effect of zinc soil contamination on selected free amino acids playing an important role in plant adaptation to stress and senescence. Ecotoxicol Environ Saf 100:166–170

Perfus-Barbeoch L, Leonhardt N, Vavaddeur A, Forestier C (2002) Heavy metal toxicity: cadmium permeates through calcium channels and disturbs the plant water status. Plant J 32:539–548

Pinto AP, Mota AM, de Varennes A, Pinto FC (2004) Influence of organic matter on the uptake of cadmium, zinc, copper and iron by Sorghum plants. Sci Total Environ 326(1–3):239–247

Polle A, Schuetzenduebel A (2003) Heavy metal signalling in plants: linking cellular and organismic responses. In: Hirt H, Shinozaki K (eds) Plant responses to abiotic stress. Springer, Berlin, pp 187–215

Rahman A, Nahar K, Hasanuzzaman M, Fujita M (2016) Manganese-induced cadmium stress tolerance in rice seedlings: coordinated action of antioxidant defense, glyoxalase system and nutrient homeostasis. C R Biol 339(11-12):462–474. https://doi.org/10.1016/j.crvi.2016.08.002

Rahman MF, Ghosal A, Alam MF, Kabir AH (2017) Remediation of cadmium toxicity in field peas (Pisum sativum L.) through exogenous silicon. Ecotoxicol Environ Saf 135:165–172

Ramesh SA, Shin R, Eide DJ, Schachtman DP (2003) Differential metal selectivity and gene expression of two zinc transporters from rice. Plant Physiol 133(1):126–134

Rizwan M, Ali S, Ur Rehman MZ, Rinklebe J, Tsang DC, Bashir A, Maqbool A, Tack FMG, Ok YS (2018) Cadmium phytoremediation potential of Brassica crop species: a review. Sci Total Environ 631:1175–1191

Romero-Puertas MC, Corpas FJ, Rodriguez-Serrano M, Gomez M, del Rio LA, Sandalio LM (2007) Differential expression and regulation of antioxidative enzymes by cadmium in pea plants. J Plant Physiol 164:1346–1357

Sakouhi L, Rahoui S, Gharsallah C, Munemasa S, El Ferjani E, Murata Y, Chaoui A (2018) Effects of calcium and EGTA on thiol homeostasis and defense-related enzymes in Cd-exposed chickpea roots. Acta Physiol Plant 40:20

Sasaki A, Yamaji N, Yokosho K, Ma JF (2012) Nramp5 is a major transporter responsible for manganese and cadmium uptake in rice. Plant Cell 24(5):2155–2167

Schaaf G, Ludewig U, Erenoglu BE, Mori S, Kitahara T, Von Wirén N (2004) ZmYS1 functions as a proton-coupled symporter for phytosiderophore- and nicotianamine-chelated metals. J Biol Chem 279(10):9091–9096

Seher Y, Filiz O, Melike B (2013) Gamma-amino butyric acid, glutamate dehydrogenase and glutamate decarboxylase levels in phylogenetically divergent plants. Plant Syst Evol 299(2): 403–412

Sharma SS, Dietz KJ (2006) The significance of amino acids and amino acid-derived molecules in plant responses and adaptation to heavy metal stress. J Exp Bot 57(4):711–726

Sharma SS, Schat H, Vooijs R (1998) In vitro alleviation of heavy metal-induced enzyme inhibition by proline. Phytochemistry 49:1531–1535

Shim D, Hwang JU, Lee J, Lee S, Choi Y, An G, Martinoia E, Lee Y (2009) Orthologs of the class A4 heat shock transcription factor HsfA4a confer cadmium tolerance in wheat and rice. Plant Cell 21:4031–4043

Song J, Feng SJ, Chen J, Zhao WT, Yang ZM (2017) A cadmium stress-responsive gene AtFC1 confers plant tolerance to cadmium toxicity. BMC Plant Biol 17(1):1–15

Sui FQ, Chang JD, Tang Z, Liu WJ, Huang XY, Zhao FJ (2018) Nramp5 expression and functionality likely explain higher cadmium uptake in rice than in wheat and maize. Plant Soil 433(1):377–389

Sun Y, Liu Z, Guo J, Zhu Z, Zhou Y, Guo C, Hu Y, Li J, Shangguan Y, Li T, Hu Y, Wu R, Li W, Rochaix J-D, Miao Y, Sun X (2020) WRKY33-PIF4 loop is required for the regulation of H_2O_2 homeostasis. Biochem Biophys Res Commun 527(4):922–928. https://doi.org/10.1016/j.bbrc.2020.05.041

Takahashi R, Ishimaru Y, Senoura T, Shimo H, Ishikawa S, Arao T, Nakanishi H, Nishizawa NK (2011) The OsNRAMP1 iron transporter is involved in Cd accumulation in rice. J Exp Bot 62: 4843–4850

Takahashi R, Ishimaru Y, Shimo H, Ogo Y, Senoura T, Nishizawa NK, Nakanishi H (2012) The OsHMA2 transporter is involved in root-to-shoot translocation of Zn and Cd in rice. Plant Cell Environ 35(11):1948–1957

Tang L, Mao B, Li Y, Lv Q, Zhang L, Chen C, He H, Wang W, Zeng X, Shao Y, Pan Y, Hu Y, Peng Y, Fu X, Li H, Xia S, Zhao B (2017) Knockout of OsNramp5 using the CRISPR/Cas9 system produces low Cd-accumulating indica rice without compromising yield. Sci Rep 7: 14438

Thomine S, Schroeder JI (2013) Plant metal transporters with homology to proteins of the NRAMP family. In: Madame curie bioscience database. Landes Bioscience, Austin

Thomine S, Wang R, Ward JM et al (2000) Cadmium and iron transport by members of a plant transporter gene family in Arabidopsis with homology to NRAMP genes. Proc Natl Acad Sci U S A 97:4991–4996

Uraguchi S, Kamiya T, Sakamoto T, Kasai K, Sato Y, Nagamura Y, Yoshida A, Kyozuka J, Ishikawa S, Fujiwara T (2011) Low-affinity cation transporter (OsLCT1) regulates cadmium transport into rice grains. Proc Natl Acad Sci U S A 108:20959–20964

Vassilev A, Lidon F (2011) Cd-induced membrane damages and changes in soluble protein and free amino acid contents in young barley plants. Emirates J Food Agric 23:130–136

Vavasseur PBLLNA, Forestier C (2002) Heavy metal toxicity: cadmium permeates through calcium channels and disturbs the plant water status. Plant J 32(4):539–548

Verret F, Gravot A, Auroy P, Leonhardt N, David P, Nussaume L, Richaud P (2004) Overexpression of AtHMA4 enhances root-to-shoot translocation of zinc and cadmium and plant metal tolerance. FEBS Lett 576(3):306–312

Vidal SM, Malo D, Vogan K, Skamene E, Gros P (1993) Natural resistance to infection with intracellular parasites: isolation of a candidate for Bcg. Cell 73(3):469–485

Wang HC, Wu JS, Chia JC, Yang CC, Wu YJ, Juang RH (2009) Phytochelatin synthase is regulated by protein phosphorylation at a threonine residue near its catalytic site. J Agric Food Chem 57: 7348–7355

Wang JW, Li Y, Zhang YX, Chai TY (2013) Molecular cloning and characterization of a Brassica juncea yellow stripe-like gene, BjYSL7, whose overexpression increases heavy metal tolerance of tobacco. Plant Cell Rep 32(5):651–662

Wani KI, Zehra A, Choudhary S, Naeem M, Khan MMA, Castroverde CDM, Aftab T (2020) Mechanistic insights into strigolactone biosynthesis, signaling, and regulation during plant growth and development. J Plant Growth Regul 88:1–17

Wani KI, Choudhary S, Zehra A, Naeem M, Weathers P, Aftab T (2021a) Enhancing artemisinin content in and delivery from Artemisia annua: a review of alternative, classical, and transgenic approaches. Planta 254(2):1–15

Wani KI, Naeem M, Castroverde CDM, Kalaji HM, Albaqami M, Aftab T (2021b) Molecular mechanisms of nitric oxide (NO) signaling and reactive oxygen species (ROS) homeostasis during abiotic stresses in plants. Int J Mol Sci 22(17):9656

Wolf AE, Dietz KJ, Schroder P (1996) Degradation of glutathione s-conjugates by a carboxypeptidase in the plant vacuole. FEBS Lett 384:31–34

Wong CKE, Cobbett CS (2009) HMA P-type ATPases are the major mechanism for root-to-shoot Cd translocation in Arabidopsis thaliana. New Phytol 181(1):71–78

Wu FB, Chen F, Wei K, Zhang GP (2004) Effect of cadmium on free amino acid, glutathione and ascorbic acid concentrations in two barley genotypes (Hordeum vulgare L.) differing in cadmium tolerance. Chemosphere 57(6):447–454

Yan JL, Wang PT, Wang P, Yang M, Lian XM, Tang Z, Huang CF, Salt DE, Zhao FJ (2016) A loss-of-function allele of OsHMA3 associated with high cadmium accumulation in shoots and grain of japonica rice cultivars. Plant Cell Environ 39:1941–1954

Yang M, Zhang YY, Zhang L, Hu J, Zhang X, Lu K, Dong H, Wang D, Zhao FJ, Huang CF, Lian XM (2014) OsNRAMP5 contributes to manganese translocation and distribution in rice shoots. J Exp Bot 65:4849–4486

Zafarzadeh A, Rahimzadeh H, Mahvi AH (2018) Health risk assessment of heavy metals in vegetables in an endemic esophageal cancer region in Iran. Health Scope 7(3):e12340. https://doi.org/10.5812/jhealthscope.12340
Zayneb C, Bassem K, Zeineb K, Grubb CD, Noureddine D, Hafedh M, Amine E (2015) Physiological responses of fenugreek seedlings and plants treated with cadmium. Environ Sci Pollut Res Int 22(14):10679–10689. https://doi.org/10.1007/s11356-015-4270-8
Zemanová V, Pavlík M, Pavlíková D, Kyjaková P (2015) Changes in the contents of amino acids and the profile of fatty acids in response to cadmium contamination in spinach. Plant Soil Environ 61:285–290
Zemanová V, Pavlík M, Pavlíková D (2017) Cadmium toxicity induced contrasting patterns of concentrations of free sarcosine, specific amino acids and selected microelements in two Noccaea species. PLoS One 12(5):e0177963
Zhang F, Li X, Wang C, Shen Z (2000) Effect of cadmium on autoxidation rate of tissue and inducing accumulation of free proline in seedlings of mung bean. J Plant Nutr 23(3):357–368
Zheng X, Chen L, Li X (2018) Arabidopsis and rice showed a distinct pattern in ZIPs genes expression profile in response to Cd stress. Bot Stud 59(1):1–10
Zhigang A, Cuijie L, Yuangang Z, Yejie D, Wachter A, Gromes R, Rausch T (2006) Expression of BjMT2, a metallothionein 2 from Brassica juncea, increases copper and cadmium tolerance in Escherichia coli and Arabidopsis thaliana, but inhibits root elongation in Arabidopsis thaliana seedlings. J Exp Bot 57:3575–3582
Zhu G, Xiao H, Guo Q, Zhang Z, Zhao J, Yang D (2018) Effects of cadmium stress on growth and amino acid metabolism in two compositae plants. Ecotoxicol Environ Saf 158:300–308

5 Natural Resistance-Associated Macrophage Proteins (NRAMPs): Functional Significance of Metal Transport in Plants

Anitha Mani and Kavitha Sankaranarayanan

Abstract

In plants, there are several families of metal transporters that play an important role in metal uptake and sequestration, thus protecting the plant from stress due to heavy metals. Crop plants grown on heavy metal-contaminated agricultural land present an indirect hazard to the health of their consumers. Hence, it is very important to study the transporters associated with metal transport in plants. Natural resistance-associated macrophage protein (NRAMP) transporters are present in a wide range of organisms. This family of NRAMP transporters has been identified and functionally characterized in several plant species like *Arabidopsis*, rice, soya bean, etc. that is, in both monocot and dicot plants. In plants, several members of the NRAMP gene family have been identified and functionally characterized. They are involved in the transport of several divalent metal ions in the plant based on their localization. The presence of NRAMP transporter genes has also been computationally predicted in the genomes of all the plants studied so far.

This chapter discusses the general properties, structure, function, and expression of NRAMP transporters in several plants based on existing literature. It mainly focuses on the functional significance of the NRAMP family of transporters in both monocot and dicot plants.

Keywords

Metal · Plant · NRAMP · Structure · Function · Expression

A. Mani · K. Sankaranarayanan (✉)
Ion Channel Biology Laboratory, AU-KBC Research Centre, Madras Institute of Technology, Anna University, Chennai, Tamil Nadu, India

K. Kumar, S. Srivastava (eds.), *Plant Metal and Metalloid Transporters*,
https://doi.org/10.1007/978-981-19-6103-8_5

Abbreviations

NRAMP	Natural resistance-associated macrophage protein
CDF	Cation diffusion facilitator
CTR	Copper transporters
YSL	Yellow stripe-like
ZIP	ZRT1-IRT1-like protein
ABC	ATP binding cassette
CAX	Cation proton exchanger

5.1 Introduction

Fourteen different minerals are required by the plants for their overall growth and development (Marschner 2012). These mineral elements present in the soil are absorbed by the roots, transported to the shoots, and then distributed to various tissues and parts of the plant based on their requirement (Marschner 2012). The quality of food, to a large extent, is determined by the uptake and sequestration of essential, as well as nonessential heavy metals, by the plants. In some types of soil, the bioavailability and abundance of essential metals can be limiting, and thus plants have developed several methods for efficient absorption.

In plants, there exist several large families of metal transporter genes. These transporters are involved in the uptake and efflux of metal ions. The families of metal transporters include ZRT1-IRT1-like protein (ZIP), ATPases, cation diffusion facilitator (CDF), copper transporters (CTR), and natural resistance-associated macrophage protein (NRAMP) homologs (Guerinot 2000). Apart from them, members of the ABC transporter family and vacuolar cation proton exchanger (CAX) are involved in plant metal homeostasis (Kushnir et al. 2001).

After the uptake of these metals, plants transport them to the cellular compartments and ultimately reach the growing organs where they are required. Metals play many important roles in photosynthesis but can cause severe oxidative damage when in excess, thus necessitating a highly regulated influx/efflux mechanism in the photosynthetic tissues (Hall and Williams 2003). Plants being sessile also have to combat toxicity of nonessential heavy metals such as lead, cadmium, and mercury or essential metals when present in excessive concentrations. The transporters function to either exclude metals at the root or sequester metals in certain cell compartments like the vacuole and thus minimize damages caused due to them.

Members of the NRAMP family of transporters are found in a wide variety of living beings like animals, plants, fungi, and bacteria (Nevo and Nelson 2006). The NRAMP family of transporters are highly conserved membrane proteins that are involved in metal ion transport. The NRAMP gene was first identified in phagosomes inside infected murine macrophages. It was believed to determine sensitivity to bacterial infection by regulating the concentrations of essential divalent metal ions. Homologs of *NRAMP1* were later characterized in plants as well (Hall and Williams 2003).

In plants, the NRAMP gene family has been identified to function on several divalent metal ions, regulating their acquisition, transportation, and homeostasis. NRAMP transporters primarily transport metal ions like Fe^{+2}, Mn^{+2}, Co^{+2}, and Zn^{+2} (Nevo and Nelson 2006). The rice NRAMPs were one of the first NRAMP families to be identified in plants with the ESTs of OsNRAMP1-3 from rice being identified and cloned (Belouchi et al. 1995). Subsequently genes from this family of transporters were identified in several higher plants. A total of six NRAMP transporters have been identified in *Arabidopsis* (Mäser et al. 2001). The NRAMP transporters in plants are divided into two subfamilies, with AtNRAMP1 and AtNRAMP6 belonging to the first subfamily and NRAMP2–5 to the second subfamily (Mäser et al. 2001). The rice OsNRAMP1 and OsNRAMP3 are grouped in the first subfamily while OsNramp 2 has been grouped into the second subfamily.

A variable number of NRAMP proteins have been identified in the genomes of several plant species. A basal angiosperm, *A. trichopoda*, has three copies of NRAMP; these copies underwent a lineage-specific expansion to 10 copies found in *Panicum virgatum*, a monocot species, and 13 in *Glycine max*, a eudicot species (Ullah et al. 2018). NRAMP genes are present in all plant families in both the grass and non-graminaceous species. These two groups of plants use different approaches to transport iron from the soil (Marschner and Romheld 1994). In non-graminaceous plants, the IRT/FRO system is a major component of the Fe uptake system (Qin et al. 2017). The main transporter in graminaceous plants responsible for uptake of Fe from siderophore-Fe complexes is YSL (Thomine and Vert 2013). Apart from the transporters involved in both these strategies, the NRAMP family represents another transporter family associated with Fe uptake and transport (Thomine and Vert 2013). The presence of plant NRAMP genes in these two groups of plants suggests their involvement in metal uptake from the soil solution and metal homeostasis. NRAMP genes also affect the remobilization of intracellular Cd (Cailliatte et al. 2009), which might also help to increase tolerance to toxic heavy metals in plants.

The current knowledge regarding the properties and functional aspects of the plant NRAMP is presented in this chapter. The plant NRAMP family of transporters has been reported in both plant genomic and plant EST databases, indicating that genes from this family are present in all the plants studied at the molecular level. However, there is a huge gap in the understanding of the molecular and physiological functions of these groups of proteins. Here, we provide an overview of the plant NRAMP systems that are thought to be involved in the acquisition, distribution, and redistribution of transition metals. Specific focus would be paid to their location, function, and known substrate specificity considering the overall view of transition metal nutrition in plants.

5.2 Genomic Analysis

The NRAMP family of transporters in yeast is represented by three genes, namely *SMF1*, *SMF2*, and *SMF3*, in fly by *MVL* (*malvolio*), and by *NRAMP1* and *NRAMP2* in mammals. The rat isoform of the human NRAMP2 is DCT1 (*d*ivalent *c*ation *t*ransporter 1 or *d*ivalent *m*etal *t*ransporter DMT1) (Gunshin et al. 1997). One of the characteristics of plant *NRAMP* genes is the comparatively high number of *NRAMP* gene homologs per species. The rice genome contains seven distinct genes which encode NRAMP homologous proteins (Bennetzen 2002). Additionally, several ESTs with homology to NRAMP in several species like *Medicago truncatula*, soya bean, maize, cotton, pine, tomato, and barley have been identified.

In higher plants like *Arabidopsis*, rice, soybean, *Medicago*, barley, peanut, mustard, tomato, and apple, the identified NRAMPs are divided into two large groups, namely, subfamily I and subfamily II, with both dicots and monocots represented in each of the groups (Qin et al. 2017). The *Arabidopsis* genome encodes six *NRAMP* homologs; it has four members in subfamily I and two members in subfamily II. The rice genome encodes seven members, two in subfamily I and five in subfamily II and correspondingly four and three from *Medicago*. In soya bean (*Glycine max*), 13 *GmNRAMP* homologs have been identified to be encoded by the genome. Of the 13 NRAMP members in soya bean, eight belong to subfamily I and five to subfamily II. While in mustard, NRAMP proteins are grouped into subfamily I alone, those in peanut, barley, and apple are grouped into subfamily II (Qin et al. 2017). There are 11 CsNRAMPs, in tea which are divided into group I (CsNRAMP3, CsNRAMP4, CsNRAMP5, and CsNRAMP8) and group II (the remaining CsNRAMPs) (Jinqiu Li et al. 2021).

The members of the group II family of NRAMPs appear to be more related to the reported animal NRAMP homologs. It has been reported that both groups of NRAMPs are required for a regulated metal homeostasis in both the plant and animal kingdoms (Thomine and Vert 2013).

5.3 Structural Analysis

The plant NRAMPs have been highly conserved throughout evolution. In rice, OsNRAMPs are 518–550 amino acid residues long and are predicted to have 11–12 transmembrane domains (Mani and Sankaranarayanan 2018a). In *Phaseolus vulgaris* (common bean), PvNRAMPs have 12 transmembrane domains, and they are 507–554 amino acid residues long (Ishida et al. 2018). In *Camellia sinensis*, CsNRAMP proteins are 279–1373 amino acid residues long and contain 3–12 transmembrane regions. This is believed to be due to a broken NRAMP domain (Jinqiu Li et al. 2021). In *Brassica napus*, BnNRAMPs are only 100–200 amino acid residues long (Meng et al. 2017).

Members of the NRAMP family have considerable protein sequence similarity of 28% (yeast), 40% (plant), and 55% (fly) with the mammalian proteins (46%, 58%, and 73% similarity, respectively) (Cellier et al. 1995). The rat isoform DCT1 shares

92% identity with NRAMP2 and 73% identity to NRAMP1 in humans (Gunshin et al. 1997). In the plant kingdom, NRAMP proteins show high amino acid sequence similarity than NRAMP proteins from other kingdoms. *Arabidopsis* NRAMP proteins share 40–50% amino acid sequence identity with NRAMP1 in mouse and about 30% with SMF1 the yeast NRAMP. The sequence similarity of the bacterial consensus transport sequence found between transmembrane domains TM8 and TM9 and the predicted transmembrane domains is even greater (Kerppola and Ames 1992).

NRAMPs generally have 11 or 12 transmembrane domains. The first ten TMs form a LeuT-fold, like in bacterial homolog structures and the model *Deinococcus radiodurans* (Dra) NRAMP (Bozzi et al. 2016b). Conserved residues in TM1 and TM6 like asparagine, methionine, and aspartate coordinate transition metal substrates (Ehrnstorfer et al. 2014). In between TM8 and TM9, there is a characteristic "consensus transport motif" (CTM) (Williams et al. 2000). This CTM has similarities with the previously studied conserved regions of numerous transport proteins in bacteria and particularly with the highly conserved part of the permeation pore of the *Shaker*-type K^+ channel in animals (Belouchi et al. 1997).

It has been reported that NRAMP proteins carry consensus residues between TMD8 and TMD9. GQSSTITGT YAG QY(/F)V(/I)MQGFLD(/E/N) is the consensus transport motif (CTM) commonly present among NRAMP proteins. In OsNRAMP1, OsNRAMP2, OsNRAMP3, OsNRAMP5, and OsNRAMP7, it is highly conserved, and in other members of the OsNRAMP family of transporter proteins, it is partially conserved. In the OsNRAMP6 protein sequence, it is least conserved (Mani and Sankaranarayanan 2018b). In *Arabidopsis*, the AtNRAMP proteins have GQSSTITGTY AGQXXMXGFLX as CTM, while in *Phaseolus vulgaris*, PvNRAMP proteins have GQSSTITGTYAGQFIMGGFLN (Ishida et al. 2018). The CsNRAMP proteins in *Camellia sinensis* also have similar consensus residues, that is, GQSSTxTGTYAGQFIMxGFLxLxxKKW (Jinqiu Li et al. 2021). For the NRAMP transporter family, the signature sequence is DPGN, and any mutation in these residues leads to defects in the transporter function (Mani and Sankaranarayanan 2018a).

X-ray crystallography which is one of the primary means of elucidating the structures of transporter proteins has the potential to provide snapshots of their structures at different stages of its functioning. Knowledge about the conformational changes taking place in the NRAMP transporter structure as it transports metal ions and protons can help us to understand its functioning. Using X-ray crystallography, Bozzi et al. (2019) studied the structure of the NRAMP transporters from bacterium *Deinococcus radiodurans*. The results hint at four distinct conformations that the protein adopts during different stages of metal transport. A unique feature of this mechanism is that it transports metal ions and protons by different pathways that has not been previously reported in transporter proteins with similar structures.

5.4 Functional Characterization

NRAMPs are a class of amino acid-polyamine-organocation (APC) superfamily transition metal transporters that are involved in the uptake of micronutrients like Mn^{2+} in plants, bacteria, and Fe^{2+} in animals (Cellier 2012). The APC superfamily of secondary transporters consists of several related transporters that transport various substrates like metabolites, neurotransmitters, and transition metals in all living organisms (Vastermark et al. 2014). Essential divalent metals like Mn^{2+}, Ni^{2+}, Fe^{2+}, Cu^{2+}, Co^{2+}, and Zn^{2+} and toxic elements like Pb^{2+}, Cd^{2+}, and Hg^{2+} are transported by NRAMPs. However alkaline metals like Mg^{2+} and Ca^{2+} which are widely available in soil are not transported by them (Bozzi et al. 2016a). Generally, metal uptake by the NRAMP transporter is triggered by acidic pH and followed by an influx of proton (Ehrnstorfer et al. 2017). No studies to date have confirmed that NRAMP is a thermodynamically coupled secondary transporter capable of using the desired gradient of metal or proton to power electrochemical uphill transport of the other substrate.

The transporters of NRAMP family have been reported in various plant species with diversified functions (Ishimaru et al. 2012). In numerous plant species like rice, soya bean, and *Arabidopsis*, several members of the NRAMP family have been identified and functionally characterized (Qin et al. 2017). Various *NRAMP* genes also have been recognized in plant species, like *TjNRAMP4* in *Thlaspi japonicum* (Mizuno et al. 2005), *MbNRAMP1* in *Malus baccata* (Xiao et al. 2008), *MtNRAMP1* in *Medicago truncatula* (Tejada-Jiménez et al. 2015), *GmNRAMPs* in soybean (Qin et al. 2017), and *NtNRAMP5* in tobacco (Tang et al. 2017).

Starting from the NRAMPs found in yeast SMF1, all the NRAMP proteins have been found to function as metal transporters. In yeast, SMF1 was first identified as a part of the Mn uptake system (Supek et al. 1996). Later in the mammalian duodenum, NRAMP2 was mainly identified as the Fe uptake transporter and was found to transport a range of heavy metals (Gunshin et al. 1997). NRAMP2 in mammalian cells plays an important role in iron uptake and recycling. Currently, it is assumed that all the *NRAMP* genes encode metal transporters with broad specificity. Plant NRAMP proteins have been demonstrated to be metal transporters by complementation studies in yeast mutants compromised in metal uptake (Thomine and Vert 2013). MtNRAMP1 in *Medicago truncatula* is the only transporter that was characterized in *M. truncatula*, as a Fe uptake protein (Tejada-Jiménez et al. 2015).

The first plant NRAMP genes were cloned from rice (Belouchi et al. 1997). In rice, there are seven Nramp transporters, namely, OsNRAMP1–OsNRAMP7. But not all of them have been functionally characterized (Nevo and Nelson 2006). In the NRAMP family, many of the transporter proteins are involved in Fe transport. Under Fe deficiency, *OsNRAMP1* is highly upregulated. The OsNRAMP1 transporter is localized on the plasma membrane and is involved in the transport of Fe and Cd. *OsNRAMP1* is involved in the cellular uptake of Cd thus leading to high Cd accumulation in rice (Takahashi et al. 2011) grown in cadmium-rich soils. OsNRAMP3 functions as a Mn-influx transporter and is involved in Mn translocation from old leaves via the phloem cells to young leaves. OsNRAMP3 is regulated posttranslationally in response to environmental nutrient availability. OsNRAMP3 is not involved in Mn uptake but is involved in Mn translocation (Yang et al. 2013; Mani and Sankaranarayanan 2018a).

OsNRAMP4 is also known as NRAMP aluminum transporter 1 (Nrat1). It is the first transporter in the NRAMP family to be identified as the trivalent Al ion transporter (Xia et al. 2010). OsNRAMP4 in rice does not transport other divalent metal ions, like Zn, Mn, and Fe. It also shares relatively low similarity with the other OsNRAMP members in rice (Xia et al. 2010). In rice Al tolerance, NRAT1 plays an important role by reducing the toxic Al level in the root cell wall and sequesters Al into the vacuole in root cells. OsNRAMP4 plays an important role in Al tolerance; hence, rice is the most Al tolerant of all the cereal crops (Famoso et al. 2010).

OsNRAMP5 is a plasma membrane protein involved in Mn and Fe transport (Ishimaru et al. 2012). When plants are under Fe or Zn deficiency, OsNRAMP5 gene expression increases slightly in the roots, but varying levels of Mn in the surrounding do not affect it (Sasaki et al. 2012). During flowering and seed development in rice plant, OsNRAMP5 plays a role in Fe and Mn transport as well and is constitutively involved in the uptake of Fe and Mn (Ishimaru et al. 2012; Mani and Sankaranarayanan 2018b). OsNRAMP5 is highly expressed in the hulls. It is also expressed in leaves, but the expression levels decline as the leaf ages. *OsNRAMP5* transporters have been reported to be expressed in the rice root and shoot vascular bundles, specifically in the parenchyma cells surrounding the xylem. It plays an important role in xylem-mediated root-to-shoot transport as it is highly expressed in stele cells especially in the xylem region. Thus OsNRAMP5 in rice plants plays a crucial role in Mn uptake, translocation, and distribution. *OsNRAMP5* is highly expressed in stele cells especially in the xylem region and plays an important role in the xylem-mediated root-to-shoot transport (Yang et al. 2014). Cd uptake in rice is also majorly orchestrated by OsNRAMP5 (Sasaki et al. 2012).

OsNRAMP6 is a plasma membrane-localized protein and has been identified to be involved in the uptake of Fe and Mn. Rice plant immunity is negatively regulated by it as loss of its function will result in increased resistance against *M. oryzae* (Peris-Peris et al. 2017).

In *Arabidopsis*, there are six NRAMP transporters (Williams et al. 2000). Heterologous expression studies of transporters AtNRAMP1, AtNRAMP3, and AtNRAMP4 in yeast mutants indicated that these proteins transport Fe, Mn, and Cd (Curie et al. 2000; Thomine et al. 2000). AtNRAMP3 and AtNRAMP4 are involved in the mobilization of Fe stores from vacuoles (Thomine et al. 2003; Lanquar et al. 2005). In adult plants, they also function in vacuolar Mn export into photosynthetic tissues (Lanquar et al. 2010). AtNRAMP6 is an intracellular Cd transporter (Cailliatte et al. 2009). AtNRAMP1 has been shown to be essential for the uptake of Mn from the soil and thus is a high-affinity Mn transporter (Cailliatte et al. 2010). It has been reported recently that the AtNRAMP2 protein is involved in the remobilization of Mn^{2+} in Golgi for root growth instead of Mn^{2+} uptake through roots (Gao et al. 2018).

AtNramp3 is also associated with sensitivity and uptake of Fe and Cd (Thomine et al. 2000). AtNramp3 is thought to function in long-distance metal transport. In

Arabidopsis, AtNRAMP3 and AtNRAMP4, to some extent, are involved in resistance against *E. chrysanthemi* a bacterial pathogen (Thomine et al. 2003). AtNRAMP3 and AtNRAMP4 function in vacuolar Fe mobilization (Mary et al. 2015). AtNRAMP6 plays an important role in intracellular Fe homeostasis and is located in the Golgi network (Li et al. 2019). AtNRAMP3 and AtNRAMP4 contribute to Fe mobilization and are located on vacuolar membranes (Mary et al. 2015).

Five NRAMP genes in cacao (*Theobroma cacao*) have been identified, namely, TcNRAMP1, TcNRAMP2, TcNRAMP3, TcNRAMP5, and TcNRAMP6. TcNRAMP5 transports essential metal ions and additionally transports Cd^{2+}. TcNRAMP3 and TcNRAMP5 are involved in the transport of Fe^{2+} and Mn^{2+}. TcNRAMP6 is specific for Mn^{2+} transport. It is suggested that TcNRAMP2 may be involved in the remobilization of metal cations rather than their uptake (Ullah et al. 2018).

MbNRAMP1 in the fruit tree *Malus baccata* is involved in the transfer of Fe, Mn, and Cd (Xiao et al. 2008). Under Fe deficiency in tomato plants (*Solanum lycopersicum*), LeNRAMP1 is believed to play a role in the distribution of Fe in the vascular parenchyma (Bereczky et al. 2003). NRAMP1 and NRAMP3 in tomato have also been suggested to be involved in Mn transport (Bereczky et al. 2003). The *NRAMP* gene *AhNRAMP1* in peanuts plays a role in Fe nutrition (Xiong et al. 2012). In wheat, *TpNRAMP3* is a Mn, Co, and Cd transporter and is potentially responsible for the accumulation of Mn, Co, and Cd (Peng et al. 2018). *NRAMP* genes in soybean are widely involved in the uptake, homeostasis regulation, and transport of Fe, Mn, Cu, and Cd metal ions. In *Phaseolus vulgaris* (common bean), *PvNRAMP1*, *PvNRAMP2*, *PvNRAMP3*, *PvNRAMP4*, and *PvNRAMP5* might be required for general metal homeostasis during all the developmental stages. While *PvNRAMP6* and *PvNRAMP7* might play a role in symbiotic associations with beneficial microorganisms (Ishida et al. 2018).

In the CSS reference genome database, 11 NRAMP genes have been identified and characterized as CsNRAMPs. CsNRAMP3, CsNRAMP4, CsNRAMP5, and CsNRAMP8 participate in the absorption of metal ions. Both CsNRAMP2 and CsNRAMP5 fusion proteins located in the plasma membrane may function in the transmembrane transport of metal ions (Jinqiu Li et al. 2021).

Members of the NRAMP family of transporters located on separate organelles show various functions. In plants, the transport selectivity of NRAMP protein is independent of the group (group I or II) they belong to in the phylogenetic tree (Thomine and Vert 2013). Therefore, the biological function of the *NRAMP family of transporters* is diverse in different plant species and needs to be further studied.

The metal uptake functions of plant NRAMP homologs have been demonstrated in mutant yeasts deficient in metal uptake; however, their roles in metal homeostasis in plants have not been fully interpreted yet. The presence of a large number of *NRAMP* genes in plant genomes has made this study very complex as it may cause functional redundancy within the NRAMP gene family.

5.5 Expression Pattern and Regulation

NRAMP proteins in plants mostly localize on intracellular membranes like the vacuolar membrane and plastid envelope (Thomine and Vert 2013). Fig. 5.1 shows different regions of cell where NRAMP proteins are localised (Jogawat et al., 2021).

AtNRAMP1 and *AtNRAMP2*, *LeNRAMP1* and *LeNRAMP3*, and *OsNRAMP1* are few members of the NRAMP family which are preferentially expressed in roots, and *AtNRAMP3* and *AtNRAMP4* and *OsNRAMP2* and *OsNRAMP3* are expressed in the shoots. *OsNRAMP3* and *AtNRAMP1* and *LeNRAMP1* despite being in the same group (group I) are preferentially expressed in different parts of the plant. In shoot, *OsNRAMP3* expression is stronger, while in the root, *AtNRAMP1* and *LeNRAMP1*

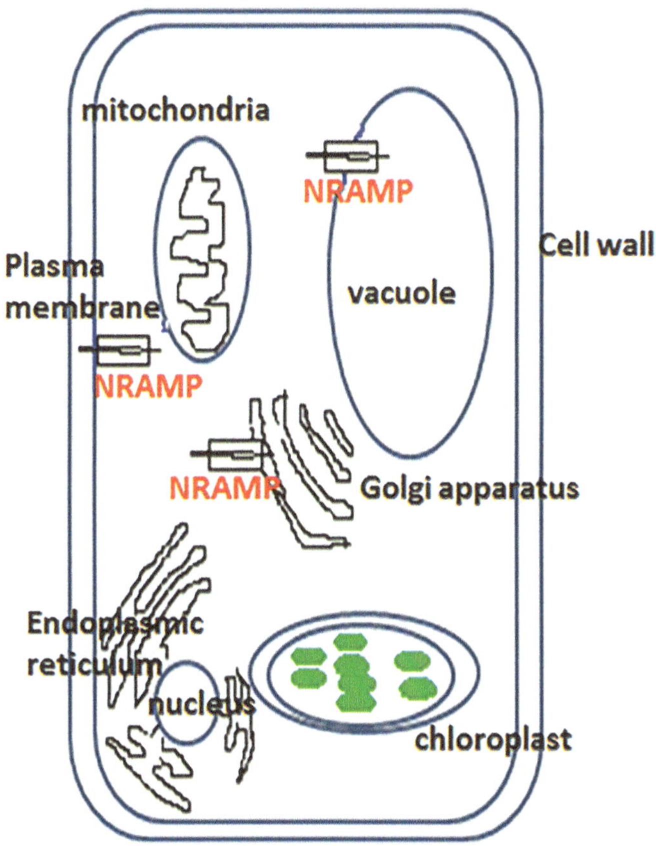

Fig. 5.1 NRAMP metal transporters expressed in different regions of the cell (plasma membrane, Golgi apparatus, vacuole) in different plant species

expressions are stronger (Curie et al. 2000; Bereczky et al. 2003). In *Arabidopsis*, the expression of AtNRAMP5 is restricted to the reproductive organs. In the vascular tissues of root and shoot, *AtNRAMP3* and *AtNRAMP4* are expressed (Thomine and Vert 2013). Thus the preferential expression of plant *NRAMP* genes in the roots or shoots does not depend on whether they belong to group I or group II of the phylogenetic tree.

During the vegetative state, *OsNRAMP1* expression is observed mainly in roots (Takahashi et al. 2011). The difference in *OsNRAMP1* expression levels in root affects Cd accumulation among different rice cultivars (Takahashi et al. 2011). During the reproductive stage, *OsNRAMP1* expression is higher in leaf blade and stem. OsNRAMP1 transporter is expressed on the plasma membrane of endodermis and pericycle cells, thus helping in the mobilization of metals from root to shoot. OsNRAMP3 is expressed in the plasma membrane and specifically in vascular bundles, particularly in phloem companion cells. In the rice node, *OsNRAMP3* is constitutively expressed (Yamaji et al. 2013). In rice plants, the expression of OsNRAMP3 in leaves slightly increases as the leaves age. To meet its minimal growth requirement during Mn deficiency, Mn from the enlarged vascular bundles to the younger tissues and panicles is transported by OsNRAMP3. However, OsNRAMP3 is internalized in vesicles and rapidly degraded when Mn is in excess. To protect the developing tissues from Mn toxicity, Mn is preferentially loaded into the older leaves and is directly connected to the enlarged vascular bundles. OsNRAMP3 undergoes post translational regulation in response to the nutrient availability in the environment (Yang et al. 2013; Mani and Sankaranarayanan 2018b).

In roots, OsNRAMP5 gene expression increases slightly when plants are under Fe or Zn deficiency but is not affected by variation in Mn level in the surrounding (Sasaki et al. 2012). It is expressed in the plasma membrane of the exodermal and endodermal layers at the mature root zone (Ishimaru et al. 2012; Sasaki et al. 2012). In rice hull, OsNRAMP5 is highly expressed. It is expressed in leaves also, but as leaves age, its expression decreases. In rice, OsNRAMP5 transporter is present in root and shoot vascular bundle, particularly the parenchyma cells around the xylem. *OsNRAMP5 is* highly expressed in stele cells especially in the xylem region and thus plays an important role in the xylem-mediated root-to-shoot transport. In rice plants, OsNRAMP5 plays an important role in the process of uptake, translocation, and distribution of Mn. *OsNRAMP5* plays an important role in the xylem-mediated root-to-shoot transport as it is highly expressed in stele cells particularly in the xylem region (Yang et al. 2014; Mani and Sankaranarayanan 2018a).

In *Arabidopsis* root, under conditions of Fe deficiency, expression of AtNramp1, AtNramp3, and AtNramp4 is upregulated (Curie et al. 2000; Thomine et al. 2000). AtNRAMP1 is located in the root plasma membrane (Meng et al. 2017). AtNRAMP3 is expressed in the vascular bundle of root, stem, and leaves (Thomine et al. 2003). It is upregulated and localized to the vacuolar membrane under conditions of Fe starvation (Thomine et al. 2003).

In tomato, *LeNramp1* is specifically expressed in the root epidermis and the cortex behind the root tip; it localizes to the root vascular parenchyma of the root hair zone and is upregulated by Fe deficiency (Bereczky et al. 2003); under conditions of Fe deficiency, *LeNramp1* is believed to mobilize Fe in the vascular tissue. In barley, when nitrogen (N) is adequate in the presence of Cd, Nramp transcript is downregulated, but under N-deficiency, it is strongly upregulated (Finkemeier et al. 2003). Some members of the NRAMP family are involved in Fe and Cd uptake and homeostasis, and other members may have different physiological functions.

In peanuts, the expression of *AhNRAMP1* is specifically higher in the roots and increased further under Fe deficiency (Xiong et al. 2012). Under Fe deficiency *LeNRAMP1*, *AtNRAMP1*, *MbNRAMP1*, and *OsNRAMP1* genes in tomato, *Arabidopsis*, *M. baccata*, and rice show higher expression in the roots (Takahashi et al. 2011). Thus these *NRAMP* genes are believed to have a conserved function in Fe homeostasis and to belong to a subclass of this family of proteins that are induced by Fe deficiency.

Expression of *TcNRAMP1* and *TcNRAMP5* in cacao root is high when compared to their expression levels in flower buds and beans. The *TcNRAMP6* gene is widely expressed in root and unopened flower buds. *TcNRAMP2* and *TcNRAMP3* are uniformly expressed across various organs and are constitutively expressed in the leaf and root tissues (Ullah et al. 2018).

In the tea plant (*Camellia sinensis*), 11 CsNRAMP genes' expression has been detected in different tissues (Jinqiu Li et al. 2021). In the root, CsNRAMP3, CsNRAMP4, and CsNRAMP5 are highly expressed, while in the leaf, CsNRAMP1, CsNRAMP2, CsNRAMP10, and CsNRAMP11 are highly expressed, and in the stem, CsNRAMP6 and CsNRAMP9 are highly expressed. CsNRAMP7 and CsNRAMP8 are highly expressed in both leaf and shoot tissues. CsNRAMP proteins are thought to play different roles in metal transport. Upon Pb treatment, expressions of CsNRAMP1, CsNRAMP2, CsNRAMP9, and CsNRAMP10 are upregulated in leaves. CsNRAMP3, CsNRAMP4, CsNRAMP5, CsNRAMP7, and CsNRAMP9 are thought to play a role in Pb transportation as their expression levels increased in the root when exposed to Pb. The expression of CsNRAMP3 increased greatly under Pb treatment (Jinqiu Li et al. 2021).

In *Arabidopsis*, AtNRAMP6 functions in young leaves and lateral roots. AtNRAMP3 and AtNRAMP4 are localized in the vacuolar membrane (Lanquar et al. 2005; Mary et al. 2015). AtNRAMP1 regulates Fe homeostasis (Curie et al. 2000) and functions as a high-affinity transporter for Mn uptake (Cailliatte et al. 2010). During seed germination, AtNRAMP3 and AtNRAMP4 participate in vacuolar Fe mobilization as both are localized on the vacuolar membrane (Lanquar et al. 2005). AtNRAMP6 functions as an intracellular metal transporter and is targeted to vesicular-shaped endomembrane compartments, and it is believed to be involved in Cd tolerance (Cailliatte et al. 2009).

A native Chinese plant species, *Sedum alfredii*, is a metal hyperaccumulator that can accumulate large amounts of Cd and Zn in the shoot without any significant effect on its growth and metabolism (Zhang et al. 2020). SaNRAMP1 transports Cd, Mn, and Zn and is highly expressed in the young shoots (Zhang et al. 2020).

TpNRAMP3 is a member of the NRAMP family identified in Polish wheat (*Triticum polonicum L.*). *TpNRAMP3* is localized on chromosome 7BL. At the jointing and booting stages, the expression of *TpNRAMP3* is very high in leaf blades and root and the first nodes during the grain filling stage. It is a plasma membrane-localized protein. The expression of *TpNRAMP3* in seedling roots is upregulated by Fe and Cu and downregulated by Mg and Ni (Peng et al. 2018).

In soybean root, under high concentration of Cu and Cd, the expressions of *GmNRAMP1a*, *GmNRAMP5a*, and *GmNRAMP3a* are highly increased, and *GmNRAMP5a* expression is increased significantly by toxic levels of metals like Cd, Cu, and Mn (Qin et al. 2017). GmNRAMP7 is the only NRAMP protein in soya bean which is expressed mainly in roots along with other monocot NRAMPs (like OsNRAMP5, OsNRAMP1, and HvNRAMP5) (Sasaki et al. 2012; Wu et al. 2016). Members of the GmNRAMPs which belong to subfamily I are predicted to localize to vacuoles, while members belonging to subfamily II localize to the plasma membrane (Table 5.1).

5.6 Conclusion

For normal plant growth and development, transition metals are essential. These metal ions are absorbed from the soil via their roots, distributed throughout the plant, and their concentrations in the organelles in specific cells of various tissues are regulated. Metal transporters play an important role in the uptake and transport of these metals. In the last decade, our knowledge regarding these metal transporters has increased tremendously, and a number of transporter families have been identified.

Similar to yeast and animal NRAMPs, *NRAMP* genes in plants also encode transition metal transporters with a broad selectivity. We have summarized the general structural information, genomic and functional analysis, and the expression pattern of these NRAMP transporters in various plant species. However, the existing knowledge regarding the functioning of NRAMP transporters in plant remains finite. Much remains to be understood specifically on the cofactors and other physiological/environmental conditions in plant cells which determine the selectivity of the NRAMP transporters in vivo. It would be fascinating to identify the variations in the substrate for each NRAMP transporter and the structural features which result in the difference in substrate selectivity. A combination of several studies like analyses of the structure-function using heterologous expression systems, molecular genetic studies in plant, cell biological and biochemical studies regarding structure stability, and imaging studies for observing metal localization will give a clear picture of the function of NRAMP transporters and their connection with other transporters and chelators associated with metal homeostasis in plants. This understanding is pivotal in harnessing the tools of genetics to develop superior varieties of heavy metal-

Table 5.1 Summary of the NRAMP transporters mentioned in the chapter

SI no.	Plant	Name of NRAMP	No. of members predicted/ identified	Predicted/identified functional role
1	Rice	OsNRAMP	7(Nevo et al., 2006)	*OsNRAMP1*—Cd uptake (Takahashi et al., 2011) OsNRAMP3—Mn translocation (Yang et al., 2013, Mani and Sankaranarayanan, 2018a, b) OsNRAMP4—Al ion transporter (Xia et al., 2010) OsNRAMP5—Mn and Fe transport (Ishimaru et al., 2012) OsNRAMP6—uptake of Fe and Mn (Peris-Peris et al., 2017)
2	*Arabidopsis*	*AtNRAMP*	6 (Williams et al., 2000).	AtNRAMP1—Mn uptake. (Cailliatte et al., 2010). AtNRAMP2—remobilization of Mn^{2+} (Gao H. et al., 2018). AtNRAMP3 and AtNRAMP4—Fe mobilization (Thomine et al., 2003; Lanquar et al., 2005) AtNRAMP6—intracellular Fe homeostasis (Cailliatte et al., 2010)
3	Glycine max	*GmNRAMP*	13	*GmNRAMP7*—acquisition of Fe in the root (Qin, L et al., 2017)
4	*Camellia sinensis*	CsNRAMP	11	CsNRAMP3, CsNRAMP4, CsNRAMP5, and CsNRAMP8—absorption of metal ions CsNRAMP2 and CsNRAMP5—transmembrane transport of metal ions (Jinqiu et al., 2021)
5	*Phaseolus vulgaris*	PvNRAMP	7	*PvNRAMP1*, *PvNRAMP2*, *PvNRAMP3*, *PvNRAMP4*, and *PvNRAMP5*—general metal homeostasis during all developmental stages of the common bean *PvNRAMP6* and *PvNRAMP7*—symbiosis with beneficial microorganism (Ishida et al., 2018)
6	*Brassica napus*	BnNRAMP	22	BnNRAMP1—transport of Cd (Meng et a., 2017)
7	*Theobroma cacao*	TcNRAMP	5	TcNRAMP2—remobilization of metal ions TcNRAMP3—transport of Fe^{2+} and Mn^{2+} TcNRAMP5—transport of Fe^{2+}, Mn^{2+}, and Cd^{2+} TcNRAMP6—Mn^{2+} transporter (Ullah et al., 2018)

(continued)

Table 5.1 (continued)

SI no.	Plant	Name of NRAMP	No. of members predicted/ identified	Predicted/identified functional role
8	Wheat	*TpNRAMP*	NA	*TpNRAMP3*—accumulation of Mn, Co, and Cd (Peng et al., 2018)
9	Tomato	*LeNRAMP*	NA	LeNRAMP1—distribution of Fe LeNramp1 and LeNramp3—Mn transport (Bereczky et al., 2003)
10	Peanut	*AhNRAMP*	NA	*AhNRAMP1—Fe nutrition* (Xiong et al., 2012)
11	*Sedum alfredii*	SaNRAMP	NA	SaNRAMP1—which transports Cd, Mn, and Zn (Zhang et al., 2020)
12	*Hordeum vulgare*	HvNRAMP	NA	HvNramp5—uptake of Mn and Cd (Wu et al., 2016)
13	*Malus baccata*	MbNRAMP	NA	MbNRAMP1—which transfers Fe, Mn, and Cd (Xiao et al., 2008)

NA clear data not available

resistant crop plants by plant breeding techniques and also in heavy metal hyperaccumulation.

Acknowledgments AM would like to thank Anna University for providing the Anna Centenary Research Fellowship (ACRF).

References

Belouchi A, Cellier M, Kwan T et al (1995) The macrophage-specific membrane protein Nramp controlling natural resistance to infections in mice has homologues expressed in the root system of plants. Plant Mol Biol 29:1181–1196

Belouchi A, Kwan T, Gros P (1997) Cloning and characterization of the OsNramp family from Oryza sativa, a new family of membrane proteins possibly implicated in the transport of metal ions. Plant Mol Biol 33:1085–1092

Bennetzen J (2002) The rice genome. Opening the door to comparative plant biology. Science 296: 60–63

Bereczky Z, Wang H-Y, Schubert V et al (2003) Differential regulation of nramp and irt metal transporter genes in wild type and iron uptake mutants of tomato. J Biol Chem 278:24697–24704

Bozzi AT, Bane LB, Weihofen WA, McCabe AL, Singharoy A, Chipot CJ, Gaudet SK (2016a) Conserved methionine dictates substrate preference in Nramp-family divalent metal transporters. PNAS 113:10310–10315

Bozzi AT, Bane LB, Weihofen WA, Singharoy A, Guillen ER, Ploegh HL, Schulten K, Gaudet R (2016b) Crystal structure and conformational change mechanism of a bacterial Nramp-family divalent metal transporter. Structure 24:2102–2114

Bozzi AT, Zimanyi CM, Nicoludis JM, Lee BK, Zhang CH, Gaudet R (2019) Structures in multiple conformations reveal distinct transition metal and proton pathways in an Nramp transporter. elife 8:e41124. https://doi.org/10.7554/eLife.41124

Cailliatte R, Lapeyre B, Briat J, Mari S, Curie C (2009) The NRAMP6 metal transporter contributes to cadmium toxicity. Biochem J 422:217–228

Cailliatte R, Schikora A, Briat J-F, Mari S, Curie C (2010) High-affinity manganese uptake by the metal transporter NRAMP1 is essential for arabidopsis growth in low manganese conditions. Plant Cell 22:904–917. https://doi.org/10.1105/tpc.109.073023

Cellier MFM (2012) Nramp: from sequence to structure and mechanism of divalent metal import. Metal Transp 69:249–293

Cellier M, Prive G, Belouchi A, Kwan T, Rodrigues V, Chia W, Gros P (1995) Nramp defines a family of membrane proteins. Proc Natl Acad Sci U S A 92(22):10089–10093

Curie C, Alonso JM, Le Jean M et al (2000) Involvement of NRAMP1 from Arabidopsis thaliana in iron transport. Biochem J 347:749–755

Ehrnstorfer IA, Geertsma ER, Pardon E, Steyaert J, Dutzler R (2014) Crystal structure of a SLC11 (NRAMP) transporter reveals the basis for transition-metal ion transport. Nat Struct Mol Biol 21:990–996

Ehrnstorfer IA, Manatschal C, Arnold FM, Laederach J, Dutzler R (2017) Structural and mechanistic basis of proton-coupled metal ion transport in the SLC11/NRAMP family. Nat Commun 8:14033. https://doi.org/10.1038/ncomms14033

Famoso AN, Clark RT, Shaff JE, Craft E, McCouch SR, Kochian LV (2010) Development of a novel aluminum tolerance phenotyping platform used for comparisons of cereal aluminum tolerance and investigations into rice aluminum tolerance mechanisms. Plant Physiol 153(4): 1678–1691

Finkemeier I, Kluge C, Metwally A, Georgi M, Grotjohann N, Dietz KJ (2003) Alterations in Cd-induced gene expression under nitrogen deficiency in Hordeum vulgare. Plant Cell Environ 26(6):821–833. https://doi.org/10.1046/j.1365-3040.2003.01014.x

Gao H et al (2018) NRAMP2, a trans-Golgi network-localized manganese transporter, is required for *Arabidopsis* root growth under manganese deficiency. New Phytol 217:179–193

Guerinot ML (2000) The ZIP family of metal transporters. Biochim Biophys Acta 1465:190–198

Gunshin H, Mackenzie B, Berger UV, Gunshin Y, Romero MF, Boron WF, Nussberger S, Gollan JL, Hediger MA (1997) Cloning and characterization of a proton-coupled mammalian metal-ion transporter. Nature 388:482–488

Hall JL, Williams LE (2003) Transition metal transporters in plants. J Exp Bot 54(393):2601–2613. https://doi.org/10.1093/jxb/erg303

Ishida JK, Caldas DG, Oliveira LR, Frederici GC, Leite LMP, Mui TS (2018) Genet Mol Biol 41: 820–833

Ishimaru Y, Bashir K, Nakanishi H, Nishizawa NK (2012) OsNRAMP5, a major player for constitutive iron and manganese uptake in rice. Plant Signal Behav **7**(7):763–766. https://doi.org/10.4161/psb.20510

Jinqiu Li Y, Duan ZH, Shang X, Zhang K, Zou Z, Ma Y, Li F, Fang W, Zhu X (2021) Genome-wide identification and expression analysis of the *NRAMP* family genes in tea plant (*Camellia sinensis*). Plan Theory 10(6):1055

Jogawat A, Yadav B, Chhaya NOP (2021) Metal transporters in organelles and their roles in heavy metal transportation and sequestration mechanisms in plants. Physiol Plant:1–17. https://doi.org/10.1111/ppl.13370

Kerppola RE, Ames GF (1992) Topology of the hydrophobic membrane-bound components of the histidine periplasmic permease. Comparison with other members of the family. J Biol Chem 267:2329–2336

Kushnir S, Babiychuk E, Storozhenko S et al (2001) A mutation of the mitochondrial ABC transporter Sta1 leads to dwarfism and chlorosis in the arabidopsis. Mut Plant Cell 13:89–100

Lanquar V, Lelièvre F, Bolte S, Hamès C, Alcon C, Neumann D, Vansuyt G, Curie C, Schröder A, Krämer U et al (2005) Mobilization of vacuolar iron by AtNRAMP3 and AtNRAMP4 is essential for seed germination on low iron. EMBO J 24:4041–4051

Lanquar V, Ramos MS, Lelièvre F, Barbier-Brygoo H, Krieger-Liszkay A, Krämer U et al (2010) Export of vacuolar manganese by AtNRAMP3 and AtNRAMP4 is required for optimal photosynthesis and growth under manganese deficiency. Plant Physiol 152:1986–1999. https://doi.org/10.1104/pp.109.150946

Li J, Wang Y, Zheng L, Li Y, Zhou X, Li J, Gu D, Xu E, Lu Y, Chen X et al (2019) The intracellular transporter AtNRAMP6 is involved in Fe homeostasis in arabidopsis. Front Plant Sci 10:1124

Mani A, Sankaranarayanan K (2018a) In silico analysis of natural resistance-associated macrophage protein (NRAMP) family of transporters in rice. Protein J 37(3):237–247. https://doi.org/10.1007/s10930-018-9773-y

Mani A, Sankaranarayanan K (2018b) Heavy metal and mineral element-induced abiotic stress in rice plant. In: Farooq S, Hayat KZ, Amjad I (eds) Rice crop – current developments. IntechOpen, London. https://doi.org/10.5772/intechopen.76080

Marschner P (2012) Mineral nutrition of higher plants, 3rd edn. Academic Press, London

Marschner H, Romheld V (1994) Strategies of plants for acquisition of iron. Plant Soil 165:261–274

Mary V, Ramos MS, Gillet C, Socha AL, Giraudat J, Agorio A, Merlot S, Clairet C, Kim SA, Punshon T et al (2015) Bypassing iron storage in endodermal vacuoles rescues the iron mobilization defect in the natural resistance associated-macrophage protein3natural resistance associated-macrophage protein4 double mutant. Plant Physiol 169:748–759

Mäser P, Thomine S, Schroeder JI, Ward JM, Hirschi K, Sze H, Talke IN, Amtmann A, Maathuis FJ, Sanders D et al (2001) Phylogenetic relationships within cation transporter families of arabidopsis. Plant Physiol 126:1646–1667

Meng JG, Zhang XD, Tan SK, Zhao KX, Yang ZM (2017) Genome-wide identification of Cd-responsive NRAMP transporter genes and analyzing expression of NRAMP 1 mediated by miR167 in Brassica napus. Biometals 30(6):917–931

Mizuno T, Usui K, Horie K, Nosaka S, Mizuno N, Obata H (2005) Cloning of three ZIP/Nramp transporter genes from a Ni hyperaccumulator plant *Thlaspi japonicum* and their Ni^{2+}-transport abilities. Plant Physiol Biochem 43:793–801. https://doi.org/10.1016/j.plaphy.2005.07.006

Nevo Y, Nelson N (2006) The NRAMP family of metal-ion transporters. Biochim Biophys Acta 1763:609–620

Peng F, Wang C, Cheng Y, Kang H, Fan X, Sha L, Zhang H, Zeng J, Zhou Y, Wang Y (2018) Cloning and characterization of *TpNRAMP3*, a metal transporter from polish wheat (*Triticum polonicum* L.). Front Plant Sci 9:1354. https://doi.org/10.3389/fpls.2018.01354

Peris-Peris C, Serra-Cardona A, Sánchez-Sanuy F, Campo S, Ariño J, Segundo BS (2017) Two NRAMP6 isoforms function as iron and manganese transporters and contribute to disease resistance in rice. Mol Plant-Microbe Interact 30:385–398. https://doi.org/10.1094/MPMI-01-17-0005-R

Qin L, Han P, Chen L, Walk TC, Li Y, Hu X, Xie L, Liao H, Liao X (2017) Genome-wide identification and expression analysis of *NRAMP* family genes in soybean (*Glycine Max* L.). Plant Sci 8:1436. https://doi.org/10.3389/fpls.2017.01436

Sasaki A, Yamaji Y, Yokosho K, Ma JF (2012) Nramp5 is a major transporter responsible for manganese and cadmium uptake in rice. Plant Cell 24:2155–2167

Supek F, Supekova L, Nelson H et al (1996) Yeast manganese transporter related to the macrophage protein involved in conferring resistance to mycobacteria. Proc Natl Acad Sci U S A 93:5105–5110

Takahashi R et al (2011) The OsNRAMP1 iron transporter is involved in Cd accumulation in rice. J Exp Bot 62:4843–4850

Tang Z, Cai H, Li J, Lv Y, Zhang W, Zhao F (2017) Allelic variation of *NtNramp5* associated with cultivar variation in cadmium accumulation in tobacco. Plant Cell Physiol 58:1583–1593. https://doi.org/10.1093/pcp/pcx087

Tejada-Jiménez M, Castro-Rodríguez R, Kryvoruchko I, Lucas MM, Udvardi M, Imperial J et al (2015) *Medicago truncatula* natural resistance-associate macrophage protein1 is required for iron uptake by rhizobia-infected nodule cells. Plant Physiol 168:258–272. https://doi.org/10.1104/pp.114.254672

Thomine S, Vert G (2013) Iron transport in plants: better be safe than sorry. Curr Opin Plant Biol 163:322–327

Thomine S, Wang R, Ward JM, Crawford NM, Schroeder JI (2000) Cadmium and iron transport by members of a plant metal transporter family in *Arabidopsis* with homology to *Nramp* genes. Proc Natl Acad Sci 97:4991–4996

Thomine S, Leliävre F, Debarbieux E et al (2003) AtNRAMP3, a multispecific vacuolar metal transporter involved in plant responses to iron deficiency. Plant J 34:685–695

Ullah I, Wang Y, Eide DJ et al (2018) Evolution, and functional analysis of natural resistance-associated macrophage proteins (NRAMPs) from *Theobroma cacao* and their role in cadmium accumulation. Sci Rep 8:14412

Vastermark A, Wollwage S, Houle ME, Rio R, Saier MH (2014) Expansion of the APC superfamily of secondary carriers. Proteins 82:2797–2811

Williams LE, Pittman JK, Hall JL (2000) Emerging mechanisms for heavy metal transport in plants. Biochim Biophys Acta 1465:104–126

Wu D, Yamaji N, Yamane M, Kashino-Fujii M, Sato K, Feng Ma J (2016) The HvNramp5 transporter mediates uptake of cadmium and manganese, but not iron. Plant Physiol 172(3): 1899–1910

Xia J, Yamaji N, Kasai T, Ma JF (2010) Plasma membrane localized transporter for aluminum in rice. Proc Natl Acad Sci U S A 107:18381–18385

Xiao H, Yin L, Xu X, Li T, Han Z (2008) The iron-regulated transporter. Ann Bot 102:881–889. https://doi.org/10.1093/aob/mcn178

Xiong H, Kobayashi T, Kakei Y, Senoura T, Nakazono M, Takahashi H et al (2012) AhNRAMP1 iron transporter is involved in iron acquisition in peanut. J Exp Bot 63:4437–4446

Yamaji N, Xia J, Mitani-Ueno N, Yokosho K, Ma JF (2013) Preferential delivery of zinc to developing tissues in rice is mediated by P-type heavy metal ATPase OsHMA2. Plant Physiol 162:927–939

Yang M, Zhang W, Dong H, Zhang Y, Lv K et al (2013) OsNRAMP3 is a vascular bundles specific manganese transporter that is responsible for manganese distribution in rice. PLoS One 8(12): e83990. https://doi.org/10.1371/journal.pone.0083990

Yang M et al (2014) OsNRAMP5 contributes to manganese translocation and distribution in rice shoots. J Exp Bot 65:4849–4861

Zhang J, Zhang M, Song H, Zhao J, Shabala S, Tian S, Yang X (2020) A novel plasma membrane-based NRAMP transporter contributes to Cd and Zn hyperaccumulation in Sedum alfredii Hance. Environ Exp Bot 176:104121

Role of Heavy Metal ATPases in Transport of Cadmium and Zinc in Plants

6

Dwaipayan Sinha, Pramod Kumar Tandon,
Gyan Prakash Srivastava, Shobhit Kumar Srivastava,
and Suchetana Mukherjee

Abstract

Anthropogenic causes have led to a constant increase in heavy metal concentration in soil. Cadmium (Cd), copper (Cu), zinc (Zn), nickel (Ni), cobalt (Co), chromium (Cr), lead (Pb), and arsenic (As) toxicity is recorded in agricultural soils throughout the world. Some of these heavy metal ions are essential micronutrients for plant metabolism but when present in excess are extremely toxic. Heavy metal toxicity in soil not only affects crop yield but may also lead to biomagnification of metals as they enter the food chain, thereby exerting detrimental effect both on human health and the environment. Plants experience oxidative stress upon exposure to heavy metals as they produce ROS (reactive oxygen species) that leads to cellular damage. Plants also accumulate metal ions that disturb cellular ionic homeostasis and several physiological processes which ultimately manifest in reduction of growth and yield. To overcome these, stress plants are equipped with different transporters which transport, sequester, and ultimately detoxify nonessential or excess heavy metals. Heavy metal ATPases need special mention in this respect. They play a significant role in the trafficking

D. Sinha (✉)
Department of Botany, Government General Degree College, Mohanpur, Paschim Medinipur, West Bengal, India

P. K. Tandon
Department of Botany, University of Lucknow, Lucknow, Uttar Pradesh, India

G. P. Srivastava
Department of Botany, University of Allahabad, Prayagraj, Uttar Pradesh, India

S. K. Srivastava
Department of Botany, Asian International University, Imphal, Manipur, India

S. Mukherjee
Department of Botany, Sripat Singh College, Jiaganj, West Bengal, India

K. Kumar, S. Srivastava (eds.), *Plant Metal and Metalloid Transporters*,
https://doi.org/10.1007/978-981-19-6103-8_6

and sequestration of hazardous metals across the membrane in plants, thereby minimizing their toxic effect. The chapter aims to highlight the role of heavy metal ATPases in cadmium and zinc transportation and stress-related toxicity in plants.

Keywords

HMA · Cadmium · Zinc · Toxicity · Translocation · Transport · Tonoplast · ROS

6.1 Introduction

Heavy metal toxicity, with cadmium (Cd), copper (Cu), zinc (Zn), nickel (Ni), cobalt (Co), chromium (Cr), lead (Pb), and arsenic (As), has been detected in agricultural soils in many parts of the world. This could be due to the usage of phosphatic fertilizers for a long time (Jayasumana et al. 2015), the application of sewage sludge (Shahbazi et al. 2017), smelting process (Jaishankar et al. 2014), industrial waste (Ahmad et al. 2021), and irrigation processes in agricultural lands (Rizvi et al. 2020). When plants are exposed to high quantities of heavy metals, their major response is to produce reactive oxygen species (ROS) (Shahid et al. 2014). Various metals either directly generate ROS through Haber–Weiss reactions and Fenton reactions (Nikalje and Suprasanna 2018) or overproduce ROS (Tiwari and Lata 2018), causing nitro-oxidative stress in plants (Georgiadou et al. 2018) as an indirect result of heavy metal toxicity. Modulation of the antioxidant system (Keyster et al. 2020), disruption of the electron transport chain (Hoque et al. 2021), and alteration of mineral nutrition (Angulo-Bejarano et al. 2021) are examples of indirect pathways. Lipid peroxidation, which can directly cause biomembrane damage (Zhao et al. 2021), is one of the most harmful impacts of heavy metal exposure in plants. One of the breakdown products of polyunsaturated fatty acids in membranes, malondialdehyde (MDA), is recognized as a good biomarker of oxidative stress (Melandri et al. 2020). Plants, on the other hand, have developed a potentially effective mechanism to resist heavy metal toxicity in the environment. Plant cells are equipped with several transporters, namely the heavy metal ATPases (HMAs), Nramps (natural resistance-associated macrophage proteins), the cation diffusion facilitator (CDF) family, the ZIP (ZRT, IRT-like proteins) family, ABC transporters (ATP-binding cassette), cation antiporters, and other putative transition metal transporters. They are involved in the detoxification of nonessential or excess metal, despite their importance in maintaining the homeostasis of essential metal micronutrients. As a unique characteristic of metal hyperaccumulators, they are overexpressed and have a higher cellular concentration (Viehweger 2014). In this chapter, the HMAs will be discussed in detail with special emphasis on their role in the detoxification of Zn and Cd in the plant system.

6.2 Heavy Metal ATPases in Alleviating Heavy Metal Toxicity

Heavy metal ATPases (HMAs) are P1B-type ATPases that play a significant role in metal trafficking in plants (Huang et al. 2020a). Heavy metal ATPases (HMAs) of the P1B type have been linked to the transport of a variety of necessary and potentially hazardous metals across cell membranes. The P-type ATPase superfamily's ion pumps share an enzymatic mechanism in which ATP hydrolysis aids in ion transport across the membrane (Lee et al. 2007). Phylogenetic analysis can segregate this HMA group into two different groups (Lee et al. 2007). Based on their metal-substrate specificity, functional tests on the HMAs have revealed that they are separated into two subgroups: a copper (Cu)/silver (Ag) group and a zinc (Zn)/cobalt (Co)/cadmium (Cd)/lead (Pb) group (Deng et al. 2013). Six to eight transmembrane (TM) helices, an ATP-binding domain (ATPBD), an actuator domain, and, in many cases, an extra soluble metal-binding domains (MBDs), often at the N terminus, make up P1BATPases. The enzyme cycles between high (E1) and low (E2) affinity metal-binding states as a result of phosphorylation of a conserved aspartate residue in the ATPBD (Purohit et al. 2018). Depending on conserved motifs in the TM domain, the presence of different types of MBDs, and biochemical and genetic data correlating transporters to specific metal ions, P1B-ATPases have been divided into seven subtypes (P1B-1–P1B-7) (Purohit et al. 2018). Of the various subclasses, P1B-2-ATPases transport Zn^{2+}, Cd^{2+}, and Pb^{2+} (Lu et al. 2016), while P1B-4-ATPase subtype transports Co^{2+}, Cd^{2+}, and Zn^{2+} (Smith et al. 2017). Although some evidence relates the P1B-5-ATPases to Ni^{2+} and Fe^{2+} (Zielazinski et al. 2013), the metal specificities of the P1B-5-, P1B-6-, and P1B-7-ATPases remain unknown (Purohit et al. 2018). Cu^{+} and Ag^{+} are recognized by the P1B-1-ATPases, which contain ATP7A and ATP7B (Beneš et al. 2018), whereas P1B-3-ATPases are attributed for transportation of Cu^{+} (Meloni et al. 2014). The two subfamilies are distinct in several respects. First, in P1B-1 ATPases, the TM helix 4 motif is CPC as well as strong conservation of a YN(X)4P motif in transmembrane helix 5 and an MXXSS motif in TM helix 6, while in P1B-3 ATPases, it is CPH (Smith et al. 2014). It is widely thought that the presence of this histidine confers selectivity for Cu^{2+} (Purohit et al. 2018). CopBs, or P1B-3-ATPases, are distinctive in that they have a histidine-rich N-terminal extension that is thought to represent an MBD and is required for maximal transportation of Cu+ (Rosenzweig and Argüello 2012). The MBDs of P1B-1-ATPases, many of which are called CopAs, normally consist of 1–6 ferredoxin-like domains with a conserved CXXC metal-binding motif that attaches a single Cu+ ion (Boal and Rosenzweig 2009). In terms of tissue distribution, subcellular localization, and metal selectivity, HMAs are diverse (Takahashi et al. 2012a). However, their primary function is to transport heavy metals within the plant with a motive of either transportation or detoxification (Table 6.1). In this chapter, the relevance and action of heavy metal ATPase in sequestration and transportation of zinc and cadmium would be discussed.

Table 6.1 Heavy metal ATPases in the selected group of plants and their action

Name of the plant	Heavy metal ATPase	Intracellular location	Cellular location	Function	References
Arabidopsis thaliana	AtHMA1	Chloroplast envelope	Guard cell	Detoxification of Zn	Kim et al. (2009)
	AtHMA2	Plasma membrane	Root pericycle	Transportation of Zn and Cd	Eren and Arguello (2004) and Wong et al. (2009)
	AtHMA3	Vacuolar membrane	Guard cells, hydathodes, vascular tissues, and the root apex	Transportation of Cd	Morel et al. (2009)
	AtHMA4	Plasma membrane	Root pericycle	Transportation of Zn and Cd	Chen et al. (2018)
	AtHMA5	Plasma membrane (Wenli et al. 2020)	Roots	Cu compartmentalization and detoxification	Andrés-Colás et al. (2006)
				Involved in loading of Cu in the xylem of root for transportation in the shoot	Huang et al. (2016)
	AtHMA6	Chloroplast envelope	The inner membrane of the chloroplast envelope	Transportation of Cu through the chloroplast envelope	Catty et al. (2011)
	AtHMA7	Golgi membrane		Transportation of Cu to ethylene receptors of Golgi membrane	Huang et al. (2016)
	AtHMA8	Chloroplast and thylakoid membrane		Transportation of Cu into the thylakoid lumen	Huang et al. (2016)
Thlaspi caerulescens	TcHMA4	Plasma membrane	Roots	Transportation of Zn and Cd	Papoyan and Kochian (2004)
Hordeum vulgare	HvHMA1	Chloroplast envelope and aleurone cells	Aerial parts and grains	Mobilization of Cu and Zn from the plastids and aleurone cells	Mikkelsen et al. (2012)
	HvHMA2		Roots and shoots	Transportation of Zn and Cd	Mills et al. (2012)

Hordeum vulgare		Plasma membrane and chloroplast			
	HvHMA3	Tonoplast	Roots and leaves	Transportation of Cd	Kaznina et al. (2014)
	HvHMA4	Plasma membrane		Upregulated during Cd stress	Zhang et al. (2021)
Oryza sativa	OsHMA2	Plasma membrane	Roots and vascular bundles	It facilitates loading of Cd and Zn in xylem and facilitates root to shoot translocation	Takahashi et al. (2012b)
	OsHMA3	Tonoplast	Roots	Transportation of Cd into the vacuole	Miyadate et al. (2011)
	OsHMA4	Tonoplast	Roots	Sequestration of Cu into the vacuoles	Huang et al. (2016)
	OsHMA5	Plasma membrane	Root pericycle cells and xylem region	Loading of Cu in the xylem	Deng et al. (2013)
	OsHMA6	Plasma membrane	Spikelets and leaf blade	Efflux of Cu	Wenli et al. (2020)
	OsHMA9		Xylem and phloem	Efflux of Cu, Zn, and lead from the cells Loading and unloading of heavy metals in the vascular tissues	Lee et al. (2007)

6.3 Cadmium Toxicity in Plants

Cadmium is a nonessential element (Wan and Zhang 2012) that has a deleterious impact on plant development and growth (Zhu et al. 2018). Power plants, heating systems, metal-working sectors, and vehicular emissions all emit it into the environment. Electroplating, pigments, plastic stabilizers, and Ni–Cd batteries all employ it. Because of its high toxicity and considerable solubility in water, it is considered a major contaminant (Benavides et al. 2005). The global average Cd concentration in soil is 0.35 mg kg^{-1} (Li et al. 2016). In natural soils, however, it ranges from 0.01 to 0.8 mg kg^{-1}, with some regions reaching 2.0–8.9 mg kg^{-1} (Wei and Twardowska 2013). Cadmium is absorbed by the soil in solid phase or by insoluble compounds when it enters the soil system, and a tiny amount is also available in soil solution in the soluble form that can be quickly taken over by plants (Hussain et al. 2021). Roots are the first organ in plants to come into touch with hazardous metal ions; hence, roots have higher metal concentrations than aerial parts (Muradoglu et al. 2015). Cadmium toxicity is manifested by limiting growth (Younis et al. 2016), affecting the mineral nutrition of plants (Yildirim et al. 2021), and affecting the activities of several enzymes (Singh et al. 2019; Haider et al. 2021). It causes complicated genetic, physiological, and biochemical changes in plants, leading to phytotoxicity, which manifests itself in leaf rolls, chlorosis (Abedi and Mojiri 2020), and a reduction in root growth (Ronzan et al. 2018). It also limits respiratory (Belyaeva 2018; Branca et al. 2020) and photosynthetic activities (Xue et al. 2014; Song et al. 2019), enzyme activities (Farooq et al. 2016), and membrane functions (Yang et al. 2016) in plants, as well as inducing lipid peroxidation (Khan et al. 2013) and decreased chlorophyll content (Szopiński et al. 2019; Zhao et al. 2021) and causing oxidative stress (Kolahi et al. 2020; Żabka et al. 2012). Among crop plants, Cd is easily absorbed by the rice, and it is transported to plants via membrane channels that transport other nutrient elements with similar physical and chemical properties. Cadmium impacts rice leaf size, plant height, development, and biomass, among other things (Hussain et al. 2021). It has been reported that Cd stress results in a reduction in the number of panicles, spikelets per panicles, seed set, and grain yield (Kanu et al. 2017). In maize, Cd stress results in the onset of oxidative stress and lipid peroxidation and ultimately decreases the grain yield (Anjum et al. 2015). Figure 6.1 diagrammatically illustrates the selected effect of cadmium toxicity in plants.

6.3.1 Transporters in Alleviating Cadmium Stress

During Cd toxicity, a plant may either avoid the metal by restricting its entry within the body or may accumulate and compartmentalize the heavy metals in various subcellular locations. The strategies of avoiding metal may be diverse. It largely occurs at the extracellular level in which the plant restricts the entry of Cd or other metals through adsorption in the cell wall (Chen et al. 2013; Riaz et al. 2021; Yu et al. 2021). The roots of the plants also secrete extracellular carbohydrates (Kintlová

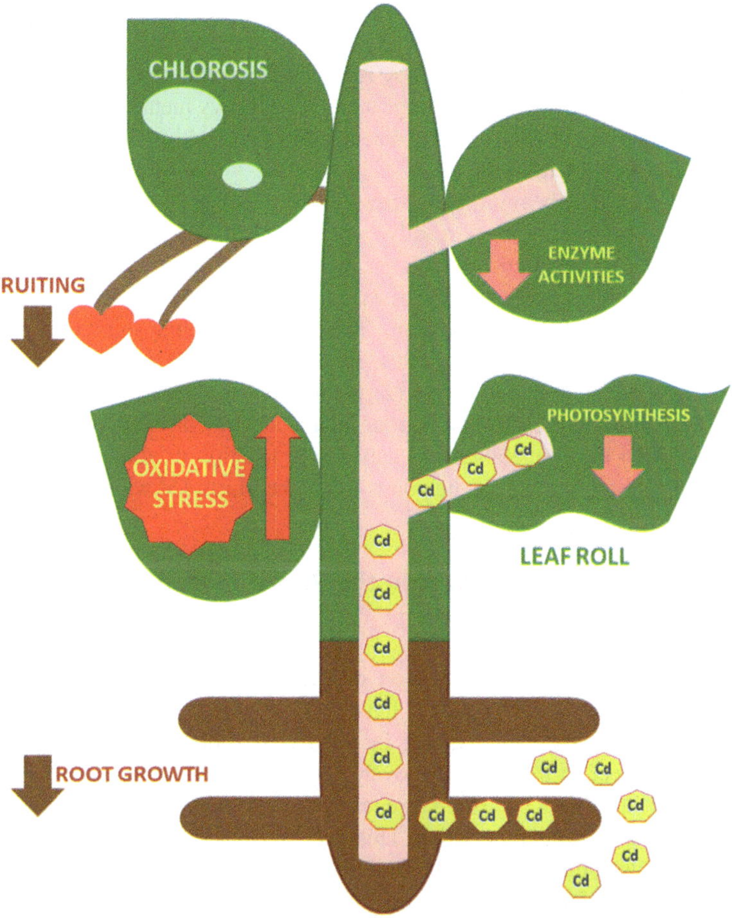

Fig. 6.1 Diagrammatic representation of cadmium toxicity in plants. Cadmium toxicity first affects the root and results in inhibition of growth. Once the cadmium is transported through the vascular bundle, it exhibits several toxicity symptoms including chlorosis and rolling of leaf, decreased fruit set, inhibition of enzyme activities, photosynthesis, and induction of oxidative stress

et al. 2021), root exudates (Sun et al. 2020), and mucilage (Lapie et al. 2019) which traps the Cd, thereby preventing its entry. However, in the case of a high concentration of Cd in the soil, some plants tend to absorb the metal, accumulate in their body, and ultimately get affected by its deleterious effects (Yang et al. 2017). However, some plants tend to accumulate exceptionally high quantities of heavy metals within their body and are known as hyperaccumulators. They are distinguished from non-hyperaccumulating species by three characteristics: a significantly higher rate of heavy metal uptake, a faster root-to-shoot translocation, and a higher ability to detoxify and sequester heavy metals in leaves (Rascio and Navari-Izzo 2010). In metal hyperaccumulators, constitutive overexpression of genes encoding transmembrane transporters, such as members of the ZIP (Liu et al. 2019), HMA (Fang et al. 2016), MATE (Li et al. 2017), YSL (Gendre et al. 2007), and MTP families (Gao et al. 2020), plays a key role in driving the uptake, translocation to leaves, and finally sequestration of large amounts of heavy metals in vacuoles. In this section, the activities of HMA will be discussed in detail with special reference to the transportation of Cd within the plant.

During Cd or any heavy metals stress, once the metal can cross the cell wall barrier of the root, it is readily uptaken by the plant especially the hyperaccumulators. As mentioned earlier, several types of transporters are involved in the transportation of the heavy metal within the plant body with an ultimate aim to sequester in an intracellular location. Thus, the entire transportation might be precisely segregated into three phases, namely (1) uptake and transportation of heavy metals within the root tissue, (2) loading of heavy metals in the xylem for onward transportation to the aboveground part, and (3) distribution and compartmentalization of heavy metals in the subcellular location of aboveground parts. In the upcoming sections, each of the points would be illustrated with respect to HMA.

6.3.2 Activities of HMA Within the Roots in Response to Cadmium Stress

Upon Cd stress or high concentration of Cd, several genes responsible for the synthesis of HMA are upregulated. It is observed that Cd stress resulted in the upregulation of HMA1 and HMA2 in the fruits, roots, and stems of different cultivars of pepper. It was also discovered that as soil Cd levels rose, so did the expression of the HMA1 and HMA2 genes, with stem expression being substantially higher than root and fruit expression. This suggests that exogenous Cd regulates the expression folds of the HMA1 and HMA2 genes and that the two genes are involved in the transport of Cd from pepper roots to the aboveground part, as well as in the buildup of Cd in the aboveground part (Hu et al. 2021). In another study, it was concluded that OsHMA3n from a low Cd-accumulating rice cultivar (Nipponbare) was found to act as a firewall by sequestering Cd into the vacuoles in the roots, keeping it out of the aboveground tissues. In contrast, OsHMA3a from a high Cd-accumulating cultivar (Anjana Dhan) has lost its function to sequester Cd into the vacuoles, resulting in excessive Cd translocation from the roots to the shoots, most likely due to a mutation of an amino acid at position 80 (Ueno et al. 2010). In another experiment, it was observed that Cd stress resulted in the upregulation of HvHMA1, HvHMA3, and HvHMA4 genes (Zhang et al. 2021). Additionally, HvHMA3 was shown to be homologous to OsHMA3 (Wu et al. 2015), a rice protein that sequesters Cd in the vacuoles of root cells, limiting the rate of Cd translocation from roots to shoots (Sasaki et al. 2014). In another study using durum wheat (*Triticum turgidum* L. subsp. *durum*), it was observed that Cd activates a complex gene network including the Cu transporter ATPase, HMA5, in the roots indicating its possible involvement in combatting the stressed condition (Aprile et al. 2018). In *Thlaspi caerulescens*, transcriptome analysis revealed activation of 20 genes under

Cd-enriched condition, out of which TcHMA3 was chosen for further study. It was observed that TcHMA3 was expressed in the tonoplast of pericycle cells of the roots indicating its involvement in sequestration and compartmentalization of Cd (Ueno et al. 2011).

6.3.3 Heavy Metal ATPase Associated with Cadmium Translocation

In the case of Cd stress, translocation from roots to shoots is a usual phenomenon with several transporters involved in the process (Rouached 2013). These transporters are in general located in the plasma membrane or the pericycle and facilitate loading of Cd in the xylem. In rice, OsHMA2 is involved in the transportation of Cd from roots to shoots. OsHMA2 is largely expressed in the roots, and its transcripts were mainly localized in vascular bundles in presence of Cd (Takahashi et al. 2012b). In *Arabidopsis*, HMA2 and HMA4 are involved in the translocation of Cd from roots to shoots. It was also observed that mutations in hma4 and hma2hma4 resulted in the decrease of Cd translocation (Wong and Cobbett 2009). Both HMA2 and HMA4 are localized in the plasma membrane of the root pericycle and pump the metal into the vascular bundle (Nouet et al. 2015; Sinclair et al. 2018). In *Noccaea caerulescens*, Cd treatment resulted in the expression of plasma membrane-bound NcHMA4 in roots and shoots. It was further observed that NcHMA4 was strongly expressed in the root tips. Moreover, in the elongation zone and hairy root region, it is expressed mostly in the xylem parenchyma indicating its association with the translocation of Cd (Craciun et al. 2012). Another study reports that TaHMA2 is involved in root-to-shoot translocation of Cd and Zn. It was observed that transgenic tobacco plants overexpressing TaHMA2 resulted in increased translocation of Cd, Zn, and some amounts of lead (Tan et al. 2013). Figure 6.2 diagrammatically illustrates the loading and transportation of heavy metals in the root zone.

6.3.4 Heavy Metal ATPase Associated with Xylem Unloading and Cadmium Distribution

Mineral ions are discharged from the xylem vessels and dispersed in the other leaf tissue after their long-distance journey in the xylem sap toward the aerial parts of the plant (Assaha et al. 2017). Cadmium xylem unloading is expected to follow the same pattern as nutritional ion unloading. Even for nutrients, however, this mechanism has yet to be fully defined. Evaporation of water causes element concentrations in the xylem sap of the leaves to rise dramatically (Sterckeman and Thomine 2020). Plants have evolved xylem unloading mechanisms in leaves to prevent hazardous ion overload. These mechanisms take place in the leaf blade's highly branched network of main and minor veins (Sterckeman and Thomine 2020). It has been observed in a study that Cd and Zn trigger the upregulation of HMA4 in roots and leaves which suggested that the transporter is involved in root-to-shoot translocation. Moreover, the expression of HMA4 is also detected in the leaves which suggests the role of

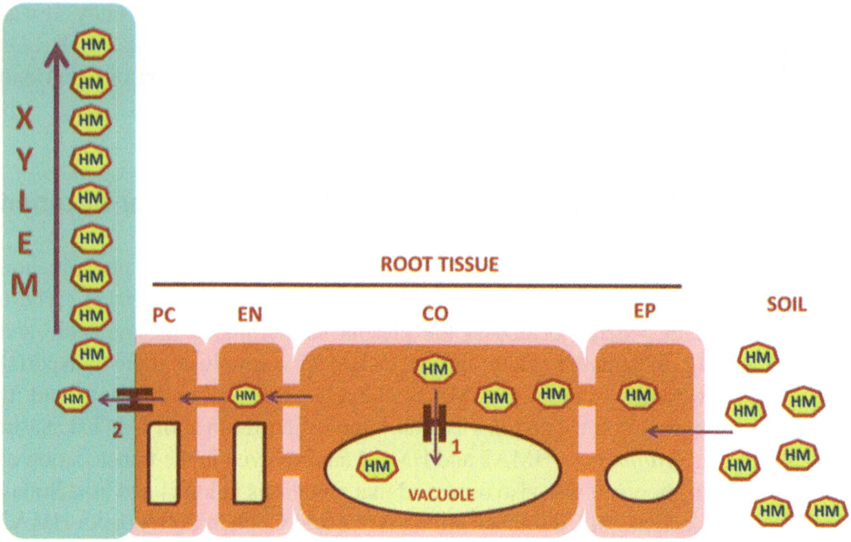

Fig. 6.2 Loading of heavy metals in the roots. The HMAs (H) are present either in the tonoplast of the cortical cells or in the plasma membrane of the pericycle. The vacuolar HMA (1) helps in the sequestration of heavy metals within the vacuoles while those located in the pericycle (2) facilitate the xylem loading process. This model is applicable to all heavy metals including cadmium. *EP* epidermis, *CO* cortex, *EN* endodermis, *HM* heavy metal, and *PC* pericycle

HMA4 in the sequestration of Cd in the aerial part (Wiyono et al. 2021). In *Noccaea caerulescens*, NcHMA4 is responsible for the transportation of Zn and Cd from the roots, while in shoots, it is involved in catering to the metals through the unloading process in the storage sites. Additionally, HMA3 is also expressed in bundle sheath and veins and also might be instrumental in the sequestration of the metals (Mishra et al. 2017).

The unloading process of Cd is also accompanied by compartmentalization of the same within the vacuole or any other intracellular location. In this case, also several HMAs are involved. In one study, it is reported that in Cd hyperaccumulator, *Sedum alfredii*, SaHMA3 is involved in the vacuolar sequestration and detoxification of Cd in the shoot (Zhang et al. 2016). Similar reports are also obtained from a study on *Sedum plumbizinicola* in which it was observed that SpHMA3 is instrumental in Cd detoxification and sequesters the metal within the vacuoles of the leaf cells (Liu et al. 2017). In *Thlaspi caerulescens*, the TcHMA3 is a tonoplast-localized Cd transporter and is responsible for the sequestration of Cd within the vacuoles of the leaf cells (Ueno et al. 2011). Figure 6.3 diagrammatically illustrates the events associated with the xylem unloading of heavy metals.

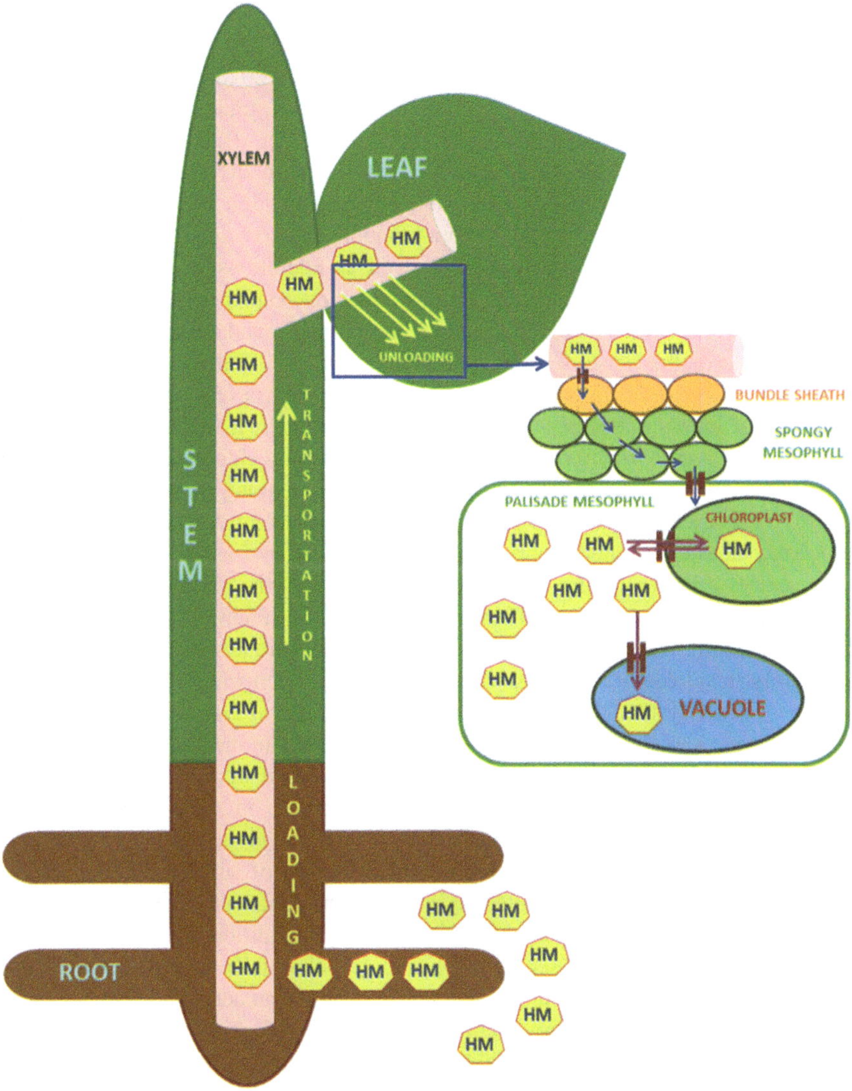

Fig. 6.3 Diagrammatic representation of the unloading of heavy metals in the aerial part of the plant. Heavy metals are unloaded from the xylem through specific transporters and travel the bundle sheath and spongy mesophyll to reach the palisade tissue. Specific HMAs are located in the intracellular organelle of the palisade mesophyll cells which facilitates sequestration of heavy metals in the intracellular compartments. This model is applicable to all heavy metals including cadmium. *HM* heavy metal

6.4 Zinc Toxicity in Plants

Increased Zn levels in soils are phytotoxic (Guarino et al. 2020), causing a variety of structural and functional problems that eventually degrade plant performance (Balafrej et al. 2020). These reactions, however, differ depending on the plant species and stage of development. Zinc toxicity results in alteration of the architecture of roots (Disante et al. 2014). Plant exposed to high metal concentrations exhibited increased root branching with a pronounced curve and a higher branching percentage in the metal contact area, as well as alterations in root morphology (Bochicchio et al. 2015). It is also reported that Zn results in a decrease in root length (Li et al. 2012). The reduction of cell proliferation and subsequent expansion was found to be the cause of root elongation suppression (Fukao and Ferjani 2011; Marichali et al. 2016). In aerial parts of the plants, high levels of Zn result in the chlorosis of leaves (Cheah et al. 2021). In *Beta vulgaris*, it was observed that Zn toxicity results in inwardly rolled leaves accompanied by chlorosis (Sagardoy et al. 2009). In a study on durum wheat, it was shown that Zn along with Cd disturbed the photosynthetic electron transport process which resulted in downregulation of the efficiency of energy transformation in photosystem II by four to fivefold (Paunov et al. 2018). Zn toxicity also has a profound impact on the photophysiology of all cyanobacterial strains, with significant inhibition of maximum photosynthetic quantum production (Sarker et al. 2021). In another experiment, it was shown that Zn and lead stress resulted in a decrease in photochemical quenching and quantum efficiency of photosystem II in young plants of *Koelreuteria paniculata* and *Zelkova schneideriana* (Huang et al. 2019). Zinc toxicity results in the generation of oxidative stress. In *Brassica napus*, it was observed that Zn treatment resulted in an increment of superoxide anion along with an increase in superoxide dismutase activity. However, it was concluded that the increased activity of superoxide dismutase failed to nullify the increased concentration of superoxide (Feigl et al. 2015). In wheat, Zn stress resulted in a significant enhancement of the levels of hydrogen peroxide and malondialdehyde contents. In addition, modulation of antioxidant enzymes was also observed in wheat in response to Zn stress (Li et al. 2013). In tomatoes, it was observed that Zn at high concentration acts synergistically with Cd and enhances oxidative stress (Cherif et al. 2011). In *Coffea arabica*, an increase in hydrogen peroxide and malondialdehyde content was observed both in deficient and stressed conditions. Modulation of the antioxidant enzymes was also observed under altered Zn conditions (Dos Santos et al. 2019). Similar results were also obtained in *Citrus reticulata* when subjected to Zn stress (Subba et al. 2014). Thus to exert its toxic effect in various levels of plants, the metal is efficiently transported across the plant body. In the next phase, the transportation of Zn would be discussed giving special emphasis on HMAs.

6.4.1 Heavy Metal ATPases in Zinc Homeostasis

Zinc is an essential micronutrient in plants and is associated with several enzymes including RNA polymerase and carbonic anhydrase (Sinha and Tandon 2020). However, at higher concentrations, the metal can pose toxicity. Land plants adjust the processes that contribute to Zn homeostasis throughout the plant in response to external supply and internal demand to keep Zn concentrations within physiological limits. The acquisition and uptake of Zn from the soil in roots is the first step (Sinclair and Krämer 2012). Zinc transporters can be found in all cellular compartments in plants, including the plasma membrane (Lee et al. 2010), chloroplasts (Ajeesh Krishna et al. 2020), and vacuoles (Menguer et al. 2013). Their main motive is to compartmentalize Zn so that it remains at the physiological level within the cell and no toxicity takes place. Though several types of transporters are involved in the transportation of Zn within a plant, in this section, the role of HMAs would be discussed. As mentioned earlier, the acquisition of Zn starts from the roots and is finally compartmentalized in the aboveground part. The uptake of Zn is associated with the radial transport of Zn with a motive of accessing the xylem. Though the first step of Zn acquisition is largely done by the zip transporters which are expressed in the epidermis (Liu et al. 2019; Huang et al. 2020b), the HMA transporters are highly instrumental in the xylem loading of Zn. Studies reveal that HMA2 (Wong et al. 2009; Satoh-Nagasawa et al. 2012) and HMA4 (Iqbal et al. 2013) transporters, which are primarily expressed in the pericycle cells close to the xylem, are responsible for Zn efflux from the root to the shoot (Chen et al. 2018). In addition, HMA9 is also involved in the loading of Zn into the xylem from the pericycle (Kawakami and Bhullar 2018). Upon reaching the aerial part of the plant, excess Zn requires to be compartmentalized so that the physiological balance is maintained. In *Arabidopsis*, HMA3 is responsible for the detoxification of biological Zn and mediates the sequestration of Zn within the vacuoles (Morel et al. 2009). In one interesting study, it was observed that in barley, the HvHMA1 is localized in the chloroplast periphery in the leaves and also in the intracellular compartments of aleurone grains. In the study, it was concluded that HvHMA1 is a broad specific exporter of metals from the chloroplast and mobilizes Cu and Zn from the plastids when the cells become deficient in those metals (Mikkelsen et al. 2012). In another study, it was shown that HvHMA2 was localized in the plasma membrane of the roots and cells of the cotyledon. It was further observed that the expression of HvHMA2 in *Arabidopsis* hma2hma4 double mutants increased levels of Cd and Zn in the plant. It was concluded that HvHMA2 is involved in the transport of Zn and Cd, and it may act similarly to AtHMA2 and AtHMA4 in *Arabidopsis* in moving these ions from roots to shoots (Mills et al. 2012). In a study on rice, it was found that OSHMA3 is a tonoplast-localized Zn transporter and is responsible for the homeostasis of the metal. It was further observed that accessions of rice with the OsHMA3 allele showed greater tolerance toward Zn in comparison to the accessions with nonfunctional allele (Cai et al. 2019).

6.5 Expression of Heavy Metal ATPases

The ATPase family is an integral membrane protein transporter responsible for translocating the divalent cation transporters coding genes representing HMAs. The ATPase protein family represents eight membranes found on different cellular compartments. HMA2 (Sinclair et al. 2018), HMA4 (Olsen et al. 2016), and HMA5 (Deng et al. 2013) are present in the cell membrane whereas the membranes of vacuoles (Miyadate et al. 2011) and Golgi body sharing HMA3 and HMA7 (Huang et al. 2016), respectively. The chloroplast itself contains three ATPases, HMA6, HMA8 (Sautron et al. 2015), and HMA1 (Boutigny et al. 2014). Functionally HMA2 helps in translocation of Zn and Cd in *Arabidopsis thaliana* (Wong and Cobbett 2009), *Hordeum vulgare* (Mills et al. 2012), *Oryza sativa* (Satoh-Nagasawa et al. 2012), and *Triticum aestivum* (Tan et al. 2013). Several studies have been undertaken to study the expression pattern of HMAs in several plants with a motive to study its activity in transgenic overexpressed plants. In *Arabidopsis*, overexpression of AtHMA4 enhanced the tolerance to Cd, Zn, and Co, resulting in more Zn and Cd in the root (Verret et al. 2004). Still, surprisingly the same is unable to be concluded in the case of tobacco. The cellular and subcellular expression patterns of AtHMA2 were similar to those of AtHMA4 (Wang et al. 2019). HMA2-GAP protein was localized as plasma membrane residential protein. The expression of the HMA2 gene in various plants suggests its possible role in making the plant capable of heavy metal uptake, thereby aiding in phytoremediation (Hussain et al. 2004). In tobacco, transgenes with HvHMA2 enhanced the Zn sensitivity. Moreover, HvHMA2 also interferes with tobacco metal Zn–Cd–Fe homeostatic along with internal mechanisms regulating metal uptake and tolerance (Barabasz et al. 2013). On the other hand, HMA3 was reported as Cd-detoxifying protein subcellular localized into the vacuoles (Zhang et al. 2018). In *Sedum plumbizincicola*, HMA3, an ATPase, acts in Cd hyper-toleration and Cd/Zn hyperaccumulation (Liu et al. 2017). Table 6.2 illustrates selected HMA genes that have been genetically engineered in other plants to check their expression and function.

6.6 Prospects and Conclusion

Plants have adapted themselves to a wide array of tolerance mechanisms against heavy metal stress. Some of them are avoiders whereas others are hyperaccumulators. In the case of hyperaccumulators, the heavy metals are largely sequestered within the vacuoles or other intracellular locations of the aboveground parts. In addition, they are also sequestered within the vacuoles of the root cells with a common motive of limiting their deleterious effects. In most cases, the transporters play a significant role in the internal distribution and localization of heavy metals. This chapter discusses the role of HMAs in the translocation of Cd and Zn. It is widely known that out of the two, Cd is highly toxic to plants. So far as the role of Zn is concerned, it often acts as a micronutrient, but its toxic effect on the plant system cannot be ruled out. Both metals are emitted from industrial sources and contaminate

Table 6.2 Illustration of selected plants in which HMA has been genetically expressed

Gene	Host plant	Target plant	Effect	References
HMA2	*Hordeum vulgare*	Tobacco	Increased sensitivity toward Zn Inhibition of iron translocation to aboveground parts	Barabasz et al. (2013)
	Triticum aestivum	Rice, tobacco, and wheat	Enhanced translocation of Zn, Cd, and lead to the aerial parts of the plants Decrease in Zn concentration and increase in iron content in rice and wheat seeds	Tan et al. (2013)
HMA3	Rice	Rice	Enhanced content of Cd and Zn in the roots	Sasaki et al. (2014)
	Sedum alfredii	Tobacco	Enhancement of Cd tolerance through an increase in Cd concentration in root vacuoles	Zhang et al. (2016)
	Rice	Rice	Higher concentration of Cd and Zn in the roots	Sasaki et al. (2014)
	Thlaspi caerulescens	*Arabidopsis thaliana*	Increased concentration of Cd in the roots	Ueno et al. (2011)
HMA4	*Arabidopsis thaliana*	Tobacco	Increased sensitivity toward Zn and tolerance toward Cd	Siemianowski et al. (2011)
	Arabidopsis thaliana	*Lycopersicon esculentum*	Reduction in iron and Zn accumulation Higher accumulation of Cd in roots and leaves Elevation in the levels of LeCHLN	Kendziorek et al. (2014)
	Arabidopsis thaliana	*Nicotiana tabacum* v. *Xanthi*	Enhanced lignification of walls of external layers of cells Enhanced hydrogen peroxide accumulation Reduced accumulation of Cd in the plant	Siemianowski et al. (2011) and Siemianowski et al. (2014)
	Populus trichocarpa	*Populus trichocarpa*	Maintenance of Zn homeostasis	Adams et al. (2011)

the atmosphere and consequently reach the plant, though the plant has its mechanism of tolerance that remains up to a certain level. At higher concentrations of the metals, they succumb to the toxic effect. This situation becomes more critical for the crop plants which manifest their toxicity largely through reduction of biomass and yield. However, there is a silver lining to every event. If the hyperaccumulating plants are considered, they are equipped with several heavy metal transporters including the HMAs. As discussed in the chapter, different HMAs perform specific functions including uptake, translocation, and unloading. Thus narrowing down of the HMAs from the tolerant or hyperaccumulating plants may prove to be of extreme relevance. The research can now proceed toward two directions, namely environmental cleanup and setting up of tolerance species. It is known that HMA2 and HMA4 proteins contribute to the translocation of Zn and Cd ions from roots and shoot resultant detoxification of these metals from plant tissues (Wu et al. 2014). The sequence similarity of HMA2, HMA3, and HMA4 indicated similar functions, but the study suggested that gene function varies in different species (Migocka et al. 2015). Plants are pretty adaptive to harsh environmental conditions, and various gene functions may be adopted due to the natural selection of ecological niches. In this case of heavy metals, tolerance may increase in HMA2, HMA3, and HMA4 substrate affinity depending on the type of metal contamination (Park et al. 2014). Thus, it can be concluded that the overexpression of HMA2, HMA3, and HMA4 with strong constituting shoot-specific promoters may improve the translocation, sequestration, and detoxification of toxic metals in root and shoot tissues. The mentioned approach strongly suggested that the expression of ATPase transporters enhanced the possibility of Cd and Zn accumulation and toleration in transgenic plants without distributing plants' normal physiology. This approach is of high relevance from both an agricultural and environmental points of view. Efforts should be made to engineer the HMA into susceptible varieties of crop plants and consequently check their expression levels. However, this exercise should not be confined to the laboratory. Serious efforts are required to check the efficacy of these transgenic crops in the field and also check their yield pattern or any physiological changes. From an environmental point of view, the HMA genes may be made to overexpress in the phytoremediation plants for a very effective uptake and translocation and consequent compartmentalization in their various parts. This will enhance the cleanup of the contaminated region, thereby making the region free from the deleterious effect of the heavy metals. Thus, systematic engineering of HMA in different plants would provide an edge intolerance as well as environmental cleanup, thereby benefitting the humans.

References

Abedi T, Mojiri A (2020) Cadmium uptake by wheat (*Triticum aestivum* L.): an overview. Plants (Basel) 9(4):500

Adams JP, Adeli A, Hsu CY, Harkess RL, Page GP, dePamphilis CW, Schultz EB, Yuceer C (2011) Poplar maintains zinc homeostasis with heavy metal genes HMA4 and PCS1. J Exp Bot 62(11):3737–3752

Ahmad W, Alharthy RD, Zubair M, Ahmed M, Hameed A, Rafique S (2021) Toxic and heavy metals contamination assessment in soil and water to evaluate human health risk. Sci Rep 11(1): 1–12

Ajeesh Krishna TP, Maharajan T, Victor Roch G, Ignacimuthu S, Antony Ceasar S (2020) Structure, function, regulation and phylogenetic relationship of ZIP family transporters of plants. Front Plant Sci 11:662

Andrés-Colás N, Sancenón V, Rodríguez-Navarro S, Mayo S, Thiele DJ, Ecker JR, Puig S, Peñarrubia L (2006) The *Arabidopsis* heavy metal P-type ATPase HMA5 interacts with metallochaperones and functions in copper detoxification of roots. Plant J 45(2):225–236

Angulo-Bejarano PI, Puente-Rivera J, Cruz-Ortega R (2021) Metal and metalloid toxicity in plants: an overview on molecular aspects. Plants (Basel) 10(4):635

Anjum SA, Tanveer M, Hussain S, Bao M, Wang L, Khan I, Ullah E, Tung SA, Samad RA, Shahzad B (2015) Cadmium toxicity in maize (*Zea mays* L.): consequences on antioxidative systems, reactive oxygen species and cadmium accumulation. Environ Sci Pollut Res 22(21): 17022–17030

Aprile A, Sabella E, Vergine M, Genga A, Siciliano M, Nutricati E, Rampino P, De Pascali M, Luvisi A, Miceli A, Negro C (2018) Activation of a gene network in durum wheat roots exposed to cadmium. BMC Plant Biol 18(1):1–6

Assaha DVM, Ueda A, Saneoka H, Al-Yahyai R, Yaish MW (2017) The role of Na^+ and K^+ transporters in salt stress adaptation in glycophytes. Front Physiol 8:509

Balafrej H, Bogusz D, Triqui ZA, Guedira A, Bendaou N, Smouni A, Fahr M (2020) Zinc hyperaccumulation in plants: a review. Plants (Basel) 9(5):562

Barabasz A, Wilkowska A, Tracz K, Ruszczyńska A, Bulska E, Mills RF, Williams LE, Antosiewicz DM (2013) Expression of HvHMA2 in tobacco modifies Zn–Fe–Cd homeostasis. J Plant Physiol 170(13):1176–1186

Belyaeva EA (2018) Respiratory complex II in mitochondrial dysfunction-mediated cytotoxicity: insight from cadmium. J Trace Elem Med Biol 50:80–92

Benavides MP, Gallego SM, Tomaro ML (2005) Cadmium toxicity in plants. Braz J Plant Physiol 17:21–34

Beneš V, Leonhardt T, Sácký J, Kotrba P (2018) Two P1B-1-ATPases of *Amanita strobiliformis* with distinct properties in Cu/Ag transport. Front Microbiol 9:747

Boal AK, Rosenzweig AC (2009) Structural biology of copper trafficking. Chem Rev 109(10): 4760–4779

Bochicchio R, Sofo A, Terzano R, Gattullo CE, Amato M, Scopa A (2015) Root architecture and morphometric analysis of *Arabidopsis thaliana* grown in Cd/Cu/Zn-gradient agar dishes: a new screening technique for studying plant response to metals. Plant Physiol Biochem 91:20–27

Boutigny S, Sautron E, Finazzi G, Rivasseau C, Frelet-Barrand A, Pilon M, Rolland N, Seigneurin-Berny D (2014) HMA1 and PAA1, two chloroplast-envelope PIB-ATPases, play distinct roles in chloroplast copper homeostasis. J Exp Bot 65(6):1529–1540

Branca JJV, Pacini A, Gulisano M, Taddei N, Fiorillo C, Becatti M (2020) Cadmium-induced cytotoxicity: effects on mitochondrial electron transport chain. Front Cell Dev Biol 8:604377

Cai H, Huang S, Che J, Yamaji N, Ma JF (2019) The tonoplast-localized transporter OsHMA3 plays an important role in maintaining Zn homeostasis in rice. J Exp Bot 70(10):2717–2725

Catty P, Boutigny S, Miras R, Joyard J, Rolland N, Seigneurin-Berny D (2011) Biochemical characterization of AtHMA6/PAA1, a chloroplast envelope Cu(I)-ATPase. J Biol Chem 286(42):36188–36197

Cheah BH, Chen YL, Lo JC, Tang IC, Yeh KC, Lin YF (2021) Divalent nutrient cations: friend and foe during zinc stress in rice. Plant Cell Environ 44(10):3358–3375

Chen G, Liu Y, Wang R, Zhang J, Owens G (2013) Cadmium adsorption by willow root: the role of cell walls and their subfractions. Environ Sci Pollut Res 20(8):5665–5672

Chen ZR, Kuang L, Gao YQ, Wang YL, Salt DE, Chao DY (2018) AtHMA4 drives natural variation in leaf Zn concentration of *Arabidopsis thaliana*. Front Plant Sci 9:270

Cherif J, Mediouni C, Ben Ammar W, Jemal F (2011) Interactions of zinc and cadmium toxicity in their effects on growth and in antioxidative systems in tomato plants (*Solanum lycopersicum*). J Environ Sci (China) 23(5):837–844

Craciun AR, Meyer CL, Chen J, Roosens N, De Groodt R, Hilson P, Verbruggen N (2012) Variation in HMA4 gene copy number and expression among *Noccaea caerulescens* populations presenting different levels of Cd tolerance and accumulation. J Exp Bot 63(11): 4179–4189

Deng F, Yamaji N, Xia J, Ma JF (2013) A member of the heavy metal P-type ATPase OsHMA5 is involved in xylem loading of copper in rice. Plant Physiol 163(3):1353–1362

Disante KB, Cortina J, Vilagrosa A, Fuentes D, Hernández EI, Ljung K (2014) Alleviation of Zn toxicity by low water availability. Physiol Plant 150(3):412–424

dos Santos JO, Andrade CA, de Souza KR, de Oliveira SM, Brandão IR, Alves JD, Santos IS (2019) Impact of zinc stress on biochemical and biophysical parameters in *Coffea arabica* seedlings. J Crop Sci Biotechnol 22(3):253–264

Eren E, Arguello JM (2004) Arabidopsis HMA2, a divalent heavy metal-transporting PIB-type ATPase, is involved in cytoplasmic Zn^{2+} homeostasis. Plant Physiol 136(3):3712–3723

Fang X, Wang L, Deng X, Wang P, Ma Q, Nian H, Wang Y, Yang C (2016) Genome-wide characterization of soybean P1B-ATPases gene family provides functional implications in cadmium responses. BMC Genomics 17:376

Farooq MA, Ali S, Hameed A, Bharwana SA, Rizwan M, Ishaque W, Farid M, Mahmood K, Iqbal Z (2016) Cadmium stress in cotton seedlings: physiological, photosynthesis and oxidative damages alleviated by glycinebetaine. S Afr J Bot 104:61–68

Feigl G, Lehotai N, Molnár Á, Ördög A, Rodríguez-Ruiz M, Palma JM, Corpas FJ, Erdei L, Kolbert Z (2015) Zinc induces distinct changes in the metabolism of reactive oxygen and nitrogen species (ROS and RNS) in the roots of two *Brassica* species with different sensitivity to zinc stress. Ann Bot 116(4):613–625

Fukao Y, Ferjani A (2011) V-ATPase dysfunction under excess zinc inhibits Arabidopsis cell expansion. Plant Signal Behav 6(9):1253–1255

Gao Y, Yang F, Liu J, Xie W, Zhang L, Chen Z, Peng Z, Ou Y, Yao Y (2020) Genome-wide identification of metal tolerance protein genes in *Populus trichocarpa* and their roles in response to various heavy metal stresses. Int J Mol Sci 21(5):1680

Gendre D, Czernic P, Conéjéro G, Pianelli K, Briat JF, Lebrun M, Mari S (2007) TcYSL3, a member of the YSL gene family from the hyper-accumulator *Thlaspi caerulescens*, encodes a nicotianamine-Ni/Fe transporter. Plant J 49(1):1–15

Georgiadou EC, Kowalska E, Patla K, Kulbat K, Smolińska B, Leszczyńska J, Fotopoulos V (2018) Influence of heavy metals (Ni, Cu, and Zn) on nitro-oxidative stress responses, proteome regulation and allergen production in basil (*Ocimum basilicum* L.) plants. Front Plant Sci 9:862

Guarino F, Improta G, Triassi M, Cicatelli A, Castiglione S (2020) Effects of zinc pollution and compost amendment on the root microbiome of a metal tolerant poplar clone. Front Microbiol 11:1677

Haider FU, Liqun C, Coulter JA, Cheema SA, Wu J, Zhang R, Wenjun M, Farooq M (2021) Cadmium toxicity in plants: impacts and remediation strategies. Ecotoxicol Environ Saf 211: 111887

Hoque MN, Tahjib-Ul-Arif M, Hannan A, Sultana N, Akhter S, Hasanuzzaman M, Akter F, Hossain MS, Sayed MA, Hasan MT, Skalicky M, Li X, Brestič M (2021) Melatonin modulates plant tolerance to heavy metal stress: morphological responses to molecular mechanisms. Int J Mol Sci 22(21):11445

Hu X, Li T, Xu W, Chai Y (2021) Distribution of cadmium in subcellular fraction and expression difference of its transport genes among three cultivars of pepper. Ecotoxicol Environ Saf 216: 112182

Huang XY, Deng F, Yamaji N, Pinson SR, Fujii-Kashino M, Danku J, Douglas A, Guerinot ML, Salt DE, Ma JF (2016) A heavy metal P-type ATPase OsHMA4 prevents copper accumulation in rice grain. Nat Commun 7(1):1–13

Huang XH, Zhu F, Yan WD, Chen XY, Wang GJ, Wang RJ (2019) Effects of Pb and Zn toxicity on chlorophyll fluorescence and biomass production of *Koelreuteria paniculata* and *Zelkova schneideriana* young plants. Photosynthetica 57(2):688–697

Huang X, Duan S, Wu Q, Yu M, Shabala S (2020a) Reducing cadmium accumulation in plants: structure–function relations and tissue-specific operation of transporters in the spotlight. Plants (Basel) 9(2):223

Huang S, Sasaki A, Yamaji N, Okada H, Mitani-Ueno N, Ma JF (2020b) The ZIP transporter family member OsZIP9 contributes to root zinc uptake in rice under zinc-limited conditions. Plant Physiol 183(3):1224–1234

Hussain D, Haydon MJ, Wang Y, Wong E, Sherson SM, Young J, Camakaris J, Harper JF, Cobbett CS (2004) P-type ATPase heavy metal transporters with roles in essential zinc homeostasis in Arabidopsis. Plant Cell 16(5):1327–1339

Hussain B, Ashraf MN, Shafeeq-Ur-Rahman AA, Li J, Farooq M (2021) Cadmium stress in paddy fields: effects of soil conditions and remediation strategies. Sci Total Environ 754:142188

Iqbal M, Nawaz I, Hassan Z, Hakvoort HW, Bliek M, Aarts MG, Schat H (2013) Expression of HMA4 cDNAs of the zinc hyperaccumulator *Noccaea caerulescens* from endogenous NcHMA4 promoters does not complement the zinc-deficiency phenotype of the *Arabidopsis thaliana* hma2hma4 double mutant. Front Plant Sci 4:404

Jaishankar M, Tseten T, Anbalagan N, Mathew BB, Beeregowda KN (2014) Toxicity, mechanism and health effects of some heavy metals. Interdiscip Toxicol 7(2):60–72

Jayasumana C, Fonseka S, Fernando A, Jayalath K, Amarasinghe M, Siribaddana S, Gunatilake S, Paranagama P (2015) Phosphate fertilizer is a main source of arsenic in areas affected with chronic kidney disease of unknown etiology in Sri Lanka. Springerplus 4:90

Kanu AS, Ashraf U, Mo Z, Fuseini I, Mansaray LR, Duan M, Pan S, Tang X (2017) Cadmium uptake and distribution in fragrant rice genotypes and related consequences on yield and grain quality traits. J Chem 2017:1405878

Kawakami Y, Bhullar NK (2018) Molecular processes in iron and zinc homeostasis and their modulation for biofortification in rice. J Integr Plant Biol 60(12):1181–1198

Kaznina NM, Titov AF, Topchieva LV, Batova YV, Laidinen GF (2014) The content of HvHMA2 and HvHMA3 transcripts in barley plants treated with cadmium. Russ J Plant Physiol 61(3): 355–359

Kendziorek M, Barabasz A, Rudzka J, Tracz K, Mills RF, Williams LE, Antosiewicz DM (2014) Approach to engineer tomato by expression of AtHMA4 to enhance Zn in the aerial parts. J Plant Physiol 171(15):1413–1422

Keyster M, Niekerk LA, Basson G, Carelse M, Bakare O, Ludidi N, Klein A, Mekuto L, Gokul A (2020) Decoding heavy metal stress signalling in plants: towards improved food security and safety. Plants (Basel) 9(12):1781

Khan MD, Mei L, Ali B, Chen Y, Cheng X, Zhu SJ (2013) Cadmium-induced upregulation of lipid peroxidation and reactive oxygen species caused physiological, biochemical, and ultrastructural changes in upland cotton seedlings. Biomed Res Int 2013:374063

Kim YY, Choi H, Segami S, Cho HT, Martinoia E, Maeshima M, Lee Y (2009) AtHMA1 contributes to the detoxification of excess Zn(II) in Arabidopsis. Plant J 58(5):737–753

Kintlová M, Vrána J, Hobza R, Blavet N, Hudzieczek V (2021) Transcriptome response to cadmium exposure in barley (*Hordeum vulgare* L.). Front Plant Sci 12:629089

Kolahi M, Mohajel Kazemi E, Yazdi M, Goldson-Barnaby A (2020) Oxidative stress induced by cadmium in lettuce (*Lactuca sativa* Linn.): oxidative stress indicators and prediction of their genes. Plant Physiol Biochem 146:71–89

Lapie C, Leglize P, Paris C, Buisson T, Sterckeman T (2019) Profiling of main metabolites in root exudates and mucilage collected from maize submitted to cadmium stress. Environ Sci Pollut Res Int 17:17520–17534

Lee S, Kim YY, Lee Y, An G (2007) Rice P1B-type heavy-metal ATPase, OsHMA9, is a metal efflux protein. Plant Physiol 145(3):831–842

Lee S, Jeong HJ, Kim SA, Lee J, Guerinot ML, An G (2010) OsZIP5 is a plasma membrane zinc transporter in rice. Plant Mol Biol 73(4–5):507–517

Li X, Yang Y, Zhang J, Jia L, Li Q, Zhang T, Qiao K, Ma S (2012) Zinc induced phytotoxicity mechanism involved in root growth of *Triticum aestivum* L. Ecotoxicol Environ Saf 86:198–203

Li X, Yang Y, Jia L, Chen H, Wei X (2013) Zinc-induced oxidative damage, antioxidant enzyme response and proline metabolism in roots and leaves of wheat plants. Ecotoxicol Environ Saf 89: 150–157

Li X, Zhang X, Yang Y, Li B, Wu Y, Sun H, Yang Y (2016) Cadmium accumulation characteristics in turnip landraces from China and assessment of their phytoremediation potential for contaminated soils. Front Plant Sci 7:1862

Li N, Meng H, Xing H, Liang L, Zhao X, Luo K (2017) Genome-wide analysis of MATE transporters and molecular characterization of aluminum resistance in *Populus*. J Exp Bot 68(20):5669–5683

Liu H, Zhao H, Wu L, Liu A, Zhao FJ, Xu W (2017) Heavy metal ATPase 3 (HMA3) confers cadmium hypertolerance on the cadmium/zinc hyperaccumulator *Sedum plumbizincicola*. New Phytol 215(2):687–698

Liu XS, Feng SJ, Zhang BQ, Wang MQ, Cao HW, Rono JK, Chen X, Yang ZM (2019) OsZIP1 functions as a metal efflux transporter limiting excess zinc, copper and cadmium accumulation in rice. BMC Plant Biol 19(1):283

Lu M, Li Z, Liang J, Wei Y, Rensing C, Wei G (2016) Zinc resistance mechanisms of P1B-type ATPases in *Sinorhizobium meliloti* CCNWSX0020. Sci Rep 6:29355

Marichali A, Dallali S, Ouerghemmi S, Sebei H, Casabianca H, Hosni K (2016) Responses of *Nigella sativa* L. to zinc excess: focus on germination, growth, yield and yield components, lipid and terpene metabolism, and total phenolics and antioxidant activities. J Agric Food Chem 64(8):1664–1675

Melandri G, AbdElgawad H, Riewe D, Hageman JA, Asard H, Beemster GTS, Kadam N, Jagadish K, Altmann T, Ruyter-Spira C, Bouwmeester H (2020) Biomarkers for grain yield stability in rice under drought stress. J Exp Bot 71(2):669–683

Meloni G, Zhang L, Rees DC (2014) Transmembrane type-2-like Cu^{2+} site in the P1B-3-type ATPase CopB: implications for metal selectivity. ACS Chem Biol 9(1):116–121

Menguer PK, Farthing E, Peaston KA, Ricachenevsky FK, Fett JP, Williams LE (2013) Functional analysis of the rice vacuolar zinc transporter OsMTP1. J Exp Bot 64(10):2871–2883

Migocka M, Papierniak A, Maciaszczyk-Dziubinska E, Posyniak E, Kosieradzka A (2015) Molecular and biochemical properties of two P1B2-ATPases, CsHMA3 and CsHMA4, from cucumber. Plant Cell Environ 38(6):1127–1141

Mikkelsen MD, Pedas P, Schiller M, Vincze E, Mills RF, Borg S, Møller A, Schjoerring JK, Williams LE, Baekgaard L, Holm PB, Palmgren MG (2012) Barley HvHMA1 is a heavy metal pump involved in mobilizing organellar Zn and Cu and plays a role in metal loading into grains. PLoS One 7(11):e49027

Mills RF, Peaston KA, Runions J, Williams LE (2012) HvHMA2, a P(1B)-ATPase from barley, is highly conserved among cereals and functions in Zn and Cd transport. PLoS One 7(8):e42640

Mishra S, Mishra A, Küpper H (2017) Protein biochemistry and expression regulation of cadmium/zinc pumping ATPases in the hyperaccumulator plants *Arabidopsis halleri* and *Noccaea caerulescens*. Front Plant Sci 8:835

Miyadate H, Adachi S, Hiraizumi A, Tezuka K, Nakazawa N, Kawamoto T, Katou K, Kodama I, Sakurai K, Takahashi H, Satoh-Nagasawa N, Watanabe A, Fujimura T, Akagi H (2011) OsHMA3, a P1B-type of ATPase affects root-to-shoot cadmium translocation in rice by mediating efflux into vacuoles. New Phytol 189(1):190–199

Morel M, Crouzet J, Gravot A, Auroy P, Leonhardt N, Vavasseur A, Richaud P (2009) AtHMA3, a P1B-ATPase allowing Cd/Zn/Co/Pb vacuolar storage in *Arabidopsis*. Plant Physiol 149(2): 894–904

Muradoglu F, Gundogdu M, Ercisli S, Encu T, Balta F, Jaafar HZ, Zia-Ul-Haq M (2015) Cadmium toxicity affects chlorophyll *a* and *b* content, antioxidant enzyme activities and mineral nutrient accumulation in strawberry. Biol Res 48(1):11

Nikalje GC, Suprasanna P (2018) Coping with metal toxicity—cues from halophytes. Front Plant Sci 9:777

Nouet C, Charlier JB, Carnol M, Bosman B, Farnir F, Motte P, Hanikenne M (2015) Functional analysis of the three HMA4 copies of the metal hyperaccumulator *Arabidopsis halleri*. J Exp Bot 6(19):5783–5795

Olsen LI, Hansen TH, Larue C, Østerberg JT, Hoffmann RD, Liesche J, Krämer U, Surblé S, Cadarsi S, Samson VA, Grolimund D (2016) Mother-plant-mediated pumping of zinc into the developing seed. Nat Plants 2(5):1–6

Papoyan A, Kochian LV (2004) Identification of *Thlaspi caerulescens* genes that may be involved in heavy metal hyperaccumulation and tolerance. Characterization of a novel heavy metal transporting ATPase. Plant Physiol 136(3):3814–3823

Park W, Feng Y, Ahn SJ (2014) Alteration of leaf shape, improved metal tolerance, and productivity of seed by overexpression of CsHMA3 in *Camelina sativa*. Biotechnol Biofuels 7:96

Paunov M, Koleva L, Vassilev A, Vangronsveld J, Goltsev V (2018) Effects of different metals on photosynthesis: cadmium and zinc affect chlorophyll fluorescence in durum wheat. Int J Mol Sci 19(3):787

Purohit R, Ross MO, Batelu S, Kusowski A, Stemmler TL, Hoffman BM, Rosenzweig AC (2018) Cu+-specific CopB transporter: revising P1B-type ATPase classification. Proc Natl Acad Sci U S A 115(9):2108–2113

Rascio N, Navari-Izzo F (2010) Heavy metal hyperaccumulating plants: how and why do they do it? And what makes them so interesting? Plant Sci 180(2):169–181

Riaz M, Kamran M, Rizwan M, Ali S, Zhou Y, Núñez-Delgado A, Wang X (2021) Boron application mitigates Cd toxicity in leaves of rice by subcellular distribution, cell wall adsorption and antioxidant system. Ecotoxicol Environ Saf 222:112540

Rizvi A, Zaidi A, Ameen F, Ahmed B, AlKahtani MD, Khan MS (2020) Heavy metal induced stress on wheat: phytotoxicity and microbiological management. RSC Adv 10(63): 38379–38403

Ronzan M, Piacentini D, Fattorini L, Rovere FD, Eiche E, Riemann M, Altamura MM, Falasca G (2018) Cadmium and arsenic affect root development in *Oryza sativa* L. negatively interacting with auxin. Environ Exp Bot 151:64–75

Rosenzweig AC, Argüello JM (2012) Toward a molecular understanding of metal transport by P (1B)-type ATPases. Curr Top Membr 69:113–136

Rouached H (2013) Recent developments in plant zinc homeostasis and the path toward improved biofortification and phytoremediation programs. Plant Signal Behav 8(1):e22681

Sagardoy RU, Morales FE, López-Millán AF, Abadía AN, Abadía JA (2009) Effects of zinc toxicity on sugar beet (*Beta vulgaris* L.) plants grown in hydroponics. Plant Biol 11(3):339–350

Sarker I, Moore LR, Tetu SG (2021) Investigating zinc toxicity responses in marine *Prochlorococcus* and *Synechococcus*. Microbiology 167(6):001064

Sasaki A, Yamaji N, Ma JF (2014) Overexpression of OsHMA3 enhances Cd tolerance and expression of Zn transporter genes in rice. J Exp Bot 65(20):6013–6021

Satoh-Nagasawa N, Mori M, Nakazawa N, Kawamoto T, Nagato Y, Sakurai K, Takahashi H, Watanabe A, Akagi H (2012) Mutations in rice (*Oryza sativa*) heavy metal ATPase 2 (OsHMA2) restrict the translocation of zinc and cadmium. Plant Cell Physiol 53(1):213–224

Sautron E, Mayerhofer H, Giustini C, Pro D, Crouzy S, Ravaud S, Pebay-Peyroula E, Rolland N, Catty P, Seigneurin-Berny D (2015) HMA6 and HMA8 are two chloroplast Cu+-ATPases with different enzymatic properties. Biosci Rep 35(3):e00201

Shahbazi F, Ghasemi S, Sodaiezadeh H, Ayaseh K, Zamani-Ahmadmahmoodi R (2017) The effect of sewage sludge on heavy metal concentrations in wheat plant (*Triticum aestivum* L.). Environ Sci Pollut Res Int 24(18):15634–15644

Shahid M, Pourrut B, Dumat C, Nadeem M, Aslam M, Pinelli E (2014) Heavy-metal-induced reactive oxygen species: phytotoxicity and physicochemical changes in plants. Rev Environ Contam Toxicol 232:1–44

Siemianowski O, Mills RF, Williams LE, Antosiewicz DM (2011) Expression of the $P(_1B)$-type ATPase AtHMA4 in tobacco modifies Zn and Cd root to shoot partitioning and metal tolerance. Plant Biotechnol J 9(1):64–74

Siemianowski O, Barabasz A, Kendziorek M, Ruszczynska A, Bulska E, Williams LE, Antosiewicz DM (2014) HMA4 expression in tobacco reduces Cd accumulation due to the induction of the apoplastic barrier. J Exp Bot 65(4):1125–1139

Sinclair SA, Krämer U (2012) The zinc homeostasis network of land plants. Biochim Biophys Acta Mol Cell Res 1823(9):1553–1567

Sinclair SA, Senger T, Talke IN, Cobbett CS, Haydon MJ, Krämer U (2018) Systemic upregulation of MTP2- and HMA2-mediated Zn partitioning to the shoot supplements local Zn deficiency responses. Plant Cell 30(10):2463–2479

Singh P, Singh I, Shah K (2019) Reduced activity of nitrate reductase under heavy metal cadmium stress in rice: an in silico answer. Front Plant Sci 9:1948

Sinha D, Tandon PK (2020) Biological interventions towards management of essential elements in crop plants. In: Mishra K, Tandon PK, Srivastava S (eds) Sustainable solutions for elemental deficiency and excess in crop plants. Springer, Singapore, pp 209–258

Smith AT, Smith KP, Rosenzweig AC (2014) Diversity of the metal-transporting P1B-type ATPases. J Biol Inorg Chem 19(6):947–960

Smith AT, Ross MO, Hoffman BM, Rosenzweig AC (2017) Metal selectivity of a Cd-, Co-, and Zn-transporting P1B-type ATPase. Biochemistry 56(1):85–95

Song X, Yue X, Chen W, Jiang H, Han Y, Li X (2019) Detection of cadmium risk to the photosynthetic performance of hybrid *Pennisetum*. Front Plant Sci 10:798

Sterckeman T, Thomine S (2020) Mechanisms of cadmium accumulation in plants. CRC Crit Rev Plant Sci 39(4):322–359

Subba P, Mukhopadhyay M, Mahato SK, Bhutia KD, Mondal TK, Ghosh SK (2014) Zinc stress induces physiological, ultra-structural and biochemical changes in mandarin orange (*Citrus reticulata* Blanco) seedlings. Physiol Mol Biol Plants 20(4):461–473

Sun L, Cao X, Tan C, Deng Y, Cai R, Peng X, Bai J (2020) Analysis of the effect of cadmium stress on root exudates of *Sedum plumbizincicola* based on metabolomics. Ecotoxicol Environ Saf 205:111152

Szopiński M, Sitko K, Gieroń Ż, Rusinowski S, Corso M, Hermans C, Verbruggen N, Małkowski E (2019) Toxic effects of Cd and Zn on the photosynthetic apparatus of the *Arabidopsis halleri* and *Arabidopsis arenosa* pseudo-metallophytes. Front Plant Sci 10:748

Takahashi R, Bashir K, Ishimaru Y, Nishizawa NK, Nakanishi H (2012a) The role of heavy-metal ATPases, HMAs, in zinc and cadmium transport in rice. Plant Signal Behav 12:1605–1607

Takahashi R, Ishimaru Y, Shimo H, Ogo Y, Senoura T, Nishizawa NK, Nakanishi H (2012b) The OsHMA2 transporter is involved in root-to-shoot translocation of Zn and Cd in rice. Plant Cell Environ 11:1948–1957

Tan J, Wang J, Chai T, Zhang Y, Feng S, Li Y, Zhao H, Liu H, Chai X (2013) Functional analyses of TaHMA2, a P(1B)-type ATPase in wheat. Plant Biotechnol J 11(4):420–431

Tiwari S, Lata C (2018) Heavy metal stress, signaling, and tolerance due to plant-associated microbes: an overview. Front Plant Sci 9:452

Ueno D, Yamaji N, Kono I, Huang CF, Ando T, Yano M, Ma JF (2010) Gene limiting cadmium accumulation in rice. Proc Natl Acad Sci U S A 107(38):16500–16505

Ueno D, Milner MJ, Yamaji N, Yokosho K, Koyama E, Clemencia Zambrano M, Kaskie M, Ebbs S, Kochian LV, Ma JF (2011) Elevated expression of TcHMA3 plays a key role in the extreme Cd tolerance in a Cd-hyperaccumulating ecotype of *Thlaspi caerulescens*. Plant J 66(5): 852–862

Verret F, Gravot A, Auroy P, Leonhardt N, David P, Nussaume L, Vavasseur A, Richaud P (2004) Overexpression of AtHMA4 enhances root-to-shoot translocation of zinc and cadmium and plant metal tolerance. FEBS Lett 576(3):306–312

Viehweger K (2014) How plants cope with heavy metals. Bot Stud 55(1):35

Wan L, Zhang H (2012) Cadmium toxicity effects on cytoskeleton, vesicular trafficking and cell wall construction. Plant Signal Behav 7(3):345–348

Wang J, Liang S, Xiang W, Dai H, Duan Y, Kang F, Chai T (2019) A repeat region from the *Brassica juncea* HMA4 gene BjHMA4R is specifically involved in Cd^{2+} binding in the cytosol under low heavy metal concentrations. BMC Plant Biol 19(1):89

Wei S, Twardowska I (2013) Main rhizosphere characteristics of the Cd hyperaccumulator *Rorippa globosa* (Turcz.) Thell. Plant Soil 372(1–2):669–681

Wenli ZO, Chang LI, Yajun ZH, Jingguang CH, Haohua HE, Guoyou YE (2020) Rice heavy metal P-type ATPase OsHMA6 is likely a copper efflux protein. Rice Sci 27(2):143–151

Wiyono CDA, Inoue C, Chien MF (2021) HMA4 and IRT3 as indicators accounting for different responses to Cd and Zn by hyperaccumulator *Arabidopsis halleri* ssp. *gemmifera*. Plant Stress 2: 100042

Wong CKE, Cobbett CS (2009) HMA P-type ATPases are the major mechanism for root-to-shoot Cd translocation in *Arabidopsis thaliana*. New Phytol 181(1):71–78

Wong CKE, Jarvis RS, Sherson SM, Cobbett CS (2009) Functional analysis of the heavy metal binding domains of the Zn/Cd-transporting ATPase, HMA2, in *Arabidopsis thaliana*. New Phytol 181(1):79–88

Wu Z, Zhao X, Sun X, Tan Q, Tang Y, Nie Z, Hu C (2014) Xylem transport and gene expression play decisive roles in cadmium accumulation in shoots of two oilseed rape cultivars (*Brassica napus*). Chemosphere 119:1217–1223

Wu D, Sato K, Ma JF (2015) Genome-wide association mapping of cadmium accumulation in different organs of barley. New Phytol 208(3):817–829

Xue Z, Gao H, Zhao S (2014) Effects of cadmium on the photosynthetic activity in mature and young leaves of soybean plants. Environ Sci Pollut Res Int 21(6):4656–4664

Yang L, Ji J, Harris-Shultz KR, Wang H, Wang H, Abd-Allah EF, Luo Y, Hu X (2016) The dynamic changes of the plasma membrane proteins and the protective roles of nitric oxide in rice subjected to heavy metal cadmium stress. Front Plant Sci 7:190

Yang Y, Ge Y, Zeng H, Zhou X, Peng L, Zeng Q (2017) Phytoextraction of cadmium-contaminated soil and potential of regenerated tobacco biomass for recovery of cadmium. Sci Rep 7(1):7210

Yildirim E, Ekinci M, Turan M, Ağar G, Dursun A, Kul R, Alim Z, Argin S (2021) Humic + fulvic acid mitigated Cd adverse effects on plant growth, physiology and biochemical properties of garden cress. Sci Rep 11(1):8040

Younis U, Malik SA, Rizwan M, Qayyum MF, Ok YS, Shah MH, Rehman RA, Ahmad N (2016) Biochar enhances the cadmium tolerance in spinach (*Spinacia oleracea*) through modification of Cd uptake and physiological and biochemical attributes. Environ Sci Pollut Res Int 21: 21385–21394

Yu H, Yang A, Wang K, Li Q, Ye D, Huang H, Zhang X, Wang Y, Zheng Z, Li T (2021) The role of polysaccharides functional groups in cadmium binding in root cell wall of a cadmium-safe rice line. Ecotoxicol Environ Saf 226:112818

Żabka A, Winnicki K, Polit JT, Wróblewski M, Maszewski J (2012) Cadmium(II)-induced oxidative stress results in replication stress and epigenetic modifications in root meristem cell nuclei of *Vicia faba*. Cell 10(3):640

Zhang J, Zhang M, Shohag MJ, Tian S, Song H, Feng Y, Yang X (2016) Enhanced expression of SaHMA3 plays critical roles in Cd hyperaccumulation and hypertolerance in Cd hyperaccumulator *Sedum alfredii* Hance. Planta 3:577–589

Zhang J, Martinoia E, Lee Y (2018) Vacuolar transporters for cadmium and arsenic in plants and their applications in phytoremediation and crop development. Plant Cell Physiol 59(7): 1317–1325

Zhang C, Yang Q, Zhang X, Zhang X, Yu T, Wu Y, Fang Y, Xue D (2021) Genome-wide identification of the HMA gene family and expression analysis under Cd stress in barley. Plants (Basel) 10(9):1849

Zhao H, Guan J, Liang Q, Zhang X, Hu H, Zhang J (2021) Effects of cadmium stress on growth and physiological characteristics of sassafras seedlings. Sci Rep 11(1):9913

Zhu G, Xiao H, Guo Q, Zhang Z, Zhao J, Yang D (2018) Effects of cadmium stress on growth and amino acid metabolism in two compositae plants. Ecotoxicol Environ Saf 30(158):300–308

Zielazinski EL, González-Guerrero M, Subramanian P, Stemmler TL, Argüello JM, Rosenzweig AC (2013) *Sinorhizobium meliloti* Nia is a P(1B-5)-ATPase expressed in the nodule during plant symbiosis and is involved in Ni and Fe transport. Metallomics 12:1614–1623

The Versatile Role of Plant Aquaglyceroporins in Metalloid Transport

7

Suhas Balasaheb Karle, Kundan Kumar, and Om Parkash Dhankher

Abstract

Metalloids are elements with intermediate chemical properties between metals and nonmetals. The metalloids are biologically important elements, ranging from essential to extremely toxic elements with contrasting effects on organisms. Plants deal with a considerable imbalance of metalloids in the environment. Plants must acquire adequate amounts of essential metalloids for metabolism or contrarily exclude toxic metalloids to avoid cellular toxicity. The process of uptake and exclusion is guided by channel proteins, which transport metalloids across cellular membranes. Major intrinsic proteins (MIPs) are a family of selective channels that includes aquaporins (water channels) and aquaglyceroporins (glycerol and other solute channels). Aquaglyceroporin facilitates the transport of small solutes, including glycerol, small uncharged solutes, and gasses across biological membranes. Plant MIPs are grouped into five subfamilies based on sequence similarity and subcellular localization. Plant MIPs are mainly categorized into five subfamilies – plasma membrane intrinsic proteins (PIPs), nodulin-26-like intrinsic proteins (NIPs), tonoplast intrinsic proteins (TIPs), small basic intrinsic proteins (SIPs), and uncharacterized intrinsic proteins (XIPs). The uptake of environmental metalloids by aquaglyceroporins explains how beneficial elements such as silicon are taken up in plants. Conversely, toxic elements such as arsenic and antimony also enter the food chain via these channel proteins. The present

S. B. Karle · K. Kumar (✉)
Department of Biological Sciences, Birla Institute of Technology & Science Pilani, K. K. Birla Goa Campus, Goa, India
e-mail: kundan@goa.bits-pilani.ac.in

O. P. Dhankher
Stockbridge School of Agriculture, University of Massachusetts Amherst, Amherst, MA, USA

K. Kumar, S. Srivastava (eds.), *Plant Metal and Metalloid Transporters*,
https://doi.org/10.1007/978-981-19-6103-8_7

review summarizes the role of various MIP homologs for transporting metalloids into and out of plant cells. This review discusses the detailed mechanism of MIPs for acquiring essential metalloids and their role in the influx and efflux in plant cells.

Keywords

Metalloids · Major intrinsic proteins · Aquaglyceroporins · Homeostasis · Toxicity

7.1 Introduction

Abiotic stress conditions such as salinity, drought, temperature extremes, nutrient deficiency, metalloid toxicity, flooding, etc. cause serious damage to plants in terms of growth, development, and productivity (Singh et al. 2020; Zulfiqar et al. 2020). For instance, tainting the soil and water bodies with metals, their unfavorable effects on plants, and retransmission to the consuming populations creates significant worries worldwide. Heavy metals and metalloid stress contribute to adverse effects on the health of plants and consumers. Metalloids are semimetals that share some common properties of metals and nonmetals. Metalloid boron (B) acts as a micronutrient. At the same time, silicon (Si) is viewed as a useful component for providing mechanical strength and protection from abiotic and biotic stresses to certain crop plants.

On the other hand, arsenic (As) is an extreme ecological danger to most living forms, including plants, while the less plentiful metalloids germanium (Ge), antimony (Sb), and tellurium (Te) show variable levels of harm (Angulo-Bejarano et al. 2021; Sharma et al. 2021). Plants in small amounts require the metalloids B, Si, and Se, while some are toxic and nonessential, namely As, Ge, Sb, and Te. Cereals like wheat, rice, maize, and barley are major sources of biomagnification of metalloid (s) in consumers (Deng et al. 2019). B is conjectured to exist as an all-important micronutrient in soil; even so, its role as an essential trace element is being questioned (Pereira et al. 2021). The boric acid (BA) form of B, which possesses no charge at acidic or neutral pH, gets transported in cells by passive transport (Hrmova et al. 2020); however, that is a half story, and there exist regulated transport mechanisms that help retain B equilibrium in plants (Miwa and Fujiwara 2010; Pereira et al. 2021). The requirement of B varies among the plant species. Some plant species like cotton, beet, and *Brassica* require B during the entire plant development cycle. On the other hand, soybean, pea, wheat, and barley need relatively low B levels and that too during the development of seeds and initiation of flowering (Pommerrenig et al. 2015).

Arsenic is considered a potent group I carcinogen and the leading cause of health issues related to cardiovascular, neurological, hematological, renal, and respiratory systems. Various natural and anthropogenic activities release As into the environment as a potentially toxic, ubiquitous entity in soil and water bodies (Deng et al. 2019). The inorganic form of As in the form of arsenite As(III) and arsenate As (V) predominantly exists in soil and is taken up by the plants (Panda et al. 2010). Roots usually take up the excess concentrations of metalloids by various transporters, including phosphate transporters, sulfate transporters, aquaglyceroporins, nodulin-26-like proteins, Si influx transporter, hexose transporter, etc. (Deng et al. 2019). The detoxification of such metalloids via conjugation with glutathione (GSH) and phytochelatins (PCs) and vacuolar sequestration in roots could be a primary strategy for reducing the metalloid contents in cereal grains (Deng et al. 2019). Si in the form of silicates is the second most plentiful component after oxygen in the environment. The expanding proofs have shown that this metalloid is advantageous to plants, particularly under stress conditions such as salinity, drought, and biotic stresses (Yan et al. 2018). The plant accessibility of Si in soil (as monosilicic acid, H_4SiO_4) differs between 0.1 and 0.6 mM (Yan et al. 2018). Si is taken up and moved through the plant to be saved as SiO_2 phytoliths (Epstein 2001; Ma and Yamaji 2008; Deshmukh et al. 2017a; Luyckx et al. 2017; Pavlovic et al. 2021). The main Si species accessible to living beings is uncharged orthosilicic acid. Like Si, the most bio-accessible Ge type is the uncharged germanic acid (Pommerrenig et al. 2015).

Plants have transporters that keep the required degree of metal and metalloid ion homeostasis inside the cells to perform distinctive functions. These transporters facilitate the entry, multi-organ, and subcellular distribution and exit of various metal(loids) ions. These transporters involve aquaporins, phosphate and sulfate transporters, hexose transporters, etc. (Pandey et al. 2019). Major intrinsic proteins (MIPs) are the membrane proteins that facilitate the transport of small polar molecules across membranes (Johanson et al. 2001). The plant MIP superfamily is highly conserved, with members ranging in size from 23 to 31 kDa (Gomes et al. 2009; Venkatesh et al. 2013; Maurel et al. 2015). The MIPs are divided into five main aquaporin subfamilies based on their sequence similarities and localization: plasma membrane intrinsic protein (PIPs), tonoplast membrane intrinsic proteins (TIPs), nodulin-26-like intrinsic membrane proteins (NIPs), uncharacterized intrinsic proteins (XIPs), and the small basic intrinsic proteins (SIPs) (Maurel et al. 2008; Sakurai et al. 2005; Kumar et al. 2018). Structurally, MIP tetramer is functional, and each monomer comprises six transmembrane alpha-helices (H1–H6) and five loops (LA–LE), of which loops B and E are hydrophobic. In contrast, loops A, C, and D are hydrophilic. The conserved motifs among all MIPs include two highly conserved NPA (asparagine–proline–alanine) motifs in loops B and E. The NPA motifs are known to regulate the substrate selectivity of the AQPs (Deshmukh et al. 2015). Moreover, the aromatic/arginine (ar/R) selectivity filter of MIPs is a tetrad in which each residue is located in H2 and H5 helices and loops LE1 and LE2. The ar/R filter forms a central pore and contributes to substrate selectivity (Afzal et al. 2016; Deshmukh et al. 2016, 2017b; Shivaraj et al. 2017).

The members of the TIP subfamily are mainly localized to the vacuole and facilitate the transport of water, glycerol, hydrogen peroxide, urea, and ammonia by different ar/R selectivity filters (Deshmukh et al. 2015). According to the current

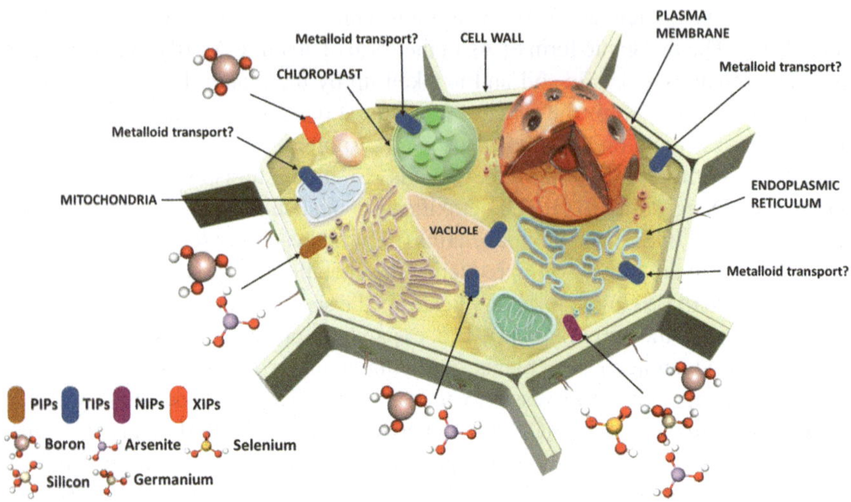

Fig. 7.1 Diagrammatic representation of different MIP localizations in plant cells and their role as various metalloid transporters

consensus, aquaporins have been major contributing entities in plant metalloid stress response and adaptations. The localization of MIPs and their role as potential metalloids (As, B, Si, Se, and Sb) are represented in Fig. 7.1. The possible mechanisms reported so far include changes in the activities of ROS-scavenging enzymes, the altered expression level of certain stress marker genes through the cross talk, hormone-mediated response, and so on (Sun et al. 2017). This chapter will discuss the role of different classes of MIPs in metalloid transport and stress responses.

7.2 PIP Members as Metalloid Transporters

PIP members are the largest subfamilies of plant MIPs and are localized to the plasma membrane. PIPs mainly regulate water homeostasis, and they are also known to be the transporters of urea, H_2O_2, CO_2, and metalloids such as, B, Sb, and Si (Kumar et al. 2018). PIPs are further grouped in two subgroups, namely PIP1 and PIP2 (Chaumont et al. 2001). Plant roots absorb essential mineral nutrients such as B and Si from the soil and transport them to other parts of the plants via various transporters, including PIPs. PIPs also provide sensitivity/tolerance to metalloids, particularly As, via uptake/influx/efflux and compartmentalization of toxic metalloids in different plant tissues. PIPs form channel-like structures that act differently in plants under various environmental conditions during metalloid transport. PIPs are involved in B uptake and support the plants to survive under B limitations. In barley (*Hordeum vulgare*), PIP members such as HvPIP1;3 and *HvPIP1;4* act as B transporters, confirmed in the heterologous system by complementation studies (Fitzpatrick and Reid 2009). The overexpression of ZmPIPs, HvPIP1;3 and HvPIP1;4, in yeast showed B influx and channel activities (Zangi and Filella 2012). Several studies in rice demonstrated that OsPIP1;3, OsPIP2;4, OsPIP2;6, and OPIP2;7 are involved in B transport and lead to influx and efflux activities in rice (Kumar et al. 2014; Mosa et al. 2016). The heterologous expression of OsPIPs genes, namely, OsPIP1;2, OsPIP1;3, OsPIP2;4, OsPIP2;7, and OsPIP2;8, in deficient strain (*Δfps1Δacr3Δycf1*) of yeast resulted in enhanced B transport and sensitivity (Mosa et al. 2016). Heterologous overexpression of OsPIP1;3 and OsPIP2;6 in *Arabidopsis* exhibited enhanced tolerance under higher B concentration (Mosa et al. 2016). Transcript accumulation of several *PIP* members in *Arabidopsis* roots was decreased under higher B concentrations (Aquea et al. 2012). The expression of *PIP1;2*, *PIP2;1*, and *PIP2;2* was higher in shoots compared to roots of *Arabidopsis* at increased B concentration (Macho-Rivero et al. 2018). In non-AM (arbuscular mycorrhizal) plants like maize, the transcript of *ZmPIP2;2* was upregulated at a higher concentration of B under drought stress, showing its role in enhancing water flow in plant roots and decreasing excessive B levels (Quiroga et al. 2020). In contrast, the AM plants under excessive B levels showed reduced *ZmPIP2;2* expressions. The authors hypothesized that AM plants might have a different process to regulate excess B. Higher phosphorous (P) concentration in tissues and interplay of boron and phosphorous reduces B toxicity (Quiroga et al. 2020). These studies concluded strongly that B toxicity caused water flux reduction to the shoot and caused reduced hydraulic conductance of shoot tissue, resulting in downregulation of aquaporin genes in roots and shoots. However, further studies suggested that repression of these *PIP* genes in the shoot is due to the excessive B accumulation, or excess B leads to activation of signaling cascades (Macho-Rivero et al. 2018). Studies involving RNA sequencing transcriptional profiling of the *Brassica napus BnaAQP*s in contrasting B-resistant genotype (Qingyou10) and B-sensitive genotype (Westar10) under sufficient and deficient B conditions suggested that *BnaPIP*s are highly expressed in roots, old leaves, and juvenile leaves (Yuan et al. 2017). In both the cultivars, *BnaPIP1;1*s, *BnaPIP1;2*s, and *BnaPIP2;2/2;3*s showed elevated transcript abundance in roots under the B-sufficient condition, whereas the same *PIPs* transcripts were downregulated in minimal B stress (Yuan et al. 2017).

To subside the metalloid toxicity, the various plants have developed potential strategies, including downregulation of the specialized transporters, metalloid exclusion through efflux channel proteins, and metalloid complexation with thiols such as GSH and PCs, followed by vacuolar sequestration. Most of the NIP subfamilies of aquaporins are identified as As(III) transporters; however, fewer studies also revealed the role of the PIP subfamily in providing metalloids such as B tolerance in plants. To reduce the metalloid accumulation in plant cells and cope with the metalloid stress in rice, PIPs act as a bidirectional pump (Kumar et al. 2014; Mosa et al. 2016). Members of rice PIPs play a potential role in As(III) and B transport pathways (Mosa et al. 2012, 2016; Kumar et al. 2014). Several studies also demonstrated downregulation of transcripts of five rice PIPs (OsPIP1;2, OsPIP1;3, OsPIP2;4, OsPIP2;6, and OsPIP2;7) and 13 PIPs of *Brassica juncea* under arsenite stress (Mosa et al. 2012; Srivastava et al. 2013). In heterologously expressed rice, PIPs (OsPIP2;4, OsPIP2;6, and OsPIP2;7) in *Xenopus* oocytes showed enhanced As (III) transport. Studies of transgenic *Arabidopsis* with constitutive expression of *OsPIP2;4*, *OsPIP2;6*, and *OsPIP2;7* exhibit enhanced tolerance under As(III) stress without affecting the levels of As in root and shoot tissues. To underpin the role of PIPs as bidirectional As(III) transporters, heterologous overexpression of OsPIPs in *Arabidopsis* roots displayed both arsenite influx and efflux processes (Mosa et al. 2012). The other members of rice PIPs need to be characterized at the plant level, and their prominent role needs to be identified. AtPIP2;2 is largely expressed with cell division-related protein (NtCyc07) and demonstrated to provide higher tolerance toward As(III) in *Arabidopsis* (Lee and Hwang 2012). *Arabidopsis atpip2;2* exhibited decreased As(III) tolerance without difference in As(III) concentration, whereas the overexpressed AtPIP2;2 exhibited increased As(III) tolerance and decreased As(III) levels in yeast and *Arabidopsis* (Modareszadeh et al. 2021). The studies potentiate the role of AtPIP2;2 in bidirectional transporter of As(III). The enhanced As(III) efflux in *Arabidopsis* is due to the higher As(III) exporter activity compared to the importer activity indicating that AtPIP2;2 provides As(III) tolerance by reducing As(III) accumulation (Modareszadeh et al. 2021).

The various transporters present in the roots help regulate Si uptake and, therefore, act as a major element to cope with biotic and abiotic stresses in plants (Rios et al. 2017). The Si-treated roots exhibited enhanced expression of PIP genes. In salt-stressed cucumber (*Cucumis sativus*), Si enhances root water uptake via upregulation of aquaporin gene expression. The expression of PIPs (PIP1;2, PIP2;1, PIP2;4, PIP2;5) was upregulated on treatment with Si in cucumber (Zhu et al. 2015). Similarly, Si supplementation in leaves of bottle gourd (*Lagenaria siceraria*) decreased transcript accumulation of *LsiPIP1-5* at 24 h, while at 72 h, *LsiPIP1-5* expression was enhanced (Kumawat et al. 2021). In stem, Si treatment enhanced transcript accumulation of *LsiPIP1-5*, *LsiPIP2-4*, and *LsiPIP4-1* at 24 h with a decrease at 72 h. The root Si application exhibits enhanced transcript accumulation of *LsiPIP1-5* and *LsiPIP2-4* (Kumawat et al. 2021). The transcript levels of *Sorghum bicolor* PIPs (SbPIP1;6, SbPIP2;2, and SbPIP2;6) show upregulation upon Si treatment in limited time salt stress (Liu et al. 2014). The improvement in root water uptake under dehydration stress may be due to the upregulation of PIP genes (Liu et al. 2014). These reports highlight that the manipulation of PIPs in many plant species exhibits an enhanced level of B and Si and controls and limits the toxic As concentration. However, further studies are needed to demonstrate the role of PIPs in metalloid transport and tolerance at the field level under natural environmental and soil conditions.

7.3 NIP Members as Metalloid Transporters

NIPs allow the diffusion of metalloids through membranes and facilitate their cellular transport in different parts of the plants (Wallace et al. 2006; Pommerrenig et al. 2015). Different forms of metalloids transported by NIPs include B, Si, As, Ge, selenium (Se), and Sb (Pommerrenig et al. 2015). For instance, Ma and Yamaji (2006) showed that Si transporter Lsi1 (OsNIP2;1) also transported As(III) and facilitated the entry of As(III) into rice roots. In addition to Lsi1, three other rice NIPs (OsNIP1;1, OsNIP2;2, and OsNIP3;1) were also able to mediate As(III) influx when expressed in *Xenopus laevis* oocytes; however, these genes were expressed at significantly low levels in rice roots (Li et al. 2009). The expression of Lsi1 in *X. laevis* oocytes also significantly increased the uptake of the methylated As species MMAV (monomethylarsonic acid) but not DMAV (dimethylarsonic acid) (Li et al. 2009). Lsi1 helped increase As(III) efflux in rice roots exposed to AsV (Zhao et al. 2010). Additionally, the heterologous expression of OsNIP2;1 and OsNIP3;2 in yeast increased sensitivity to As(III) and As accumulation (Bienert et al. 2008). *Arabidopsis thaliana* NIP1;1, NIP2;1, NIP3;1, NIP5;1, NIP6;1, and NIP7;1 were permeable to As(III) in the yeast expression system (Isayenkov and Maathuis 2008; Bienert et al. 2008), and AtNIP1;1 was capable of transporting As(III) when expressed in *X. laevis* oocytes (Kamiya and Fujiwara 2009). Other NIPs such as LjNIP5;1 and LjNIP6;1 from *Lotus japonicas* are also permeable to As(III) (Bienert et al. 2008). It is clear that members of NIP aquaporins are metalloid channels that transport Si, As, and Sb.

Although the genetic regulation of the B accumulation is less known, a recent study demonstrated that HvNIP2;2/HvLsi6 appear within quantitative trait loci (QTL) largely responsible for B-level dynamics in barley. Further, complementation studies in yeast Δ*atr1* mutant (B efflux transporter deficient) showed that the expression of HvNIP2;2/HvLsi6 induces growth suppression in Δ*atr1* yeast cells, indicating that these NIPs function as B transporters. Also, its heterologous expression in *X. laevis* oocytes showed a high rate of B uptake (Jia et al. 2021). As HvNIP2;2/HvLsi6 is previously reported to function as Si transporter, studies with B transport system show the versatile role of NIP aquaporins in the metalloid transport. The rapeseed's QTL mapping and differential gene expression (DGE) analysis has identified putative NIP transporters that may transport B (Hua et al. 2016). The QTL mapping approach encouraged the identification of the B-efficient gene *BnaA3.NIP5;1* which induces the root tip growth under B deficiency in *Brassica napus*. Additionally, the 5′-untranslated region (UTR) of *BnaA3.NIP5;1* contains "CTTTC" repeats which contribute to the differential expression of the genes involved in plant growth and seed setting (He et al. 2021). Hence, not only the key functional residues but the untranslated regions of the NIPs may also function in the regulatory mechanisms involved in metalloid transport-associated mechanisms in plants. The increased expression of *CiNIP5* under B-deficient conditions has been reported in trifoliate orange and *Carrizo citrange*; in this case, the gene's predominant expression was found in root tissues. The function of *CiNIP5* has similarities with AtNIP5;1 gene from the *Arabidopsis* (An et al. 2012). Thus, the tissue-specific NIP expression reflects their particular roles in plants.

Moreover, certain functions can be seen conserved among different plant species. The site-directed mutagenesis in the H2 and H5 regions of AtNIP5;1 and OsLsi1 (OsNIP2;1) shows that the amino acid residues in the H5 region of the ar/R filter are critical for the transport of B and Si (Mitani-Ueno et al. 2011). The QTL analysis revealed *HvNIP2;1* marker gene associated with B and Ge toxicity in barley. Further site-directed mutagenesis in the amino acid residues of H2 and H5 regions and yeast complementation assay showed that the mutant's B, Ge, and As transport activity was altered (Hayes et al. 2013). Thus, the conserved structural features are important in deciding the transport substrates for NIP aquaporins.

The transcriptome analysis has shown that *Arabidopsis NIP5;1* gene expression increases under B-deficient conditions. Studies with the promoter-GUS fusion showed that NIP5;1 expression is upregulated in the root extension and the root hair zone under B-deficient conditions. Heterologous expression in *X. laevis* oocytes showed that NIP5;1 transported boric acid. Further, T-DNA mutants of NIP5;1 exhibited impaired boric acid uptake by roots, less growth of plants, and overall increased sensitivity to B inadequacy in both roots and shoots. Thus, it may be stated that under B-deficient stress conditions, NIPs could notably influence plant growth and also influence the transport processes in plants (Takano et al. 2006). Similarly, Gómez-Soto et al. (2019) proposed that the *AtNIP5;1* promoter is largely regulated by phytohormones, including abscisic acid (ABA) and ethylene. When applied exogenously, the influence of ABA and ethylene results in the induced expression of the *AtNIP5;1* gene. Further, the *AtNIP5;1* exhibits ABA-induced B uptake and induces root growth.

Sb is toxic to all living beings, including plants, if taken up by plants in the rhizosphere system (Bienert et al. 2008; Kamiya and Fujiwara 2009). The toxicity and tolerance studies in yeast showed that the *AtNIP1;1* transport antimonite (SbIII) and determine consequent sensitivity. Although selenium (Se) is not essential for plants, selenocysteine in plants is involved in the defense against oxidative stress. Vegetables and fruits are the rich sources of Se (Arnér 2010). Examples of the Se transporters in plants include sulfate transporters and OsNIP2;1 (Sors et al. 2005; Zhang et al. 2010, 2012; Zhao et al. 2010). Geranium and silica share similarities in their chemical properties. Ge has been proposed as a biomarker and has been useful for identifying *OsNIP2;1* responsible for the silica accumulation (Ma and Yamaji 2006). Also, the radiolabeled Ge has been shown to trace silica accumulation in heterologous expression systems. A similar biomarker principle has been applied to assess B uptake in barley mediated by NIP aquaporins (Takahashi et al. 1976; Mitani-Ueno et al. 2011; Bárzana et al. 2014).

7.4 XIP Members as Metalloid Transporters

In the genome of different plant species, including *Physcomitrella patens*, *Nicotiana benthamiana*, grape, cotton, tomato, and poplar, various XIP subfamilies were identified (Shelden et al. 2009; Park et al. 2010; Lopez et al. 2012; Ampah-Korsah et al. 2016). The XIPs are absent in monocots and found only in diversified dicot species, except Brassicaceae (Danielson and Johanson 2008). The NPARC motif, the signature sequence for XIPs, showed cysteine residue, followed after the second NPA motif (Danielson and Johanson 2008; Gupta and Sankararamakrishnan 2009). XIPs exhibit discrepancies in amino acids of the first NPA motif and ar/R filter. XIPs are subdivided into four subclasses based on demarcating ar/R filters. In some plants, NIPs showed similarity with the two subclasses having an ar/R signature; instead, the other two subclasses had hydrophobic signature amino acids (Noronha et al. 2016). The XIP cDNA was cloned from morning glory, potato, tobacco, and tomato and showed to be localized in the plasma membrane (Bienert et al. 2011).

Bienert et al. (2011) showed that in a yeast mutant (δfps1), XIPs belonging to Solanales species transport boric acid. In the Solanales' XIPs mutants, the expression of the various splice variants showed impaired and no growth at 10 mM and 20 mM boric acid, respectively. The role of XIPs to facilitate B transport was proved by increased sensitivity to boric acid (Bienert et al. 2011). Overexpression of NtXIP1;1 in *Nicotiana tabacum* with *Nicotiana plumbaginifolia PMA4* promoter fused with 35S enhancer (En2pPMA4) resulted in severe B deficiency symptoms. Overexpression of NtXIP1;1 in *Arabidopsis*, which does not contain any *XIP* gene, by CaMV 35S promoter, resulted in severe B deficiency symptoms. The expression of the two splice variants NtXIP1;1α and NtXIP1;1β in *Xenopus* oocytes showed increased boric acid uptake (Bienert et al. 2019). However, heterologous expression of *NtXIP1;1* with *Arabidopsis* AtNIP5;1 promoter rescued the B deficiency symptoms of *Atnip5;1* mutant, indicating that NtXIP1;1 is a functional boric acid channel in *N. tabacum* and provides B homeostasis and its distribution in tobacco through the tissue-specific expression (Bienert et al. 2019). Some studies on the heterologous expression of grapevine *Vitis vinifera* VvXIP1 in yeast showed their inclusion in metalloids (B and As) and heavy metals (copper and nickel) transport and reduced yeast cell growth in the presence of 40 mM B, 0.5 mM nickel, and 5 mM copper, whereas the growth of the transformed yeast was improved in the presence of 0.45 mM As (Noronha et al. 2016). The knowledge of XIPs and their role in metals and metalloid transport in plants is very limited. Further studies are needed to decipher the exact role of the XIP subfamily in plants.

7.5 Role of TIPs in Metalloid Transport and Tolerance

TIPs are a subfamily of MIPs and are detected to be predominantly located at the tonoplast membrane (Jauh et al. 1998, 1999; Srivastava et al. 2013). However, some TIPs such as AtTIP2;1 and AtTIP1;2 are also localized in endosomal membrane compartments and plasma membrane (Liu et al. 2003). GFP fusion experiments confirmed that AtTIP5;1 is localized to mitochondria (Soto et al. 2010). PvTIP4;1 from *Pteris vittata*, an arsenite hyperaccumulator, has been shown to mediate As(III) transport by yeast functional complementation assay and heterologous expression in *Arabidopsis* (He et al. 2016). Srivastava et al. (2013), through the transcriptomic studies in *B. juncea* under arsenate As(V) stress, found that the *TIP2* gene is responsive to different As(V) treatment periods in both root and shoot parts. The TIP function in the B transport has been investigated either in vivo by utilizing transgenic plants or in vitro by heterologous expression analysis (Bienert and Bienert 2017; Bienert et al. 2019; Pang et al. 2010). *AtTIP5;1* overexpression builds the resilience to high B in the transgenic *Arabidopsis* plants, probably occurring through the vacuolar compartmentation (Pang et al. 2010). Porcel et al. (2018) identified a novel gene, *BvCOLD1*, in *Beta vulgaris* whose sequence shares similarities with the TIP aquaporins. However, it was localized to the endoplasmic reticulum during the subcellular localization analysis with the GFP-fused construct. In yeast cells, *BvCOLD1* showed B homeostasis. When overexpressed, it exhibited enhanced tolerance against the B deficiency in *Arabidopsis*. However, mere sequence similarity may not determine its aquaporin type and role, and further structural studies are imperative to determine the same. Pang et al. (2017) stated that AtTIP5;1 manages elongation of hypocotyl under high convergences of B in *A. thaliana* by unknown mechanisms. Studies with two mutants of *AtTIP5;1* showed that the elongation of cells is hampered when given high B or gibberellin (GA3) compared to the wild-type plants. Further, paclobutrazol (GA synthesis inhibitor) hindered the upregulation of *AtTIP5;1* expression influenced by high B. In the case of GA3 treatment, the expression of *AtTIP5;1* was upregulated compared to wild-type plants. Further, treatment with high B stimulated the transcript level increase of the GA biosynthesis genes in WT seedlings. Also, double mutant studies with the *DELLA* genes showed no upregulation of the *AtTIP5;1* gene under high B stress. Altogether, these outcomes recommend that *AtTIP5;1* function downstream to GA signaling in high B stress (Pang et al. 2017). Hence, TIP aquaporins may cross talk with phytohormones and participate in various stress regulatory mechanisms under metalloid stress. Moreover, they may stimulate growth and development under these stress conditions.

Rivera-Serrano et al. (2012) proposed that the localization of the TIP aquaporins may occur by the vesicular trafficking that involves the Golgi-dependent and Golgi-independent transport through the endoplasmic reticulum. However, the underlying mechanisms are less known; besides, the role of vesicular trafficking under metalloid stress conditions has not been elucidated. Gattolin et al. (2011) reported the localization of *Arabidopsis* TIP3;1 and TIP3;2 to the tonoplast and the plasma membrane; however, their significance to the metalloid stress response is not known. Some reports suggest the differential localization of the TIPs to the subcellular organelles other than the vacuole (Sudhakaran et al. 2021). Further, the localization of the TIPs to other cellular organelles has also been reported; for instance, AtTIP2;1, AtTIP1;2, and AtTIP1;1 localized to the chloroplast membrane (Ferro et al. 2010), GmTIP3;3 and AtTIP5;1 to the mitochondria, GmTIP1;1 to the endoplasmic reticulum, and GmTIP2;6, GmTIP2;7, GmTIP1;7, GmTIP2;3, and GmTIP2;1 to the plasma membrane (Deshmukh et al. 2013). Also, the cytoplasmic localization of the TIPs has been reported in the case of the flax (Shivaraj et al. 2017), nonetheless, their importance in the transport of metalloids is yet to be characterized. The representative plant MIPs' role as metalloid transporters determined by functional studies and the studies conducted to determine functionality are listed in Table 7.1.

Table 7.1 The role of representative MIPs from different plant species acts as metalloid transporters as determined by functional studies

S. no.	Plant species	MIPS (gene/ protein)	Functional studies employed	References
	Boron			
1	*Hordeum vulgare*	*HvPIP1;3* and *HvPIP1;4*	Yeast complementation studies	Fitzpatrick and Reid (2009)
2	*Oryza sativa*	*OsPIP2;4*, *OsPIP2;6*, *OSPIP2;7*, and *OsPIP1;3*	Yeast complementation and overexpression studies in *Arabidopsis thaliana*	Kumar et al. (2014) and Mosa et al. (2016)
3	*Nicotiana tabacum*	*NtXIP1;1*	*Xenopus* oocyte assay, yeast complementation, and overexpression in *Nicotiana* and *Arabidopsis*	Bienert et al. (2011, 2019)
4	*Vitis vinifera*	*VvXIP1*	Overexpression and transformation in yeast	Noronha et al. (2016)
5	*Hordeum vulgare*	*HvNIP2;2*	Yeast complementation studies Heterologous expression in *X. laevis* oocytes	Jia et al. (2021)
6	*Arabidopsis thaliana*	AtNIP5;1 and OsLsi1 (OsNIP2;1)	Site-directed mutagenesis	Mitani-Ueno et al. (2011)
7	*Hordeum vulgare*	HvNIP2;1	Site-directed mutagenesis and yeast complementation studies	Hayes et al. (2013)
8	*Arabidopsis thaliana*	*NIP5;1*	Heterologous expression in *X. laevis* oocytes T-DNA mutant analysis	Takano et al. (2006)
	Arsenic			
9	*Oryza sativa*	*OsPIP2;4*, *OsPIP2;6*, and *OSPIP2;7*	*Xenopus* oocyte assay and overexpression studies in *Arabidopsis thaliana*	Mosa et al. (2012)
10	*Arabidopsis thaliana*	*AtPIP2;2*	Overexpression studies in yeast and *Arabidopsis*. Knockout studies in *Arabidopsis*	Modareszadeh et al. (2021)
11	*Oryza sativa*	*OsNIP2;1* *OsNIP1;1*, *OsNIP2;2*, and *OsNIP3;1*	Transport in *Xenopus laevis* oocytes	Li et al. (2009)
12	*Arabidopsis thaliana*	*NIP1;1*, *NIP2;1*, *NIP5;1*, *NIP6;1*, and *NIP7;1*	Transport studies in yeast cells	Isayenkov and Maathuis (2008)
13	*Arabidopsis thaliana*	*AtNIP1;1*	Transport in *Xenopus laevis* oocytes	Kamiya and Fujiwara (2009)
14	*Lotus japonicas*	*LjNIP5;1* and *LjNIP6;1*	Transport studies in yeast cells	Bienert et al. (2008)

(continued)

Table 7.1 (continued)

S. no.	Plant species	MIPS (gene/protein)	Functional studies employed	References
15	*Hordeum vulgare*	HvNIP2;1	Site-directed mutagenesis and yeast complementation studies	Hayes et al. (2013)
16	*Pteris vittata*	*PvTIP4;1*	Heterologous expression in yeast and *Arabidopsis*	He et al. (2016)
	Silicon			
17	*Arabidopsis thaliana*	AtNIP5;1 and OsLsi1 (OsNIP2;1)	Site-directed mutagenesis	Mitani-Ueno et al. (2011)
	Germanium			
18	*Hordeum vulgare*	HvNIP2;1	Site-directed mutagenesis and yeast complementation studies	Hayes et al. (2013)
	Antimony			
19	*Arabidopsis thaliana*	*AtNIP1;1*	Toxicity and tolerance assay in yeast	Kamiya and Fujiwara (2009)

7.6 Future Perspectives

Metalloid contamination, its stress to plants, and accumulation in the edible plant parts have emerged as serious food production and safety problems. Cereals like wheat, rice, maize, and barley are major sources of biomagnification of metalloid (s) in consumers. For instance, rice is one of the major sources of As exposure. Compared to other crop plants, the translocation factor (TF) for As is higher in rice (0.8), indicating its ten times higher concentration in rice than in other cereals. Being a staple food in most parts of the world, As accumulation in rice grains poses a great concern to consumers. Thus major efforts are needed to reduce the As content in rice grains up to at least maximum permissible limits (safe for consumption), 200 μg/kg for white rice and 300 μg/kg for brown rice as suggested by the World Health Organization (WHO 2014). Brown rice contains 80% more inorganic As than white rice because of the germ layer in brown rice, which retains a considerable amount of As. Hence, it is imperative to reduce the deposition of this metalloid into the edible plant parts. Although notable studies have helped to understand the role of different types of transporters, including MIPs in metalloid transport in plants, many research gaps need to be identified and addressed. As stated above, certain MIPs (e.g., TIPs) have been shown to cross talk with multiple plant hormones in metalloid stress conditions. This indicates that MIPs could directly or indirectly influence plant growth and development. However, the scarcity of knowledge about the underlying signaling mechanisms may limit to employ different MIPs for the crop improvement programs. The selective transport of specific metalloid types by the aquaporins denotes their functions. However, their versatility in transporting other metalloid types directly or indirectly needs to be assessed.

As of now, only PIPs, TIPs, NIPs, and XIPs are involved in metalloid stress tolerance/sensitivity in plants. The literature shows that most of the known metalloids that influence plant growth and productivity are transported by NIP and PIP aquaporins showing their versatile role in metalloid stress response. However, some reports studied their differential regulation at the gene expression level, and functional characterization of them either by homologous or heterologous systems will certainly help uncover many unknown aspects of the NIP- and PIP-mediated metalloid stress signaling in plants. Few reports have proved the metalloid transportability of the MIPs through the heterologous expression studies, which could further be scaled up in many other model systems. When it comes to the function, the sensitivity or tolerance to the metalloid stress responses can be seen as variable among each MIP type from species to species; this discrepancy and dynamic functionality of the MIPs should be addressed at the individual level. Also, distinct MIP signaling pathways regarding the corresponding morphological and physiological modifications in plants are yet to be studied. The inconsistencies among the current data may result from the sort of plant species, the stress treatment systems, etc.

Consequently, the exploratory examinations on the model plant species could give a baseline to future research. It is well known that the MIPs contain significant key amino acid residues in certain conserved regions which take part in the metalloid transport process. Identifying such critical amino acid residues through in silico and mutant analysis approach will help understand the nature of aquaporins to show differential preference to various metalloid substrates for transport. Moreover, it may help engineer modified and improved crop plants exhibiting multimetalloid stress tolerance and biomagnification in permissible limits in the food chain (Mittler and Blumwald 2010). Additionally, studies with other biologically significant metalloids/metal-induced stress in plants will be useful to expand the spectrum of functional characterization of the MIPs.

References

Afzal Z, Howton TC, Sun Y, Mukhtar MS (2016) The roles of aquaporins in plant stress responses. J Dev Biol 4(1):9

Ampah-Korsah H, Anderberg HI, Engfors A, Kirscht A, Norden K, Kjellstrom S, Kjellbom P, Johanson U (2016) The aquaporin splice variant NbXIP1;1α is permeable to boric acid and is phosphorylated in the N-terminal domain. Front Plant Sci 7:862

An JC, Liu YZ, Yang CQ, Zhou GF, Wei QJ, Peng SA (2012) Isolation and expression analysis of CiNIP5, a citrus boron transport gene involved in tolerance to boron deficiency. Sci Hortic 142: 149–154

Angulo-Bejarano PI, Puente-Rivera J, Cruz-Ortega R (2021) Metal and metalloid toxicity in plants: an overview on molecular aspects. Plants 10(4):1–28

Aquea F, Federici F, Moscoso C, Vega A, Jullian P, Haseloff JI, Arce-Johnson PA (2012) A molecular framework for the inhibition of *Arabidopsis* root growth in response to boron toxicity. Plant Cell Environ 35(4):719–734

Arnér ES (2010) Selenoproteins—what unique properties can arise with selenocysteine in place of cysteine? Exp Cell Res 316(8):1296–1303

Bárzana G, Aroca R, Bienert GP, Chaumont F, Ruiz-Lozano JM (2014) New insights into the regulation of aquaporins by the arbuscular mycorrhizal symbiosis in maize plants under drought stress and possible implications for plant performance. Mol Plant Microbe Interact 27(4): 349–363

Bienert MD, Bienert GP (2017) Plant aquaporins and metalloids. In: Plant aquaporins. Springer, Cham, pp 297–332

Bienert GP, Thorsen M, Schüssler MD, Nilsson HR, Wagner A, Tamás MJ, Jahn TP (2008) A subgroup of plant aquaporins facilitate the bi-directional diffusion of As(OH)3 and Sb(OH)3 across membranes. BMC Biol 6(1):1–5

Bienert GP, Bienert MD, Jahn TP, Boutry M, Chaumont F (2011) Solanaceae XIPs are plasma membrane aquaporins that facilitate the transport of many uncharged substrates. Plant J 66(2): 306–317

Bienert MD, Muries B, Crappe D, Chaumont F, Bienert GP (2019) Overexpression of X intrinsic protein 1;1 in *Nicotiana tabacum* and *Arabidopsis* reduces boron allocation to shoot sink tissues. Plant Direct 3(6):1–16

Chaumont F, Barrieu F, Wojcik E, Chrispeels MJ, Jung R (2001) Aquaporins constitute a large and highly divergent protein family in maize. Plant Physiol 125:1206–1215

Danielson JÅ, Johanson U (2008) Unexpected complexity of the aquaporin gene family in the moss *Physcomitrella patens*. BMC Plant Biol 8(1):1–5

Deng F, Yu M, Martinoia E, Song WY (2019) Ideal cereals with lower arsenic and cadmium by accurately enhancing vacuolar sequestration capacity. Front Genet 10(322):1–7

Deshmukh RK, Vivancos J, Guérin V, Sonah H, Labbé C, Belzile F, Bélanger RR (2013) Identification and functional characterization of silicon transporters in soybean using comparative genomics of major intrinsic proteins in *Arabidopsis* and rice. Plant Mol Biol 83(4–5): 303–315

Deshmukh RK, Vivancos J, Ramakrishnan G, Guérin V, Carpentier G, Sonah H, Labbé C, Isenring P, Belzile FJ, Bélanger RR (2015) A precise spacing between the NPA domains of aquaporins is essential for silicon permeability in plants. Plant J 83(3):489–500

Deshmukh RK, Sonah H, Bélanger RR (2016) Plant aquaporins: genome-wide identification, transcriptomics, proteomics, and advanced analytical tools. Front Plant Sci 7:1896

Deshmukh RK, Ma JF, Bélanger RR (2017a) Role of silicon in plants. Front Plant Sci 8(1858):5–7

Deshmukh RK, Nguyen HT, Belanger RR (2017b) Aquaporins: dynamic role and regulation. Front Plant Sci 8:1420

Epstein E (2001) Silicon in plants: facts vs. concepts. In: Studies in plant science, vol 8. Elsevier, Amsterdam, pp 1–15

Ferro M, Brugière S, Salvi D, Seigneurin-Berny D, Moyet L, Ramus C, Miras S, Mellal M, Le Gall S, Kieffer-Jaquinod S, Bruley C (2010) AT_CHLORO, a comprehensive chloroplast proteome database with subplastidial localization and curated information on envelope proteins. Mol Cell Proteome 9(6):1063–1084

Fitzpatrick KL, Reid RJ (2009) The involvement of aquaglyceroporins in transport of boron in barley roots. Plant Cell Environ 32(10):1357–1365

Gattolin S, Sorieul M, Frigerio L (2011) Mapping of tonoplast intrinsic proteins in maturing and germinating *Arabidopsis* seeds reveals dual localization of embryonic TIPs to the tonoplast and plasma membrane. Mol Plant 4(1):180–189

Gomes D, Agasse A, Thiébaud P, Delrot S, Gerós H, Chaumont F (2009) Aquaporins are multifunctional water and solute transporters highly divergent in living organisms. Biochim Biophys Acta Biomembr BBA Biomembr 1788(6):1213–1228

Gómez-Soto D, Galván S, Rosales E, Bienert P, Abreu I, Bonilla I, Bolaños L, Reguera M (2019) Insights into the role of phytohormones regulating pAtNIP5;1 activity and boron transport in *Arabidopsis thaliana*. Plant Sci 287:1–13

Gupta AB, Sankararamakrishnan R (2009) Genome-wide analysis of major intrinsic proteins in the tree plant *Populus trichocarpa*: characterization of XIP subfamily of aquaporins from evolutionary perspective. BMC Plant Biol 9(1):1–28

Hayes JE, Pallotta M, Baumann U, Berger B, Langridge P, Sutton T (2013) Germanium as a tool to dissect boron toxicity effects in barley and wheat. Funct Plant Biol 40(6):618–627

He Z, Yan H, Chen Y, Shen H, Xu W, Zhang H, Shi L, Zhu YG, Ma M (2016) An aquaporin PvTIP 4;1 from *Pteris vittata* may mediate arsenite uptake. New Phytol 209(2):746–761

He M, Wang S, Zhang C, Liu L, Zhang J, Qiu S, Wang H, Yang G, Xue S, Shi L, Xu F (2021) Genetic variation of BnaA3. NIP5;1 expressing in the lateral root cap contributes to boron deficiency tolerance in *Brassica napus*. PLoS Genet 17(7):1–20

Hrmova M, Gilliham M, Tyerman SD (2020) Plant transporters involved in combating boron toxicity: beyond 3D structures. Biochem Soc Trans 48(4):1683–1696

Hua Y, Zhang D, Zhou T, He M, Ding G, Shi L, Xu F (2016) Transcriptomics-assisted quantitative trait locus fine mapping for the rapid identification of a nodulin 26-like intrinsic protein gene regulating boron efficiency in allotetraploid rapeseed. Plant Cell Environ 39(7):1601–1618

Isayenkov SV, Maathuis FJ (2008) The *Arabidopsis thaliana* aquaglyceroporin AtNIP7;1 is a pathway for arsenite uptake. FEBS Lett 582(11):1625–1628

Jauh GY, Fischer AM, Grimes HD, Ryan CA, Rogers JC (1998) δ-Tonoplast intrinsic protein defines unique plant vacuole functions. PNAS 95(22):12995–12999

Jauh GY, Phillips TE, Rogers JC (1999) Tonoplast intrinsic protein isoforms as markers for vacuolar functions. Plant Cell 11(10):1867–1882

Jia Z, Bienert MD, von Wirén N, Bienert GP (2021) Genome-wide association mapping identifies HvNIP2;2/HvLsi6 accounting for efficient boron transport in barley. Physiol Plant 171(4): 809–822

Johanson U, Karlsson M, Johansson I, Gustavsson S, Sjövall S, Fraysse L, Weig AR, Kjellbom P (2001) The complete set of genes encoding major intrinsic proteins in Arabidopsis provides a framework for a new nomenclature for major intrinsic proteins in plants. Plant Physiol 126(4): 1358–1369

Kamiya T, Fujiwara T (2009) *Arabidopsis* NIP1;1 transports antimonite and determines antimonite sensitivity. Plant Cell Physiol 50(11):1977–1981

Kumar K, Mosa KA, Chhikara S, Musante C, White JC, Dhankher OP (2014) Two rice plasma membrane intrinsic proteins, OsPIP2;4 and OsPIP2;7, are involved in transport and providing tolerance to boron toxicity. Planta 239(1):187–198

Kumar K, Mosa KA, Meselhy AG, Dhankher OP (2018) Molecular insights into the plasma membrane intrinsic proteins roles for abiotic stress and metalloids tolerance and transport in plants. Indian J Plant Physiol 23(4):721–730

Kumawat S, Khatri P, Ahmed A, Vats S, Kumar V, Jaswal R, Wang Y, Xu P, Mandlik R, Shivaraj SM, Deokar A (2021) Understanding aquaporin transport system, silicon and other metalloids uptake and deposition in bottle gourd (*Lagenaria siceraria*). J Hazard Mater 409:124598

Lee MS, Hwang S (2012) Cyc07 enhances arsenite tolerance by reducing As levels in *Nicotiana tabacum* and *Arabidopsis thaliana*. Plant Biotech Rep 6(4):391–405

Li RY, Ago Y, Liu WJ, Mitani N, Feldmann J, McGrath SP, Ma JF, Zhao FJ (2009) The rice aquaporin Lsi1 mediates uptake of methylated arsenic species. Plant Physiol 150(4):2071–2080

Liu LH, Ludewig U, Gassert B, Frommer WB, von Wirén N (2003) Urea transport by nitrogen-regulated tonoplast intrinsic proteins in *Arabidopsis*. Plant Physiol 133(3):1220–1228

Liu P, Yin L, Deng X, Wang S, Tanaka K, Zhang S (2014) Aquaporin-mediated increase in root hydraulic conductance is involved in silicon-induced improved root water uptake under osmotic stress in *Sorghum bicolor* L. J Exp Bot 65(17):4747–4756

Lopez D, Bronner G, Brunel N, Auguin D, Bourgerie S, Brignolas F, Carpin S, Tournaire-Roux C, Maurel C, Fumanal B, Martin F, Sakr S, Label P, Julien JL, Gousset-Dupont A, Venisse JS (2012) Insights into Populus XIP aquaporins: evolutionary expansion, protein functionality, and environmental regulation. J Exp Bot 63(5):2217–2230

Luyckx M, Hausman JF, Lutts S, Guerriero G (2017) Silicon and plants: current knowledge and technological perspectives. Front Plant Sci 8(411):1–8

Ma JF, Yamaji N (2006) Silicon uptake and accumulation in higher plants. Trends Plant Sci 11(8): 392–397

Ma JF, Yamaji N (2008) Functions and transport of silicon in plants. Cell Mol Life Sci 65(19): 3049–3057

Macho-Rivero MA, Herrera-Rodríguez MB, Brejcha R, Schäffner AR, Tanaka N, Fujiwara T, González-Fontes A, Camacho-Cristóbal JJ (2018) Boron toxicity reduces water transport from root to shoot in *Arabidopsis* plants. Evidence for a reduced transpiration rate and expression of major PIP aquaporin genes. Plant Cell Physiol 59(4):841–849

Maurel C, Verdoucq L, Luu DT, Santoni V (2008) Plant aquaporins: membrane channels with multiple integrated functions. Annu Rev Plant Biol 59:595–624

Maurel C, Boursiac Y, Luu DT, Santoni V, Shahzad Z, Verdoucq L (2015) Aquaporins in plants. Physiol Rev 95(4):1321–1358

Mitani-Ueno N, Yamaji N, Zhao FJ, Ma JF (2011) The aromatic/arginine selectivity filter of NIP aquaporins plays a critical role in substrate selectivity for silicon, boron, and arsenic. J Exp Bot 62(12):4391–4398

Mittler R, Blumwald E (2010) Genetic engineering for modern agriculture: challenges and perspectives. Annu Rev Plant Biol 61:443–462

Miwa K, Fujiwara T (2010) Boron transport in plants: coordinated regulation of transporters. Ann Bot 105(7):1103–1108

Modareszadeh M, Bahmani R, Kim D, Hwang S (2021) Decreases in arsenic accumulation by the plasma membrane intrinsic protein PIP2;2 in *Arabidopsis* and yeast. Environ Pollut 275:1–10

Mosa KA, Kumar K, Chhikara S, Mcdermott J, Liu Z, Musante C, White JC, Dhankher OP (2012) Members of rice plasma membrane intrinsic proteins subfamily are involved in arsenite permeability and tolerance in plants. Transgenic Res 21(6):1265–1277

Mosa KA, Kumar K, Chhikara S, Musante C, White JC, Dhankher OP (2016) Enhanced boron tolerance in plants mediated by bidirectional transport through plasma membrane intrinsic proteins. Sci Rep 6(1):1–4

Noronha H, Araújo D, Conde C, Martins AP, Soveral G, Chaumont F, Delrot S, Gerós H (2016) The grapevine uncharacterized intrinsic protein 1 (VvXIP1) is regulated by drought stress and transports glycerol, hydrogen peroxide, heavy metals but not water. PLoS One 11(8):1–18

Panda SK, Upadhyay RK, Nath S (2010) Arsenic stress in plants. J Agron Crop Sci 196(3):161–174

Pandey AK, Gautam A, Dubey RS (2019) Transport and detoxification of metalloids in plants in relation to plant-metalloid tolerance. Plant Gene 17:1–26

Pang Y, Li L, Ren F, Lu P, Wei P, Cai J, Xin L, Zhang J, Chen J, Wang X (2010) Overexpression of the tonoplast aquaporin AtTIP5;1 conferred tolerance to boron toxicity in *Arabidopsis*. J Genet Genomics 37(6):389–397

Pang Y, Li J, Qi B, Tian M, Sun L, Wang X, Hao F (2017) Aquaporin AtTIP5;1 as an essential target of gibberellins promotes hypocotyl cell elongation in *Arabidopsis thaliana* under excess boron stress. Funct Plant Biol 45(3):305–314

Park W, Scheffler BE, Bauer PJ, Campbell BT (2010) Identification of the family of aquaporin genes and their expression in upland cotton (*Gossypium hirsutum* L.). BMC Plant Biol 10(1): 1–7

Pavlovic J, Kostic L, Bosnic P, Kirkby EA, Nikolic M (2021) Interactions of silicon with essential and beneficial elements in plants. Front Plant Sci 12:1–19

Pereira GL, Siqueira JA, Batista-Silva W, Cardoso FB, Nunes-Nesi A, Araújo WL (2021) Boron: more than an essential element for land plants? Front Plant Sci 11:1–10

Pommerrenig B, Diehn TA, Bienert GP (2015) Metalloido-porins: essentiality of nodulin 26-like intrinsic proteins in metalloid transport. Plant Sci 238:212–227

Porcel R, Bustamante A, Ros R, Serrano R, Mulet Salort JM (2018) BvCOLD1: a novel aquaporin from sugar beet (*Beta vulgaris* L.) involved in boron homeostasis and abiotic stress. Plant Cell Environ 41(12):2844–2857

Quiroga G, Erice G, Aroca R, Ruiz-Lozano JM (2020) Elucidating the possible involvement of maize aquaporins in the plant boron transport and homeostasis mediated by Rhizophagus irregularis under drought stress conditions. Int J Mol Sci 21(5):1–21

Rios JJ, Martínez-Ballesta MC, Ruiz JM, Blasco B, Carvajal M (2017) Silicon-mediated improvement in plant salinity tolerance: the role of aquaporins. Front Plant Sci 8:948

Rivera-Serrano EE, Rodriguez-Welsh MF, Hicks GR, Rojas-Pierce M (2012) A small molecule inhibitor partitions two distinct pathways for trafficking of tonoplast intrinsic proteins in *Arabidopsis*. PLoS One 7(9):1–11

Sakurai J, Ishikawa F, Yamaguchi T, Uemura M, Maeshima M (2005) Identification of 33 rice aquaporin genes and analysis of their expression and function. Plant Cell Physiol 46(9): 1568–1577

Sharma SS, Kumar V, Dietz KJ (2021) Emerging trends in metalloid-dependent signaling in plants. Trends Plant Sci 26(5):452–471

Shelden MC, Howitt SM, Kaiser BN, Tyerman SD (2009) Identification and functional characterisation of aquaporins in the grapevine, *Vitis vinifera*. Funct Plant Biol 36(12): 1065–1078

Shivaraj SM, Deshmukh RK, Rai R, Bélanger R, Agrawal PK, Dash PK (2017) Genome-wide identification, characterization, and expression profile of aquaporin gene family in flax (*Linum usitatissimum*). Sci Rep 7(1):1–7

Singh S, Kumar A, Panda D, Modi MK, Sen P (2020) Identification and characterization of drought responsive miRNAs from a drought tolerant rice genotype of Assam. Plant Gene 21:1–8

Sors TG, Ellis DR, Salt DE (2005) Selenium uptake, translocation, assimilation and metabolic fate in plants. Photosynth Res 86(3):373–389

Soto G, Fox R, Ayub N, Alleva K, Guaimas F, Erijman EJ, Mazzella A, Amodeo G, Muschietti J (2010) TIP5;1 is an aquaporin specifically targeted to pollen mitochondria and is probably involved in nitrogen remobilization in *Arabidopsis thaliana*. Plant J 64(6):1038–1047

Srivastava S, Srivastava AK, Suprasanna P, D'Souza SF (2013) Quantitative real-time expression profiling of aquaporins-isoforms and growth response of *Brassica juncea* under arsenite stress. Mol Biol Rep 40(4):2879–2886

Sudhakaran S, Thakral V, Padalkar G, Rajora N, Dhiman P, Raturi G, Sharma Y, Tripathi DK, Deshmukh R, Sharma TR, Sonah H (2021) Significance of solute specificity, expression, and gating mechanism of tonoplast intrinsic protein during development and stress response in plants. Physiol Plant 172(1):258–274

Sun H, Li L, Lou Y, Zhao H, Yang Y, Wang S, Gao Z (2017) The bamboo aquaporin gene PeTIP4;1–1 confers drought and salinity tolerance in transgenic *Arabidopsis*. Plant Cell Rep 36(4):597–609

Takahashi E, Syo S, Miyake Y (1976) Effect of germanium on the growth of plants with special reference to the silicon nutrition. 1. Comparative studies on the silica nutrition in plants. J Sci Soil Manure 2:191–197

Takano J, Wada M, Ludewig U, Schaaf G, Von Wirén N, Fujiwara T (2006) The *Arabidopsis* major intrinsic protein NIP5;1 is essential for efficient boron uptake and plant development under boron limitation. Plant Cell 18(6):1498–1509

Venkatesh J, Yu JW, Park SW (2013) Genome-wide analysis and expression profiling of the *Solanum tuberosum* aquaporins. Plant Physiol Biochem 73:392–404

Wallace IS, Choi WG, Roberts DM (2006) The structure, function and regulation of the nodulin 26-like intrinsic protein family of plant aquaglyceroporins. Biochim Biophys Acta (BBA) Biomembr 1758(8):1165–1175

WHO (2014) Codex Alimentarius Commission, 37th Session, Geneva, 14–18 July 2014. World Health Organization, Geneva

Yan GC, Nikolic M, Ye MJ, Xiao ZX, Liang YC (2018) Silicon acquisition and accumulation in plant and its significance for agriculture. J Integr Agric 17(10):2138–2150

Yuan D, Li W, Hua Y, King GJ, Xu F, Shi L (2017) Genome-wide identification and characterization of the aquaporin gene family and transcriptional responses to boron deficiency in *Brassica napus*. Front Plant Sci 8:1–17

Zangi R, Filella M (2012) Transport routes of metalloids into and out of the cell: a review of the current knowledge. Chem Biol Interact 197(1):47–57

Zhang L, Yu F, Shi W, Li Y, Miao Y (2010) Physiological characteristics of selenite uptake by maize roots in response to different pH levels. J Plant Nutr Soil Sci 173(3):417–422
Zhang H, Feng X, Zhu J, Sapkota A, Meng B, Yao H, Qin H, Larssen T (2012) Selenium in soil inhibits mercury uptake and translocation in rice (*Oryza sativa* L.). Environ Sci Technol 46(18): 10040–10046
Zhao XQ, Mitani N, Yamaji N, Shen RF, Ma JF (2010) Involvement of silicon influx transporter OsNIP2;1 in selenite uptake in rice. Plant Physiol 153(4):1871–1877
Zhu YX, Xu XB, Hu YH, Han WH, Yin JL, Li HL, Gong HJ (2015) Silicon improves salt tolerance by increasing root water uptake in *Cucumis sativus* L. Plant Cell Rep 34(9):1629–1646
Zulfiqar F, Akram NA, Ashraf M (2020) Osmoprotection in plants under abiotic stresses: new insights into a classical phenomenon. Planta 251(1):1–7

The Multidrug and Toxic Compound Extrusion (MATE) Family in Plants and Their Significance in Metal Transport

8

Debojyoti Kar, Ajar Anupam Pradhan, Arka Dutta, Lavanya Bhagavatula, and Sourav Datta

Abstract

The mineral and nutrient status of the soil is a critical factor determining optimum growth and development of a plant. The varying composition of the soil with respect to the availability of various metals poses a developmental challenge to plants. High concentration of toxic metals in the soil hinders plant growth. The ability of plants to withstand such stress depends on its cellular detoxification mechanisms. In this chapter, we elaborate the role of the multidrug and toxic compound extrusion (MATE) transporter family in regulating the transport of various metals like iron and aluminum into and out of the plant. This ubiquitous family of transporters generally consists of 12 transmembrane helices and function as cationic effluxers. They are mostly present in the plasma membrane of a cell of various tissues in a plant. They are reported to perform a wide array of functions and efflux various classes of substrates. MATEs are mostly known for their ability to efflux citrate and chelate external free aluminum ions to prevent aluminum toxicity. They also help in efficient iron solubilization and translocation of iron to the xylem sap. Some of the other roles of MATE proteins, like transporting secondary metabolites, xenobiotic detoxification, and biotic interactions, and roles in developmental pathways are also discussed here.

Keywords

MATE transporters · Aluminum toxicity · Iron homeostasis · Xenobiotic detoxification · Biotic stress · Secondary metabolites

D. Kar · A. A. Pradhan · A. Dutta · L. Bhagavatula · S. Datta (✉)
Department of Biological Sciences, Indian Institute of Science Education and Research (IISER) Bhopal, Bhopal, Madhya Pradesh, India
e-mail: dkar@iiserb.ac.in; ajar19@iiserb.ac.in; sdatta@iiserb.ac.in

K. Kumar, S. Srivastava (eds.), *Plant Metal and Metalloid Transporters*,
https://doi.org/10.1007/978-981-19-6103-8_8

Abbreviations

Al	Aluminum
EDS5	Enhanced disease susceptibility 5
Fe	Iron
HCAA	Hydroxycinnamic acid amide
MATE	Multidrug and toxic compound extrusion
PCA	Protocatechuic acids
TMACl	Tetramethylammonium chloride
TNT	2,4,6-trinitro toluene

8.1 Introduction

Every living organism faces multiple environmental challenges in their lifetime. These adverse situations have helped them to develop resistance mechanisms and exhibit adaptive response, which is crucial for their survival. Animals have an advantage of mobility, which helps them to escape from certain environmental stress factors, for example, they can move to cooler places under extreme sunlight to avoid the heat or escape from the high UV light exposure associated with sunlight. Plants, being sessile, do not have the option of mobility; instead, they have developed various other adaptive mechanisms for their survival. However, a common mechanism that exists in both animals and plants is the cellular protection machinery against toxic compounds. One such common mechanism is the multidrug efflux transporters, which provide a high degree of resistance to antimicrobial compounds, xenobiotics, genotoxins, metals, etc. They mediate efflux of toxic compounds from cytosol to extracellular space or detoxify by internalizing them into vacuoles (Du et al. 2018). In this particular chapter, we will be focusing mainly on a class of plant-specific multidrug transporters, known as the multidrug and toxic compound extrusion (MATE) transporter family and how they help plants to survive under various stress conditions.

There are six classes of identified bacterial multidrug efflux transporters. They are ATP-binding cassette (ABC) superfamily, major facilitator superfamily (MFS), small multidrug resistance (SMR) family, resistance nodulation cell division (RND) family, proteobacterial antimicrobial compound efflux (PACE) family, and the multidrug and toxic compound extrusion (MATE) transporter family (Du et al. 2018). Of these, ABC transporters are the only class of active transporters, while others are secondary active transporters. The first identified MATE transporter, NorM, was found in *Vibrio parahaemolyticus* along with its homologue YdhE from *Escherichia coli* and *Haemophilus influenzae* (Morita et al. 1998). The *E. coli* cells transformed with a plasmid harboring the *NorM* gene showed a higher energy-dependent norfloxacin efflux. The elevated norfloxacin efflux after addition of a H^+ conductor, CCCP, in these transformed cells indicated the efflux machinery to be driven by a H^+ electrochemical potential (Morita et al. 1998). Gene and protein sequence studies revealed NorM consists of 456 amino acid residues, and hydropathy index identified it to be a 12 transmembrane helix containing membrane protein. However, sequence homology with other members of reported MFS family did not show any sequence similarity, which led to the identification of a novel class of multidrug efflux family protein (Morita et al. 1998; Brown et al. 1999). Brown et al. (1999) have identified that NorM and YdhE shared sequence homology and transmembrane topology with members of an unidentified family, which was ubiquitously present across all life forms; hence, they named this new prototype of multidrug efflux family as the multidrug and toxic compound extrusion (MATE) transporter family.

Though the identified bacterial MATEs were found to be confer drug resistance, the substrate specificity varies greatly among other MATEs. They provide resistance to cations, fluoroquinolones, aminoglycosides, etc. MATEs were categorized under multidrug/oligosaccharidyl-lipid/polysaccharide (MOP) exporter superfamily in bacteria (Hvorup et al. 2003). They have undergone rapid expansion and duplication over time and are present in all life forms starting from unicellular protists, bacteria, and archaea to multicellular complex organisms such as fungi, animals, and plants. The broad substrate spectrum could be the reason for its positive selection over the course of evolution and being ubiquitously present in all life-forms. Recent study has shown that plants being sessile have undergone larger expansion/duplication of MATEs than animals. The sequence to species ratio of MATEs is less in prokaryotes as compared to eukaryotes, mainly plants. In plants, the MATE paralog number increases as we move from water to land plants. The chlorophyta contains approximately 15–20 MATEs, and bryophytes have 20–40 *MATE* genes. In Arabidopsis, there are 58 MATE transporters, 117 in *Glycine max*, and 70 in *Triticum aestivum*. MATEs typically contain 12 transmembrane helices, share a weak sequence similarity among other MATEs, and are membrane localized (Morita et al. 1998; Brown et al. 1999; Hvorup et al. 2003).

They can be identified using the MATE domain (Pfam 01554). In humans, MATE transporters have a 13 transmembrane helix and show elevated expression in brush-border membranes of kidney and in the liver (Otsuka et al. 2005a; Omote et al. 2006). In plants, apart from plasma membrane, MATEs can also be found in other subcellular membranes, for example, the tonoplast, Golgi, and chloroplast membranes across most of the tissues (Park et al. 2011; Zhang et al. 2014, 2017).

They generally efflux cationic drugs, for example, cimetidine, metformin, and acyclovir in animals (Masuda et al. 2006). In plants, they exhibit a wide substrate spectrum including nicotine, flavonoids, proanthocyanidins, and organic acids (Upadhyay et al. 2019). This wide category of substrate efflux capacity by MATEs has rendered them with multiple crucial functions in plants. Briefly, MATEs function to provide resistance against genotoxins, xenobiotics, antimicrobial compounds, transport of secondary metabolites, resistance to biotic stress, tolerance to metals, and response to aluminum toxicity and have roles in phytohormone transport as well as several developmental pathways in plants (Fig. 8.1, Table 8.1) (Upadhyay et al. 2019).

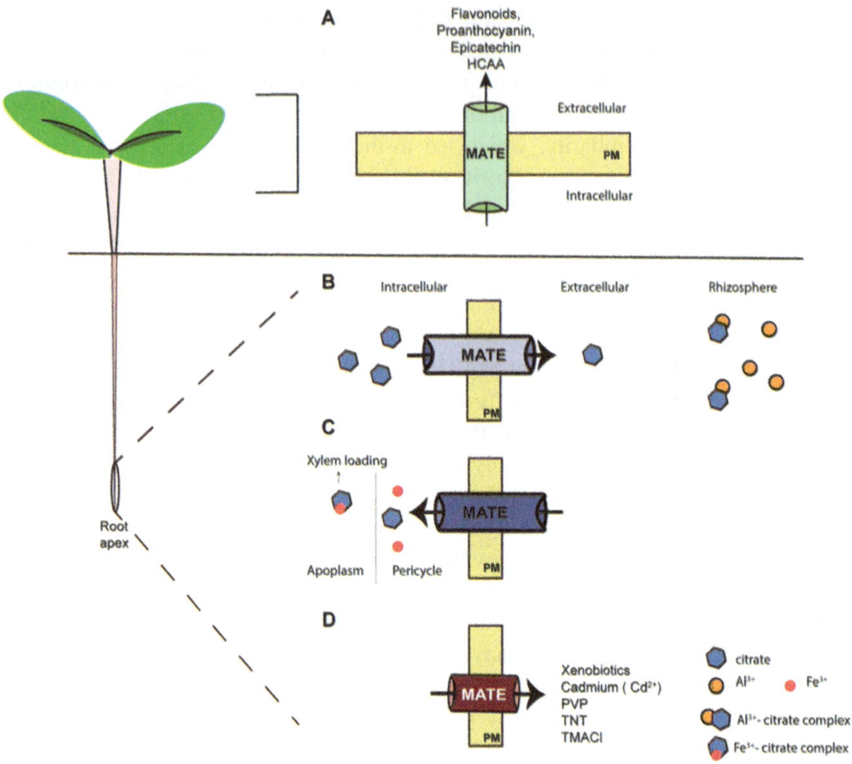

Fig. 8.1 Schematic representation of diverse substrate transport functions exhibited by MATE transporters in a plant. (**a**) transport of secondary metabolites to outside of the cell mostly in the aerial tissues, *HCAA* Hydroxycinnamic acid amides (**b**) citrate exudation from epidermal cell plasma membrane to the root rhizosphere to chelate external Al^{3+} ions thereby alleviating Al toxicity (**c**) loading of citrate in pericycle cells resulting in iron-citrate complex formation in the apoplast for efficient iron translocation into the xylem (**d**) exudation of toxic compounds such as xenobiotics, cadmium ions, *PVP* polyvinylpyrollidone, *TNT* 2,4,6-trinitrotoluene, *TMACl* tetramethylammonium chloride. The different colours of MATE genes indicate different MATE transporters involved in transporting different substrates

8.2 Structure of MATEs

MATE proteins generally contain 400–500 amino acid residues. Owing to their weak sequence similarity, they can be classified into three different structural subfamilies, namely, the NorM, DinF, and the eukaryotic MATEs (eMATEs). The topology exclusive to MATEs is composed of N-terminal and C-terminal lobes, which are related by intramolecular pseudo twofold symmetry. The NorM type has conserved acidic residues at the N and C-lobe, whereas the DinF and eMATE have conserved acidic residues at N- and C-lobes, respectively, for transport (He et al. 2010; Lu et al. 2013; Tanaka et al. 2013; Miyauchi et al. 2017). Generally,

Table 8.1 List of MATE transporters involved in iron homeostasis, aluminum toxicity, secondary metabolite transport, developmental roles, biotic stress

Organism	MATE genes	Tissue expressed	Subcellular localization	Function	Reference
MATE transporters involved in iron homeostasis					
Arabidopsis thaliana	*AtFRD3*	Root pericycle	Plasma membrane	Effluxes citrate into root vasculature, iron solubilization	Durrett et al. (2007) and Roschzttardtz et al. (2011)
	BCD1	Higher in root, induced by Fe	Golgi complex	Reallocation of iron	Seo et al. (2012)
	AtDTX25	All over the plant	Tonoplast	Regulation of vacuolar Fe export	Thi et al. (2021)
Oryza sativa	*OsPEZ1*	Stele of the roots	Plasma membrane	Efflux of PCA into xylem to solubilize the apoplasmic precipitated iron in the stele	Ishimaru et al. (2011)
	OsPEZ2	Root	Plasma membrane	Efflux of CA and PCA into xylem and apoplasmic Fe solubilization	Bashir et al. (2011)
	OsFRDL1	Roots	Plasma membrane	Citrate transport; Fe solubilization	Yokosho et al. (2009)
Medicago truncatula	*MtMATE66*		Plasma membrane	Citric acid transport	Wang et al. (2017)
	MtMATE55	Hypocotyls, vegetative buds and leaves	N. A	Iron homeostasis	
	MtMATE69	Roots and stems	Plasma membrane	Citrate transporter	
	MtMATE67	Nodules	Plasma membrane	Fe activated citrate transport	Kryvoruchko et al. (2018)
Arachis hypogea	*AhFRDL1*	Al-induced expression in the pericycle of root tip	Plasma membrane	Citrate transport, Fe translocation and homeostasis	Qiu et al. (2019)
Secale cereal	*ScFRDL1*	All over plant		Efflux of citrate into the xylem for Fe translocation from roots to shoots	Yokosho et al. (2010)

(continued)

Table 8.1 (continued)

Organism	MATE genes	Tissue expressed	Subcellular localization	Function	Reference
MATE transporters involved in aluminum toxicity					
Arabidopsis thaliana	*AtMATE*	Roots	Plasma membrane	Citrate exudation	Liu et al. (2009)
	AtDTX30	Mainly in roots, slight transcript levels observed in rosette leaf, bud, dry seeds	Plasma membrane	Al-responsive citrate exudation	Upadhyay et al. (2020)
Sorghum bicolor	*SbMATE*	Roots	Plasma membrane (GFP onion epidermis); tonoplast in vacuolated cells	Al exclusion by citrate efflux	Magalhaes et al. (2007)
Hordeum vulgare	*HvAACT*	Root tip epidermis (in situ hybridization; antibody staining)	Plasma membrane (GFP, onion epidermis)	Citrate exudation (*Xenopus oocyte*)	Furukawa et al. (2007)
Brachypodium distachyon	*BdMATE1*, *BdMATE2*	Root	N. A	Citrate exudation	Contreras et al. (2014)
Zea maize	*ZmMATE1*	Roots; more at the root tip	Plasma membrane (GFP)	Citrate exudation (*Xenopus oocyte*)	Maron et al. (2010)
	ZmMATE6	Roots and leaves		Citrate efflux (transgenic Arabidopsis)	Du et al. (2021)
Fagopyrum esculentum	*FeMATE1*	Al-induced upregulation in roots	Plasma membrane (GFP; onion epidermal cell)	Al-induced citrate efflux	Lei et al. (2017)
	FeMATE2	Roots and leaves	Golgi complex	Internal detoxification of Al by citrate secretion into the Golgi	
Oryza sativa	*OsFRDL4*	Root	Plasma membrane	Al-induced citrate efflux (transgenic)	Yokosho et al. (2011)
	OsFRDL2	Root, leaves and shoot. Al induced upregulation in root	Vesicles in cytosol (GFP; onion epidermis)	Al-induced citrate efflux	Yokosho et al. (2016)
Secale cereale	*ScFRDL2*	Roots; root tip and basal root regions		Al-activated citrate secretion	Yokosho et al. (2010)

Glycine soja	*GsMATE*	All over the plant, higher in roots; Al induced increase in root tip	Plasma membrane	Al-induced citrate secretion	Ma et al. (2018)
Glycine max	*GmMATE13*	Root tips	Plasma membrane	Citrate exudation and enhanced Al resistance	Wang (2019)
	GmMATE47				
	GmMATE75	Constitutively expressed in soybean roots	Plasma membrane	Al-induced citrate secretion	Zhou et al. (2019)
	GmMATE87				
	GmMATE79				
Brassica oleracea	*BoMATE*	Very low in shoot; higher in root, inducible by aluminum	Plasma membrane	Al-induced citrate efflux	Wu et al. (2014)
Medicago truncatula	*MtMATE66*	Leaves, stems, nodules roots, stems and pods	Plasma membrane	Citrate efflux and Al tolerance	Wang et al. (2017)
	MtMATE69	Roots and stems	Plasma membrane		
Gossypium hirsutum	*GhMATE1*	NA	NA	Al-activated citrate secretion	Kundu and Ganesan (2020)
Arachis hypogea	*AhFRDL1*	Al induced expression in the pericycle of root tip	Plasma membrane	Citrate efflux and Al tolerance	Qiu et al. (2019)
Eucalyptus camaldulensis	*EcMATE1*	Roots	Plasma membrane	Al-responsive citrate excretion	Sawaki et al. (2013)
Populus trichocarpa	*PtrMATE1*	Central cylinder of mature roots; Al-responsive expression in root apex	Plasma membrane (onion epidermis)	Al-induced enhanced citrate efflux	Li et al. (2017)
Vigna umbellata	*VuMATE1*	Root apex (up to 1 cm in the central cylinder)	Plasma membrane	Al-induced citrate exudation and Al tolerance	Liu et al. (2013)
	VuMATE2	Al induced expression in root apex	Plasma membrane	Root citrate secretion and Al resistance	Liu et al. (2018)
MATE transporters in secondary metabolite transport					
Nicotiana tabacum	Nt*MATE1*	Expressed all over the plant, abundant in root tissue	Vacuolar membrane	Transport of H^+/nicotine	Takase et al. (2008)

(continued)

Table 8.1 (continued)

Organism	MATE genes	Tissue expressed	Subcellular localization	Function	Reference
Arabidopsis thaliana	At*DTX46*	Flowers and a silique	Chloroplast	Transportation of phenolic compounds	Parinthawong et al. (2015)
Medicago truncatula	*MtMATE1* *MtMATE2*	Seed coat	Tonoplast membrane	Transport of epicatechin 3′-*O*-glucoside	Zhao and Dixon (2009)
Arabidopsis thaliana	*AtDTX41*	Seed-coat endothelium	Ovules and developing seeds	Flavonoid/H+- antiporter	Debeaujon et al. (2007) and Marinova et al. (2007)
Vitis vinifera	*VvMATE*		Vacuole	Transport of proanthocyanin into the vacuole	Gomez et al. (2009) and Pérez-Díaz et al. (2014)
Fragaria ananassa	*FaTT12*	Abundant in seeds	Vacuole	Transport of proanthocyanin into the vacuole	Chen et al. (2018)
Capsicum annuum	*CaMATE*	Leaf, flower, pericarp, placenta, or seed	Vacuole and plastids	Transport of secondary metabolites such as proanthocyanin, flavonoids and nicotine	Chen et al. (2020)
Developmental roles					
Arabidopsis thaliana	*AtDTX33* *AtDTX35*	Root hairs and guard cells	Tonoplast	Act as chloride channels	Zhang et al. (2017)
Arabidopsis thaliana	*DTX31*	Root hairs	Plasma membrane	Helps in root hair elongation	Won et al. (2009)
Arabidopsis thaliana	*DTX34*	Pollen grain	–	Helps in male gametophyte development	Bock (2006))

Zea mays	*BIGE1*	Lateral root (LR) zone, elongation zone, and root tip	Trans-Golgi	Accelerated root and leaf initiation as well as enlargement of embryo scutellum	Suzuki et al. (2015)
Lupinus albus	*LaMATE*	Root hairs	Plasma membrane	Role in this modification under phosphate deficiency	Uhde-Stone et al. (2005)
Malus domestica	*MdMATE*	Mostly in flowers and pollen	Fruit flesh and peel	Role in maturation of fruits	Zhang et al. (2021)

transmembrane domain 1–6 form the N-lobe, and transmembrane 7–12 form the C-lobe. Depending on the protonation of the acidic residues, MATEs exist in two conformations: straight and bent, which helps them extrude substrate in a rocker-switcher mechanism (Kusakizako et al. 2020). The cations (H^+/Na^+) bind to the center resulting in a change into an inward conformation from an outer conformation, which directs ion coordination across the membrane. MATEs mainly use Asp and Glu residues for substrate binding (He et al. 2010; Nie et al. 2016; Kusakizako et al. 2020).

Crystallographic studies have helped elucidating crystal structures of MATE proteins across all of its structural groups. The *NorM* gene from *Vibrio cholerae* was the first crystallized MATE protein, which revealed a topology exclusive to the MATEs (He et al. 2010). The Glu255 and Asp371 at the center of C-lobe form a negatively charged pocket which is responsible for substrate/cation binding. These acidic residues are conserved among the NorM family members (He et al. 2010). The significance of this Asp residue was proved by mutational studies in the *NorM* gene from *Vibrio parahaemolyticus*, where the Asp 367, corresponding to Asp371 of NorM *Vibrio cholerae* was mutated to Asn367 and Ala367 (D367N and D367A). These mutations completely abolished the transport activity of the NorM (Otsuka et al. 2005b). The DinF subfamily also transports substrates from the intracellular to the extracellular side of the membrane using a proton or Na^+ electrochemical gradient (Tanaka et al. 2013; Miyauchi et al. 2017; Kusakizako et al. 2020). The first crystal structure of this DinF subclass type was obtained from *Pyrococcus furiosus* H^+-coupled MATE (*PfMATE*) (Tanaka et al. 2013). This structure was obtained at two conformations under different pH conditions: straight and bent. The transmembrane 1 at the N-lobe either had a straight helix or a bent due to a kink at Pro26 and Gly30. The Asp41 residue on the TM1 is well conserved among the DinF and NorM subfamilies. Mutational studies on this particular Asp41 residue have shown reduced drug/H^+ transport activity (Tanaka et al. 2013; Kusakizako et al. 2020). All of these structures were captured on the outward-facing conformation. A recent study has resolved the first inward-facing state structure of *Pyrococcus furiosus* MATE (Zakrzewska et al. 2019). It is noteworthy that TM1 is likely the reason for the conformational flexibility between outward-facing and inward-facing states of DinF subfamily transporters (Kusakizako et al. 2020).

The eMATEs and NorM have highly conserved acidic residues at the C-lobe (Miyauchi et al. 2017). Among the eMATES, plant MATEs are structurally most resolved. The *Camelina sativa* (CsMATE) and *Arabidopsis thaliana* (AtDTX14) are among the resolved plant MATE structures (Miyauchi et al. 2017; Tanaka et al. 2017). CsMATE also has similar TM positioning with pseudo-symmetry similar to bacterial proteins. There are three amino acids connecting every TM except for TM6 and TM7, which are connected by 12 amino acids located at the cytoplasmic end of the N-lobe. The N-lobe side is narrower than C-lobe resulting in an asymmetric pocket size. This asymmetry is mainly due to the presence of the bulky amino acid residues, Y82, F216, and Y219 on TM2 and TM6 as compared to TM8 and TM12 (Tanaka et al. 2017). The internal pocket of CsMATE is completely negatively charged, and the residues are ordered similar to NorM MATEs. The pockets are large to move out plant substrates like berberine. Even the positioning of certain residues is similar to NorM-Vc and CsMATE shares around 83% similarity with AtDTX1 (responsible for cadmium transport), it is likely that CsMATE could also be involved in metal ion transport (Tanaka et al. 2017).

Another plant MATE transporter, AtDTX14, was structurally resolved at 2.6 Å (Miyauchi et al. 2017). This protein showed 32% sequence identity with hMATE1 and 50% amino acid sequence similarity with AtDTX1, and it also exhibits similar norfloxacin export activity. The AtDTX14 also had 12 TMs and a V-shaped structure with TM1–6 forming the N-lobe and TM7–12 forming the C-lobe similar to bacterial NorM subfamily proteins (Miyauchi et al. 2017). In contrast to bacterial MATEs, the cavities in both lobes are absent in AtDTX14. Also, a narrower extracellular entrance of AtDTX14 than bacterial MATEs represented a partially outward-open structure (Miyauchi et al. 2017). Distinctly, AtDTX14 differs from other MATEs at the TM7 in C-lobe, which assumes a 50° bent conformation around Cys263 forming a hinge region. Also, an intense hydrogen bonding network is formed around this hinge region with Glu265 and Asp383 protonation likely to be the cause for this hydrogen bonding (Kusakizako et al. 2020).

Among the characterized eMATE structures, a distinct conformational change lies in the straight and bent form of TM7 of CsMATE and AtDTX14, respectively. When compared with bacterial MATEs, eMATEs do not harbor the conserved aspartate residue at TM1 for transport activity. Hydrogen bonding network in the N-lobes of DinF subfamilies are unconserved in eMATEs. The C-lobe of eMATE and NorM shows conserved acidic residues but is absent in DinF subfamily depicting the significance of C-lobe of eMATE and NorM in substrate transport activity (Kusakizako et al. 2020). The readers are redirected to (Kusakizako et al. 2020) for a detailed structural comparison among the different MATE subfamilies.

8.3 Function of MATE Transporters in Metal Toxicity Tolerance

MATEs are reported to perform several crucial roles with respect to metal toxicity tolerance. They include xenobiotic detoxification, iron homeostasis, and transport of organic acid in response to aluminum ions present in the root rhizosphere. The following section details about the role of MATEs specifically in metal toxicity tolerance.

8.3.1 Role of MATE Transporters in Xenobiotic Toxicity Tolerance

The first identified Arabidopsis MATE gene *AtDTX1* was functionally characterized by complementation in KAM3 bacterial mutant (Li et al. 2002). The survival of the complemented KAM3 mutants in norfloxacin media identified the toxic efflux carrier function of this transporter. AtDTX1 contains 12 putative transmembrane domains, localized to the plasma membrane, and requires proton motif force to efflux the toxic compounds from the cytoplasm. It is able to detoxify a wide range of toxic substances such as the Cd^{2+}, EtBr, allelo-chemicals, and plant alkaloids such as berberine and palmatine (Li et al. 2002). A synthetic pollutant 2,4,6-trinitro toluene (TNT), which is highly phytotoxic, can induce the expression of *AtDTX1 and AtDTX3* at low concentrations, indicating the involvement of these genes in TNT toxicity tolerance and a probable role in bioremediation (Gandia-herrero et al. 2008). The transcript levels of *AtDTX3* also increase in response to genotoxins such as bleomycin and mitomycin-C after 6 h of treatment, identified using a high-density colony array (HDCA) method of genome-wide transcriptome profiling (Chen et al. 2003). PVP and pyrrolidinone are rhizo-inhibitory substances. *AtDTX19/ALF5*, which is expressed in the root epidermis of Arabidopsis, protects the root from these rhizo-inhibitory substances. The *alf5* mutant is sensitive to these compounds, but in the absence of these, the root growth of the mutant is similar to that of wild type. Expression of *ALF5* in Yeast *(Saccharomyces cerevisiae)* provides resistance against a toxic cation, tetramethylammonium chloride (TMACl), and also against the growth inhibitory concentrations of NaCl and KCl. These indicate the toxic cation extrusion function of AtDTX19/ALF5 in plants (Diener et al. 2001). Atrazine is an herbicide that induces ROS production in plants by blocking the electron transport chain. Exogenous application of the herbicide atrazine induces *ATDTX21*, which might be involved in the detoxification of glutathione-conjugated xenobiotic and their transport from cytoplasm to vacuole (Ramel et al. 2007). Expression of *AtDTX14* in *E. coli* drug exporter mutant strain (6-KO) provided resistance against the antibiotic norfloxacin indicating the norfloxacin export activity of AtDTX14 (Miyauchi et al. 2017). AtDTX18/LAL5 is strongly expressed in response to *Phytophthora infestans*. It exports hydroxycinnamic acid amides (HCAA) such as p-coumaroylagmatine and feruloyltyramine, which are exported to the apoplast and incorporated into the cell wall in response to the pathogen attack (Dobritzsch et al. 2016). Arsenate (As [V]) induces the expression of two mate genes in rice *OsMATE1* and *OsMATE2*. Transgenic Arabidopsis lines overexpressing *OsMATE1* and *OsMATE2* are hypersensitive to arsenate and lack the ability of ROS scavenging due to reduced flavonoid content (Tiwari et al. 2014). In blueberry, the two MATE family transporters, VcMATE1 and VcMATE4, have been reported to be involved in xenobiotic efflux or in the export of toxic cations (Chen et al. 2015). In potato (*Solanum tuberosum* L.), the expression analysis of the MATE genes in response to the bivalent heavy metals was performed, and eight MATE genes were identified to be heavy metal responsive. *StMATE33, StMATE5, StMATE61, StMATE18, StMATE60*, and *StMATE40* were upregulated in the leaves in response to cadmium (Cd^{2+}), copper (Cu^{2+}), lead (Pb^{2+}), nickel (Ni^{2+}), and zinc (Zn^{2+}), while the expression of *StMATE45* and *StMATE59* was lowered in response to these heavy metals (Huang et al. 2021). In pigeon pea, *CcMATE4*, *CcMATE34*, and *CcMATE45* are induced under metal stress (Dong et al. 2019). *CcMATE45* is a homolog of *AtALF5*, and *CcMATE34* is a homolog of *AtDTX1* and is therefore predicted to perform a similar function in pigeon pea, that is, efflux harmful substances or alkaloids (Dong et al. 2019). The cation efflux activity of the MATE transporters plays important role in toxicity, antibiotic, and heavy metal stress tolerance in different species as a multitasker and versatile transporter family.

8.3.2 Effect of Aluminum on Plants

Plants grown under acidic soil conditions often experience aluminum (Al^{3+}) toxicity. Under low soil pH conditions, that is, below pH 5, the trivalent cationic form of aluminum is prevalent as the aluminum salts $AlCl_3$, $Al(OH)_3$, dissociate into free Al^{3+} ions. A number of studies carried out have mentioned about the effect of aluminum toxicity in tissue and organelle-specific manner. Aluminum toxicity primarily affects the root system architecture of the plants by inhibiting the primary root elongation (Sun et al. 2010; Kopittke 2016). Al toxicity can induce alteration in metabolic processes by generating higher reactive oxygen species (ROS). It also can cause cell wall damage by binding to the pectin matrix, hemicellulose, and xyloglucans. Lipid peroxidation is induced by Al toxicity causing plasma membrane and other subcellular membrane damage. Aluminum can replace calcium in its cellular signaling network and can cause a disturbance of calcium homeostasis. Aluminum can cause nutrient imbalance, induce ROS accumulation, and can also damage DNA, resulting in an overall growth reduction and toxic responses (Bojórquez-Quintal et al. 2017; Kar et al. 2021).

There are two major strategies used by plants to mitigate the Al stress, that is, (1) Al exclusion and (2) Al tolerance. In the Al exclusion strategy, the plants secrete organic anions (e.g., citrate, malate, oxalate, etc.) into the rhizosphere to externally chelate the Al^{3+} before it enters the plant. While in Al tolerance strategy, Al enters the plant and is detoxified and sequestered inside the plant (Kochian et al. 2015). MATE transporter-mediated Al toxicity by effluxing citrate into the rhizosphere is categorized under the Al exclusion strategy (Fig. 8.1b). Substrate specificity of MATE transporters has been characterized by expressing them in *Xenopus* oocytes. A recent study showed the presence of a conserved amino acid, aspartate, within a motif, citrate exuding motif (CEM) in the citrate exuding MATEs from various plants (Upadhyay et al. 2019). This motif is absent in non-citrate exuding MATEs shown to function in aluminum toxicity tolerance (Upadhyay et al. 2019). The following section describes the role of MATE transporters involved in Al toxicity tolerance.

8.3.2.1 MATE Transporters Exude Citrate in Response to Aluminum Toxicity

The first reported Al-responsive citrate exuding plant MATE gene *SbMATE* was identified in sorghum. It underlies an Al-tolerance locus Alt_{SB} and encodes a plasma membrane-localized citrate transporter that belongs to the MATE family. *SbMATE* is expressed in root apices of Al-tolerant lines and is responsible for the Al-activated root citrate exudation. Overexpression of *SbMATE* in transgenic *Arabidopsis* conferred increased Al tolerance and root citrate exudation (Magalhaes et al. 2007). Following this discovery, there are many MATE transporters identified, which are homologous to *SbMATE* and provide Al stress tolerance.

In *Arabidopsis thaliana*, Al stress induces a lower level of root citrate exudation. AtMATE and AtDTX30 have been reported to provide aluminum toxicity tolerance (Liu et al. 2009; Upadhyay et al. 2020). *AtMATE*, which is a homolog of *SbMATE*, is expressed in the roots upon Al stress induction. It is responsible for citrate exudation into the rhizosphere to protect the roots from Al toxicity (Liu et al. 2009). AtDTX30 is localized to plasma membrane of the root epidermis. It is induced by Al in a concentration-dependent manner and localizes to the DTZ in response to Al toxicity and indirectly modulates the citrate exudation through the regulation of AtMATE and AtSTOP1 (Upadhyay et al. 2020). In Barley, Al resistance is mainly controlled by the secretion of citrate. *HvAACT1* shares 59% identity and 86% similarity with *AtMATE/FRDL* of Arabidopsis (Furukawa et al. 2007). Its protein product localizes to the plasma membrane of root tip epidermis. It is highly expressed in the roots, and its expression positively correlates with Al-induced citrate secretion. Heterologous expression of *HvAACT1* in the *Xenopus* oocytes showed citrate efflux activity. Tobacco plants overexpressing *HvAACT1* showed higher citrate efflux in response to Al (Furukawa et al. 2007). In *Brachypodium distachyon,* the expression of *BdMATE1* and *BdMATE2* was induced by Al, which can be appropriate candidates for Al-induced citrate efflux in Brachypodium (Contreras et al. 2014). Transgenic *Setaria viridis* plants overexpressing *BdMATE* show sustained root growth and higher root citrate exudation from the root under Al stress. They exclude Al from the root apex confirmed by hematoxylin assay (Ribeiro et al. 2017). An EST in wheat identified as *TaMATE1* is highly expressed in genotypes with high citrate efflux. *TaMATE1* is homologous to *HvAACT1* and identified as a candidate controlling the citrate efflux phenotype in wheat (Ryan et al. 2009). In maize, the MATE transporter family member *ZmMATE1* localize to the Al tolerance QTL. It is a functional homolog of the Al tolerance genes characterized in sorghum, barley, and Arabidopsis. It encodes a plasma membrane-localized citrate efflux transporter. The expression of *ZmMATE1* is mostly concentrated in root tissues and is upregulated by Al. Heterologous expression of *ZmMATE1* in transgenic Arabidopsis confers significantly increased Al tolerance (Maron et al. 2010). Recently, another MATE transporter in Maize, ZmMATE6, has been found to be induced by Al treatment, and transgenic Arabidopsis lines expressing *ZmMATE6* were found to display a greater Al-activated citrate secretion from the roots and were significantly resistant to Al toxicity (Du et al. 2021). Two MATE transporters in Buckwheat, *FeMATE1* and *FeMATE2*, are induced in response to Al stress. FeMATE1 is localized to the plasma membrane and is involved in Al-activated citrate efflux for Al detoxification in the roots, whereas FeMATE2 is localized to trans-Golgi and is involved in transporting citrate into the Golgi system for the internal detoxification of Al^{3+} in the roots and leaves of buckwheat (Lei et al. 2017). In rice, two MATE transporters, OsFRDL4 and OsFRDL2, act as a citrate transporters induced by Al and localize to the plasma membrane of rice root cells (Yokosho et al. 2011, 2016). In rye, *ScFRDL2* is induced by Al in the roots, and the expression pattern is consistent with the citrate secretion pattern. ScFRDL2 is involved in Al-activated citrate secretion (Yokosho et al. 2010). In soybean (*Glycine max*), GmMATE75, GmMATE79, and GmMATE87 are plasma-membrane-localized citrate transporters and have capabilities to increase Al-induced citrate efflux (Zhou et al. 2019). GmMATE13 and GmMATE47 also mediate citrate exudation as characterized in *Xenopus* oocytes and enhanced Al resistance as shown in transgenic Arabidopsis (Wang 2019). In *Glycine soja,*

GsMATE is expressed mainly in roots and is specifically upregulated by Al with higher expression level in the root tips. The GsMATE protein is also localized to the plasma membrane and involved in Al-induced citrate secretion (Ma et al. 2018). In cotton, *GhMATE1* is found to be induced by Al, and the *ghmate1* RNA-i mutants showed a negligible amount of citrate exudation and highly reduced primary and secondary growth in response to Al stress. But in the transgenic Arabidopsis plants, overexpressing *GhMATE1* showed higher citrate efflux and enhanced root growth compared to the wild type in Al stress condition. This indicates that GhMATE1 can be involved in Al stress-induced citrate efflux and Al tolerance (Kundu and Ganesan 2020). In rice bean (*Vigna umbellata*), the Al-induced citrate efflux from the root apex is biphasic. The early phase of low citrate secretion is mediated by VuMATE2, which is a proton-coupled citrate permeable channel and regulates citrate efflux after 1.5–3 h of Al treatment (Liu et al. 2018). But the later phase of higher citrate secretion is mediated by VuMATE1. There is a 6 h lag between Al treatment and the increase in expression of *VuMATE1*, while the citrate efflux increases 9 h after Al treatment indicating that the function of VuMATE1 is subjected to both transcriptional and post-transcriptional regulation (Liu et al. 2013). In *Eucalyptus camaldulensis*, *EcMATE1, 2, 3,* and *4* are induced by Al. EcMATE1 is localized to plasma membrane and abundantly expressed in roots. Al-responsive citrate excretion through EcMATE1 involves protein phosphorylation/dephosphorylation processes. *EcMATE1* and *EcMATE3* have been characterized in tobacco hairy roots to enhance the Al-responsive citrate excretion (Sawaki et al. 2013). In Cabbage, *BoMATE* is highly expressed in the roots and localizes to the plasma membrane. It is induced by Al and characterized to mediate Al-induced citrate transport in *Xenopus* oocyte. Overexpression of *BoMATE* in transgenic Arabidopsis enhances citrate exudation and provides tolerance to Al^{3+} toxicity (Wu et al. 2014). In *Medicago truncatula*, *MtMATE66* is a plasma membrane citric acid transporter primarily expressed in root epidermal cells. It takes part in Al toxicity tolerance as the *mtmate66* mutants showed less root growth compared to the wild type under Al^{3+} stress, whereas overexpression of *MtMATE66* resulted in more tolerant hairy root phenotype. Another MATE transporter, *MtMATE69*, is localized to the plasma membrane, and upon exposure to Al, it can induce enhanced citrate efflux (Wang et al. 2017). In *Populus trichocarpa*, *PtrMATE1* expression was specifically high in the central cylinder of mature roots and was significantly increased under Al treatment. Overexpression of *PtrMATE1* in *Populus* resulted in increased Al tolerance and root citrate exudation (Li et al. 2017). In peanut, *AhFRDL1* is expressed throughout the root tip and contributes to Al tolerance by promoting citrate exudation. The knockdown *ahfrdl1* plants are susceptible to Al, and expression of *AhFRDL1* in *atmate* knockout plants restored their Al tolerance (Qiu et al. 2019). In pigeon pea (*Cajanus cajan*), *CcMATE34, CcMATE45*, and *CcMATE4* are induced under Al stress, but *CcMATE4* is induced only in the roots and shares homology with *MtMATE*. It is predicted to play a role in Al toxicity tolerance similar to *MtMATE* (Dong et al. 2019). Thus, MATE transporters present in the root plasma membrane are aluminum inducible and function to exude citrate to the root rhizosphere, thereby chelating external free Al^{3+} ions resulting in Al toxicity alleviation.

8.3.3 Role of MATE Transporters in Iron Homeostasis

Iron (Fe) is an important micronutrient for plants. It acts as a cofactor of the enzymes in many physio-biochemical processes such as nitrogen fixation, electron transport system photosynthesis, and redox reactions. The chloroplasts, mitochondria, peroxisomes, and the endoplasmic reticulum are rich in iron-sulfur proteins, for example, the photosystems, iron-sulfur complexes, hem groups, and peroxidases with iron as a cofactor. Iron deficiencies can lead to chlorosis causing the loss of chlorophyll and turning the leaves in yellow color. Iron is available as ferric (Fe^{3+}) ions in the rhizosphere. Plants uptake iron by utilizing two strategies: (1) reducing and (2) chelating (Romheld and Marschner 1986). MATE transporters play an important role in Fe uptake by secreting organic acids such as citrate into the rhizosphere, which can chelate the ferric ions. MATE proteins are also expressed in the pericycle where they efflux citrate into the xylem as well as the apoplast surrounding the stele for solubilization of Fe by chelating them with the organic acids. The Fe-citrate complex is then loaded into the xylem sap and distributed inside the plant.

In Arabidopsis, the *atfrd3* knockout mutants are deficient in Fe uptake, and the leaves are chlorotic (Rogers and Guerinot 2002; Green and Rogers 2004). They accumulate high level of ferric ions (Fe^{3+}) in their root vasculature, and the expression of the Fe deficiency-induced genes is very high. *AtFRD3* is identified as a plasma membrane-localized transporter found in the pericycle that effluxes citrate into root vasculature, which helps in the formation of Fe-citrate complexes and helps in Fe transport to the shoot (Durrett et al. 2007). It also helps in Fe solubilization in the apoplastic space separating two tissues without symplastic connections (Roschzttardtz et al. 2011). Another MATE transporter BCD1 also plays role in Fe homeostasis. It is induced by excessive iron but repressed by iron deficiency. Iron levels are elevated in the knockout mutant *bcd1–1*, but the overexpressor plants show chlorosis and other iron deficiency symptoms. It localizes to the Golgi complex and reallocates the excessive iron released from stress-induced cellular damage (Park et al. 2011). ELS1is a homolog of BCD1 and is induced by iron (Wang et al. 2016). The overexpressor plants, *els1-D*, show leaf chlorosis, reduced iron content, and accelerated senescence. This dark-induced senescence phenotype can be recovered by supplementing *els1-D* with 400 μM $FeSO_4$. These findings indicate that ELS1 plays a role in iron homeostasis and leaf senescence (Wang et al. 2016). *AtDTX25* is an ascorbate transporter localized to the vacuolar membrane. This is highly expressed in the early phase of germination and required for the remobilization of Fe during seedling development. The *atdtx25–1* mutants are highly sensitive to Fe deficiency and accumulate Fe in their vacuoles (Thi et al. 2021). In rice, *OsPEZ1* is localized to the stele of the roots and transports protocatechuic acids (PCA), which help in solubilizing the apoplastic iron. In *ospez1* mutants, the apoplastic Fe precipitates, while in the overexpressor lines, accumulation of more iron is observed (Ishimaru et al. 2011). *OsPEZ2* is expressed in root tips and transports PCA and CA (caffeic acid) to solubilize the apoplasmic Fe in rice. The *ospez2* mutants contain reduced PCA and CA in xylem sap and with low xylem Fe concentration (Bashir et al. 2011). *OsFRDL1* is a citrate transporter localized to the root pericycle. The knockout mutants *osfrdl1* show leaf chlorosis, lower citrate, and ferric concentration in xylem sap and precipitation of Fe in stele (Yokosho et al. 2009). *OsFRDL1* localizes to plasma membrane and is involved in long-distance iron transport (Inoue et al. 2004). LjMATE1 is a nodule-specific iron transporter in the model legume *Lotus japonica* that effluxes citrate and assists in iron translocation from the roots to the nodule (Takanashi et al. 2013). In *Medicago truncatula*, *MtMATE66* is a citrate transporter localized to the plasma membrane in root epidermis. *mtmate66* mutants display chlorosis in iron-deficient conditions. *MtMATE55* is involved in iron homeostasis as both the overexpressor and mutants for this gene show Fe accumulation and mutants display chlorosis. *MtMATE69* is another citric acid effluxing MATE transporter, which is induced under iron deficiency conditions. Overexpressing *MtMATE69* resulted in altered iron homeostasis (Wang et al. 2017). *MtMATE67* is a citrate transporter localized to the nodule and the plasma membrane of the symbiosome surrounding the bacteroids. It effluxes citrate in an iron-activated manner from nodule cells, which is necessary to ensure ferric solubility and mobility in to the apoplasm and uptake into nodule cells (Kryvoruchko et al. 2018). In soybean, two MATE genes, *GmFRD3a* and *GmFRD3b*, are homologous to the *AtFRD3*. Both are induced under iron deficiency in an iron-efficient cultivar. Higher expression levels of *GmFRD3b* correspond to higher increased xylem citrate levels in an iron-efficient cultivar, indicating the involvement of both these genes in citrate-mediated iron homeostasis. In rye, *ScFRDL1* is induced under iron deficiency. This citrate transporter is localized to the central cylinder and endodermis. It mediates Fe translocation from roots to shoot by effluxing citrate into the xylem (Yokosho et al. 2010). In peanuts, the expression of *AhFRDL* is localized to root stele and induced by iron deficiency. Transgenic *atfrd3* mutants overexpressing *AhFRDL1* recovered Fe deficiency. Downregulation of *AhFRDL1* knockdown in roots resulted in reduced citrate in xylem and low active Fe in young leaves, indicating that it plays a role in iron uptake and homeostasis (Qiu et al. 2019). Apart from aluminum, MATEs also function under iron-deficient conditions and help in efficient iron translocation to maintain iron homeostasis within a plant.

8.4 Other Functions of MATE Transporters in Plants

MATEs exhibit other versatile functions apart from its role in metal toxicity. Most of the MATEs perform transport of secondary metabolites in aerial tissues of the plant. They play critical roles in multiple developmental processes within a plant. The following section entails the functions of MATE transporters in secondary metabolite transport, biotic stress, and developmental roles.

8.4.1 Secondary Metabolite Transport

Plants have been studied to produce different types of secondary metabolites essential for their day-to-day life purposes, which are further classified into alkaloids, terpenoids, and phenols (Yazaki 2005). The biosynthesis and transort of secondary metabolites regulate many physiological and stress responses in plants, for example, protection against pathogens, prevention against damage caused by exposure to UV and also important for pollination in plants due to their role in coloration of different organs (Chen et al. 2015). However, the overproduction of secondary metabolites can also turn out to be detrimental to plants as they are often highly reactive and have the potential to denature proteins. Thus, through the course of evolution, plants have devised certain means to contain these detrimental effects. One of the common ways employed by them is to store these metabolites inside compartments and vesicles after their primary processing (oxidation or conjugation with other compounds) or to directly dislodge them from the cell (Freeman and Beattie 2008). Studies reveal that the spatiotemporal distribution and formation of secondary metabolites is often observed in some of the differentiated tissues in plants, where, for example, the production of a secondary metabolite occurs in root tissues in *Nicotiana tabacum*, but their site of action is in the aerial parts of plants where they act as an anti-herbivore chemical (Takase et al. 2008). The MATE transporters stand out to facilitate the localization of secondary metabolites across different membranes. Characterization of such MATE transporters involved in these types of localization has also been reported thoroughly using many experimental approaches. Heterologous expression studies in yeast have suggested a possible mechanism for the transport of H^+/nicotine via Nt*MATE1* present in the vacuolar membrane (Takase et al. 2008). Exogenous treatment of nicotine or jasmonate in the Nt*MATE1* overexpressing cells resulted in lowering of cellular pH levels leading to rapid acidification inside the cells proving the vacuolar influx of nicotine using outward proton gradient via Nt*MATE1* to be true (Takase et al. 2008). Other class of MATE transporters, namely, AtDTX46, is studied to be involved in the transportation of phenolic compounds in *Arabidopsis thaliana* (Parinthawong et al. 2015). MATE transporters mainly Mt*MATE1* and Mt*MATE2* (in tonoplast membrane) in *Medicago truncatula* are involved in the transport of epicatechin 3′-O-glucoside(Zhao and Dixon 2009) and malonylated flavonoids (Zhao et al. 2011), respectively. A class of MATE transporter proteins, namely, AtDTX41 or transparent testa 12 (TT12), is mainly expressed in the ovules, and developing seeds play a major role as flavonoid/H^+-antiporter administering vacuolar sequestration of glycosylated flavan-3-ol monomers in the seed-coat endothelium (Debeaujon et al. 2007; Marinova et al. 2007). A homologue of AtTT12, MTP77, has been studied to transport proanthocyanin in tomato leaves (Mathews et al. 2003). Interestingly, VvMATE in grapevine (*Vitis vinifera*) shows transport of proanthocyanin into the vacuole, whereas MdMATE1 and MdMATE2 serve as flavonoid/H^+ transporter in proanthocyanidin accumulating cells of apple (*Malus domestica*) as reported by (Zhao and Dixon 2009; Gomez et al. 2009, 2011; Frank et al. 2011; Pérez-Díaz et al. 2014; Chen et al. 2018). A recent study characterized the role of 11 *CaMATEs*

(*Capsicum annuum*) and 21 *StMATEs* (*Solanum tuberosum)* in transport of secondary metabolites such as proanthocyanin, flavonoids, and nicotine (Chen et al. 2020). Two of the MATE genes in apple (*MdMATE1 and MdMATE2*) known to be close homologs of TT12 in *Arabidopsis* play an important role in proanthocyanidin accumulation in cells and flavonoid transport (Zhang et al. 2021). The ability of MATE transporters to transport secondary metabolites across various tissues can be used as potential targets to modulate a plant response under abiotic stress.

8.4.2 Developmental Roles

MATEs have been classified as a class of transporter proteins involved in efflux and influx of various molecules across different membranes in plants. However, with the advancement in research, their potential in regulating the growth and development of plants has got significant attention. Two of the members of MATE family transporters, AtDTX33 and AtDTX35, have been reported to play an important role in regulating the turgor pressure in the guard cells of stomata allowing cell expansion and also regulating gaseous exchange. AtDTX33 and AtDTX35 mainly act as chloride channels localized in the tonoplast of the root hairs and guard cells (Zhang et al. 2017). At the early root developmental phase, the double mutants were shown to have reduced root hair elongation compared to the wild type. Their mutant study also revealed that the double mutant of *dtx33dtx35* is impaired in stomatal opening and is less likely to lose water in the adult stages, hence is drought-tolerant compared to wild type and either of the single mutants (Zhang et al. 2017). Another mate transporter, *DTX31*, localized to the plasma membrane might play some important role in root hair elongation. Root hair elongation (RHE) element is also found in its promoter region. *dtx31* mutants show a root hair-defective phenotype compared to the wild type (Won et al. 2009). *DTX34* might have a crucial role to play in male gametophyte development as it was found to be expressed in the pollen grain, indicating its possible role in tip growth in plants (Bock 2006). A loss-of-function mutation in a MATE gene *Big embryo1* (*bige1*) in maize results in accelerated root and leaf initiation as well as enlargement of embryo scutellum. BIGE1 is found in the trans-Golgi, suggesting that it may have a function in the secretion of a signaling molecule (Suzuki et al. 2015). Interaction autophagy-related protein (ATG8) present at the endosomal membrane with AtDTX51 promotes senescence in *Arabidopsis thaliana* in natural and carbon-deprived conditions. This ATG8-ABS3 interaction is hypothesized to be conserved across monocots and dicots (Jia et al. 2019). White lupin (*Lupinus albus*) has been widely accepted as a model system for studying adaptation of plants due to its extreme tolerance of low phosphate stress. To tolerate low phosphate stress, it adapts certain modification in root architecture resulting in short and densely clustered roots (Uhde-Stone et al. 2005). *LaMATE* in lupin is seen to be playing an important role in this modification under phosphate deficiency (Uhde-Stone et al. 2005). MATE gene in *Capsicum annum*, namely, *CaMATE*, might play an important role in the reproductive development of the plant as they have been observed to be expressed in the reproductive tissues including the placenta, pericarp, and the seeds (Chen et al. 2020). Flowers were observed to have high expression of *CaMATE18* and *CaMATE39*, while the expression of *CaMATE30* was found to be upregulated during the G11 stage of pericarp development, suggesting the specific and narrow function roles of *CaMATEs* in the reproductive tissues (Chen et al. 2020). The peak expression of *CaMATE37* was observed at the F7 stage in flowers, while its expression in other tissues was highly downregulated suggesting their role in the specification of floral organs at specific stages (Chen et al. 2020). Among the 55 *MATE* genes identified in *Malus domestica*, expression of various *MATEs* (*MDMATE23, MdMATE49, and MdMATE58*) was found to be increased during the formation and maturation stages of the fruit, indicating their possible role in development (Zhang et al. 2021). To conclude, MATE/DTX transporters have a variety of roles in plant development, from cell/organ growth initiation to later processes like senescence, and are highly conserved across a wide range of plant species.

8.4.3 Biotic Stress

Plant species growing in the wild are often exposed to numerous sorts of stress conditions. Among these, the existence of diverse microorganisms in the soil biosphere, namely, bacteria, fungi, and viruses, often turns out to be a major threat and have detrimental effects on the survival of plants. Thus, plants have co-evolved along with these microorganisms developing various innate immune defense responses, which induce systemic and local resistance (Upadhyay et al. 2019). Recent studies reveal the role of MATE transporters in response to such biotic stresses. The production and accumulation of salicylic acid (SA) plays an important role in the innate immune response that provides resistance to infected and neighboring cells via transcriptional regulation (Serrano et al. 2013). Characterization of AtDTX47 or EDS5, a homologue of MATE, revealed the role of these transporters in exporting salicylic acid (SA) from chloroplast to cytoplasm (site of action), thereby facilitating the defense response (Nawrath et al. 2002). It is also hypothesized that the induction SA biosynthesis might require the expression of EDS5. As reported by Nawrath et al. (2002), EDS5 also acts as a major regulator in age-related resistance (ARR) pathway. A virulent strain of *Pseudomonas syringae*, pathogenic to plant species, increases the expression level of EDS5, thereby conferring resistance to the plant body (Nawrath et al. 2002). It was also observed that the activity of one of the MATE transporters AtDTX21 is coregulated with type III effectors in response to *P. syringae* infection activating the ABA signaling pathway (De Torres-Zabala et al. 2007). A MATE protein AtDTX51/ADS1, which was earlier reported to be a negative regulator of plant defense resistance, was later identified to play a substantial role in SA signaling/synthesis pathway (Sun et al. 2011). The efflux of hydroxycinnamic acid amide (HCAA) by DTX18 is also proven to be a part of plant defense response preventing late blight in potato (*Solanum tuberosum*) caused by *Phytophthora infestans* (Dobritzsch et al. 2016). Reportedly 66 MATE (*MdMATE*) transporters were identified in apple (*Malus domestica*) genome, and the study of cis-regulatory elements in their promoter suggested their role in various stress responses (Zhang et al. 2021). *MdMATE* genes are implicated in biotic stress response, as evidenced by variations in gene expression levels in response to various pathogen infections (Zhang et al. 2021). It is not only metal ions but also MATEs can come to aid in response to biotic stress. Thus, the ability of MATE transporters to perform a variety of functions is noteworthy to take them into account as potential targets for plant adaptation and survivability to adverse conditions.

8.5 Conclusion and Future Perspectives

The function of transporters in plants is indispensable for nutrient transport, cell communication and signaling, growth, defense, and stress responses (Tang et al. 2020). Current advances in genomics, sequencing, and gene editing technologies have provided with approaches to modulate transporter activity for a particular physiological response. The intricate cellular details of a transporter including phosphorylation, sumoylation, and ligand binding structural changes have brought about changes in approaches to fight stress (Tang et al. 2020). The MATE transporters perform a multitude of functions in a plant such as hormonal transport, plant secondary metabolite transport, and regulation of plant development. These transporters provide insights of a plant defense mechanism to biotic and abiotic stress (Upadhyay et al. 2019). MATEs are highly recognized to play a role in aluminum toxicity where they exude citrate to chelate Al^{3+} ions externally in the root rhizosphere. Several studies identify the role of MATE transporter in iron transport. Thus, MATE transporters play crucial roles in uptake of important metals as well as exclusion of hazardous metals at toxic concentrations. Their role in plant development has only been characterized in some model plants and opens a wide field for their application in food crops and economically important plants. Transporter engineering provides a good opportunity to engineer the MATE transporters in cereals, oilseed plants, and cash crops. These can be engineered to regulate the transport of beneficial as well as hazardous metals together with the possibility of enhancing secondary metabolites. The multitasking abilities of MATE transporters in a plant system provide opportunities to study and explore them further for crop improvement in the future.

Acknowledgments D.K acknowledges CSIR, Government of India, for his fellowship. A.A.P acknowledges PMRF fellowship, Ministry of Education, Government of India. S.D. would like to thank MHRD-STARS grant (STARS/APR2019/BS/245/FS), Government of India for funding. Authors acknowledge PCDB lab members for suggestions and IISER Bhopal for infrastructure and facilities. The authors sincerely apologize omission of any reference, which might have been excluded owing to lack of space. The authors declare no conflict of interest.

References

Bashir K, Ishimaru Y, Shimo H et al (2011) Rice phenolics efflux transporter 2 (PEZ2) plays an important role in solubilizing apoplasmic iron. Soil Sci Plant Nutr 57:803–812. https://doi.org/10.1080/00380768.2011.637305

Bock KW (2006) Integrating membrane transport with male gametophyte development and function through transcriptomics. Plant Physiol 140:1151–1168. https://doi.org/10.1104/pp.105.074708

Bojórquez-Quintal E, Escalante-Magaña C, Echevarría-Machado I, Martínez-Estévez M (2017) Aluminum, a friend or foe of higher plants in acid soils. Front Plant Sci 8:1767

Brown MH, Paulsen IT, Skurray RA (1999) The multidrug efflux protein NorM is a prototype of a new family of transporters. Mol Microbiol 31:394–395

Chen IP, Haehnel U, Altschmied L et al (2003) The transcriptional response of Arabidopsis to genotoxic stress—a high-density colony array study (HDCA). Plant J 35:771–786. https://doi.org/10.1046/j.1365-313X.2003.01847.x

Chen L, Liu Y, Liu H et al (2015) Identification and expression analysis of MATE genes involved in flavonoid transport in blueberry plants. PLoS One 10:e0118578. https://doi.org/10.1371/journal.pone.0118578

Chen SY, Tang YM, Hu YY et al (2018) FaTT12-1, a multidrug and toxin extrusion (MATE) member involved in proanthocyanidin transport in strawberry fruits. Sci Hortic 231:158–165. https://doi.org/10.1016/j.scienta.2017.12.032

Chen Q, Wang L, Liu D et al (2020) Identification and expression of the multidrug and toxic compound extrusion (mate) gene family in capsicum annuum and *Solanum tuberosum*. Plan Theory 9:1–23. https://doi.org/10.3390/plants9111448

Contreras R, Figueiras AM, Gallego FJ, Benito C (2014) Brachypodium distachyon: a model species for aluminium tolerance in Poaceae. Funct Plant Biol 41:1270. https://doi.org/10.1071/FP13362

De Torres-Zabala M, Truman W, Bennett MH et al (2007) Pseudomonas syringae pv. Tomato hijacks the *Arabidopsis* abscisic acid signalling pathway to cause disease. EMBO J 26:1434–1443. https://doi.org/10.1038/sj.emboj.7601575

Debeaujon I, Peeters AJM, Leon-Kloosterziel KM, Koornneef M (2007) The TRANSPARENT TESTA12 gene of *Arabidopsis* encodes a multidrug secondary transporter-like protein required for flavonoid sequestration in vacuoles of the seed coat endothelium. Plant Cell 19:2023–2038. https://doi.org/10.2307/3871345

Diener AC, Gaxiola RA, Fink GR (2001) *Arabidopsis* ALF5, a multidrug efflux transporter gene family member, confers resistance to toxins. Plant Cell 13:1625. https://doi.org/10.2307/3871390

Dobritzsch M, Lübken T, Eschen-Lippold L et al (2016) MATE transporter-dependent export of hydroxycinnamic acid amides. Plant Cell 28:583–596. https://doi.org/10.1105/tpc.15.00706

Dong B, Niu L, Meng D et al (2019) Genome-wide analysis of MATE transporters and response to metal stress in Cajanus cajan. J Plant Interact 14(1):265–275. https://doi.org/10.1080/17429145.2019.1620884

Du D, Wang-Kan X, Neuberger A et al (2018) Multidrug efflux pumps: structure, function and regulation. Nat Rev Microbiol 16:523–539. https://doi.org/10.1038/s41579-018-0048-6

Du H, Ryan PR, Liu C et al (2021) ZmMATE6 from maize encodes a citrate transporter that enhances aluminum tolerance in transgenic *Arabidopsis thaliana*. Plant Sci 311:1–21. https://doi.org/10.1016/j.plantsci.2021.111016

Durrett TP, Gassmann W, Rogers EE (2007) The FRD3-mediated efflux of citrate into the root vasculature is necessary for efficient iron translocation. Plant Physiol 144:197–205. https://doi.org/10.1104/pp.107.097162

Frank S, Keck M, Sagasser M et al (2011) Two differentially expressed MATE factor genes from apple complement the *Arabidopsis* transparent testa12 mutant. Plant Biol 13:42–50. https://doi.org/10.1111/j.1438-8677.2010.00350.x

Freeman BC, Beattie GA (2008) An overview of plant defenses against pathogens and herbivores. Plant Health Instr 94. https://doi.org/10.1094/phi-i-2008-0226-01

Furukawa J, Yamaji N, Wang H et al (2007) An aluminum-activated citrate transporter in barley. Plant Cell Physiol 48:1081–1091. https://doi.org/10.1093/pcp/pcm091

Gandia-herrero F, Lorenz A, Larson T et al (2008) Detoxification of the explosive 2,4,6-trinitrotoluene in *Arabidopsis*: discovery of bifunctional O-and C-glucosyltransferases. Plant J 56:963–974. https://doi.org/10.1111/j.1365-313X.2008.03653.x

Gomez C, Terrier N, Torregrosa L et al (2009) Grapevine MATE-type proteins act as vacuolar H+-dependent acylated anthocyanin transporters. Plant Physiol 150:402–415. https://doi.org/10.1104/pp.109.135624

Gomez C, Conejero G, Torregrosa L et al (2011) In vivo grapevine anthocyanin transport involves vesicle-mediated trafficking and the contribution of anthoMATE transporters and GST. Plant J 67:960–970. https://doi.org/10.1111/j.1365-313X.2011.04648.x

Green LS, Rogers EE (2004) FRD3 controls iron localization in *Arabidopsis*. Plant Physiol 136: 2523–2531. https://doi.org/10.1104/pp.104.045633

He X, Szewczyk P, Karyakin A et al (2010) Structure of a cation-bound multi drug and toxic compound extrusion transporter. Nature 467:991–994. https://doi.org/10.1038/nature09408

Huang Y, He G, Tian W et al (2021) Genome-wide identification of MATE gene family in potato (Solanum tuberosum L.) and expression analysis in heavy metal stress. Front Genet 12:1–14. https://doi.org/10.3389/fgene.2021.650500

Hvorup RN, Winnen B, Chang AB et al (2003) The multidrug/oligosaccharidyl-lipid/polysaccharide (MOP) exporter superfamily. Eur J Biochem 270(5):799–813

Inoue H, Mizuno D, Takahashi M et al (2004) A rice FRD3-like (OsFRDL1) gene is expressed in the cells involved in long-distance transport a Rice FRD3-like (OsFRDL1) gene is expressed in the cells. Soil Sci Plant Nutr 50:1133–1140. https://doi.org/10.1080/00380768.2004.10408586

Ishimaru Y, Kakei Y, Shimo H et al (2011) A rice phenolic efflux transporter is essential for solubilizing precipitated apoplasmic iron in the plant stele. J Biol Chem 286:24649–24655. https://doi.org/10.1074/jbc.M111.221168

Jia M, Liu X, Xue H et al (2019) Noncanonical ATG8-ABS3 interaction controls senescence in plants. Nat Plants 5:212–224. https://doi.org/10.1038/s41477-018-0348-x

Kar D, Pradhan AA, Datta S (2021) The role of solute transporters in aluminum toxicity and tolerance. Physiol Plant 171:638–652. https://doi.org/10.1111/ppl.13214

Kochian LV, Piñeros MA, Liu J, Magalhaes JV (2015) Plant adaptation to acid soils: the molecular basis for crop aluminum resistance. Annu Rev Plant Biol 66:571–598. https://doi.org/10.1146/annurev-arplant-043014-114822

Kopittke PM (2016) Role of phytohormones in aluminium rhizotoxicity. Plant Cell Environ 39: 2319–2328. https://doi.org/10.1111/pce.12786

Kryvoruchko IS, Routray P, Sinharoy S et al (2018) An iron-activated citrate transporter, MtMATE67, is required for symbiotic nitrogen fixation. Plant Physiol 176:2315–2329. https://doi.org/10.1104/pp.17.01538

Kundu A, Ganesan M (2020) GhMATE1 expression regulates Aluminum tolerance of cotton and overexpression of GhMATE1 enhances acid soil tolerance of *Arabidopsis* ☆. Curr Plant Biol 24:100160. https://doi.org/10.1016/j.cpb.2020.100160

Kusakizako T, Miyauchi H, Ishitani R, Nureki O (2020) Structural biology of the multidrug and toxic compound extrusion superfamily transporters. Biochim Biophys Acta Biomembr 1862: 183154. https://doi.org/10.1016/j.bbamem.2019.183154

Lei GJ, Yokosho K, Yamaji N, Ma JF (2017) Two MATE transporters with different subcellular localization are involved in Al tolerance in buckwheat. Plant Cell Physiol 58:2179–2189. https://doi.org/10.1093/pcp/pcx152

Li L, He Z, Pandey GK et al (2002) Functional cloning and characterization of a plant efflux carrier for multidrug and heavy metal detoxification. J Biol Chem 277:5360–5368. https://doi.org/10.1074/jbc.M108777200

Li N, Meng H, Xing H et al (2017) Genome-wide analysis of MATE transporters and molecular characterization of aluminum resistance in Populus. J Exp Bot 68:5669–5683. https://doi.org/10.1093/jxb/erx370

Liu J, Magalhaes JV, Shaff J, Kochian LV (2009) Aluminum-activated citrate and malate transporters from the MATE and ALMT families function independently to confer *Arabidopsis* aluminum tolerance. Plant J 57:389–399. https://doi.org/10.1111/j.1365-313X.2008.03696.x

Liu MY, Chen WW, Xu JM et al (2013) The role of VuMATE1 expression in aluminium-inducible citrate secretion in rice bean (Vigna umbellata) roots. J Exp Bot 64:1795–1804. https://doi.org/10.1093/jxb/ert039

Liu MY, Lou HQ, Chen WW et al (2018) Two citrate transporters coordinately regulate citrate secretion from rice bean root tip under aluminum stress. Plant Cell Environ 41:809–822. https://doi.org/10.1111/pce.13150

Lu M, Radchenko M, Symersky J et al (2013) Structural insights into H +-coupled multidrug extrusion by a MATE transporter. Nat Struct Mol Biol 20:1310–1317. https://doi.org/10.1038/nsmb.2687

Ma Q, Yi R, Li L et al (2018) GsMATE encoding a multidrug and toxic compound extrusion transporter enhances aluminum tolerance in *Arabidopsis thaliana*. BMC Plant Biol 18:212. https://doi.org/10.1186/s12870-018-1397-z

Magalhaes JV, Liu J, Guimarães CT et al (2007) A gene in the multidrug and toxic compound extrusion (MATE) family confers aluminum tolerance in sorghum. Nat Genet 39:1156–1161. https://doi.org/10.1038/ng2074

Marinova K, Pourcel L, Weder B et al (2007) The *Arabidopsis* MATE transporter TT12 acts as a Vacuolar flavonoid/H^+-antiporter active in Proanthocyanidin-accumulating cells of the seed coat. Plant Cell 19:2023–2038. https://doi.org/10.1105/tpc.106.046029

Maron LG, Piñeros MA, Guimarães CT et al (2010) Two functionally distinct members of the MATE (multi-drug and toxic compound extrusion) family of transporters potentially underlie two major aluminum tolerance QTLs in maize. Plant J 61:728–740. https://doi.org/10.1111/j.1365-313X.2009.04103.x

Masuda S, Terada T, Yonezawa A et al (2006) Identification and functional characterization of a new human kidney-specific H+/organic cation antiporter, kidney-specific multidrug and toxin extrusion 2. J Am Soc Nephrol 17:2127–2135. https://doi.org/10.1681/ASN.2006030205

Mathews H, Clendennen SK, Caldwell CG et al (2003) Activation tagging in tomato identifies a transcriptional regulator of anthocyanin biosynthesis, modification, and transport. Plant Cell 15:1689–1703. https://doi.org/10.1105/tpc.012963

Miyauchi H, Moriyama S, Kusakizako T et al (2017) Structural basis for xenobiotic extrusion by eukaryotic MATE transporter. Nat Commun 21:1633. https://doi.org/10.1038/s41467-017-01541-0

Morita Y, Kodama K, Shiota S et al (1998) NorM, putative multidrug efflux protein, of Vibrio parahaemolyticus and its homolog in Escherichia coli. Antimicrob Agents Chemother 42:1778–1782

Nawrath C, Heck S, Parinthawong N et al (2002) EDS5, an essential component of salicylic acid-dependent signaling for disease resistance in *Arabidopsis*, is a member of the MATE transporter family. Plant Cell 14:275–286. https://doi.org/10.1105/tpc.010376

Nie L, Grell E, Malviya VN et al (2016) Identification of the high-affinity substrate-binding site of the multidrug and toxic compound extrusion (MATE) family transporter from pseudomonas stutzeri. J Biol Chem 291:15503–15514. https://doi.org/10.1074/jbc.M116.728618

Omote H, Hiasa M, Matsumoto T et al (2006) The MATE proteins as fundamental transporters of metabolic and xenobiotic organic cations. Trends Pharmacol Sci 27:587–593. https://doi.org/10.1016/j.tips.2006.09.001

Otsuka M, Matsumoto T, Morimoto R et al (2005a) A human transporter protein that mediates the final excretion step for toxic organic cations. Proc Natl Acad Sci U S A 102:17923–17928. https://doi.org/10.1073/pnas.0506483102

Otsuka M, Yasuda M, Morita Y et al (2005b) Identification of essential amino acid residues of the NorM Na +/multidrug antiporter in Vibrio parahaemolyticus. J Bacteriol 187:1552–1558. https://doi.org/10.1128/JB.187.5.1552-1558.2005

Parinthawong N, Cottier S, Buchala A et al (2015) Localization and expression of EDS5H a homologue of the SA transporter EDS5. BMC Plant Biol 15:1. https://doi.org/10.1186/s12870-015-0518-1

Park J, Park M-J, Seo PJ et al (2011) A Golgi-localized MATE transporter mediates iron homoeostasis under osmotic stress in *Arabidopsis*. Biochem J 442:551–561. https://doi.org/10.1042/bj20111311

Pérez-Díaz R, Ryngajllo M, Pérez-Díaz J et al (2014) VvMATE1 and VvMATE2 encode putative proanthocyanidin transporters expressed during berry development in Vitis vinifera L. Plant Cell Rep 33:1147–1159. https://doi.org/10.1007/s00299-014-1604-9

Qiu W, Wang N, Dai J et al (2019) AhFRDL1-mediated citrate secretion contributes to adaptation to iron deficiency and aluminum stress in peanuts. J Exp Bot 70:2873–2886. https://doi.org/10.1093/jxb/erz089

Ramel F, Sulmon C, Cabello-hurtado F et al (2007) Genome-wide interacting effects of sucrose and herbicide-mediated stress in *Arabidopsis thaliana*: novel insights into atrazine toxicity and sucrose-induced tolerance. BMC Genomics 20:1–20. https://doi.org/10.1186/1471-2164-8-450

Ribeiro AP, de Souza WR, Martins PK et al (2017) Overexpression of BdMATE gene improves Aluminum tolerance in Setaria viridis. Front Plant Sci 8:1–12. https://doi.org/10.3389/fpls.2017.00865

Rogers EE, Guerinot ML (2002) FRD3, a member of the multidrug and toxin efflux family, controls iron deficiency responses in *Arabidopsis*. Plant Cell 14:1787–1799. https://doi.org/10.1105/tpc.001495

Romheld V, Marschner H (1986) Evidence for a specific uptake system for iron Phytosiderophores in roots of grasses. Plant Physiol 80:175–180

Roschzttardtz H, Séguéla-Arnaud M, Briat JF et al (2011) The FRD3 citrate effluxer promotes iron nutrition between symplastically disconnected tissues throughout *Arabidopsis* development. Plant Cell 23:2725–2737. https://doi.org/10.1105/tpc.111.088088

Ryan PR, Raman H, Gupta S et al (2009) A second mechanism for aluminum resistance in wheat relies on the constitutive efflux of citrate from roots. Plant Physiol 149:340–351. https://doi.org/10.1104/pp.108.129155

Sawaki Y, Kihara-Doi T, Kobayashi Y et al (2013) Characterization of Al-responsive citrate excretion and citrate-transporting MATEs in Eucalyptus camaldulensis. Planta 237(4):979–989. https://doi.org/10.1007/s00425-012-1810-z

Seo PJ, Park J, Park M et al (2012) A Golgi-localized MATE transporter mediates iron homoeostasis under osmotic stress in *Arabidopsis*. Biochem J 442:551–561. https://doi.org/10.1042/BJ20111311

Serrano M, Wang B, Aryal B et al (2013) Export of salicylic acid from the chloroplast requires the multidrug and toxin extrusion-like transporter EDS5. Plant Physiol 162:1815–1821. https://doi.org/10.1104/pp.113.218156

Sun P, Tian QY, Chen J, Zhang WH (2010) Aluminium-induced inhibition of root elongation in *Arabidopsis* is mediated by ethylene and auxin. J Exp Bot 61:347–356. https://doi.org/10.1093/jxb/erp306

Sun X, Gilroy EM, Chini A et al (2011) ADS1 encodes a MATE-transporter that negatively regulates plant disease resistance. New Phytol 192:471–482. https://doi.org/10.1111/j.1469-8137.2011.03820.x

Suzuki M, McCarty DR, Sato Y et al (2015) Conserved functions of the MATE transporter BIG EMBRYO1 in regulation of lateral organ size and initiation rate. Plant Cell 27:2288–2300. https://doi.org/10.1105/tpc.15.00290

Takanashi K, Yokosho K, Saeki K et al (2013) LjMATE1: a citrate transporter responsible for iron supply to the nodule infection zone of lotus japonicus. Plant Cell Physiol 54:585–594. https://doi.org/10.1093/pcp/pct019

Takase H, Matsuoka K, Yazaki K et al (2008) Multidrug and toxic compound extrusion-type transporters implicated in vacuolar sequestration of nicotine in tobacco roots. Plant Physiol 149: 708–718. https://doi.org/10.1104/pp.108.132811

Tanaka Y, Hipolito CJ, Maturana AD et al (2013) Structural basis for the drug extrusion mechanism by a MATE multidrug transporter. Nature 496:247–251. https://doi.org/10.1038/nature12014

Tanaka Y, Iwaki S, Tsukazaki T (2017) Crystal structure of a plant multidrug and toxic compound extrusion family protein. Structure 25:1455–1460. https://doi.org/10.1016/j.str.2017.07.009

Tang RJ, Luan M, Wang C et al (2020) Plant membrane transport research in the post-genomic era. Plant Commun 1:1–18. https://doi.org/10.1016/j.xplc.2019.100013

Thi M, Hoang T, Almeida D et al (2021) AtDTX25, a member of the multidrug and toxic compound extrusion family, is a vacuolar ascorbate transporter that controls intracellular iron cycling in *Arabidopsis*. New Phytol 231:1956–1967. https://doi.org/10.1111/nph.17526

Tiwari M, Sharma D, Singh M et al (2014) Expression of OsMATE1 and OsMATE2 alters development, stress responses and pathogen susceptibility in *Arabidopsis*. Sci Rep 4:3964. https://doi.org/10.1038/srep03964

Uhde-Stone C, Liu J, Zinn KE et al (2005) Transgenic proteoid roots of white lupin: a vehicle for characterizing and silencing root genes involved in adaptation to P stress. Plant J 44:840–853. https://doi.org/10.1111/j.1365-313X.2005.02573.x

Upadhyay N, Kar D, Deepak Mahajan B et al (2019) The multitasking abilities of MATE transporters in plants. J Exp Bot 70:4643–4656. https://doi.org/10.1093/jxb/erz246

Upadhyay N, Kar D, Datta S (2020) A multidrug and toxic compound extrusion (MATE) transporter modulates auxin levels in root to regulate root development and promotes aluminium tolerance. Plant Cell Environ 43:745–759. https://doi.org/10.1111/pce.13658

Wang H (2019) Interactions between hydrogen sulphide and nitric oxide regulate two soybean citrate transporters during the alleviation of aluminium toxicity. Plant Cell Environ 42:2340–2356. https://doi.org/10.1111/pce.13555

Wang Z, Qian C, Guo X et al (2016) ELS1, a novel MATE transporter related to leaf senescence and iron homeostasis in *Arabidopsis thaliana*. Biochem Biophys Res Commun 476(4): 319–325. https://doi.org/10.1016/j.bbrc.2016.05.121

Wang JQH, Li P, Yang L et al (2017) Diverse functions of multidrug and toxin extrusion (MATE) transporters in citric acid efflux and metal homeostasis in Medicago truncatula. Plant J 90:79–95. https://doi.org/10.1111/tpj.13471

Won S-K, Lee Y-J, Lee H-Y et al (2009) Cis-element- and transcriptome-based screening of root hair-specific genes and their functional characterization in *Arabidopsis*. Plant Physiol 150: 1459–1473. https://doi.org/10.1104/pp.109.140905

Wu X, Li R, Shi J et al (2014) Brassica oleracea MATE encodes a citrate transporter and enhances aluminum tolerance in *Arabidopsis thaliana*. Plant Cell Physiol 55:1426–1436. https://doi.org/10.1093/pcp/pcu067

Yazaki K (2005) Transporters of secondary metabolites. Curr Opin Plant Biol 8(3):301–307. https://doi.org/10.1016/j.pbi.2005.03.011

Yokosho K, Yamaji N, Ueno D et al (2009) OsFRDL1 is a citrate transporter required for efficient translocation of iron in rice. Plant Physiol 149:297–305. https://doi.org/10.1104/pp.108.128132

Yokosho K, Yamaji N, Ma JF (2010) Isolation and characterisation of two MATE genes in rye. Funct Plant Biol 37:296–303. https://doi.org/10.1071/fp09265

Yokosho K, Yamaji N, Ma JF (2011) An Al-inducible MATE gene is involved in external detoxification of Al in rice. Plant J 68:1061–1069. https://doi.org/10.1111/j.1365-313X.2011.04757.x

Yokosho K, Yamaji N, Fujii-Kashino M, Ma JF (2016) Functional analysis of a MATE gene OsFRDL2 revealed its involvement in Al-induced secretion of citrate, but a lower contribution to Al tolerance in Rice. Plant Cell Physiol 57:976–985. https://doi.org/10.1093/pcp/pcw026

Zakrzewska S, Mehdipour AR, Malviya VN et al (2019) Inward-facing conformation of a multidrug resistance MATE family transporter. Proc Natl Acad Sci U S A 116:12275–12284. https://doi.org/10.1073/pnas.1904210116

Zhang H, Zhu H, Pan Y et al (2014) A DTX/MATE-type transporter facilitates abscisic acid efflux and modulates ABA sensitivity and drought tolerance in *Arabidopsis*. Mol Plant 7:1522–1532. https://doi.org/10.1093/mp/ssu063

Zhang H, Zhao FG, Tang RJ et al (2017) Two tonoplast MATE proteins function as turgor-regulating chloride channels in *Arabidopsis*. Proc Natl Acad Sci U S A 114:E2036–E2045. https://doi.org/10.1073/pnas.1616203114

Zhang W, Liao L, Xu J et al (2021) Genome-wide identification, characterization and expression analysis of MATE family genes in apple (Malus × domestica Borkh). BMC Genomics 22:1–14. https://doi.org/10.1186/s12864-021-07943-1

Zhao J, Dixon RA (2009) MATE transporters facilitate vacuolar uptake of Epicatechin 3'-O-glucoside for Proanthocyanidin biosynthesis in Medicago truncatula and *Arabidopsis*. Plant Cell 21:2323–2340. https://doi.org/10.1105/tpc.109.067819

Zhao J, Huhman D, Shadle G et al (2011) MATE2 mediates vacuolar sequestration of flavonoid glycosides and glycoside malonates in Medicago truncatula. Plant Cell 23:1536–1555. https://doi.org/10.1105/tpc.110.080804

Zhou Y, Wang Z, Gong L, Chen A (2019) Functional characterization of three MATE genes in relation to aluminum-induced citrate efflux from soybean root Green fluorescent protein. Plant and Soil 443:121–138

Molecular Mechanism of Aluminum Tolerance in Plants: An Overview

9

Dwaipayan Sinha and Soumi Datta

Abstract

Aluminum (Al) is a metal that is abundantly available in the earth's crust in various forms. Though Al has some beneficial role in selected plants, its toxicity and stress symptoms are a matter of concern to agriculturists. Aluminum at high concentration can severely damage a crop by affecting its root system and limiting the uptake of nutrients. As a consequence, the productivity of the plant is diminished. The plant, however, has devised several tolerance mechanisms through which the can combat stress. These mechanisms primarily involve restriction of Al either outside the plant's body or compartmentalization of the metal in a subcellular location thereby limiting its reactivity. A wide array of organic acids are involved in this process, and all are exuded by their transporters. These transporter proteins are synthesized from their respective genes, each of which has several transcription factors. This chapter is an attempt to overview the Al tolerance mechanism of a plant at the molecular level. Efforts have been made to highlight the functions of various transporters involved in the process.

Keywords

Aluminum · Tolerance · Toxicity · Stress · Compartmentalization · Productivity

D. Sinha
Department of Botany, Government General Degree College, Mohanpur, Paschim Medinipur, West Bengal, India

S. Datta (✉)
Department of Pharmacognosy and Phytochemistry, School of Pharmaceutical Education and Research, Jamia Hamdard, New Delhi, India

K. Kumar, S. Srivastava (eds.), *Plant Metal and Metalloid Transporters*,
https://doi.org/10.1007/978-981-19-6103-8_9

9.1 Aluminum Toxicity and Tolerance in Plants: An Introduction

Toxicity caused by aluminum (Al) is the main agricultural difficulty on acidic soil reported globally, mostly because of enhanced Al solubility at low pH and its inhibitory role in the growth of a plant. Most crops are tolerant to acidic soils except a few crops like maize, which are less tolerant and thus yield less. Aluminum is an abundantly available metal and ranks third on the earth's crust (Tietz et al. 2019). Aluminum and its compounds are important components of the Earth's crust, accounting for up to 8% of the planet's surface. It is found in mixtures with oxygen, fluorine, silicon, sulfur, and other elements; it does not exist in its elemental form. Aluminum is found in silicates such as feldspars and micas, as well as in cryolite, a compound of sodium and fluorine, and in bauxite rock (comprising hydrous Al oxides, Al hydroxides, and impurities such as free silica) (Krewski et al. 2007). The pH and chemical environment of the solution affect the total Al content in the soil and Al speciation (Kisnierienė and Lapeikaitė 2015). The toxic effect of various forms of Al on the growth of plants reduces in the following order: Al, $Al(OH)^{2+}$, $Al(OH)_2^{+}$, $Al(OH)_4^{-}$ (Bojórquez-Quintal et al. 2017). Trivalent aluminum (Al^{3+}) is the most prevalent form at a low pH (about 4.3) and has the largest impact on plant growth. Plants are not poisoned by Al that has been precipitated or chelated with organic substances (Nogueirol et al. 2015).

Aluminum toxicity, even at micromolar concentrations, can limit root growth, affecting water and nutrient absorption and finally resulting in agricultural output decrease (Du et al. 2020). Elongation of the root is dependent on cell division, and this process is hampered by Al toxicity. The metal inhibits cell division resulting in brittleness of the root and stunted development. The roots become stunted with scanty root hair development. The apices of the roots become swollen and ultimately get damaged (Panda et al. 2009). Aluminum is thought to cause toxicity in the apoplast by interacting with the cell walls' negative binding sites, primarily pectin in root epidermal and cortical cells (Jiang et al. 2009). Aluminum exposure also results in rapid depolarization of plasma membrane in plants (Illés et al. 2006). In soils and plant roots, Al tends to combine with phosphorus (P) in a less accessible and insoluble form, resulting in a P shortage for plant growth. Aluminum also reduces root respiration (Ward et al. 2011), interferes with enzymes that control polysaccharide deposition in cell walls, reduces cytokinin synthesis and transport, and alters the structure and function of plasma membranes, meddling with the uptake, transport, and use of multiple elements (Ca, Mg, P, and K), as well as uptake of water (Bojórquez-Quintal et al. 2017). Aluminum toxicity can reduce agricultural yields by 30–40% in rice and other crops (Panhwar et al. 2015). However, a combination of modern genetics and molecular biology, sustainable agricultural practices, and alteration of acidic soil will help to improve the crop yields resulting from Al lethality. Plants have devised several mechanisms to combat Al toxicity. Some do so by immobilization of the metal, while others internalize and compartmentalize the metal in their subcellular location thereby rendering them inactive. This chapter is an attempt to overview the tolerance mechanism of Al by the plants. Various transporters involved in the process have also been highlighted.

9.2 Effect of Aluminum Stress in Plants

Aluminum poses a major hindrance in the overall growth and development of a plant in acidic soil where the root, especially the apex of the root, is the main target. Aluminum toxicity is predominant in acidic soils, and thus plants growing in acid soils not only need to overcome Al toxicity but also nitrogen limitation as acidic soil has less nitrification with comparatively higher levels of Al (Zhao and Shen 2018). When soil pH falls below 5.5, Al becomes soluble and changes its hydroxide form to hazardous forms, primarily Al^{3+}, which is considered phytotoxic to the majority of plants (Bonomelli and Artacho 2021). Thickening of cell wall accompanied by callose (Gao et al. 2019) and lignin deposition (Szurman-Zubrzycka et al. 2021), structural alterations and depolarization of the plasma membrane (Jaskowiak et al. 2018), alterations in the cytoskeleton (Blancaflor et al. 1998), alterations in cell shape and vacuolization, disruption of cytosolic Ca^{2+} homeostasis, inhibition of cation uptake by channel protein blocking, generation of reactive oxygen species (Ranjan et al. 2021b), lipid peroxidation (Liang et al. 2021), and mitochondrial dysfunction (Yamamoto et al. 2002) are all toxic effects of Al (Bonomelli and Artacho 2021). In this section, the toxic effects of Al in various plant organs, as well as physiological level, are briefly illustrated.

Aluminum toxicity primarily affects the growth of roots, which leads to the hampered acquisition of nutrients thereby affecting the growth of plants (Yu et al. 2016). The most prominent symptoms of retarded root growth are root stunting and a change in root morphology (Ambachew and Blair 2021). The root also becomes brittle with swollen deformed root tips (Jung and McCouch 2013) and stubby (Jaskowiak et al. 2018) with lateral roots, and root tips appear thick and turn brown (Shetty et al. 2021). Aluminum's primary toxicity is restricted to the root tip's distal transition zone. Aluminum's primary toxicity is restricted to the root tip's distal transition zone. Meristematic cells in the root zone exit division and prepare for F-actin-dependent fast cell elongation. Aluminum has a major impact on cell division in the meristem and cell elongation in the elongation zone (Panda et al. 2009). It is also reported that Al depolymerizes cortical microtubules in a plant cell thereby affecting cell division and growth (Sivaguru et al. 2003).

The primary symptom of Al toxicity is inhibition of shoot growth (Pan et al. 1988). Other visible symptoms include chlorosis and necrosis of leaves, reduction in leaf number and size, and a decrease in shoot biomass (Mossor-Pietraszewska 2001). Another effect of Al toxicity is a reduction in stomatal density. This is possibly due to the inhibitory effect of the metal on protodermal sister cell division, which results in the formation of guard cells (Smirnov et al. 2014). Aluminum toxicity also results in decreased stomatal conductance (Cárcamo et al. 2019). Aluminum toxicity results in a gradual buildup of Al in the leaves of *Oryza sativa*, thereby increasing its temperature. In addition, there is an increase in chlorophyll degradation, restricted assimilation of carbon dioxide, and increased crop water stress index (Phukunkamkaew et al. 2021). In *Eucalyptus*, Al stress results in degradation of chloroplast envelope and decreased chlorophyll content. Along with it, the photosynthetic rate also decreased with an increase in stress (Yang et al. 2015).

Induction of oxidative stress is an important biochemical manifestation of the plant in response to Al stress (Ma et al. 2012). Aluminum stress results in the peroxidation of lipids (Awasthi et al. 2017). In *Arachis hypogea*, Al treatment increases mitochondrial superoxide and hydrogen peroxide content. This was accompanied by tie-dependent increase in malondialdehyde content (He et al. 2019). In *Medicago sativa,* Al stress resulted in upregulation in the activities of catalase, peroxidase, and glutathione reductase (Min et al. 2019). In *Citrus grandis* and *Citrus sinensis*, treatment with Al resulted in leakage of electrolytes along with increased production of hydrogen peroxide and superoxide ions. In addition, the content of malondialdehyde also increased (Guo et al. 2018). In *Allium cepa*, Al induced damage to DNA and initiate cell death (Murali Achary and Panda 2010).

Aluminum interacts with calcium within a plant cell specifically in acidic soil conditions. There are three mechanisms by which Al interacts with calcium in a plant cell. Firstly, Al inhibits the symplastic transport of calcium. Secondly, it disrupts the calcium homeostasis within the cytoplasm. Thirdly, Al displaces calcium from the apoplasm (Meriño-Gergichevich et al. 2010). One of the reasons for the displacement of calcium from the cell wall is that Al binds to pectin more strongly than calcium. This makes the cell wall more rigid, thereby reducing the extensibility that is necessary for the cell elongation process (Brunner and Sperisen 2013). In many plant species, Al-exposed roots are known to induce callose (1,3-β-D-glucan) (Eticha et al. 2005; Tahara et al. 2005; Jones et al. 2006), and root tips with Al-induced callose are identified as Al sensibility indicator (Hirano et al. 2004; Tahara et al. 2005).

9.3 Aluminum Tolerance Mechanism

Plants that thrive in acidic soil have developed adaptations to detoxify Al, which includes external exclusion and internal tolerance (Yan et al. 2020; Wang et al. 2021). The secretion of Al chelators, elevations in rhizosphere pH, secretion of mucilage, immobilization of Al by the cell wall, and Al efflux are all possible mechanisms implicated in external exclusion. Internal tolerance, on the other hand, may be mediated by complexation, compartmentalization, and sequestration of internal Al (Yang et al. 2019). Several methodologies have recently been employed to investigate the processes of Al tolerance in plants. The secretion of organic acids for Al tolerance has been linked to the transcriptome, proteome, metabolome and mutant breeding techniques, and various Al-tolerant genes (Wei et al. 2021). These organic acids are largely responsible for chelation of Al, modification of cell wall, and change in pH, which aids in the tolerance of the metal by the plant. Table 9.1 illustrates the genes responsible for the tolerance of Al in plants.

Table 9.1 Aluminum-tolerant genes were identified in selected plant species

Plant species	Genes	Function	Reference
Arabidopsis thaliana	AtALMT1 (aluminum-activated malate transporter 1)	Transports malate	Hoekenga et al. (2006)
	AtALS3	Transport Al away from sensitive tissue	Larsen et al. (2005)
	AtMATE (*Arabidopsis thaliana*- multidrug and toxic compound extrusion)	Transport citrate	Liu et al. (2009)
	AtNIP1;2 (Nod26-like intrinsic protein 1–2)	Transported Al-malate into the cytoplasm	Wang et al. (2017)
	AtSTOP1 (*Arabidopsis thaliana*- sensitive to proton rhizotoxicity 1)	Regulate Al tolerance genes	Iuchi et al. (2007)
	AtWRKY46	Regulate Al tolerance genes	Ding et al. (2013)
Arabidopsis thaliana	AtXET31	Function in cell wall extension	Zhu et al. (2012)
Brassica napus	BnALMT1/2	Transport malate	Ligaba et al. (2007)
Cajanus cajan	CcSTOP1	A homolog of AtSTOP1	Daspute et al. (2018)
Fagopyrum esculentum Moench.	FeALS1.1 (aluminum sensitive 1)	A homolog of OsALS1	Lei et al. (2017)
	FeALS1.2	A homolog of OsALS1	Lei et al. (2017)
Glycine max	GmALMT1	Transports malate	Liang et al. (2013)
	GmSTOP1–1	A homolog of AtSTOP1	Wu et al. (2018b)
	GmSTOP1–3	A homolog of AtSTOP1	Wu et al. (2018b)
Holcus lanatus	HlALMT1	Transports malate	Ohyama et al. (2013)
	HlART1	A homolog of OsART1	Chen et al. (2013b)
Hydrangea macrophylla	HmPALT1	Transport Al into the cytoplasm	Negishi et al. (2012)
	HmVALT1 (*Hydrangea macrophylla*- vacuolar Al transporter 1)	Sequester Al into the vacuoles	Negishi et al. (2012)
Hordeum vulgare L.	HvAACT1 (anthocyanin 5-aromatic acyltransferase 1)	Transport citrate	Zhou et al. (2013)
Medicago sativa	MsALMT1	Transport malate	Chen et al. (2013a)

(continued)

Table 9.1 (continued)

Plant species	Genes	Function	Reference
Nicotiana tabacum	NtSTOP1	A homolog of AtSTOP1	Ohyama et al. (2013)
Oryza sativa	OsALS1	Sequester Al into the vacuoles	Huang et al. (2021b)
	OsART1	Regulation of Al tolerance genes	Yamaji et al. (2009)
	OsFRDL4 (*Oryza sativa*-ferric reductase defective3-like 4)	Transports citrate	Yokosho et al. (2016a)
	OsFRDL2	Transports citrate	Yokosho et al. (2016b)
	OsEXPA10	Mediate cell wall loosening	Che et al. (2016)
	OsCDT3	Bind Al	Xia et al. (2013)
	OsASR5	Regulation of Al tolerance genes	Arenhart et al. (2012)
	OsART2	Regulation of Al tolerance genes	Che et al. (2018)
	OsASR1	Regulation of Al tolerance genes	Arenhart et al. (2016)
	OsMGT1	Transportation of mg into the cytoplasm	Chen et al. (2012)
	OsWRKY22	Regulation of Al tolerance genes	Li et al. (2018b)
	OsSTAR2	Transport UDP-glucose to the cell wall	Huang et al. (2009)
	OsSTAR1	Transport UDP-glucose to the cell wall	Huang et al. (2009)
	OsNrat1	Transport Al into the cytoplasm	Xia et al. (2010)
Sorghum bicolor L.	SbMATE1	Transport citrate	Magalhaes et al. (2007)
	SbZNF1	Regulate Al tolerance genes	Melo et al. (2019)
	SbWRKY1	Regulate Al tolerance genes	Melo et al. (2019)
	SbSTOP1	A homolog of AtSTOP1	Huang et al. (2018)
Secale cereale L.	ScALMT1	Transports malate	Collins et al. (2008)
	ScFRDL2	Transports citrate	Yokosho et al. (2010)
Triticum aestivum L.	TaALMT1	Transports malate	Sasaki et al. (2004)

(continued)

Table 9.1 (continued)

Plant species	Genes	Function	Reference
	TaSTOP1	A homolog of AtSTOP1	Garcia-Oliveira et al. (2013)
	TaMATE1B	Transports citrate	Tovkach et al. (2013)
	TaMATE1	Transports citrate	Garcia-Oliveira et al. (2014)
Vigna umbellata	VuSTOP1	A homolog of AtSTOP1	Fan et al. (2015)
Zea mays	ZmMATE1	Transports citrate	Maron et al. (2010)

9.3.1 External Tolerance Mechanism

The plants secrete a wide array of organic acids from the roots in response to Al stress (Wu et al. 2018a). Aluminum enhances citrate production by altering the gene expression of citrate metabolic pathways. Aluminum activates transcription and anion channels in the plasma membrane, resulting in citrate secretion by roots, via heterotrimeric G-proteins, phospholipase C, inositol triphosphate, diacylglycerol, calcium ions, and protein kinases (Jiang et al. 2018). Other organic acids secreted in response to Al include malate and oxaloacetate (Yang et al. 2019). One of the most common tolerance mechanisms in plants is the exclusion of Al from the root apex by root exudation of organic acids such as malate and citrate. The aluminum-activated malate transporter (ALMT) and multidrug and toxic compound extrusion (MATE) families of anion channels that confer Al tolerance have been widely documented in the literature. MATE transporters are known to exude citrate in the rhizosphere to chelate Al, whereas ALMTs are involved in the extrusion of Al via malate exudation (Ribeiro et al. 2021). The ALMT family is present universally in sequenced genomes throughout the plant kingdom (Palmer et al. 2016). The ALMT gene encodes anion transporter and transports organic acids through vacuolar and plasma membranes in cells (Ma et al. 2020). The localization of ALMT among various plants is illustrated in Table 9.2.

TaALMT1 was the first member of the family to be identified in wheat root tips and was determined to be involved in Al resistance through malate exudation into the soil (Palmer et al. 2016). TaALMT1 is expressed in the root apices all of the time and is activated by the Al cations found in acid soils to release malate anions into the apoplast. Malate chelates Al, protecting the cell wall, membranes, and other cellular components (Liu and Zhou 2018). Rye is the most Al-resistant of cereal crops. Segregating populations and wheat-rye addition lines were used to map Al resistance loci on chromosomes 3R (Alt2), 4RL (Alt3), 6RS (Alt1), and 7RS (Alt4) (Liu and Zhou 2018). A 1359 bp cDNA termed ScALMT1 was cloned from rye using TaALMT1 gene primers, and it co-segregated with the Al resistance locus Alt4

Table 9.2 Illustration of subcellular location of selected ALMT transporters

Subcellular location	Cellular location	ALMT transporter	Plant species	Reference
Plasma membrane	Root cell	AtALMT1	*Arabidopsis thaliana*	Kovermann et al. (2007)
	Root cell	BnALMT1	*Brassica napus*	Ligaba et al. (2007)
	Root cell	BnALMT2	*Brassica napus*	Ligaba et al. (2007)
	Root cell	GmALMT1	*Glycine max*	Liang et al. (2013)
	Root and shoot cells	HlALMT1	*Holcus lanatus*	Chen et al. (2013b)
	Root cell	ScALMT1	*Secale cereal*	Palmer et al. (2016)
	Root cell	TaALMT1	*Triticum aestivum*	Ramesh et al. (2015)
	Throughout the plant	ZmALMT1	*Zea mays*	Sharma et al. (2016)
	Root cell	ZmALMT2	*Zea mays*	Sharma et al. (2016)
	Guard cells and certain root tissues	HvALMT1	*Hordeum vulgare*	Xu et al. (2015)
	Guard cell	AtALMT12	*Arabidopsis thaliana*	Meyer et al. (2010)
Tonoplast	Guard cell	AtALMT6	*Arabidopsis thaliana*	Meyer et al. (2011)
	Guard cell	AtALMT9	*Arabidopsis thaliana*	De Angeli et al. (2013a)
	Berry mesocarp tissue	VvALMT9	*Vitis vinifera*	De Angeli et al. (2013b)

(Fontecha et al. 2007). ScALMT1 is predominantly expressed in roots, and exogenous Al treatment increases its expression. On chromosome 7R, Al-tolerant rye genotypes have five ScALMT1 gene groups, two of which are highly expressed in the root tip, whereas sensitive genotypes have only two copies, one of which is highly expressed in the root tip (Ma et al. 2014). There are nine members of the ALMT rice family (Heng et al. 2018) out of which OsALMT has been studied thoroughly. OsALMT4 is a plasma membrane protein that is widely expressed in roots and shoots, particularly in the vasculature. Malate was discharged from the roots of transgenic plants overexpressing OsALMT4, which is consistent with OsALMT4 encoding a malate-permeable anion channel (Liu et al. 2017). The *Arabidopsis* ALMT gene family has 14 members (Sharma et al. 2016). Nine of the 14 AtALMT members in *Arabidopsis* have seven transmembrane domains, while the other five AtALMT members have just six (Peng et al. 2018). AtALMT is involved in the production of malate in response to Al toxicity in *Arabidopsis* (Sharma et al. 2016; Zhang et al. 2019).

Multidrug and toxic compound extrusion (MATE) is a type of transporter that is involved in the detoxication of endogenous secondary metabolites as well as exogenous substances in both animal and plant cells (Dong et al. 2019). MatE, a common domain found in most plant MATE proteins, is shared by all MATE transporters. MATE genes range in length from 400 to 550 amino acids on average, with yeast having the longest genes at 700 residues and archaeal having the shortest. These transporters are split into bacterial NorM and DinF and eukaryotic MATE (eMATE) subfamilies based on amino acid sequence similarity (Upadhyay et al. 2019). MATE genes have been discovered in *Arabidopsis* (Wang et al. 2015b) and other plant species including cabbage (Wu et al. 2014), *Eucalyptus* (Sawaki et al. 2013), soybean (Rogers et al. 2009), potato (Huang et al. 2021a), and sorghum (Melo et al. 2019). In *Populus trichocarpa*, PtrMATE1 expression was induced 12 hours after Al stress, while PtrMATE2 expression was induced 24 h later, demonstrating that both proteins work together in response to Al stress in poplar (Li et al. 2017). In *Vigna umbellata*, there is Al stress-induced expression of VuMATE1 and was followed by secretion of citrate (Liu et al. 2013, 2018). In *Glycine max*, it was observed that there was an increase in Al-induced citrate exudation through upregulation of GmMATE and increased phosphorylation of the plasma membrane H + -ATPase (Wang et al. 2016). In *Glycine soja*, GsMATE acts as a citrate transporter and is responsible for the secretion of citrate and reduction of Al content in the root (Ma et al. 2018). In a study, BdMATE from *Brachypodium distachyon* is made to overexpress in *Setaria viridis* and resulted in improved Al tolerance through the exclusion of the metal from the root apex (Ribeiro et al. 2017).

9.3.2 Internal Tolerance Mechanism

Internal tolerance mechanisms play a major role in Al detoxification, absorption, translocation, and storage of non-phytotoxic Al complexes in plant organs particularly in vacuoles (Grisel et al. 2010). The genus exhibits high tolerance to Al. It is reported that *Fagopyrum tataricum* and *Fagopyrum homotropicum* secrete oxalic acid from the root. It was further observed that both species accumulated Al in the leaves upon short-term exposure. Aluminum was stored as Al oxalate in the leaves and roots but in the form of Al citrate in the xylem sap indicating detoxification in the form of chelation accompanied by transportation to the aerial part of the plant (Wang et al. 2015a). In *Fagopyrum esculentum*, two half-size ABC transporters, namely, FeALS1.1 and FeALS1.2, participate in the internal detoxification of Al in the roots and leaves, respectively, through sequestration of the metal in the vacuoles. It was also observed that Al increased the expression of FeALS1.1 in both leaves and roots, and the level of expression in the roots was six times higher than that of its *Arabidopsis* homolog gene (AtALS1). FeALS1.2 expression, on the other hand, was unaffected by Al and showed a 39-fold higher level of expression in the leaves than AtALS1 (Lei et al. 2017). In *Hydrangea macrophylla*, there is a distinct vacuolar and plasma membrane-localized Al transporter that facilitates localization of Al in the vacuole, and its concentration may well reach a level of 15 mM.

Moreover, vacuolar Al transporters (VALT) and plasma membrane Al transporter 1 (PALT1) are identified in the pant and are involved in Al tolerance mechanism (Negishi et al. 2012). An additional Al transporter gene, namely, HmPALT2, has also been identified in the pant and belongs to the member of anion permease (Negishi et al. 2013). In tea, Al exposure results in the accumulation of the metal in walls of the leaf epidermal cells (Tolrà et al. 2011). The absorption, sequestration, and movement of Al from roots to aboveground sections of plants have been linked to several transporters. OsNrat1, a plasma membrane-located transporter in rice, is a member of the natural resistance-associated macrophage protein (Nramp) family and shares little in common with other Nramp members and is responsible for the transportation of Al. Except for epidermal cells, Nrat1 is found in the plasma membranes of all cells in the root tip (Xia et al. 2010). Moreover, OsALS1, which encodes a half-size ABC transporter in rice, has been discovered to be localized at the tonoplast and is essential for Al detoxification via vacuole sequestration (Huang et al. 2012). AtALS3 (AtABCI16) is an ABC transporter found in the plasma membrane of *Arabidopsis* that is involved in plant tolerance to Al toxicity. It only has a TMD domain and is implicated in plant tolerance to Al toxicity (Huang et al. 2021b). In *Andropogon virginicus*, treatment with Al resulted in the induction of half-type ABCG transporter AvABCG1. The AvABCG1 transporter gene was activated in roots and/or shoots of *Andropogon virginicus* by Al, Cu, Zn, and diamide, with Al stress causing the most expression (Ezaki et al. 2015). In barley, ATP-binding cassette transporter HvABCB25 is responsible for Al detoxification. The tissue expression pattern analysis revealed that HvABCB25 is mostly expressed in the leaf and stem in the control condition, however, that Al only strongly stimulated HvABCB25 expression in the root tip. Furthermore, as compared to other metals (such as Cd, La) and low pH, Al stresses specifically upregulated HvABCB25. The expression level of HvABCB25 increases with the exposure period to Al stress, according to time-course investigations (Liu et al. 2021). In Arabidopsis, NIP1;2 is a bidirectional Al transporter that allows removal of Al from the root cell wall by trans-plasma membrane transport into the cytosol and then transportation from root to shoot through loading of the metal in the xylem. It was further observed that the transport substrate for NIP1;2 is the Al-malate complex, and NIP1;2-mediated root Al uptake necessitates a functional ALMT1-mediated root malate exudation system (Wang et al. 2017). The presence of other elements also alleviates Al toxicity in plants (Rahman et al. 2018). A study reported that silicon ameliorates the toxic effect of Al in potatoes concerning the number of roots, branches, and leaves (Dorneles et al. 2016). In another study, it was reported that Boron stimulates polar auxin transport, which is mediated by the auxin efflux transporter PIN2 and leads to the regulation of the plasma membrane-H^+-ATPase, leading to a raised root surface pH, which is required to reduce Al accumulation in this Al-targeted apical root zone (Li et al. 2018a). Magnesium can also alleviate Al toxicity through improved compartmentalization of carbon from shoots to roots, upregulation in the synthesis of organic acids, increased activity of phosphatase, proton-ATPase activity maintenance, and cytoplasmic pH modulation and counteracting oxidative stress (Bose et al. 2011). In rice, upregulation of

OsMGT1, a magnesium transporter in roots, is instrumental in providing Al tolerance through increased magnesium accumulation in the cells (Chen et al. 2012). Similarly, overexpression of AtMGT1 in *Arabidopsis* confers Al tolerance to the plant (Deng et al. 2006). Another study reported that magnesium-mediated root development and Al tolerance in *Arabidopsis* is linked to changes in nitric oxide production (Li et al. 2019). In *Zea mays*, Al induces activation of ZmAT6, which provides Al tolerance by scavenging reactive oxygen species and boosting antioxidant enzymes (Du et al. 2021). In rice, ectopic expression of HtNHX1 or HtNHX2 (from *Helianthus tuberosus*) resulted in the improvement of growth in rice in an Al-stressed condition. It does so through alteration of potassium and hydrogen ion fluxes and structure of cell wall (Li et al. 2020a). In Al-stressed *Lolium perenne*, supplementation with sulfate resulted in inhibition of lipid peroxidation and upregulation of superoxide dismutase activity and can offer short-term protection against Al toxicity (Vera-Villalobos et al. 2020).

9.3.3 Transcription Factors Involved in Combatting Aluminum Stress

Transcription factors are the primary regulators of gene expression. Transcription factors are proteins that regulate gene expression by binding to certain DNA regions. A variety of variables, including epigenetic mechanisms, gene regulatory elements, and chemical cofactors, influence their evolution (Mitsis et al. 2020). Sensitive to proton rhizotoxicity, STOP1 is a type of C2H2 zinc finger transcription factor that plays an important role in regulating the downstream expression of genes involved in tolerance of stresses (Fan et al. 2016). In *Arabidopsis*, AtSTOP1 is involved in the positive regulation of three major Al resistance genes, namely, AtALMT1, AtMATE, and AtALS3 (Iuchi et al. 2007; Sawaki et al. 2009). In *Nicotiana tabacum*, NtSTOP1, regulating Al tolerance gene, has been identified. The GGNVS motif of the NtMATE promoter is found in the NtSTOP1 binding site. Although the STOP1/NtSTOP1 motif that leads them to bind to AtALMT1/NtMATE was similar, AtALMT1 and NtMATE are essentially different genes (Ito et al. 2019). In *Glycine max*, GmSTOP1 homologs have been identified. A homolog search revealed three STOP1 homologs in the soybean genome: GmSTOP1–1 (Glyma.10G215200), GmSTOP1–2 (Glyma.16G156400), and GmSTOP1–3 (Glyma.16G156400). All three GmSTOP1 proteins shared similar characteristics, including the presence of four putative zinc finger domains, nuclear localization, and transactivation activity. H+ stress altered transcription of all three GmSTOP1s marginally, but Al substantially upregulated GmSTOP1–1 and GmSTOP1–3 in root apexes and GmSTOP1–3 in basal root areas. It was also observed that GmSTOP1–1 and GmSTOP1–3 could partially mend Al tolerance in the Atstop1 mutant of *Arabidopsis* (Wu et al. 2018b). GhSTOP1 from *Gossypium hirsutum* is also reported to be responsible for the regulation of GhALMT1, GhMATE, and GhALS3 (Kundu et al. 2019). In *Sorghum bicolor*, four STOP1-like proteins, namely, SbSTOP1a, SbSTOP1b, SbSTOP1c, and SbSTOP1d, have been identified.

All the STOP proteins were localized in the nucleus. It was also shown in the study that apart from the known SbMATE-mediated Al exclusion mechanism, SbSTOP1d demonstrated transcriptional regulation of SbSTAR2 and SbMATE, indicating the possibility of additional SbSTOP1 and SbSTAR2-dependent Al tolerance mechanism in sorghum (Huang et al. 2018). In addition, AtHB7 and AtHB12 are two members of the HD-ZIP I subfamily and play an important role in the regulation of tolerance toward Al toxicity. A study reported that Al stress resulted in 13-fold and 206-fold upregulation of AtHB7 and AtHB12, respectively. AtHB7 and AtHB12 had varied dynamic expression patterns in response to Al stress. Although both AtHB7 and AtHB12 positively regulate root growth in the absence of Al stress, our findings reveal that in the presence of Al stress, AtHB7 antagonizes AtHB12 to restrict root growth. AtHB7 and AtHB12 influence the cell wall's ability to bind Al in opposite ways (Liu et al. 2020b). One of the factors of Al tolerance in plants are the changes in cell wall characteristics. In *Arabidopsis thaliana*, the transcription factor WRKY47 is involved in root growth and Al tolerance. It was further observed that extensin-like protein (ELP) and xyloglucan endotransglucosylase-hydrolases17 (XTH17), which are involved in cell wall modification, are directly regulated by WRKY47 (Li et al. 2020b). In *Oryza sativa*, the nuclear-located OsWRKY22 is a transcriptional activator that binds to the promoter of OsFRDL4 via W-box elements. Thus, OsWRKY22 enhances Al-induced citrate secretion and Al tolerance in rice via promoting Al-induced increases in OsFRDL4 expression (Li et al. 2018b). In a separate investigation, GsGIS3 was identified from Al-tolerant wild soybean, and gene expression profiles were described in *Arabidopsis*. GsGIS3 is a nuclear protein with one zinc finger motif (C2H2). It was observed that Al tolerance was increased in transgenic lines with more root development, higher proline, and decreased malondialdehyde accumulation when GsGIS3 was overexpressed in *Arabidopsis*. This suggests that GsGIS3, as a C2H2 ZFP, may improve tolerance to Al toxicity by regulating Al-tolerant-related genes positively (Liu et al. 2020a). In *Glycine soja*, the GsMAS1 gene from the wild soybean BW69 line encodes a MADS-box transcription factor. The nucleus was found to contain the putative GsMAS1 protein. The GsMAS1 gene was abundant in soybean roots, with a constitutive expression pattern and a concentration time-specific pattern caused by Al stress. Furthermore, six genes resistant to Al stress were elevated, whereas Al stress and GsMAS1 overexpression dramatically activated AtALMT1 and STOP2 (Zhang et al. 2020). In *Hordeum vulgare*, the HvATF1, an Al tolerance factor has been identified. HvATF1 is a transcriptional activator that is expressed in the nucleus. HvATF1 transcription was found to be constitutive in several tissues and to be unaffected by Al stress. Transcriptomics analysis revealed 64 genes that were differentially expressed in the RNAi lines compared to the wild type, and these were identified as potential downstream HvATF1 regulators (Wu et al. 2020). In *Hordeum vulgare*, HvHOX9, a novel homeobox-leucine zipper transcription factor, was discovered to be a miR166b target gene and functionally described. Al stress increased HvHOX9 expression, which was exclusively found in the root tip. It is suggested that miR166b/HvHOX9 plays a key role in Al tolerance by reducing Al binding to the root cell wall and boosting apoplastic pH, allowing for Al detoxification in the root (Feng et al. 2020). In *Vigna umbellata*, the NAC-type transcription factor VuNAR1 has been studied in terms of the Al stress response. VuNAR1 is a nuclear-localized transcriptional activator whose expression was elevated in roots but not in shoots under Al stress. Aluminum exclusion improves Al resistance in *Arabidopsis* plants overexpressing VuNAR1. VuNAR1 controls the expression of genes involved in protein phosphorylation and cell wall modification in *Arabidopsis*, according to a comparative transcriptomic study. The direct transcriptional activation of cell wall-associated receptor kinase 1 (WAK1) by VuNAR1 was demonstrated in a transient expression test, and the result of the study indicated that VuNAR1 controls Al resistance through regulation of cell wall pectin metabolism via binding to the promoter of WAK1 and inducing its expression (Lou et al. 2020).

9.3.4 Plant Hormones Involved in Aluminum Stress Adaptation

Phytohormones are also known to play a function in the control of plant tolerance to a variety of biotic and abiotic stimuli (Ranjan et al. 2021a). The bulk of abiotic stress tolerance mechanisms in plants is regulated by the ABA signaling system (Sah et al. 2016). Following heavy metal exposure, ABA concentrations were shown to be higher in numerous plant tissues, implying that it may have a role in reducing heavy metal stress (Vishwakarma et al. 2017). In soybean (Hou et al. 2010) and barley (Ahmed et al. 2016), Al stress-induced ABA accumulation has been demonstrated to modulate organic acid efflux. In one study, it was observed that in *Fagopyrum esculentum*, there is an increase in ABA content upon treatment with Al. It was further observed that the increase in ABA content is correlated with the expression of ABC-like gene ALS3 (FeALS3) (Reyna-Llorens et al. 2015). In *Arabidopsis*, it was reported that ABA induced expression of AtALMT, thereby indicating its role in detoxification of Al (Kobayashi et al. 2013). In *Vigna umbellata*, it was shown that ABA-responsive and SA-responsive cis-acting elements are present in the promoter region of VuMATE indicating its probable role in the Al-regulated expression of VuMATE (Liu et al. 2016). OsSTAR1 and other Al tolerance genes in rice have been demonstrated to be regulated by members of the ABA, stress, and ripening (ASR) gene family (Arenhart et al. 2016). The expression of rice ASR family genes in response to Al treatment was investigated, and it was discovered that ASR5 functions as a transcription factor to control the expression of STAR1, Nrat1, and FRDL4 (Arenhart et al. 2013). STAR1 (sensitive to aluminum rhizotoxicity 1) belongs to the ABC transporter family, and in rice, it promotes the transport of UDP-glucose to the cell wall, as well as masking Al-binding sites, with the help of STAR2 (Huang et al. 2009). Apart from ABA, the other phytohormones also play their respective roles during Al stress. It was observed that in Maize, treatment with Al influences basipetal transport of auxin within the root tissue (Hasenstein and Evans 1988). In *Arabidopsis*, it was shown that Al inhibited root elongation and resulted in the generation of ethylene, and this possibly acts as a signal for change in the distribution of auxin through disruption of AUX1- and PIN2-mediated auxin polar transport. This accounts for inhibition of root elongation (Sun et al. 2010). In wheat, it is shown that Al-induced malic acid outflow was aided by indole acetic acid (IAA) (Yang et al. 2011). This suggests that, in response to Al toxicity, auxin alters anion channels to regulate the outflow of organic acids such as malate. Cytokinins are adenine-derived small-molecule plant growth regulators that regulate practically every aspect of plant development and growth. Cytokinins are important regulators of plant cell division, nutrient allocation, and photosynthetic activity on the inside, as well as detecting and signaling agents for plant responses to external stressors (Emery and Kisiala 2020). In *Arabidopsis,* it is shown that through modulation of Al-induced auxin signaling, cytokinin increases root growth inhibition under stress. In the root-apex transition zone, Al stress produces a local cytokinin response that is dependent on IPTs, which encode adenosine phosphate isopentenyl transferases and regulate cytokinin production. In response to Al stress, IPTs are upregulated primarily in the root-apex transition zone, where they enhance local cytokinin production and root development inhibition. Ethylene signaling, which operates upstream of auxin, is also involved in root growth suppression (Yang et al. 2017b). Jasmonic acid (JA) is a natural hormone regulator that plays a role in development, wound healing, and pathogen attack. When pathogens are detected, JA is produced and activates a signaling cascade that triggers a variety of defense responses (Antico et al. 2012). It was reported from a study that the exogenous jasmonic acid-enhanced Al induced root growth suppression. In response to Al stress, the expression of the JA receptor coronatine insensitive1 (COI1) and the essential JA signaling regulator MYC2 was upregulated in the root tips. Ethylene was in charge of this mechanism, as well as COI1-mediated Al-induced root development suppression under Al stress. Furthermore, transcriptomic research demonstrated that jasmonic acid signaling regulated numerous responsive genes under Al stress (Yang et al. 2017a). In *Lycopersicon esculentum*, the link between the WRKY and ALMT genes as well as their role in JA-mediated root development suppression in tomatoes under Al stress was demonstrated (Wang et al. 2020). AtWRKY46 acts as a negative regulator of AtALMT1 in *Arabidopsis*, binding to its promoter at the W-box and causing increased malate efflux and lower root Al buildup in Atwrky46 mutants (Ding et al. 2013).

9.4 Manipulation of Aluminum-Tolerant Genes Using Transgenic Approaches

Transgenic breeding is a promising approach for improving crop abiotic stress tolerance, as it can generate novel and important sources of resistance with rapid replication potential. In most plant species, Al-induced organic acid secretion is a key mechanism for modulating the degree of resistance to Al toxicity (Singh and Chauhan 2011; Liu et al. 2018). As discussed, ALMT and MATE transporters are instrumental in conferring Al resistance in plants. They are responsible for the secretion of citrate and malate, which bring about Al tolerance (Liu et al. 2009). Thus upregulation of organic acid secretion is the prime focus of biotechnologists for increasing Al tolerance in crop plants. This can be achieved by the genetic transformation of crop plants with genes conferring Al tolerance. Table 9.3 represents selected Al-tolerant genes that are made of express in other plants to confer Al tolerance.

9.5 Conclusion and Future Perspective

Aluminum stress is a matter of grave concern as it not only affects the physiological status of a plant but also possesses a threat to food security through reduction of yield. Some plants are highly sensitive to Al toxicity and succumb to it. Others are however tolerant and rely on several intricate processes to overcome the toxicity. These processes are commonly referred to as tolerance mechanisms, and depending upon the way the plant fights the Al toxicity, it can be external and internal. External tolerance involves the secretion of organic acids by the plant, which immobilizes the metal, thereby making it inaccessible to the plant system. The internal tolerance mechanism is adopted when the metal can reach the plant tissues. In this case, the metal is generally transported and localized in the subcellular location primarily the vacuoles. In both cases, various Al transporters come into action to localize the metal, thereby minimizing the deleterious effect. Thus, there has been a constant effort to identify these transporters in the resistant varieties of the plant. In this sphere, success has been achieved to genetically engineer the Al-tolerant genes especially those of the transporters into the susceptible varieties of plants to achieve Al resistance. The ALMT has been instrumental in conferring Al tolerance to several plants. They actively transport malates across the membrane, which then chelates the metals thereby rendering it inactive. Because many ALMTs are not involved in Al tolerance, the historical protein family name "ALMT" may be misleading. Their physiological responsibilities are much broader, and we've only scratched the surface of this diversity thus far. Future research will undoubtedly benefit from the availability of trustworthy structural data that allows for the correlation of structural motifs with functional features. We should expect additional surprises from this protein family in the future (Sharma et al. 2016). The MATE transporters also perform the transportation of citric acids and confer tolerance toward Al. Although much has been known about MATE transporters, many questions about their architecture and activities remain unresolved. They appear to be accomplishing a lot more in plants than had been expected. This is consistent with the extended MATE family of proteins found in plants. Only a few of these MATE proteins have been fully functionalized. The spectrum of MATE-mediated plant responses has been broadened thanks to MATE transporters acting as channels (Upadhyay et al. 2019).

One of the major domains, which are yet to be investigated in full flow, is the perception of Al by the plants. Plant sensing of Al stress and transmission of the signal to trigger multiple downstream Al-tolerant mechanisms are key frontier concerns in plant science. Heterotrimeric G-proteins, phospholipase C, inositol triphosphate, diacylglycerol, Ca^{2+}, and protein kinases were found to be involved

Table 9.3 Selected aluminum -tolerant genes, which are made of express in other plants to confer aluminum tolerance

S. No.	Gene	Source plant	Recipient plant	Function	Reference
1.	GsMATE	*Glycine soja*	*Arabidopsis thaliana*	Aluminum induced citrate transporter	Ma et al. (2018)
2.	SbMATE	*Sorghum bicolor*	*Saccharum* spp.	Exudation of citrate in the rhizosphere sustained root growth in the presence of Al	Ribeiro et al. (2021)
3.	SbMATE	*Sorghum bicolor*	*Hordeum vulgare*	Higher citrate efflux from root apex and greater tolerance toward Al	Zhou et al. (2014)
4.	GhMATE1	*Gossypium hirsutum*	*Arabidopsis thaliana*	Enhancement of citrate release and regulation of root growth in overexpressed plants	Kundu and Ganesan (2020)
5.	BdMATE	*Brachypodium distachyon*	*Setaria viridis*	Transgenic plants had increased root citrate exudation into the rhizosphere, implying that the chelation of the metal by the organic acid anion improved Al tolerance in these plants	Ribeiro et al. (2017)
6.	ZmMATE6	*Zea mays*	*Arabidopsis thaliana*	Higher release of Al-induced citrate	Du et al. (2021)
7.	AtALMT1	*Arabidopsis thaliana*	*Vigna mungo*	Increase in malate exudation and adaptation to Al toxicity in acidic soil	Saha et al. (2020)
8.	BoALMT1	*Brassica oleracea*	*Arabidopsis thaliana*	Increase in malate secretion and proton efflux	Zhang et al. (2018)
9.	TaALMT1	*Triticum aestivum*	*Triticum aestivum* bob White 26 "SH9826" line (BW26)	Greater malate efflux and Al tolerance	Pereira et al. (2010)
10.	MsALMT1	*Medicago sativa*	*Nicotiana tabacum*	Enhanced malate efflux and better Al tolerance	Chen et al. (2013a)
11.	HvALMT1	*Hordeum vulgare*	*Hordeum vulgare*	Greater efflux of malate and succinate from the roots	Gruber et al. (2011)

in Al-induced citrate exudation signaling (Poot-Poot and Teresa Hernandez-Sotomayor 2011; Jiang et al. 2018; Han et al. 2020). However, reports of Al transmembrane signal transduction are scarce. The most essential topic in plant Al stress research is the identification of the Al receptor. Using Ca^{2+} imaging-based forward genetic screening, salt and H_2O_2 sensors have been discovered, which leads to the discovery of Al receptors (Wei et al. 2021). The rapid advancement of whole-genome sequencing and genome editing technology has increased the number of opportunities to uncover the mechanisms of Al tolerance and identify novel Al-tolerant genes, allowing current biotechnology to be used to create Al-tolerant plant varieties (Wei et al. 2021). Thus, the development of Al-tolerant plant variety is extremely relevant to combat Al stress.

Even though Al is not required by all plants, it can boost development and is necessary for specific plant taxa, depending on environmental conditions, element concentration, and plant species (Bojórquez-Quintal et al. 2017). The application of Al as a biostimulant is desirable as it stimulates plant growth (e.g., tea) (Sun et al. 2020) and may fix floral colors (e.g., *Hydrangea*) (Schreiber et al. 2011). High levels of Al, on the other hand, can represent a major threat to agricultural productivity by inhibiting root elongation and plant growth via a variety of mechanisms involving Al, such as interactions in the symplast, plasma membrane, and cell wall (Kochian et al. 2015). However, it is critical to understand the concentration range at which a helpful element turns lethal, particularly when using it as a fertilizer to boost crop production under stress and/or improve the nutritional value of food plants (Pilon-Smits et al. 2009; Kaur et al. 2016). On a different front, it is also important to study in detail the physicochemical properties of Al and its salt, which are present in the earth. Proper soil testing with the motive of estimating the Al content is also required. Proper knowledge about the Al content of the soil would make the farmers alert and would enable them to use Al-containing chemicals judiciously. Moreover, it is also important to understand the environmental factors that favor the accumulation of Al in plants and improvise strategies for reduction in the accumulation of the metal within the plant body (de Silva et al. 2016). An overall environmental monitoring system that can formulate strategies to keep the levels of Al contamination under control can prove to be of immense relevance. Governmental strategies and effective enforcement of laws can further fortify these approaches, which in synchronization with an effective environmental plan can very well give ancillary support for combatting Al stress. Thus, a holistic approach needs to be taken for effective management of Al stress and consequent betterment of crop yields and ensuring food security.

References

Ahmed IM, Nadira UA, Cao F, He X, Zhang G, Wu F (2016) Physiological and molecular analysis on root growth associated with the tolerance to aluminum and drought individual and combined in Tibetan wild and cultivated barley. Planta 243(4):973–985

Ambachew D, Blair MW (2021) Genome wide association mapping of root traits in the andean genepool of common bean (*Phaseolus vulgaris* L.) grown with and without aluminum toxicity. Front Plant Sci 12:825

Antico CJ, Colon C, Banks T, Ramonell KM (2012) Insights into the role of jasmonic acid-mediated defenses against necrotrophic and biotrophic fungal pathogens. Front Biol 7(1):48–56

Arenhart RA, Margis R, Margis-Pinheiro M (2012) The rice ASR5 protein: a putative role in the response to aluminum photosynthesis disturbance. Plant Signal Behav 7:1263–1266

Arenhart RA, Lima JC, Pedron M, Carvalho FE, Silveira JA, Rosa SB, Caverzan A, Andrade CM, Schünemann M, Margis R, Margis-Pinheiro M (2013) Involvement of ASR genes in aluminium tolerance mechanisms in rice. Plant Cell Environ 36(1):52–67

Arenhart RA, Schunemann M, Bucker Neto L, Margis R, Wang ZY, Margis-Pinheiro M (2016) Rice ASR1 and ASR5 are complementary transcription factors regulating aluminium responsive genes. Plant Cell Environ 39(3):645–651

Awasthi JP, Saha B, Regon P, Sahoo S, Chowra U, Pradhan A, Roy A, Panda SK (2017) Morpho-physiological analysis of tolerance to aluminum toxicity in rice varieties of north East India. PLoS One 12(4):e0176357

Blancaflor EB, Jones DL, Gilroy S (1998) Alterations in the cytoskeleton accompany aluminum-induced growth inhibition and morphological changes in primary roots of maize. Plant Physiol 118(1):159–172

Bojórquez-Quintal E, Escalante-Magaña C, Echevarría-Machado I, Martínez-Estévez M (2017) Aluminum, a friend or foe of higher plants in acid soils. Front Plant Sci 8:1767

Bonomelli C, Artacho P (2021) Aluminum toxicity in sweet cherry trees grown in an acidic volcanic soil. Agronomy 11(6):1259

Bose J, Babourina O, Rengel Z (2011) Role of magnesium in alleviation of aluminium toxicity in plants. J Exp Bot 62(7):2251–2264

Brunner I, Sperisen C (2013) Aluminum exclusion and aluminum tolerance in woody plants. Front Plant Sci 4:172

Cárcamo MP, Reyes-Díaz M, Rengel Z, Alberdi M, Omena-Garcia RP, Nunes-Nesi A, Inostroza-Blancheteau C (2019) Aluminum stress differentially affects physiological performance and metabolic compounds in cultivars of highbush blueberry. Sci Rep 9(1):1–3

Che J, Yamaji N, Shen RF, Ma JF (2016) An Al-inducible expansin gene, OsEXPA10 is involved in root cell elongation of rice. Plant J 88:132–142

Che J, Tsutsui T, Yokosho K, Yamaji N, Ma JF (2018) Functional characterization of an aluminum (Al)-inducible transcription factor, ART2, revealed a different pathway for Al tolerance in rice. New Phytol 220:209–218

Chen ZC, Yamaji N, Motoyama R, Nagamura Y, Ma JF (2012) Up-regulation of a magnesium transporter gene OsMGT1 is required for conferring aluminum tolerance in rice. Plant Physiol 159(4):1624–1633

Chen Q, Wu KH, Wang P, Yi J, Li KZ, Yu YX, Chen LM (2013a) Overexpression of MsALMT1, from the aluminum-sensitive *Medicago sativa*, enhances malate exudation and aluminum resistance in tobacco. Plant Mol Biol Rep 31(3):769–774

Chen ZC, Yokosho K, Kashino M, Zhao FJ, Yamaji N, Ma JF (2013b) Adaptation to acidic soil is achieved by increased numbers of cis-acting elements regulating ALMT1 expression in *Holcus lanatus*. Plant J 76(1):10–23

Collins NC, Shirley NJ, Saeed M, Pallotta M, Gustafson JP (2008) An ALMT1 gene cluster controlling aluminum tolerance at the Alt4 locus of rye (*Secale cereale* L). Genetics 179:669–682

Daspute AA, Kobayashi Y, Panda SK, Fakrudin B, Kobayashi Y, Tokizawa M, Iuchi S, Choudhary AK, Yamamoto YY, Koyama H (2018) Characterization of CcSTOP1; a C2H2-type transcription factor regulates Al tolerance gene in pigeon pea. Planta 247:201–214

De Angeli A, Zhang J, Meyer S, Martinoia E (2013a) AtALMT9 is a malate-activated vacuolar chloride channel required for stomatal opening in *Arabidopsis*. Nat Commun 4(1):1

De Angeli A, Baetz U, Francisco R, Zhang J, Chaves MM, Regalado A (2013b) The vacuolar channel VvALMT9 mediates malate and tartrate accumulation in berries of *Vitis vinifera*. Planta 238(2):283–291

de Silva J, Tuwei G, Zhao FJ (2016) Environmental factors influencing aluminium accumulation in tea (*Camellia sinensis* L.). Plant Soil 400(1–2):223–230

Deng W, Luo K, Li D, Zheng X, Wei X, Smith W, Thammina C, Lu L, Li Y, Pei Y (2006) Overexpression of an *Arabidopsis* magnesium transport gene, AtMGT1, in *Nicotiana benthamiana* confers Al tolerance. J Exp Bot 57(15):4235–4243

Ding ZJ, Yan JY, Xu XY, Li GX, Zheng SJ (2013) WRKY46 functions as a transcriptional repressor of ALMT1, regulating aluminum-induced malate secretion in Arabidopsis. Plant J 76(5):825–835

Dong B, Niu L, Meng D, Song Z, Wang L, Jian Y, Fan X, Dong M, Yang Q, Fu Y (2019) Genome-wide analysis of MATE transporters and response to metal stress in Cajanus cajan. J Plant Interact 14(1):265–275

Dorneles AO, Pereira AS, Rossato LV, Possebom G, Sasso VM, Bernardy K, Sandri RD, Nicoloso FT, Ferreira PA, Tabaldi LA (2016) Silicon reduces aluminum content in tissues and ameliorates its toxic effects on potato plant growth. Cienc Rural 46:506–512

Du H, Huang Y, Qu M, Li Y, Hu X, Yang W, Li H, He W, Ding J, Liu C, Gao S, Cao M, Lu Y, Zhang S (2020) A maize ZmAT6 gene confers aluminum tolerance via reactive oxygen species scavenging. Front Plant Sci 11:1016

Du H, Ryan PR, Liu C, Li H, Hu W, Yan W, Huang Y, He W, Luo B, Zhang X, Gao S, Zhou S, Zhang S (2021) ZmMATE6 from maize encodes a citrate transporter that enhances aluminum tolerance in transgenic *Arabidopsis thaliana*. Plant Sci 311:111016

Emery RJN, Kisiala A (2020) The roles of cytokinins in plants and their response to environmental stimuli. Plants (Basel) 9(9):1158

Eticha D, The C, Welcker C, Narro L, Staß A, Horst WJ (2005) Aluminium-induced callose formation in root apices: inheritance and selection trait for adaptation of tropical maize to acid soils. Field Crops Res 93(2–3):252–263

Ezaki B, Takahashi K, Utsumi K, Higashi A (2015) A half-type AvABCG1 transporter derived from *Andropogon virginicus* L. confers aluminum tolerance. Environ Exp Bot 118:21–31

Fan W, Lou HQ, Gong YL, Liu MY, Cao MJ, Liu Y, Yang JL, Zheng SJ (2015) Characterization of an inducible C2H2-type zinc finger transcription factor VuSTOP1 in rice bean (*Vigna umbellata*) reveals differential regulation between low pH and aluminum tolerance mechanisms. New Phytol 208:456–468

Fan W, Lou HQ, Yang JL, Zheng SJ (2016) The roles of STOP1-like transcription factors in aluminum and proton tolerance. Plant Signal Behav 11(2):e1131371

Feng X, Liu W, Dai H, Qiu Y, Zhang G, Chen ZH, Wu F (2020) HvHOX9, a novel homeobox leucine zipper transcription factor, positively regulates aluminum tolerance in Tibetan wild barley. J Exp Bot 71(19):6057–6073

Fontecha G, Silva-Navas J, Benito C, Mestres MA, Espino FJ, Hernández-Riquer MV, Gallego FJ (2007) Candidate gene identification of an aluminum-activated organic acid transporter gene at the Alt4 locus for aluminum tolerance in rye (*Secale cereale* L.). Theor Appl Genet 114(2): 249–260

Gao J, Yan S, Yu H, Zhan M, Guan K, Wang Y, Yang Z (2019) Sweet sorghum (Sorghum bicolor L.) SbSTOP1 activates the transcription of a β-1,3-glucanase gene to reduce callose deposition under Al toxicity: a novel pathway for Al tolerance in plants. Biosci Biotechnol Biochem 83(3): 446–455

Garcia-Oliveira AL, Benito C, Prieto P, Menezes RDA, Rodrigues-Pousada C, Guedes-Pinto H, Martins-Lopes P (2013) Molecular characterization of TaSTOP1 homoeologues and their response to aluminium and proton (H^+) toxicity in bread wheat (*Triticum aestivum* L.). BMC Plant Biol 13:134

Garcia-Oliveira AL, Martins-Lopes P, Tolrá R, Poschenrieder C, Tarquis M, Guedes-Pinto H, Benito C (2014) Molecular characterization of the citrate transporter gene TaMATE1 and expression analysis of upstream genes involved in organic acid transport under Al stress in bread wheat (*Triticum aestivum*). Physiol Plant 152(3):441–452

Grisel N, Zoller S, Künzli-Gontarczyk M, Lampart T, Münsterkötter M, Brunner I, Bovet L, Métraux JP, Sperisen C (2010) Transcriptome responses to aluminum stress in roots of aspen (*Populus tremula*). BMC Plant Biol 10(1):1–5

Gruber BD, Delhaize E, Richardson AE, Roessner U, James RA, Howitt SM, Ryan PR (2011) Characterisation of HvALMT1 function in transgenic barley plants. Funct Plant Biol 38(2): 163–175

Guo P, Qi YP, Cai YT, Yang TY, Yang LT, Huang ZR, Chen LS (2018) Aluminum effects on photosynthesis, reactive oxygen species and methylglyoxal detoxification in two *citrus* species differing in aluminum tolerance. Tree Physiol 38(10):1548–1565

Han R, Wei Y, Xie Y, Liu L, Jiang C, Yu Y (2020) Quantitative phosphoproteomic analysis provides insights into the aluminum-responsiveness of Tamba black soybean. PLoS One 15(8): e0237845

Hasenstein KH, Evans ML (1988) Effects of cations on hormone transport in primary roots of *Zea mays*. Plant Physiol 86(3):890–894

He H, Oo TL, Huang W, He LF, Gu M (2019) Nitric oxide acts as an antioxidant and inhibits programmed cell death induced by aluminum in the root tips of peanut (*Arachis hypogaea* L.). Sci Rep 9(1):1–2

Heng Y, Wu C, Long Y, Luo S, Ma J, Chen J, Liu J, Zhang H, Ren Y, Wang M, Tan J, Zhu S, Wang J, Lei C, Zhang X, Guo X, Wang H, Cheng Z, Wan J (2018) OsALMT7 maintains panicle size and grain yield in rice by mediating malate transport. Plant Cell 30(4):889–906

Hirano Y, Pannatier EG, Zimmermann S, Brunner I (2004) Induction of callose in roots of Norway spruce seedlings after short-term exposure to aluminum. Tree Physiol 24(11):1279–1283

Hoekenga OA, Maron LG, Piñeros MA, Cancado GMA, Shaff J, Kobayashi Y, Ryan PR, Dong B, Delhaize E, Sasaki T et al (2006) AtALMT1, which encodes a malate transporter, is identified as one of several genes critical for aluminum tolerance in *Arabidopsis*. Proc Natl Acad Sci U S A 103:9738–9743

Hou N, You J, Pang J, Xu M, Chen G, Yang Z (2010) The accumulation and transport of abscisic acid in soybean (*Glycine max* L.) under aluminum stress. Plant Soil 330(1):127–137

Huang CF, Yamaji N, Mitani N, Yano M, Nagamura Y, Ma JF (2009) A bacterial-type ABC transporter is involved in aluminum tolerance in rice. Plant Cell 21(2):655–667

Huang CF, Yamaji N, Chen Z, Ma JF (2012) A tonoplast-localized half-size ABC transporter is required for internal detoxification of aluminum in rice. Plant J 69(5):857–867

Huang S, Gao J, You J, Liang Y, Guan K, Yan S, Zhan M, Yang Z (2018) Identification of STOP1-like proteins associated with aluminum tolerance in sweet sorghum (*Sorghum bicolor* L.). Front Plant Sci 9:258

Huang Y, He G, Tian W, Li D, Meng L, Wu D, He T (2021a) Genome-wide identification of MATE gene family in potato (*Solanum tuberosum* L.) and expression analysis in heavy metal stress. Front Genet 12:650500

Huang J, Li X, Chen X, Guo Y, Liang W, Wang H (2021b) Genome-wide identification of soybean abc transporters relate to aluminum toxicity. Int J Mol Sci 22(12):6556

Illés P, Schlicht M, Pavlovkin J, Lichtscheidl I, Baluska F, Ovecka M (2006) Aluminium toxicity in plants: internalization of aluminium into cells of the transition zone in Arabidopsis root apices related to changes in plasma membrane potential, endosomal behaviour, and nitric oxide production. J Exp Bot 57(15):4201–4213

Ito H, Kobayashi Y, Yamamoto YY, Koyama H (2019) Characterization of NtSTOP1-regulating genes in tobacco under aluminum stress. Soil Sci Plant Nutr 65(3):251–258

Iuchi S, Koyama H, Iuchi A, Kobayashi Y, Kitabayashi S, Kobayashi Y, Ikka T, Hirayama T, Shinozaki K, Kobayashi M (2007) Zinc finger protein STOP1 is critical for proton tolerance in Arabidopsis and coregulates a key gene in aluminum tolerance. Proc Natl Acad Sci U S A 104(23):9900–9905

Jaskowiak J, Tkaczyk O, Slota M, Kwasniewska J, Szarejko I (2018) Analysis of aluminum toxicity in *Hordeum vulgare* roots with an emphasis on DNA integrity and cell cycle. PLoS One 13(2): e0193156

Jiang HX, Tang N, Zheng JG, Chen LS (2009) Antagonistic actions of boron against inhibitory effects of aluminum toxicity on growth, CO_2 assimilation, ribulose-1, 5-bisphosphate carboxylase/oxygenase, and photosynthetic electron transport probed by the JIP-test, of Citrus grandis seedlings. BMC Plant Biol 9(1):1–6

Jiang C, Liu L, Li X, Han R, Wei Y, Yu Y (2018) Insights into aluminum-tolerance pathways in Stylosanthes as revealed by RNA-seq analysis. Sci Rep 8(1):1–9

Jones DL, Blancaflor EB, Kochian LV, Gilroy S (2006) Spatial coordination of aluminium uptake, production of reactive oxygen species, callose production and wall rigidification in maize roots. Plant, Cell and Environ 29(7):1309–1318

Jung JK, McCouch SR (2013) Getting to the roots of it: genetic and hormonal control of root architecture. Front Plant Sci 4:186

Kaur S, Kaur N, Siddique KH, Nayyar H (2016) Beneficial elements for agricultural crops and their functional relevance in defence against stresses. Arch Agron Soil Sci 62(7):905–920

Kisnierienė V, Lapeikaitė I (2015) When chemistry meets biology: the case of aluminium–a review. Chemija 26(3):148–158

Kobayashi Y, Kobayashi Y, Sugimoto M, Lakshmanan V, Iuchi S, Kobayashi M, Bais HP, Koyama H (2013) Characterization of the complex regulation of AtALMT1 expression in response to phytohormones and other inducers. Plant Physiol 162(2):732–740

Kochian LV, Piñeros MA, Liu J, Magalhaes JV (2015) Plant adaptation to acid soils: the molecular basis for crop aluminum resistance. Annu Rev Plant Biol 66:571–598

Kovermann P, Meyer S, Hörtensteiner S, Picco C, Scholz-Starke J, Ravera S, Lee Y, Martinoia E (2007) The Arabidopsis vacuolar malate channel is a member of the ALMT family. Plant J 52(6):1169–1180

Krewski D, Yokel RA, Nieboer E, Borchelt D, Cohen J, Harry J, Kacew S, Lindsay J, Mahfouz AM, Rondeau V (2007) Human health risk assessment for aluminium, aluminium oxide, and aluminium hydroxide. J Toxicol Environ Health B Crit Rev 10(Suppl 1):1–269. Erratum in: J Toxicol Environ Health B Crit Rev (2008) 11(2):147.

Kundu A, Ganesan M (2020) GhMATE1 expression regulates Aluminum tolerance of cotton and overexpression of GhMATE1 enhances acid soil tolerance of *Arabidopsis*. Curr Plant Biol 24: 100160

Kundu A, Das S, Basu S, Kobayashi Y, Kobayashi Y, Koyama H, Ganesan M (2019) GhSTOP1, a C2H2 type zinc finger transcription factor is essential for aluminum and proton stress tolerance and lateral root initiation in cotton. Plant Biol (Stuttg) 21(1):35–44

Larsen PB, Geisler MJB, Jones CA, Williams KM, Cancel JD (2005) ALS3 encodes a phloem-localized ABC transporter-like protein that is required for aluminum tolerance in *Arabidopsis*. Plant J 41:353–363

Lei GJ, Yokosho K, Yamaji N, Fujii-Kashino M, Ma JF (2017) Functional characterization of two half-size ABC transporter genes in aluminium-accumulating buckwheat. New Phytol 215(3): 1080–1089

Li N, Meng H, Xing H, Liang L, Zhao X, Luo K (2017) Genome-wide analysis of MATE transporters and molecular characterization of aluminum resistance in *Populus*. J Exp Bot 68(20):5669–5683

Li X, Li Y, Mai J, Tao L, Qu M, Liu J, Shen R, Xu G, Feng Y, Xiao H, Wu L (2018a) Boron alleviates aluminum toxicity by promoting root alkalization in transition zone via polar auxin transport. Plant Physiol 177(3):1254–1266

Li GZ, Wang ZQ, Yokosho K, Ding B, Fan W, Gong QQ, Li GX, Wu YR, Yang JL, Ma JF, Zheng SJ (2018b) Transcription factor WRKY 22 promotes aluminum tolerance via activation of Os FRDL 4 expression and enhancement of citrate secretion in rice (*Oryza sativa*). New Phytol 219(1):149–162

Li D, Ma W, Wei J, Mao Y, Peng Z, Zhang J, Kong X, Han Q, Fan W, Yang Y, Chen J (2019) Magnesium promotes root growth and increases aluminum tolerance via modulation of nitric oxide production in *Arabidopsis*. Plant Soil 2019:1–3

Li W, Du J, Feng H, Wu Q, Xu G, Shabala S, Yu L (2020a) Function of NHX-type transporters in improving rice tolerance to aluminum stress and soil acidity. Planta 251(3):71

Li CX, Yan JY, Ren JY, Sun L, Xu C, Li GX, Ding ZJ, Zheng SJ (2020b) A WRKY transcription factor confers aluminum tolerance via regulation of cell wall modifying genes. J Integr Plant Biol 62(8):1176–1192

Liang C, Piñeros MA, Tian J, Yao Z, Sun L, Liu J, Shaff J, Coluccio A, Kochian LV, Liao H (2013) Low pH, aluminum, and phosphorus coordinately regulate malate exudation through GmALMT1 to improve soybean adaptation to acid soils. Plant Physiol 161(3):1347–1361

Liang X, Ou Y, Zhao H, Zhou W, Sun C, Lin X (2021) Lipid peroxide-derived short-chain aldehydes are involved in aluminum toxicity of wheat (*Triticum aestivum*) roots. J Agric Food Chem 69(36):10496–10505

Ligaba A, Katsuhara M, Sakamoto W, Matsumoto H (2007) The BnALMT1 protein that is an aluminum-activated malate transporter is localized in the plasma membrane. Plant Signal Behav 2(4):255–257

Liu J, Zhou M (2018) The ALMT gene family performs multiple functions in plants. Agronomy 8(2):20

Liu J, Magalhaes JV, Shaff J, Kochian LV (2009) Aluminum-activated citrate and malate transporters from the MATE and ALMT families function independently to confer *Arabidopsis* aluminum tolerance. Plant J 57(3):389–399

Liu MY, Chen WW, Xu JM, Fan W, Yang JL, Zheng SJ (2013) The role of VuMATE1 expression in aluminium-inducible citrate secretion in rice bean (*Vigna umbellata*) roots. J Exp Bot 64(7): 1795–1804

Liu M, Xu J, Lou H, Fan W, Yang J, Zheng S (2016) Characterization of VuMATE1 expression in response to iron nutrition and aluminum stress reveals adaptation of rice bean (*Vigna umbellata*) to acid soils through cis regulation. Front Plant Sci 7:511

Liu J, Zhou M, Delhaize E, Ryan PR (2017) Altered expression of a malate-permeable anion channel, OsALMT4. Disrupts Mineral Nutrition Plant Physiol 175(4):1745–1759

Liu MY, Lou HQ, Chen WW, Piñeros MA, Xu JM, Fan W, Kochian LV, Zheng SJ, Yang JL (2018) Two citrate transporters coordinately regulate citrate secretion from rice bean root tip under aluminum stress. Plant Cell Environ 41(4):809–822

Liu YT, Shi QH, Cao HJ, Ma QB, Nian H, Zhang XX (2020a) Heterologous expression of a Glycine soja C2H2 Zinc finger gene improves aluminum tolerance in Arabidopsis. Int J Mol Sci 21(8):2754. https://doi.org/10.3390/ijms21082754

Liu Y, Xu J, Guo S, Yuan X, Zhao S, Tian H, Dai S, Kong X, Ding Z (2020b) AtHB7/12 regulate root growth in response to aluminum stress. Int J Mol Sci 21(11):4080

Liu W, Feng X, Cao F, Wu D, Zhang G, Vincze E, Wang Y, Chen ZH, Wu F (2021) An ATP binding cassette transporter HvABCB25 confers aluminum detoxification in wild barley. J Hazard Mater 401:123371

Lou HQ, Fan W, Jin JF, Xu JM, Chen WW, Yang JL, Zheng SJ (2020) A NAC-type transcription factor confers aluminium resistance by regulating cell wall-associated receptor kinase 1 and cell wall pectin. Plant Cell Environ 43(2):463–478

Ma B, Gao L, Zhang H, Cui J, Shen Z (2012) Aluminum-induced oxidative stress and changes in antioxidant defenses in the roots of rice varieties differing in Al tolerance. Plant Cell Rep 31(4): 687–696

Ma JF, Chen ZC, Shen RF (2014) Molecular mechanisms of Al tolerance in gramineous plants. Plant Soil 381(1):1–2

Ma Q, Yi R, Li L, Liang Z, Zeng T, Zhang Y, Huang H, Zhang X, Yin X, Cai Z, Mu Y, Cheng Y, Zeng Q, Li X, Nian H (2018) GsMATE encoding a multidrug and toxic compound extrusion transporter enhances aluminum tolerance in *Arabidopsis thaliana*. BMC Plant Biol 18(1):212

Ma X, An F, Wang L, Guo D, Xie G, Liu Z (2020) Genome-wide identification of aluminum-activated malate transporter (ALMT) gene family in rubber trees (*Hevea brasiliensis*) highlights their involvement in aluminum detoxification. Forests 11(2):142

Magalhaes JV, Liu J, Guimarães CT, Lana UG, Alves VM, Wang YH, Schaffert RE, Hoekenga OA, Piñeros MA, Shaff JE, Klein PE, Carneiro NP, Coelho CM, Trick HN, Kochian LV (2007) A gene in the multidrug and toxic compound extrusion (MATE) family confers aluminum tolerance in sorghum. Nat Genet 39:1156–1161

Maron LG, Piñeros MA, Guimarães CT, Magalhaes JV, Pleiman JK, Mao C, Shaff J, Belicuas SNJ, Kochian LV (2010) Two functionally distinct members of the MATE (multi-drug and toxic compound extrusion) family of transporters potentially underlie two major aluminum tolerance QTLs in maize. Plant J 61:728–740

Melo JO, Martins LG, Barros BA, Pimenta MR, Lana UG, Duarte CE, Pastina MM, Guimaraes CT, Schaffert RE, Kochian LV, Fontes EP (2019) Repeat variants for the SbMATE transporter protect sorghum roots from aluminum toxicity by transcriptional interplay in cis and trans. Proc Natl Acad Sci 116(1):313–318

Meriño-Gergichevich C, Alberdi M, Ivanov AG, Reyes-Díaz M (2010) Al 3+-Ca2+ interaction in plants growing in acid soils: al-phytotoxicity response to calcareous amendments. J Soil Sci Plant Nutr 10(3):217–243

Meyer S, Mumm P, Imes D, Endler A, Weder B, Al-Rasheid KA, Geiger D, Marten I, Martinoia E, Hedrich R (2010) AtALMT12 represents an R-type anion channel required for stomatal movement in *Arabidopsis* guard cells. Plant J 63(6):1054–1062

Meyer S, Scholz-Starke J, De Angeli A, Kovermann P, Burla B, Gambale F, Martinoia E (2011) Malate transport by the vacuolar AtALMT6 channel in guard cells is subject to multiple regulation. Plant J 67(2):247–257

Min CW, Khan I, Lee BH (2019) Aluminum stress inhibits root growth and alters physiological and antioxidant enzyme responses in alfalfa (*Medicago sativa* L.) roots. J Kor Grassl Forage Sci 39(4):298–302

Mitsis T, Efthimiadou A, Bacopoulou F, Vlachakis D, Chrousos GP, Eliopoulos E (2020) Transcription factors and evolution: an integral part of gene expression. World Acad Sci J 2(1):3–8

Mossor-Pietraszewska T (2001) Effect of aluminium on plant growth and metabolism. Acta Biochim Pol 48(3):673–686

Murali Achary VM, Panda BB (2010) Aluminium-induced DNA damage and adaptive response to genotoxic stress in plant cells are mediated through reactive oxygen intermediates. Mutagenesis 25(2):201–209

Negishi T, Oshima K, Hattori M, Kanai M, Mano S, Nishimura M, Yoshida K (2012) Tonoplast- and plasma membrane-localized aquaporin-family transporters in blue hydrangea sepals of aluminum hyperaccumulating plant. PLoS One 7(8):e43189

Negishi T, Oshima K, Hattori M, Yoshida K (2013) Plasma membrane-localized Al-transporter from blue hydrangea sepals is a member of the anion permease family. Genes Cells 18(5): 341–352

Nogueirol RC, Monteiro FA, Azevedo RA (2015) Tropical soils cultivated with tomato: fractionation and speciation of Al. Environ Monit Assess 187(4):160

Ohyama Y, Ito H, Kobayashi Y, Ikka T, Morita A, Kobayashi M, Imaizumi R, Aoki T, Komatsu K, Sakata Y, Iuchi S, Koyama H (2013) Characterization of AtSTOP1 orthologous genes in tobacco and other plant species. Plant Physiol 162:1937–1946

Palmer AJ, Baker A, Muench SP (2016) The varied functions of aluminium-activated malate transporters-much more than aluminium resistance. Biochem Soc Trans 44(3):856–862

Pan WL, Hopkins AG, Jackson WA (1988) Aluminum-inhibited shoot development in soybean: a possible consequence of impaired cytokinin supply. Commun Soil Sci Plant Anal 19(7–12): 1143–1153

Panda SK, Baluska F, Matsumoto H (2009) Aluminum stress signaling in plants. Plant Signal Behav 4(7):592–597

Panhwar QA, Naher UA, Radziah O, Shamshuddin J, Razi IM (2015) Eliminating aluminum toxicity in an acid sulfate soil for rice cultivation using plant growth promoting bacteria. Molecules 20(3):3628–3646

Peng W, Wu W, Peng J, Li J, Lin Y, Wang Y, Tian J, Sun L, Liang C, Liao H (2018) Characterization of the soybean GmALMT family genes and the function of GmALMT5 in response to phosphate starvation. J Integr Plant Biol 60(3):216–231

Pereira JF, Zhou G, Delhaize E, Richardson T, Zhou M, Ryan PR (2010) Engineering greater aluminium resistance in wheat by over-expressing TaALMT1. Ann Bot 106(1):205–214

Phukunkamkaew S, Tisarum R, Pipatsitee P, Samphumphuang T, Maksup S, Cha-Um S (2021) Morpho-physiological responses of indica rice (*Oryza sativa* sub. *Indica*) to aluminum toxicity at seedling stage. Environ Sci Pollut Res Int 28(23):29321–29331

Pilon-Smits EA, Quinn CF, Tapken W, Malagoli M, Schiavon M (2009) Physiological functions of beneficial elements. Curr Opin Plant Biol 12(3):267–274

Poot-Poot W, Teresa Hernandez-Sotomayor SM (2011) Aluminum stress and its role in the phospholipid signaling pathway in plants and possible biotechnological applications. IUBMB Life 63(10):864–872

Rahman MA, Lee SH, Ji HC, Kabir AH, Jones CS, Lee KW (2018) Importance of mineral nutrition for mitigating aluminum toxicity in plants on acidic soils: current status and opportunities. Int J Mol Sci 19(10):3073

Ramesh SA, Tyerman SD, Xu B, Bose J, Kaur S, Conn V, Domingos P, Ullah S, Wege S, Shabala S, Feijó JA (2015) GABA signalling modulates plant growth by directly regulating the activity of plant-specific anion transporters. Nat Commun 6(1):1

Ranjan A, Sinha R, Lal SK, Bishi SK, Singh AK (2021a) Phytohormone signalling and cross-talk to alleviate aluminium toxicity in plants. Plant Cell Rep 40(8):1331–1343

Ranjan A, Sinha R, Sharma TR, Pattanayak A, Singh AK (2021b) Alleviating aluminum toxicity in plants: implications of reactive oxygen species signaling and crosstalk with other signaling pathways. Physiol Plant 2021:1–20

Reyna-Llorens I, Corrales I, Poschenrieder C, Barcelo J, Cruz-Ortega R (2015) Both aluminum and ABA induce the expression of an ABC-like transporter gene (FeALS3) in the Al-tolerant species *Fagopyrum esculentum*. Environ Exp Bot 111:74–82

Ribeiro AP, de Souza WR, Martins PK, Vinecky F, Duarte KE, Basso MF, da Cunha BADB, Campanha RB, de Oliveira PA, Centeno DC, Cançado GMA, de Magalhães JV, de Sousa CAF, Andrade AC, Kobayashi AK, Molinari HBC (2017) Overexpression of *BdMATE* gene improves aluminum tolerance in *Setaria viridis*. Front Plant Sci 8:865

Ribeiro AP, Vinecky F, Duarte KE, Santiago TR, Casari RA, Hell AF, da Cunha BA, Martins PK, da Cruz Centeno D, de Oliveira Molinari PA, de Almeida Cançado GM (2021) Enhanced aluminum tolerance in sugarcane: evaluation of SbMATE overexpression and genome-wide identification of ALMTs in *saccharum spp*. BMC Plant Biol 21(1):1–5

Rogers EE, Wu X, Stacey G, Nguyen HT (2009) Two MATE proteins play a role in iron efficiency in soybean. J Plant Physiol 166(13):1453–1459

Sah SK, Reddy KR, Li J (2016) Abscisic acid and abiotic stress tolerance in crop plants. Front Plant Sci 7:571

Saha B, Swain D, Borgohain P, Rout GR, Koyama H, Panda SK (2020) Enhanced exudation of malate in the rhizosphere due to AtALMT1 overexpression in blackgram (*Vigna mungo* L.) confers increased aluminium tolerance. Plant Biol (Stuttg) 22(4):701–708

Sasaki T, Yamamoto Y, Ezaki B, Katsuhara M, Ahn SJ, Ryan PR, Delhaize E, Matsumoto H (2004) A wheat gene encoding an aluminum-activated malate transporter. Plant J 37:645–653

Sawaki Y, Iuchi S, Kobayashi Y, Kobayashi Y, Ikka T, Sakurai N, Fujita M, Shinozaki K, Shibata D, Kobayashi M, Koyama H (2009) STOP1 regulates multiple genes that protect arabidopsis from proton and aluminum toxicities. Plant Physiol 150(1):281–294

Sawaki Y, Kihara-Doi T, Kobayashi Y, Nishikubo N, Kawazu T, Kobayashi Y, Koyama H, Sato S (2013) Characterization of Al-responsive citrate excretion and citrate-transporting MATEs in Eucalyptus camaldulensis. Planta 237(4):979-89

Schreiber HD, Jones AH, Lariviere CM, Mayhew KM, Cain JB (2011) Role of aluminum in red-to-blue color changes in *Hydrangea macrophylla* sepals. Biometals 24(6):1005–1015

Sharma T, Dreyer I, Kochian L, Piñeros MA (2016) The ALMT family of organic acid transporters in plants and their involvement in detoxification and nutrient security. Front Plant Sci 7:1488

Shetty R, Vidya CS, Prakash NB, Lux A, Vaculík M (2021) Aluminum toxicity in plants and its possible mitigation in acid soils by biochar: a review. Sci Total Environ 765:142744

Singh D, Chauhan SK (2011) Organic acids of crop plants in aluminium detoxification. Curr Sci 100(10):1509–1515

Sivaguru M, Pike S, Gassmann W, Baskin TI (2003) Aluminum rapidly depolymerizes cortical microtubules and depolarizes the plasma membrane: evidence that these responses are mediated by a glutamate receptor. Plant Cell Physiol 44(7):667–675

Smirnov OE, Kosyan AM, Kosyk OI, Taran NY (2014) Buckwheat stomatal traits under aluminium toxicity. Mod Phytomorphol 6:15–18

Sun P, Tian QY, Chen J, Zhang WH (2010) Aluminium-induced inhibition of root elongation in *Arabidopsis* is mediated by ethylene and auxin. J Exp Bot 61(2):347–356

Sun L, Zhang M, Liu X, Mao Q, Shi C, Kochian LV, Liao H (2020) Aluminium is essential for root growth and development of tea plants (Camellia sinensis). J Integr Plant Biol 62(7):984–997

Szurman-Zubrzycka M, Chwiałkowska K, Niemira M, Kwaśniewski M, Nawrot M, Gajecka M, Larsen PB, Szarejko I (2021) Aluminum or low pH–which is the bigger enemy of barley? Transcriptome analysis of barley root meristem under Al and low pH stress. Front Genet 12:712

Tahara K, Norisada M, Hogetsu T, Kojima K (2005) Aluminum tolerance and aluminum-induced deposition of callose and lignin in the root tips of *Melaleuca* and *Eucalyptus* species. J For Res 10(4):325–333

Tietz T, Lenzner A, Kolbaum AE, Zellmer S, Riebeling C, Gürtler R, Jung C, Kappenstein O, Tentschert J, Giulbudagian M, Merkel S (2019) Aggregated aluminium exposure: risk assessment for the general population. Arch Toxicol 93(12):3503–3521

Tolrà R, Vogel-Mikuš K, Hajiboland R, Kump P, Pongrac P, Kaulich B, Gianoncelli A, Babin V, Barceló J, Regvar M, Poschenrieder C (2011) Localization of aluminium in tea (Camellia sinensis) leaves using low energy X-ray fluorescence spectro-microscopy. J Plant Res 124(1): 165–172

Tovkach A, Ryan PR, Richardson AE, Lewis DC, Rathjen TM, Ramesh S, Tyerman SD, Delhaize E (2013) Transposon-mediated alteration of TaMATE1B expression in wheat confers constitutive citrate efflux from root apices. Plant Physiol 161:880–892

Upadhyay N, Kar D, Deepak Mahajan B, Nanda S, Rahiman R, Panchakshari N, Bhagavatula L, Datta S (2019) The multitasking abilities of MATE transporters in plants. J Exp Bot 70(18): 4643–4656

Vera-Villalobos H, Lunario-Delgado L, Pérez-Retamal D, Román D, Leiva JC, Zamorano P, Mercado-Seguel A, Gálvez AS, Benito C, Wulff-Zottele C (2020) Sulfate nutrition improves short-term Al^{3+}-stress tolerance in roots of Lolium perenne L. Plant Physiol Biochem 148:103–113

Vishwakarma K, Upadhyay N, Kumar N, Yadav G, Singh J, Mishra RK, Kumar V, Verma R, Upadhyay RG, Pandey M, Sharma S (2017) Abscisic acid signaling and abiotic stress tolerance in plants: a review on current knowledge and future prospects. Front Plant Sci 8:161

Wang H, Chen RF, Iwashita T, Shen RF, Ma JF (2015a) Physiological characterization of aluminum tolerance and accumulation in tartary and wild buckwheat. New Phytol 205(1): 273–279

Wang R, Liu X, Liang S, Ge Q, Li Y, Shao J, Qi Y, An L, Yu F (2015b) A subgroup of MATE transporter genes regulates hypocotyl cell elongation in *Arabidopsis*. J Exp Bot 66(20): 6327–6343

Wang P, Yu W, Zhang J, Rengel Z, Xu J, Han Q, Chen L, Li K, Yu Y, Chen Q (2016) Auxin enhances aluminium-induced citrate exudation through upregulation of GmMATE and activation of the plasma membrane H+-ATPase in soybean roots. Ann Bot 118(5):933–940

Wang Y, Li R, Li D, Jia X, Zhou D, Li J, Lyi SM, Hou S, Huang Y, Kochian LV, Liu J (2017) NIP1;2 is a plasma membrane-localized transporter mediating aluminum uptake, translocation, and tolerance in *Arabidopsis*. Proc Natl Acad Sci U S A 114(19):5047–5052

Wang Z, Liu L, Su H, Guo L, Zhang J, Li Y, Xu J, Zhang X, Guo YD, Zhang N (2020) Jasmonate and aluminum crosstalk in tomato: identification and expression analysis of WRKYs and ALMTs during JA/Al-regulated root growth. Plant Physiol Biochem 154:409–418

Wang P, Dong Y, Zhu L, Hao Z, Hu L, Hu X, Wang G, Cheng T, Shi J, Chen J (2021) The role of γ-aminobutyric acid in aluminum stress tolerance in a woody plant, *Liriodendron chinense*× *tulipifera*. Hort Res 8(1):1–15

Ward CL, Kleinert A, Scortecci KC, Benedito VA, Valentine AJ (2011) Phosphorus-deficiency reduces aluminium toxicity by altering uptake and metabolism of root zone carbon dioxide. J Plant Physiol 168(5):459–465

Wei Y, Han R, Xie Y, Jiang C, Yu Y (2021) Recent advances in understanding mechanisms of plant tolerance and response to aluminum toxicity. Sustainability 13(4):1782

Wu X, Li R, Shi J, Wang J, Sun Q, Zhang H, Xing Y, Qi Y, Zhang N, Guo YD (2014) *Brassica oleracea* MATE encodes a citrate transporter and enhances aluminum tolerance in *Arabidopsis thaliana*. Plant Cell Physiol 55(8):1426–1436

Wu L, Kobayashi Y, Wasaki J, Koyama H (2018a) Organic acid excretion from roots: a plant mechanism for enhancing phosphorus acquisition, enhancing aluminum tolerance, and recruiting beneficial rhizobacteria. Soil Sci Plant Nutr 64(6):697–704

Wu W, Lin Y, Chen Q, Peng W, Peng J, Tian J, Liang C, Liao H (2018b) Functional conservation and divergence of soybean GmSTOP1 members in proton and aluminum tolerance. Front Plant Sci 9:570

Wu L, Guo Y, Cai S, Kuang L, Shen Q, Wu D, Zhang G (2020) The zinc finger transcription factor ATF1 regulates aluminum tolerance in barley. J Exp Bot 71(20):6512–6523

Xia J, Yamaji N, Kasai T, Ma JF (2010) Plasma membrane-localized transporter for aluminum in rice. Proc Natl Acad Sci U S A 107(43):18381–18385

Xia JX, Yamaji N, Ma JF (2013) A plasma membrane-localized small peptide is involved in rice aluminum tolerance. Plant J 76:345–355

Xu M, Gruber BD, Delhaize E, White RG, James RA, You J, Yang Z, Ryan PR (2015) The barley anion channel, HvALMT1, has multiple roles in guard cell physiology and grain metabolism. Physiol Plant 153(1):183–193

Yamaji N, Huang CF, Nagao S, Yano M, Sato Y, Nagamura Y, Ma JF (2009) A zinc finger transcription factor ART1 regulates multiple genes implicated in aluminum tolerance in rice. Plant Cell 21:3339–3349

Yamamoto Y, Kobayashi Y, Devi SR, Rikiishi S, Matsumoto H (2002) Aluminum toxicity is associated with mitochondrial dysfunction and the production of reactive oxygen species in plant cells. Plant Physiol 128(1):63–72

Yan L, Riaz M, Liu J, Yu M, Cuncang J (2020) The aluminum tolerance and detoxification mechanisms in plants; recent advances and prospects. Crit Rev Environ Sci Technol 10:1–37

Yang Y, Wang QL, Geng MJ, Guo ZH, Zhao Z (2011) Effect of indole-3-acetic acid on aluminum-induced efflux of malic acid from wheat (*Triticum aestivum* L.). Plant Soil 346(1):215–230

Yang M, Tan L, Xu Y, Zhao Y, Cheng F, Ye S, Jiang W (2015) Effect of low pH and aluminum toxicity on the photosynthetic characteristics of different fast-growing *Eucalyptus* vegetatively propagated clones. PLoS One 10(6):e0130963

Yang ZB, He C, Ma Y, Herde M, Ding Z (2017a) Jasmonic acid enhances Al-induced root growth inhibition. Plant Physiol 173(2):1420–1433

Yang ZB, Liu G, Liu J, Zhang B, Meng W, Müller B, Hayashi KI, Zhang X, Zhao Z, De Smet I, Ding Z (2017b) Synergistic action of auxin and cytokinin mediates aluminum-induced root growth inhibition in *Arabidopsis*. EMBO Rep 18(7):1213–1230

Yang JL, Fan W, Zheng SJ (2019) Mechanisms and regulation of aluminum-induced secretion of organic acid anions from plant roots. J Zhejiang Univ Sci B 20(6):513–527

Yokosho K, Yamaji N, Ma JF (2010) Isolation and characterisation of two MATE genes in rye. Funct Plant Biol 37:296–303

Yokosho K, Yamaji N, Kashino-Fujii M, Ma JF (2016a) Retro transposon-mediated aluminum tolerance through enhanced expression of the citrate transporter OsFRDL4. Plant Physiol 172(4):2327–2336

Yokosho K, Yamaji N, Miho KF, Ma JF (2016b) Functional analysis of a MATE gene OsFRDL2 revealed its involvement in Al-induced secretion of citrate, but less contribution to Al tolerance in rice. Plant Cell Physiol 57:976–985

Yu Y, Jin C, Sun C, Wang J, Ye Y, Zhou W, Lu L, Lin X (2016) Inhibition of ethylene production by putrescine alleviates aluminium-induced root inhibition in wheat plants. Sci Rep 6(1):1

Zhang L, Wu XX, Wang J, Qi C, Wang X, Wang G, Li M, Li X, Guo YD (2018) BoALMT1, an Al-induced malate transporter in cabbage, enhances aluminum tolerance in *Arabidopsis thaliana*. Front Plant Sci 8:2156

Zhang Y, Zhang J, Guo J, Zhou F, Singh S, Xu X, Xie Q, Yang Z, Huang CF (2019) F-box protein RAE1 regulates the stability of the aluminum-resistance transcription factor STOP1 in *Arabidopsis*. Proc Natl Acad Sci 116(1):319–327

Zhang X, Li L, Yang C, Cheng Y, Han Z, Cai Z, Nian H, Ma Q (2020) *GsMAS1* encoding a MADS-box transcription factor enhances the tolerance to aluminum stress in *Arabidopsis thaliana*. Int J Mol Sci 21(6):2004

Zhao XQ, Shen RF (2018) Aluminum–nitrogen interactions in the soil–plant system. Front Plant Sci 18(9):807

Zhou G, Delhaize E, Zhou M, Ryan PR (2013) The barley MATE gene, HvAACT1, increases citrate efflux and Al3+ tolerance when expressed in wheat and barley. Ann Bot 112:603–612

Zhou G, Pereira JF, Delhaize E, Zhou M, Magalhaes JV, Ryan PR (2014) Enhancing the aluminium tolerance of barley by expressing the citrate transporter genes SbMATE and FRD3. J Exp Bot 65(9):2381–2390

Zhu XF, Shi YZ, Lei GJ, Fry SC, Zhang BC, Zhou YH, Braam J, Jiang T, Xu XY, Mao CZ, Pan YJ, Yang JL, Wu P, Zheng SJ (2012) XTH31, encoding an in vitro XEH/XET-active enzyme, regulates aluminum sensitivity by modulating in vivo XET action, cell wall xyloglucan content, and aluminum binding capacity in *Arabidopsis*. Plant Cell 24:4731–4747

Functional, Structural, and Transport Aspects of ZIP in Plants

10

T. P. Ajeesh Krishna, T. Maharajan, S. Ignacimuthu,
and S. Antony Ceasar

Abstract

Zn is an essential micronutrient for plant growth and development. The unavailability of Zn may lead to impaired physiological and metabolic processes of plants. Zinc-regulated, iron-regulated transporter-like protein (ZIP) family members are actively involved in Zn uptake and transport in plants. Zn homeostasis mechanism has been studied some extend in the model plant *Arabidopsis*. The basic-region leucine zipper (bZIP) is a key transcription factor (TF) involved in the regulation of Zn transporter expression and maintains the Zn homeostasis in the cell. The transcriptional regulation of the downstream and upstream processes of the Zn homeostasis network is poorly understood in plants. The functional, structural, and transport aspects of the ZIP are very crucial for understanding the homeostasis mechanism in plants under Zn deficiency conditions. In this chapter, we discuss the details on the structural and functional insight of plant ZIPs and Zn uptake and transport aspects of ZIPs in plants. We also give the details on ZIP genes analyzed in various plants. This chapter is very useful for plant molecular biologists and physiologists to understand the role and function of ZIP in plants.

Keywords

Homeostasis · Transcription factors · Transporters · Zinc

T. P. Ajeesh Krishna · T. Maharajan · S. Antony Ceasar (✉)
Division of Plant Molecular Biology and Biotechnology, Department of Biosciences, Rajagiri College of Social Sciences, Kochi, Kerala, India

S. Ignacimuthu
Xavier Research Foundation, St Xavier's College, Palayamkottai, Tirunelveli, Tamil Nadu, India

K. Kumar, S. Srivastava (eds.), *Plant Metal and Metalloid Transporters*,
https://doi.org/10.1007/978-981-19-6103-8_10

Abbreviations

bZIP	Basic-region leucine zipper
CA	Carbonic anhydrase
CAX	Cation exchanger
CDF	Diffusion facilitator
HMA	Heavy metal ATPase
IAA	Indole-3-acetic acid
NA	Nicotianamine
NAS	Nicotianamine synthase
NRAMP	Natural resistance-associated macrophage protein
PCR	Plant cadmium resistance
P-type ATPase	Plasma membrane-type ATPase
TFs	Transcription factors
TM	Transmembrane
ZDRE	Zinc deficiency-responsive elements
ZIP	Zinc-regulated, iron-regulated transporter-like protein
ZRT	Zinc-regulated transporter
ZSM	Zinc sensor motif

10.1 Introduction

Zinc (Zn) is an essential micronutrient for plants (Krishna et al. 2017) and is involved in a wide range of physiological and molecular processes. It acts as a functional, structural, or regulatory factor for many enzymes and proteins (Gondal et al. 2021). Unavailability of Zn significantly affects the growth of the plant at any stage. The optimum concentration of Zn is critical for the normal cellular functions of plants. Zn needs to be transported from the soil solution into the root system and then distributed to the whole plant organelles, crossing both cellular and intercellular membranes (Pasricha et al. 2021). It is a metabolically controlled plant mechanism. This process is driven through carrier proteins, electrochemical gradient via ion channels, or against the electrochemical gradient via electrogenic pumps. Usually, the soil solution contains only a small amount of trace nutrients. In plants, Zn uptake and transport are very crucial under low Zn soil. Plants must use a high-affinity metal transport system to accumulate Zn ions from the soil. Many metal transporter family members are involved in cellular uptake of Zn, intercellular transport, and distribution in plants.

The zinc-regulated, iron-regulated transporter-like protein (ZIP) family members are actively involved in Zn uptake, trafficking, and detoxification in plants. The ZIP transporters are found in various cell organelles (plasma membrane, vacuolar membrane, epidermis, chloroplast, and endoplasmic reticulum) and are actively involved in metal (Zn^{2+}) homeostasis. Understanding the mechanism of Zn uptake and transport and metal homeostasis is very crucial in the plant system. The structural and functional characterization of the plant ZIP transporter is still lacking. In bacterium, the high-resolution crystal structure of the *Bordetella bronchiseptica* ZIP (BbZIP) protein was deduced, and its metal (Zn/Cd) uptake and transport mechanism were also predicted by Zhang et al. (2017). Plant ZIP transporters revealed greater variation at the Zn^{2+} binding site (functional amino acid residues) when compared to BbZIP as analyzed through in silico approaches (Krishna et al. 2020). The plant physiologist and molecular biologist need to understand the structural and functional mechanism of plant ZIPs to improve the Zn use efficiency. In this chapter, we provide the functional, structural, and transport aspects of ZIP in the plant. This chapter will help plant physiologists and molecular biologists to understand the importance ZIP transporters in plants.

10.2 Role of Zn in Plants

Plant physiology depends on various metabolic processes involving many enzymes (Escudero-Almanza et al. 2012). Zn is one of the essential micronutrients or metal elements for plants. Zn plays an important role in plants (Fig. 10.1). Plant requires

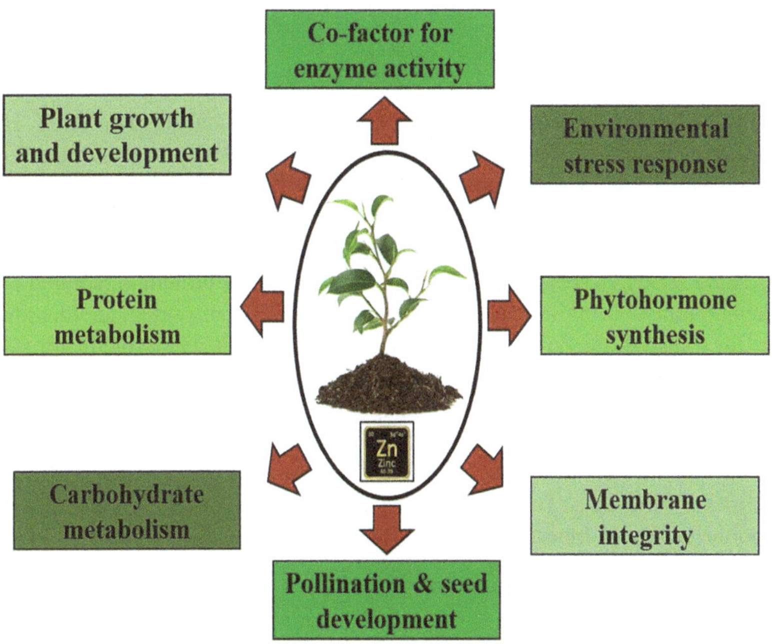

Fig. 10.1 Role of Zn in plants. The plant requires adequate levels of Zn for normal physiological and metabolic functions. Zn is involved in various metabolic processes such as protein and carbohydrate metabolism. It is only one metal required for the activity of 300 enzymes, covering all six classes of enzymes. Zn is catalyzing the synthesis of many phytohormones, which is essential for plant growth and development. Zn also plays an important role in plant stresses, and Zn-dependent protein is involved in the plant stress defense mechanism

adequate levels of Zn for normal physiological functions (Krishna et al. 2020; Hafeez et al. 2013), which is involved in many plant metabolic activities and enzymatic reactions (Hänsch and Mendel 2009). It is a catalytic metal element required for more than 300 enzymes (Vatansever et al. 2016). For example, a metalloenzyme like carbonic anhydrase (CA) requires Zn as a co-factor (Escudero-Almanza et al. 2012). In plants, CA is involved in many biological processes like ion exchange, pH regulation, respiration, CO_2 transfer, and photosynthetic CO_2 fixation (Tiwari et al. 2005). Zn helps for the stability of cellular tissues, protein amalgamation, and phytohormone production. The indole-3-acetic acid (IAA) is a natural auxin (phytohormones) usually occurring in vascular plants. The Zn is directly involved in tryptophan biosynthesis, which in turn acts as the precursor of IAA (Krishna et al. 2020). Horak and Trčka (1976) reported that Zn^{2+} ions enhance the tryptophan production in a 14-day-old shoot of pea (*Pisum sativum* L.) plants (Horak and Trčka 1976). Zn is an essential substance for the synthesis of other phytohormones such as gibberellins, cytokinin, and abscisic acid (Kumar et al. 2016). Zn deficiency decreases the level of these phytohormones in plant tissue, which causes impairment of plant growth. Zn also plays an important role in plant resistance against disease and other stresses, since Zn-dependent protein is involved in plant defense mechanisms under biotic and abiotic stresses (Cabot et al. 2019). So, Zn is a very essential micronutrient for plant growth and development.

10.3 Zn Transport Protein in Plants

In plants, Zn is absorbed in the form of divalent cation (Kumar et al. 2016). Transport of Zn^{2+} from epidermal and cortical cells to xylem vessels occurs through the symplastic or apoplastic pathways (Fig. 10.2). The root-to-shoot Zn^{2+} transport occurs via xylem loading, and it is retranslocated via the phloem cells (Clemens 2001). Charged Zn^{2+} ions cannot freely enter into cells through the plasma membrane (Eide 2005). A special type of membrane or carrier protein is required for Zn uptake and transport in plants. Many membrane family transporters are identified in plants for transporting Zn^{2+}. These include ZIP, cation diffusion facilitator (CDF), plasma membrane-type ATPase (P-type ATPase), natural resistance-associated macrophage protein (NRAMP), cation exchanger (CAX), and plant cadmium resistance (PCR) transporters. These transporters are also facilitating the transport of other metal ions such as cobalt (Co), cadmium (Cd), copper (Cu), iron (Fe), manganese (Mn), nickel (Ni), etc. (Table 10.1). The roles and functions of these membrane transporter family members are reviewed by many researchers (Kumar et al. 2016; Krishna et al. 2017, 2020; Kambe et al. 2019). Very few metal transport members have been identified in plants, and among these, many ZIP family members are identified in plants. In plants, the identification of Zn transporters is essential for understanding the molecular mechanism of Zn uptake and transport. So, plant molecular biologists need to focus on the identification and characterization of metal transporters in plants.

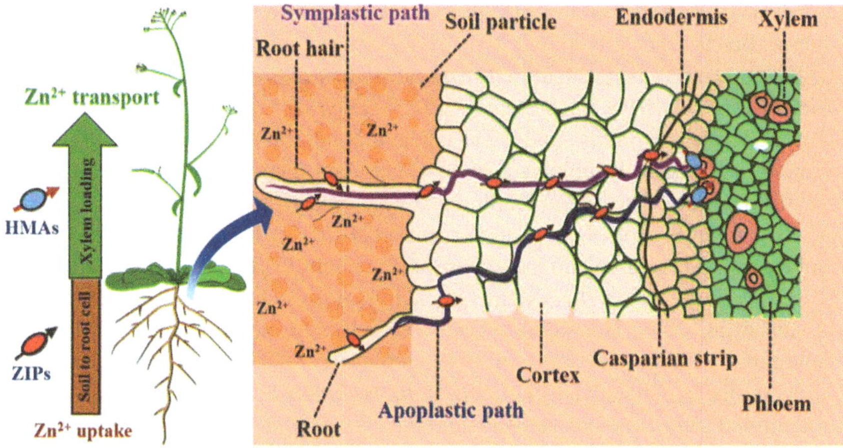

Fig. 10.2 Uptake and translocation of Zn^{2+} ions from root to shoot. Zn ions enter the root cell wall free space by a diffusion process. Uptake and transport of Zn^{2+} into the root cortex take place via a symplastic or apoplastic pathway with the help of membrane-bound low and high-affinity ZIP transporters. Zn ions are translocated into all the parts of the plants by xylem loading by HMA transporters

Table 10.1 Details of metal transporter family proteins identified in plants. The name of the transporter family and their role and functions with references are provided

Name of the family	Function	Role	Reference
ZIP	Metal homeostasis	Zn, Mn, Fe, Cd, Co, Cu, and Ni transport	Pedas and Husted (2009)
CDF	Metal homeostasis	Zn, Fe, Co, Ni, Cd, and Mn transport	Gaither and Eide (2001)
P-type ATPase	Metal homeostasis	Zn, Cu, Co, Cd, Pb, and Ca transport	Pedersen et al. (2012)
NRAMP	Metal homeostasis	Zn and Mn transport	Wang et al. (2019)
CAX	Metal homeostasis	Zn and Ca transport	Socha and Guerinot (2014)
PCR	Metal homeostasis	Zn, Cd and Ca transport	Lin et al. (2020)

10.3.1 Zn Uptake and Transport in Plants

Root Zn uptake involves different transport systems, which include a high-affinity (K_m = 0.6–2 nM) and low-affinity (K_m = 2–5 μM) membrane transporter system (Kumar et al. 2016). Most likely the high-affinity membrane transport system dominantly operates under Zn starvation (Hacisalihoglu et al. 2001). In plants, high-affinity and low-affinity ZIP family transporter members are identified, and their role and functions have been characterized. The ZIP family transporters are facilitating Zn uptake across the cellular membrane into the cytoplasm of the cell.

(Eide et al. 1996; Krishna et al. 2020). The ZIP transporters are expressed in various tissues of plants especially in root tissues under Zn starvation. For example, a total of 13 ZIP transporters were identified in *Poncirus trifoliata*. The six *PtZIP* (*PtZIP1*, *PtZIP2*, *PtZIP3*, *PtZIP5*, *PtZIP6*, and *PtZIP9*) genes are highly expressed in the root tissue of *Poncirus trifoliate* under Zn-deficient condition (Fu et al. 2017). The ZIP transporters are actively involved in Zn uptake and distribution within the cell of *Poncirus trifoliate*. The xylem loading is a very crucial step for translocation of Zn from root to whole parts of the plant (Krishna et al. 2017). The P-type ATPase family members play a crucial role in metal transport in plants (Takahashi et al. 2012). They are actively involved in the xylem (parenchymatous cells) loading of metal ions (Hussain et al. 2004; Hanikenne et al. 2008). Heavy metal ATPase (HMA) is a member of the P-type ATPase family, which plays a crucial role in xylem loading of Zn ions in plants (Chaudhary et al. 2018). In *Arabidopsis*, many HMA transporters are identified, and they are actively involved in Zn transport. The AtHMA2 and AtHMA4 are found to be located in the plasma membrane and actively involved in Zn transport efflux from cells in *Arabidopsis* (Morel et al. 2009; Eren and Arguello 2004; Mills et al. 2005; Verret et al. 2005). *AtHMA2* and *AtHMA4* are expressed in vascular vessels of the root tissue and contributed to Zn loading to the xylem (Hussain et al. 2004; Verret et al. 2004). AtHMA1 is localized in the chloroplast, and it is involved in the detoxification of Zn ions (Kim et al. 2009). AtHMA3 is found in the vacuolar membrane of the cell and also plays an important role in detoxifying Zn through vacuolar sequestration (Morel et al. 2009). The HMA transporters are also involved in metal hyperaccumulation in *Arabidopsis*. For example, the *AhHMA4* is expressed in the root tissue of *Arabidopsis halleri*; it is actively involved in a high amount of Zn translocation from root to shoot (Talke et al. 2006; Courbot et al. 2007; Hanikenne et al. 2008). The ZIP and P-type ATPase family transporters are very important for the uptake, transport, and distribution of Zn in plants. The identification and characterization of Zn transporters are very helpful for understanding the Zn transport mechanism in plants.

10.4 ZIP in Plants

The ZIP family is a metal transporter, and they are actively involved in the uptake and transport of various citations such as Cd, Fe, Mn, and Zn. The ZIP genes showed a dynamic role in plants under the Zn deficiency condition. For example, OsZIPs are involved in root to shoot Zn transportation. The OsZIP1 and OsZIP3 are involved in Zn uptake from soil solutions; OsZIP4, OsZIP5, and OsZIP8 are involved in root to shoot translocation; and OsZIP4 and OsZIP8 contribute to the accumulation of Zn in grains (Bashir et al. 2012). The ZIP genes are found to be expressed in the whole part (roots, shoot, node, leaves, culms, spikelets, and grain) of the rice plant under Zn-deficient conditions. OsZIPs are actively involved in Zn uptake, transport, and distribution in rice. Chen et al. (2008) observed that the *OsZIP1*, *OsZIP3*, and *OsZIP4* are highly expressed in the root tissue of rice under Zn-deficient conditions. The *OsZIP1* is expressed only in the root of rice under Zn-deficient conditions (Ramesh et al. 2003). *OsZIP1* is specifically involved in Zn uptake from the soil solution. *OsZIP3* gene is upregulated in root and soot tissue of rice under both Zn-deficient and Zn-sufficient conditions (Ramesh et al. 2003). It showed that the *OsZIP* genes play a very crucial role in Zn uptake and transport under both Zn-sufficient and Zn-deficient conditions in rice. In barely, 13 HvZIPs were identified, and their expression pattern was analyzed in various tissues under Zn-deficient condition (Tiong et al. 2015). In this study, six *HvZIP* genes such as *HvZIP3*, *HvZIP5*, *HvZIP7*, *HvZIP8*, *HvZIP10*, and *HvZIP13* were highly upregulated under Zn-deficient condition (Tiong et al. 2015). All these six HvZIP transporter proteins are found in the plasma membrane of the cell (Tiong et al. 2015). The *HvZIP* genes play a very crucial role in Zn uptake under Zn-deficiency conditions in wheat. Chandra et al. (2020) analyzed the expression pattern of the *EcZIP1* gene in various tissue (root, shoot, root-shoot zone, and flag leaf) of six finger millet genotypes (VHC 3582, IE 3618, IE 6240, VL 330, GE 724, and VHC 3893). This study revealed that the expression level was higher in the root, shoot, root-shoot zone, and flag leaf of GE 724 compared to the other finger millet genotypes (Chandra et al. 2020). The study also showed that the expression level of the *EcZIP1* gene contributes to the genetic variation of finger millet genotypes. ZIP family members have been identified in many plants, and their tissue-specific expression patterns were analyzed under Zn-deficient condition (Table 10.2). ZIP genes are highly expressed under Zn-deficient condition in plants to maintain the Zn homeostasis.

The functional characterization is important for understanding the role of ZIP in plants. For example, the AtZIP1, AtZIP2, and AtZIP3 facilitated Zn transport based on the yeast complementation assay (Eide 1998; Grotz et al. 1998; Guerinot 2000). AtZIP1 and AtZIP3 were confirmed to be the low-affinity and AtZIP2 as high-affinity membrane transporters (Grotz et al. 1998). These results could help understand the upregulation and downregulation of ZIP genes under both sufficient and deficient conditions. Only a few ZIP family transporters are functionally characterized in plants like *Arabidopsis* (Grotz et al. 1998; Vert et al. 2009), rice (Lee et al. 2010a; Ishimaru et al. 2005; Yang et al. 2020; Menguer et al. 2013), maize (Mondal et al. 2014), wheat (Durmaz et al. 2011; Hacisalihoglu et al. 2001), barley (Pedas and Husted 2009), and soybean (Moreau et al. 2002). The Zn-sensitive yeast (DY1457; *Saccharomyces cerevisiae*) mutant (*zrt1*/*zrt2*), which is defective in both the high-affinity (Zn-regulated transporter 1; ZRT1) and the low-affinity (ZRT2) uptake transporters system and is susceptible to Zn starvation conditions, was employed for functional characterization of plant ZIP transporters (Zhao and Eide 1996a, b). Functional characterization of all ZIP in plants will help to understand the mechanism of Zn uptake. Genetic modification of crops with ZIP genes improved the Zn uptake efficiency in plants. The overexpression of ZIP genes in crops such as *Arabidopsis* (Li et al. 2015), rice (Lee et al. 2010a,b; Ishimaru et al. 2007), barely (Tiong et al. 2014; Ramesh et al. 2004), finger millet (Ramegowda et al. 2013), cassava (Gaitán-Solís et al. 2015), and tobacco (Ramegowda et al. 2013) showed higher content of Zn in edible parts. So, the genetic modification/overexpression of ZIP in plants may contribute to improving the Zn content in the edible part of the crops.

Table 10.2 Details of ZIP genes identified in plants. The plant name, localization, and site of expression under Zn deficiency are presented with reference

Plant name	Common name	Genes identified	Localization in plant cell	Expression under Zn deficiency	Reference
Oryza sativa	Rice	*OsZIP1*	Vascular bundle	Shoot and root	Chen et al. (2008), Ishimaru et al. (2005), Yang et al. (2009) Lee et al. (2010a), Lee et al. (2010b), Ramesh et al. (2004)
		OsZIP2	Vascular bundle	Shoot and root	
		OsZIP3	Vascular bundle and epidermis	Shoot, leaves, and root	
		OsZIP4	Epidermal cell	Shoot and root	
		OsZIP5	–	Shoot and root	
		OsZIP6	–	Shoot and root	
		OsZIP7	Plasma membrane	Roots, culms, leaves spikelets, and shoot	
		OsZIP8	Plasma membrane	Roots, culms, leaves, and spikelets	
		OsZIP9-OsZIP16	–	–	
Zea mays	Maize	*ZmZIP1*	ER and plasma membrane	Shoot and root	Li et al. (2013), Kou et al. (2014)
		ZmZIP2	ER and plasma membrane	Shoot and root	
		ZmZIP3	ER and plasma membrane	Shoot and root	
		ZmZIP4	ER and plasma membrane	Shoot and root	
		ZmZIP5	ER and plasma membrane	Shoot and root	
		ZmZIP6	ER and plasma membrane	Shoot and root	
		ZmZIP7	ER and plasma membrane	Shoot and root	

(continued)

Table 10.2 (continued)

Plant name	Common name	Genes identified	Localization in plant cell	Expression under Zn deficiency	Reference
		ZmZIP8	ER and plasma membrane	Shoot and root	
		ZmZIP1	Plasma membrane	Highly expressed in flag leaf	
		ZmZIP2	Plasma membrane	Highly expressed in flag leaf	
		ZmZIP3	Plasma membrane	Kernel	
		ZmZIP4	Plasma membrane	Highly expressed in flag leaf	
		ZmZIP5	Chloroplast and plasma membrane	Highly expressed in flag leaf	
		ZmZIP6	Plasma membrane	Highly expressed in flag leaf	
		ZmZIP7	Chloroplast and plasma membrane	Highly expressed in flag leaf	
		ZmZIP8	Plasma membrane	Highly expressed in flag leaf	
		ZmZIP9	Plasma membrane	Highly expressed in flag leaf	
		ZmZIP10	Plasma membrane	Highly expressed in flag leaf	
		ZmZIP11	Plasma membrane	Highly expressed in flag leaf	
		ZmZIP12	Plasma membrane	Kernel	
Triticum aestivum	Common wheat	*TaZIP1*	–	–	Evens et al. (2017)
		TaZIP2	–	–	
		TaZIP3	–	Shoot and root	
		TaZIP5	–	–	
		TaZIP6	–	Shoot and root	

(continued)

Table 10.2 (continued)

Plant name	Common name	Genes identified	Localization in plant cell	Expression under Zn deficiency	Reference
		TaZIP7	–	Shoot and root	
		TaZIP9	–	Shoot and root	
		TaZIP10	–	–	
		TaZIP11	–	–	
		TaZIP13	–	Shoot and root	
		TaZIP14	–	–	
		TaZIP16	–	–	
Triticum durum	Durum wheat/ Macaroni wheat	*TdZIP1*	Plasma membrane	Flag, non-flag leaves, stem and spike	Deshpande et al. (2018), Durmaz et al. (2011)
		TdZIP3	Plasma membrane	Flag, non-flag leaves, stem and spike	
		TdZIP7	Plasma membrane	Flag, non-flag leaves, stem and spike	
		TdZIP10	Plasma membrane	Flag, non-flag leaves, stem and spike	
		TdZIP15	Plasma membrane	Flag, non-flag leaves, stem and spike	
		TdZIP1	Endoplasmic reticulum	Root	
Hordeum vulgare	Barely	*HvZIP1*	–	–	Tiong et al. (2014), Tiong et al. (2015), Pedas et al. (2009)
		HvZIP2	–	Shoot	
		HvZIP3	Plasma membrane	Shoot and root	
		HvZIP5	Plasma and vacuolar membrane	Shoot and root	

(continued)

Table 10.2 (continued)

Plant name	Common name	Genes identified	Localization in plant cell	Expression under Zn deficiency	Reference
		HvZIP6	–	–	
		HvZIP7	Plasma membrane	Shoot and root	
		HvZIP8	Plasma membrane	Shoot and root	
		HvZIP10	Plasma membrane	Shoot and root	
		HvZIP11	–	–	
		HvZIP13	Plasma membrane	Shoot and root	
		HvZIP14	–	–	
		HvZIP16	–	–	
Glycine max	Soybean	*GmZIP1*	Peribacteroid membrane	Nodules	Moreau et al. (2002)
		GmZIP4	–	–	
		GmZIP6	–	–	
		GmZIP10	–	–	
		GmZIP11	–	–	
Phaseolus vulgaris	Common bean	*PvZIP1-PvZIP11*	–	–	Astudillo et al. (2013)
		PvZIP12	–	Root, leaf and pod	
		PvZIP13	–	Root, leaf and pod	
		PvZIP14	–	–	
		PvZIP15	–	–	
		PvZIP16	–	Root, leaf and pod	
		PvZIP17-PvZIP19	–	–	
Poncirus trifoliate	Hardy orange	*PtZIP1*	Plasma membrane	Leaf and root	Fu et al. (2017)
		PtZIP2	Plasma membrane	Leaf and root	
		PtZIP3	Plasma membrane	Leaf and root	
		PtZIP5	Plasma membrane	Leaf and root	
		PtZIP6	Plasma membrane	Leaf and root	
		PtZIP7	Plasma membrane	Leaf and root	

(continued)

Table 10.2 (continued)

Plant name	Common name	Genes identified	Localization in plant cell	Expression under Zn deficiency	Reference
		PtZIP9	Plasma membrane	Leaf and root	
		PtZIP11	Plasma membrane	Leaf and root	
		PtZIP12	Plasma membrane	Leaf and root	
		PtZIP13	Plasma membrane	Leaf and root	
		PtZIP14	Plasma membrane	Leaf and root	
Citrus sinensis	Sweet orange	*CsZIP1*	–	Leaf	Fei et al. (2016)
		CsZIP2	–	Leaf	
		CsZIP3	–	Leaf	
		CsZIP4	–	Leaf	
Setaria italica	Foxtail millet	*SiZIP1*	–	Root, leaf, stem and spica tissue	Alagarasan et al. (2017)
		SiZIP2	–	Root, leaf, stem and spica tissue	
		SiZIP3	–	Root, leaf, stem, and spica tissue	
		SiZIP4	–	Root, leaf, stem, and spica tissue	
		SiZIP5	–	Root, leaf, stem, and spica tissue	
		SiZIP6	–	Root, leaf, stem, and spica tissue	
		SiZIP7	–	Root, leaf, stem, and spica tissue	
Eleusine coracana	Finger millet	*EcZIP1*	–	Root, shoot, root-shoot zone, and flag leaf	Chandra et al. (2020)

10.4.1 Structural and Functional Aspect of ZIP in Plants

Very little information is available on the structural and functional aspects of ZIP in plants. The ZIP transporters have six to nine transmembrane (TM) domains, while eight TM is the most prevalent form. The high-resolution crystal structure is very essential for understanding the molecular mechanism of Zn uptake and transport in plants. The ZIP crystal structure is not yet available in plants. The crystal structure of ZIP has been deducted for a bacterium *Bordetella bronchiseptica*, and its metal (Cd^{2+} and Zn^{2+}) uptake and transport mechanism are also proposed by Zhang et al. (2017).

The bacterium BbZIP transporter has eight TM domains (TM1-TM8). The metal-binding residues are located in the TM4 and TM5. Several functional amino acid residues (His177, Asn178, Pro180, Glu181, and Gly182) and metal-binding amino acid residues (Gln207, Asp/Asn208, Pro210, Glu211, and Gly212) are found in TM4 and TM5 domains of the BbZIP (Zhang et al. 2017). The structure of the BbZIP found seven Zn^{2+} metal-binding sites in the TM domains. The Ser106 on TM2 is located on the bottom of the entrance cavity, and it is guiding metals into the metal transport pathway. The Asp113 and Asp305 (metal-chelating) residues are found in the entrance cavity, which is important for recruiting metal substrates. Also, multiple metal-binding sites found in the metal exist cavity, and it is indicating that these contribute a pathway of metal releases to the cytoplasm. The metal-chelating amino acid residues such as His177, Glu276, His275, Pro180, Pro210, and Asp144 contribute to metal release into the cytoplasm. In BbZIP, the Zn^{2+} binding is penta-coordinated by three Glu residues (Glu181, Gln207, and Glu211) and two molecules of water. The detailed structural and functional aspect of BbZIP is reviewed by Krishna et al. (2020). The BbZIP is actively involved in metal (Cd^{2+} and Zn^{2+}) transport in *B. bronchiseptica*. The high-resolution BbZIP structure provides a model to analyze the functional residues of the plant ZIP transporters.

Krishna et al. (2020) predicted the metal-binding residues of plant ZIPs using the BbZIP as a template by in silico approaches (Krishna et al. 2020). It revealed that plant ZIP transporters are partially conserved with BbZIP (Krishna et al. 2020). Most of the metal-binding and transport residues are not homologous in plant ZIPs compared with BbZIP. Some of the functional residues are conserved in plant ZIPs. For example, the functional residues His177 and Gly182 are actively involved in the metal (Cd/Zn) release from the metal-binding site of the BbZIP. Most of the plant ZIP protein sequences have conserved His117 residue found in the BbZIP (Krishna et al. 2020). Also, homology modeling revealed that the plant ZIP transporters (AtZIP1, AtZIP2, AtZIP8, OsZIP13, OsZIP16, ZmZIP6, and ZmZIP11) show greater difference at the Zn^{2+} binding amino acid residues when compared to BbZIP (Krishna et al. 2020). Further studies are needed to confirm the functional role of amino acid residues in plant ZIPs. This may help for the identification of functional amino acid residues involved in Zn binding and transport in plant ZIPs. Detecting the crystal structure of plant ZIP may shed more light on Zn transport mechanism in plants. It will give an understanding of the specific Zn^{2+} transport mechanism in plants.

10.4.2 Regulation of ZIP in Plants

Plants have developed a fine Zn homeostasis mechanism to ensure optimum zinc supply to tissues throughout their lifetime under Zn-deficient or toxicity conditions. Many ZIPs are hypothesized to be responsible for Zn cellular uptake and influx into the cytoplasm of the plant cell. The sensing and signaling of the Zn status are very important for Zn homeostasis. Zn homeostasis is a complex cellular function in plant cells involved in Zn uptake, accumulation, transport, trafficking, sequestration, remobilization, and detoxification (Krishna et al. 2017, 2020). The transcriptional regulation of the downstream and upstream processes of the Zn homeostasis network is poorly understood in plants. The transcription factors (TFs) are very crucial for transcriptional regulation of target genes and contribute to Zn homeostasis. The basic-region leucine zipper (bZIP) is a TF actively involved in the regulatory mechanism of plants that underwent any stress (Corrêa et al. 2008). The TFs bZIP19 and bZIP23 are the central regulators of the Zn-deficiency response and belong to F-bZIP group family. The Cys/His-rich motif is conserved in bZIP19 and bZIP23 TFs, and it acts as a Zn sensor (Lilay et al. 2021). The Cys/His-rich amino acids are considered as a Zn-sensor motif (ZSM) of the bZIPs. The bZIP TFs are actively involved in the upregulation and/or downregulation of target ZIP genes under both Zn-sufficient and Zn-deficient conditions. In *Arabidopsis*, the TFs AtbZIP19 and AtbZIP23 are the key regulators known for Zn homeostasis. The AtbZIP19 and AtbZIP23 are considered to be the essential regulators of target ZIP genes under Zn deficiency. Recently, Lilay et al. (2021) proposed the molecular mechanism of TFs bZIP19 and bZIP23 in *Arabidopsis*. Under Zn-deficiency conditions, the TFs bZIP19 and bZIP23 bind to the Zn deficiency-responsive element (RTGTCGACAY) in the promoter region of the target genes and activate their transcription and contribute for Zn uptake and distribution (Fig. 10.3). The TFs AtbZIP19 and AtbZIP23 induce the expression of the *AtZIP4* and *nicotianamine synthase 2* (*NAS2*) gene under Zn deficiency (Lilay et al. 2021). These genes contribute to Zn uptake and nicotianamine (NA)-mediated Zn distribution in *Arabidopsis* under Zn deficiency. The AtbZIP19/23 directly bind with the Zn^{2+} ions into their Cys/His-rich motif under Zn-sufficient conditions. The Cys/His-rich residues are very crucial for detecting the Zn status in the plant cells. It provides plants with a one-step link between the cellular statuses of Zn and helps for the transcriptional regulation of target genes. Under Zn-sufficient condition, the freely available cellular Zn^{2+} ions bind with the Cys/His- rich motif. The binding of Zn^{2+} ions to the protein does not bind with the Zn deficiency-responsive elements (ZDRE) in the target gene and represses the activity of the transcription factors (Fig. 10.3). Lilay et al. (2021) demonstrated the role of Cys/His amino acid residues in the ZSM of the bZIP19 via mutation/deletion of Zn binding residue/motif. The TF bZIP19 variants, including deletion of both motif (bZIP19 [del1 and del2]), deletion of half of the motif (bZIP19 [del1] and bZIP19 [del2]), and substitution of amino acid residues Cys and His with Ala, substituting of all Cys amino acid residue (bZIP19 (Cys44Ala, Cys50Ala, and Cys63Ala), only one Cys amino acid residue in the first half motif (bZIP19 [Cys44Ala]) and two His residue in each half motif (bZIP19

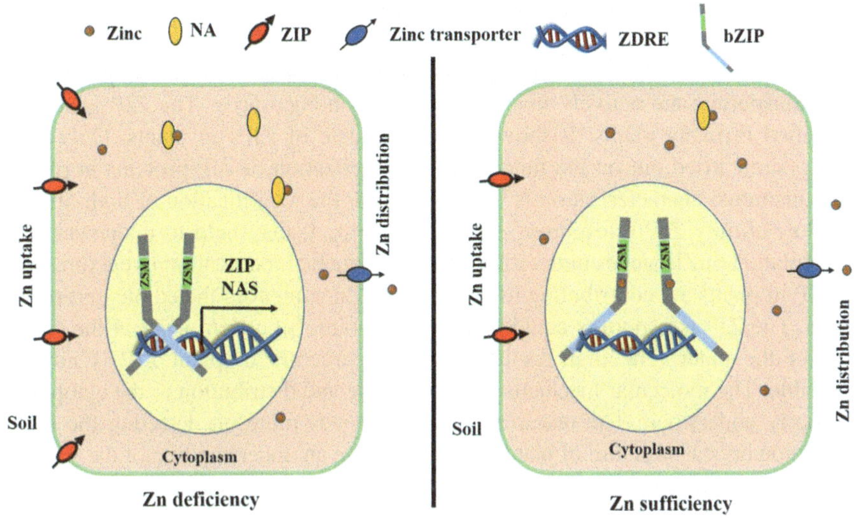

Fig. 10.3 Regulation of *AtZIP and AtNAS* gene in the model plant *Arabidopsis* under Zn deficiency and sufficiency conditions. The TFs AtbZIP19 and AtbZIP23 are considered to be the essential regulators of the adaptation to various Zn status. Under Zn deficiency conditions, the TFs bZIP19 and bZIP23 bind to the Zn deficiency-responsive element (RTGTCGACAY) in the promoter region of the target genes and activating their transcription and contribute for Zn uptake and distribution. The TFs AtbZIP19 and AtbZIP23 induce the expression of the *AtZIP* and *AtNAS* gene under Zn deficiency (Lilay et al. 2021). These genes contribute to Zn uptake and nicotianamine (NA)-mediated Zn distribution in *Arabidopsis* under Zn deficiency. Under Zn-sufficient conditions, freely available cellular Zn ions bind to the Cys/His-rich ZSM of bZIP19 and bZIP23, and this binding of Zn to the protein inhibits the regulation of the transcription factors (Lilay et al. 2021). This TF is also involved in the expression of a specific subset of genes such as *AtZIP1*, *AtZIP2*, *AtZIP4*, *AtZIP5*, *AtZIP9*, *AtZIP10*, and *AtZIP12* (Inaba et al. 2015; Lilay et al. 2019)

(His46Ala and His48Ala) abolished the Zn binding. This study revealed that the each Cys/His amino acid residues in the motif are actively involved in Zn binding. Even single Cys/His amino acid residues are essential for Zn deficiency-dependent activation of the target gene (*ZIP4*) by bZIP19 (Lilay et al. 2021). So, the Cys/His-rich motif of the bZIP19 is essential for detecting the Zn status and influence the transcriptional regulation of the target genes. AtbZIP19 and AtbZIP23 TFs are also involved in the activation of many ZIP transporters (AtZIP1, AtZIP2, AtZIP4, AtZIP5, AtZIP9, AtZIP10, and AtZIP12) (Assunção et al. 2010; Inaba et al. 2015; Lilay et al. 2019). So far, very little information is available on bZIP TFs that are responsible for the activation of target genes under Zn deficiency. More research is needed on the identification and functional characterization of Zn-responsive TFs in plants. It could help to understand the molecular mechanism of homeostasis under Zn-deficiency conditions.

10.5 Conclusion and Future Prospectus

The ZIP transporters play an important role in Zn uptake and transport in plants. The ZIP transporters are actively involved in the Zn homeostasis. The *ZIP* genes are identified in many plants. It shows the importance of ZIPs in plants. Only little studies are carried out on the functional characterization of ZIP proteins in plants. The functional characterization is very crucial for the identification of high-affinity and low-affinity ZIP transporter systems in plants. It may help to understand the upregulation of ZIP genes under Zn starvation. Genetic modification/overexpression of ZIP in plants are contributing to improve the Zn content in the edible part of the crops. The ZIP transporters could help with the biofortification of Zn in the edible part of the plant. The high-resolution crystal structure of plant ZIP is not yet available. The molecular mechanism of Zn uptake and distribution to the cytoplasm is poorly understood. The researchers need to focus more on detecting the high-resolution crystal structure of plant ZIP. It will give an understanding of the specific Zn^{2+} transport mechanism in plants. The bZIP19/23 TFs act as a Zn sensor and help to regulate the target genes under both Zn-deficient and Zn-sufficient conditions and contribute to fine Zn homeostasis. More research is needed on the identification and functional characterization of other low Zn-responsive TFs in plants. It could help to understand the molecular mechanism of homeostasis under Zn-deficiency conditions.

References

Alagarasan G, Dubey M, Aswathy KS, Chandel G (2017) Genome wide identification of orthologous ZIP genes associated with zinc and iron translocation in *Setaria italica*. Front Plant Sci 8:775

Assunção AG, Herrero E, Lin Y-F, Huettel B, Talukdar S, Smaczniak C, Immink RG, Van Eldik M, Fiers M, Schat H (2010) *Arabidopsis thaliana* transcription factors bZIP19 and bZIP23 regulate the adaptation to zinc deficiency. Proc Natl Acad Sci 107(22):10296–10301

Astudillo C, Fernandez A, Blair MW, Cichy KA (2013) The *Phaseolus vulgaris* ZIP gene family: identification, characterization, mapping, and gene expression. Front Plant Sci 4:286

Bashir K, Ishimaru Y, Nishizawa NK (2012) Molecular mechanisms of zinc uptake and translocation in rice. Plant Soil 361(1):189–201

Cabot C, Martos S, Llugany M, Gallego B, Tolrà R, Poschenrieder C (2019) A role for zinc in plant defense against pathogens and herbivores. Front Plant Sci 10:1171

Chandra AK, Pandey D, Tiwari A, Sharma D, Agarwal A, Sood S, Kumar A (2020) An omics study of iron and zinc homeostasis in finger millet: biofortified foods for micronutrient deficiency in an era of climate change? OMICS Int J Integr Biol 24(12):688–705

Chaudhary K, Agarwal S, Khan S (2018) Role of phytochelatins (PCs), metallothioneins (MTs), and heavy metal ATPase (HMA) genes in heavy metal tolerance. In: Mycoremediation and environmental sustainability. Springer, pp 39–60

Chen W, Feng Y, Chao Y (2008) Genomic analysis and expression pattern of *OsZIP1*, *OsZIP3*, and *OsZIP4* in two rice (*Oryza sativa* L.) genotypes with different zinc efficiency. Russ. J Plant Physiol 55(3):400–409

Clemens S (2001) Molecular mechanisms of plant metal tolerance and homeostasis. Planta 212(4): 475–486

Corrêa LGG, Riaño-Pachón DM, Schrago CG, Vicentini dos Santos R, Mueller-Roeber B, Vincentz M (2008) The role of bZIP transcription factors in green plant evolution: adaptive features emerging from four founder genes. PLoS One 3(8):e2944

Courbot M, Willems G, Motte P, Arvidsson S, Roosens N, Saumitou-Laprade P, Verbruggen N (2007) A major quantitative trait locus for cadmium tolerance in *Arabidopsis halleri* colocalizes with HMA4, a gene encoding a heavy metal ATPase. Plant Physiol 144(2):1052–1065

Deshpande P, Dapkekar A, Oak M, Paknikar K, Rajwade J (2018) Nanocarrier-mediated foliar zinc fertilization influences expression of metal homeostasis related genes in flag leaves and enhances gluten content in durum wheat. PLoS One 13(1):e0191035

Durmaz E, Coruh C, Dinler G, Grusak MA, Peleg Z, Saranga Y, Fahima T, Yazici A, Ozturk L, Cakmak I (2011) Expression and cellular localization of ZIP1 transporter under zinc deficiency in wild emmer wheat. Plant Mol Biol Rep 29(3):582–596

Eide DJ (1998) The molecular biology of metal ion transport in *Saccharomyces cerevisiae*. Annu Rev Nutr 18(1):441–469

Eide DJ (2005) The ZIP family of zinc transporters. In: Zinc finger proteins. Springer, pp 261–264

Eide D, Broderius M, Fett J, Guerinot ML (1996) A novel iron-regulated metal transporter from plants identified by functional expression in yeast. Proc Natl Acad Sci 93(11):5624–5628

Eren E, Arguello JM (2004) *Arabidopsis* HMA2, a divalent heavy metal-transporting PIB-type ATPase, is involved in cytoplasmic Zn2+ homeostasis. Plant Physiol 136(3):3712–3723

Escudero-Almanza DJ, Ojeda-Barrios DL, Hernández-Rodríguez OA, Chávez ES, Ruíz-Anchondo T, Sida-Arreola JP (2012) Carbonic anhydrase and zinc in plant physiology. Chil J Agric Res 72(1):140

Evens NP, Buchner P, Williams LE, Hawkesford MJ (2017) The role of ZIP transporters and group F bZIP transcription factors in the Zn-deficiency response of wheat (*Triticum aestivum*). Plant J 92(2):291–304

Fei X, Fu X-z, Wang N-q, XI J-l, Huang Y, Wei Z, Ling L-l, Peng L-z (2016) Physiological changes and expression characteristics of ZIP family genes under zinc deficiency in navel orange (*Citrus sinensis*). J Integr Agric 15(4):803–811

Fu X-Z, Zhou X, Xing F, Ling L-L, Chun C-P, Cao L, Aarts MG, Peng L-Z (2017) Genome-wide identification, cloning and functional analysis of the zinc/iron-regulated transporter-like protein (ZIP) gene family in trifoliate orange (*Poncirus trifoliata* L. Raf.). Front Plant Sci 8:588

Gaitán-Solís E, Taylor NJ, Siritunga D, Stevens W, Schachtman DP (2015) Overexpression of the transporters AtZIP1 and AtMTP1 in cassava changes zinc accumulation and partitioning. Front Plant Sci 6:492

Gaither LA, Eide DJ (2001) Eukaryotic zinc transporters and their regulation. In: Zinc biochemistry, physiology, and homeostasis, pp 65–84

Gondal AH, Zafar A, Zainab D, Toor M, Sohail S, Ameen S, Ijaz A, Ch B, Hussain I, Haider S (2021) A detailed review study of zinc involvement in animal, plant and human nutrition. Indian J Pure Appl Biosci 9(2):262–271

Grotz N, Fox T, Connolly E, Park W, Guerinot ML, Eide D (1998) Identification of a family of zinc transporter genes from *Arabidopsis* that respond to zinc deficiency. Natl Acad Sci 95(12): 7220–7224

Guerinot ML (2000) The ZIP family of metal transporters. BBA-Biomembranes 1465(1–2): 190–198

Hacisalihoglu G, Hart JJ, Kochian LV (2001) High-and low-affinity zinc transport systems and their possible role in zinc efficiency in bread wheat. Plant Physiol 125(1):456–463

Hafeez B, Khanif Y, Saleem M (2013) Role of zinc in plant nutrition-a review. J Exp Agric Int 2(2): 374–391

Hanikenne M, Talke IN, Haydon MJ, Lanz C, Nolte A, Motte P, Kroymann J, Weigel D, Krämer U (2008) Evolution of metal hyperaccumulation required cis-regulatory changes and triplication of HMA4. Nature 453(7193):391–395

Hänsch R, Mendel RR (2009) Physiological functions of mineral micronutrients (cu, Zn, Mn, Fe, Ni, Mo, B, cl). Curr Opin Plant Biol 12(3):259–266

Horak V, Trčka I (1976) The influence of Zn^{2+} ions on the tryptophan biosynthesis in plants. Biol Plant 18(5):393–396

Hussain D, Haydon MJ, Wang Y, Wong E, Sherson SM, Young J, Camakaris J, Harper JF, Cobbett CS (2004) P-type ATPase heavy metal transporters with roles in essential zinc homeostasis in *Arabidopsis*. Plant Cell 16(5):1327–1339

Inaba S, Kurata R, Kobayashi M, Yamagishi Y, Mori I, Ogata Y, Fukao Y (2015) Identification of putative target genes of bZIP19, a transcription factor essential for *Arabidopsis* adaptation to Zn deficiency in roots. Plant J 84(2):323–334

Ishimaru Y, Suzuki M, Kobayashi T, Takahashi M, Nakanishi H, Mori S, Nishizawa NK (2005) OsZIP4, a novel zinc-regulated zinc transporter in rice. J Exp Bot 56(422):3207–3214

Ishimaru Y, Masuda H, Suzuki M, Bashir K, Takahashi M, Nakanishi H, Mori S, Nishizawa NK (2007) Overexpression of the OsZIP4 zinc transporter confers disarrangement of zinc distribution in rice plants. J Exp Bot 58(11):2909–2915

Kambe T, Suzuki E, Komori T (2019) Zinc transporter proteins: a review and a new view from biochemistry. In: Zinc signaling, pp 23–56

Kim YY, Choi H, Segami S, Cho HT, Martinoia E, Maeshima M, Lee Y (2009) AtHMA1 contributes to the detoxification of excess Zn (II) in *Arabidopsis*. Plant J 58(5):737–753

Kou X, Wang S, Wu M, Guo R, Xue Z, Meng N, Tao X, Chen M, Zhang Y (2014) Molecular characterization and expression analysis of NAC family transcription factors in tomato. Plant Mol Biol Rep 32(2):501–516

Krishna T, Ceasar S, Maharajan T, Ramakrishnan M, Duraipandiyan V, Al-Dhabi N, Ignacimuthu S (2017) Improving the zinc-use efficiency in plants: a review. SABRAO J Breed Genet 49(3): 211–230

Krishna T, Maharajan T, Victor Roch G, Ignacimuthu S, Antony Ceasar S (2020) Structure, function, regulation and phylogenetic relationship of ZIP family transporters of plants. Front Plant Sci 11:662

Kumar L, Meena NL, Singh U (2016) Zinc transporter: mechanism for improving Zn availability. In: Biofortification of food crops, pp 129–146

Lee S, Jeong HJ, Kim SA, Lee J, Guerinot ML, An G (2010a) OsZIP5 is a plasma membrane zinc transporter in rice. Plant Mol Bio 73(4–5):507–517

Lee S, Kim SA, Lee J, Guerinot ML, An G (2010b) Zinc deficiency-inducible OsZIP8 encodes a plasma membrane-localized zinc transporter in rice. Mole Cells 29(6):551–558

Li S, Zhou X, Huang Y, Zhu L, Zhang S, Zhao Y, Guo J, Chen J, Chen R (2013) Identification and characterization of the zinc-regulated transporters, iron-regulated transporter-like protein (ZIP) gene family in maize. BMC Plant Bio 13(1):1–14

Li S, Zhou X, Li H, Liu Y, Zhu L, Guo J, Liu X, Fan Y, Chen J, Chen R (2015) Overexpression of ZmIRT1 and ZmZIP3 enhances iron and zinc accumulation in transgenic *Arabidopsis*. PLoS One 10(8):e0136647

Lilay GH, Castro PH, Campilho A, Assunção AG (2019) The *Arabidopsis* bZIP19 and bZIP23 activity requires zinc deficiency–insight on regulation from complementation lines. Front Plant Sci 9:1955

Lilay GH, Persson DP, Castro PH, Liao F, Alexander RD, Aarts MG, Assunção AG (2021) *Arabidopsis* bZIP19 and bZIP23 act as zinc sensors to control plant zinc status. Nat Plants 7(2):137–143

Lin J, Gao X, Zhao J, Zhang J, Chen S, Lu L (2020) Plant cadmium resistance 2 (SaPCR2) facilitates cadmium efflux in the roots of hyperaccumulator *Sedum alfredii* Hance. Front Plant Sci 11:568887

Menguer PK, Farthing E, Peaston KA, Ricachenevsky FK, Fett JP, Williams LE (2013) Functional analysis of the rice vacuolar zinc transporter OsMTP1. J Exp Bot 64(10):2871–2883

Mills RF, Francini A, da Rocha PSF, Baccarini PJ, Aylett M, Krijger GC, Williams LE (2005) The plant P1B-type ATPase AtHMA4 transports Zn and cd and plays a role in detoxification of transition metals supplied at elevated levels. FEBS Lett 579(3):783–791

Mondal TK, Ganie SA, Rana MK, Sharma TR (2014) Genome-wide analysis of zinc transporter genes of maize (*Zea mays*). Plant Mol Biol Rep 32(2):605–616

Moreau S, Thomson RM, Kaiser BN, Trevaskis B, Guerinot ML, Udvardi MK, Puppo A, Day DA (2002) GmZIP1 encodes a symbiosis-specific zinc transporter in soybean. J Biol Chem 277(7): 4738–4746

Morel M, Crouzet J, Gravot A, Auroy P, Leonhardt N, Vavasseur A, Richaud P (2009) AtHMA3, a P1B-ATPase allowing Cd/Zn/co/Pb vacuolar storage in *Arabidopsis*. Plant Physiol 149(2): 894–904

Pasricha S, Mathur V, Garg A, Lenka S, Verma K, Agarwal S (2021) Molecular mechanisms underlying heavy metal uptake, translocation and tolerance in hyperaccumulators-an analysis: heavy metal tolerance in hyperaccumulators. Environ Challeng 4:100197

Pedas P, Husted S (2009) Zinc transport mediated by barley ZIP proteins are induced by low pH. Plant Signal Behav 4(9):842–845

Pedas P, Schjoerring JK, Husted S (2009) Identification and characterization of zinc-starvation-induced ZIP transporters from barley roots. Plant Physiol Biochem 47(5):377–383

Pedersen CN, Axelsen KB, Harper JF, Palmgren MG (2012) Evolution of plant P-type ATPases. Front Plant Sci 3:31

Ramegowda Y, Venkategowda R, Jagadish P, Govind G, Hanumanthareddy R-R, Makarla U, Guligowda SA (2013) Expression of a rice Zn transporter, OsZIP1, increases Zn concentration in tobacco and finger millet transgenic plants. Plant Biotechnol Rep 7(3):309–319

Ramesh SA, Shin R, Eide DJ, Schachtman DP (2003) Differential metal selectivity and gene expression of two zinc transporters from rice. Plant Physiol 133(1):126–134

Ramesh SA, Choimes S, Schachtman DP (2004) Over-expression of an Arabidopsis zinc transporter in *Hordeum vulgare* increases short-term zinc uptake after zinc deprivation and seed zinc content. Plant Mol Biol 54(3):373–385

Socha AL, Guerinot ML (2014) Mn-euvering manganese: the role of transporter gene family members in manganese uptake and mobilization in plants. Front Plant Sci 5:106

Takahashi R, Bashir K, Ishimaru Y, Nishizawa NK, Nakanishi H (2012) The role of heavy-metal ATPases, HMAs, in zinc and cadmium transport in rice. Plant Signal Behav 7(12):1605–1607

Talke IN, Hanikenne M, Krämer U (2006) Zinc-dependent global transcriptional control, transcriptional deregulation, and higher gene copy number for genes in metal homeostasis of the hyperaccumulator *Arabidopsis halleri*. Plant Physiol 142(1):148–167

Tiong J, McDonald GK, Genc Y, Pedas P, Hayes JE, Toubia J, Langridge P, Huang CY (2014) H v ZIP 7 mediates zinc accumulation in barley (*Hordeum vulgare*) at moderately high zinc supply. New Phytol 201(1):131–143

Tiong J, McDonald G, Genc Y, Shirley N, Langridge P, Huang CY (2015) Increased expression of six ZIP family genes by zinc (Zn) deficiency is associated with enhanced uptake and root-to-shoot translocation of Zn in barley (*Hordeum vulgare*). New Phytol 207(4):1097–1109

Tiwari A, Kumar P, Singh S, Ansari S (2005) Carbonic anhydrase in relation to higher plants. Photosynthetica 43(1):1–11

Vatansever R, Özyiğit İİ, Filiz E (2016) Comparative and phylogenetic analysis of zinc transporter genes/proteins in plants. Turk J Biol 40(3):600–611

Verret F, Gravot A, Auroy P, Leonhardt N, David P, Nussaume L, Vavasseur A, Richaud P (2004) Overexpression of AtHMA4 enhances root-to-shoot translocation of zinc and cadmium and plant metal tolerance. FEBS Lett 576(3):306–312

Verret F, Gravot A, Auroy P, Preveral S, Forestier C, Vavasseur A, Richaud P (2005) Heavy metal transport by AtHMA4 involves the N-terminal degenerated metal binding domain and the C-terminal His11 stretch. FEBS Lett 579(6):1515–1522

Vert G, Barberon M, Zelazny E, Séguéla M, Briat J-F, Curie C (2009) *Arabidopsis* IRT2 cooperates with the high-affinity iron uptake system to maintain iron homeostasis in root epidermal cells. Planta 229(6):1171–1179

Wang N, Qiu W, Dai J, Guo X, Lu Q, Wang T, Li S, Liu T, Zuo Y (2019) AhNRAMP1 enhances manganese and zinc uptake in plants. Front Plant Sci 10:415

Yang X, Huang J, Jiang Y, Zhang H-S (2009) Cloning and functional identification of two members of the ZIP (Zrt, Irt-like protein) gene family in rice (*Oryza sativa* L.). Mol Bio Reps 36(2): 281–287
Yang M, Li Y, Liu Z, Tian J, Liang L, Qiu Y, Wang G, Du Q, Cheng D, Cai H (2020) A high activity zinc transporter OsZIP9 mediates zinc uptake in rice. Plant J 103(5):1695–1709
Zhang T, Liu J, Fellner M, Zhang C, Sui D, Hu J (2017) Crystal structures of a ZIP zinc transporter reveal a binuclear metal center in the transport pathway. Sci Adv 3(8):e1700344
Zhao H, Eide D (1996a) The yeast ZRT1 gene encodes the zinc transporter protein of a high-affinity uptake system induced by zinc limitation. Proc Natl Acad Sci 93(6):2454–2458
Zhao H, Eide D (1996b) The ZRT2 gene encodes the low affinity zinc transporter in Saccharomyces cerevisiae. J Biol Chem 271(38):23203–23210

The Function of HAK as K^+ Transporter and AKT as Inward-Rectifying Agent in the K^+ Channel

11

Moumita Chatterjee, Pallabi Ghosh, Supatra Sen, Dwaipayan Sinha, and Sharmistha Ganguly

Abstract

Potassium (K^+), a vital macronutrient regulates different physiological processes such as osmoregulation, pH balance, maintenance of turgor pressure, activation of an enzyme, membrane electric potential regulation, and expansion of the cell. A huge amount of K^+ is accumulated by plants in their cell vacuole. In higher plants, K^+ transportation regulated by transporter protein of two classes depending on the concentration of K^+ varies from micromoles (μmol) to millimoles (mmol). If the K^+ concentration is high in soil, then K^+ crosses the cell membrane through the K^+ channel, whereas at lower K^+ concentration, active transport system is essential for the influx of K^+ against cell electrochemical gradient. Salinity and K^+ deficiency reduced cellular K^+ content, and drought induced the K^+ concentration inside the cell vacuole. Not only dicotyledonous but also monocotyledonous plants possess HAK transporter gene, expressed mainly in roots, and have high affinity toward K^+ showing enhanced expression during K^+ starvation condition. The C terminal binding site, that is, AKT and nucleotide binding site of some K^+ transporter protein, regulates the process of binding with

M. Chatterjee (✉)
V. Sivaram Research Foundation, Bangalore, Karnataka, India

P. Ghosh
ICMR, National Institute of Pathology, New Delhi, India

S. Sen
Department of Botany, Asutosh College, Kolkata, West Bengal, India

D. Sinha
Department of Botany, Government General Degree College, Mohanpur, Paschim Medinipur, West Bengal, India

S. Ganguly
University Department of Botany, Ranchi University, Ranchi, Jharkhand, India

K. Kumar, S. Srivastava (eds.), *Plant Metal and Metalloid Transporters*,
https://doi.org/10.1007/978-981-19-6103-8_11

cytoskeleton without helping in the interaction between the subunits of K^+ channel protein. In the tissue of the root, the AKT acts as an inward-rectifying agent, and it has sensory property toward the K^+ channel and thereby plays a vital role in the translocation of K^+ through the vascular bundle. AKT-mediated K^+ absorption gets hampered due to the presence of nitric oxide (NO), which gets accumulated during salinity stress. CO_2 assimilates, and light upregulates AKT expression in the plant. These transporter proteins help plants to overcome unfavorable environmental situations like drought, salinity, and deficiency of potassium. These two proteins play a pivotal role in salt tolerance, thereby controlling different metabolisms of plants. So, the authors have made an effort to bring all the possible and relevant information in the present chapter, related to these proteins, which will help in better understanding of the physiological role and structure of the two transporter proteins.

Keywords

K^+ transportation · HAK · AKT · Potassium deficiency · Salinity · Stress response

11.1 Introduction

Potassium (K^+), an essential phytonutrient, regulates different physiological processes such as maintenance of turgor pressure, osmoregulation, activation of an enzyme, pH balance, membrane electric potential regulation, and expansion of cell (Sinha and Tandon 2020; Rodríguez-Navarro and Rubio 2006; Véry and Sentenac 2003; Maathuis et al. 1997). According to Busch and Saier Jr (2002), the K^+ acquisition pathway in a plant has two classes of transporter, which are Class 1 (channels/porins) and Class 2 (electrochemical potential-driven transporters). Potassium uptake by plants from soil occurs passively (with the help of AKT1) (Hirsch et al. 1998) as well as actively (via KUP/HAK, H^+-coupled carriers) (Qi et al. 2008). Fernando et al. (2011) reported that if external concentrations of K^+ are <10 μM, 10 and 200 μM, and >500 μM, then AtHAK5, both AtHAK5 and AKT, AKT1 respectively, can uptake. When the K^+ concentrations varied from intermediate to high, then AKT1 expressed in the epidermal cell of root helps in K^+ uptake. The external supply of K^+ does not regulate the AKT1expression (Pilot et al. 2003a, b; Pyo et al., 2010).

Buschmann et al. (2000) reported that in wheat (*Triticum aestivum*), homologous TaAKT1 expression increased if the applied growth solution is devoid of K^+. Application of phytohormone (cytokinin benzyladenine and 2,4 D) (Pilot et al. 2003a, b) and NaCl (Kaddour et al. 2009) decreases the expression of AKT1. Long-term NaCl treatment (Kaddour et al. 2009) decreases AKT1 expression, whereas short-term NaCl application (Maathuis et al. 2003; Pilot et al. 2003a, b) does not alter AKT1 expression.

AKT1 regulation is done by protein complex formation between calcineurin B-like (CBL) and a CBL-interacting protein kinase (CIPK) (Albrecht et al. 2001; Shi et al. 1999). Xu et al. (2006) reported that AKT1 is activated by the presence of

CIPK23/CBL9 or CIPK23/CBL1 pairs, AKT1 required physical interaction with CIPK23/CBL1-9 pair, while CBL1 phosphorylates CIPK23, which gives stabilization of the complex (Hashimoto et al. 2012). AKT1's ankyrin (ANK) domain interacts with CIPK23's kinase domain and suggests CIPKs' specificity toward attachment with AKT1 channel (Lee et al. 2007) to harmonize K^+ homeostasis (Grefen and Blatt 2012).

In hypersaline environment, the responses come from diverse signaling cascades as plant cell communicates with each other to overcome stress. The cell membrane is permeable to lipophilic molecules (e.g., steroid hormones) but impermeable toward hydrophilic molecules like water, ions, and several macromolecules. High-affinity K^+ transporters (HKTs) and K^+ channel transporter (AKT1) and high-affinity K^+ uptake transporter (HAK) are responsible for Na^+ influx, which further inhibits inward-rectifying K^+ channels and activates K^+ outward-rectifying channels (KOR). Saline environment exposure produces responses within a few seconds sometimes taking hours (Shah et al. 2021). The AKT1 transgene elevates salt tolerance property and also enhances the efficacy of K^+ utilization (Liu et al. 2015; Ardie et al. 2010). Potassium starvation and extrinsic ABA induce AKT2 expression level in phloem cells (Pilot et al. 2003a, b; Deeken et al. 2002).

HAK potassium transporters are carriers of amino acid-polyamine-organocation type.

HAK transporter regulation can be done at either transcriptional or posttranslational levels. At different growth conditions, root cell membrane potential alters that leads to regulation of HAK transporters at the transcriptional level (Li et al. 2018). HAK transporter is also present in bacteria.

It helps absorbed nutrients when environment possesses less K^+. Then, organism goes for nutrition absorption mode, and at that time, there is a presence of HAK transporters, whereas organism grown in K^+-rich medium is devoid of HAK. In the plant kingdom, HAK shows evolutionary diversification as it is present in the lower group like Chlorophyceae to the most advanced group, that is, flowering plants.

According to Bañuelos et al. (2002), there are Clusters I and II of the KT/KUP/HAK family, whereas *Arabidopsis* has Cluster III, and rice has Clusters III and IV. Cluster I plays a vital role in K^+ acquisition when K^+ concentration is very low. Cluster II present in the tonoplast facilitates K^+ efflux from the vacuole, which is crucial for K^+ homeostasis sustentation in the K^+-deficit medium. AtHAK6 and AtHAK2 genes of Cluster II, as well as AtHAK11 gene of Cluster III, regulate the salinity stress in *Arabidopsis* (Maathuis 2006).

11.2 HAK-AKT Transporters Present in Various Plants

Modern agriculture relies on mineral fertilization. Unlike other major macronutrients, K^+ is not incorporated into organic matter but remains as a soluble ion in the cell sap contributing up to 10% of the dry organic matter. Consequently, K^+ constitutes a chief osmoticum and plays a major function in cellular expansion such as stomatal aperture. Moreover, K^+ transport is critical for the control of cytoplasmic and lumenal pH in endosomes, regulation of membrane potential, and enzyme activity. Not surprisingly, plants have evolved a large ensemble of K^+ transporters with defined functions in nutrient uptake by roots, storage in vacuoles, and ion translocation between tissues and organs. This review describes the distribution of critical transport proteins governing K^+ nutrition in various plant species. We hope to generate new discussion by bringing together information about K^+ transporters present in different plants and their sequence homology.

K^+ is of paramount importance in plant cell physiology. The K^+ concentration in the soil solution may vary widely from 0.01 to 20 mm, and plant cells maintain a relatively constant concentration of 80–100 mM in the cytoplasm (Rodríguez-Navarro 2000). Thus, plants accumulate large amounts of K^+ in their vacuoles, surpassing purely nutritional requirements. Hence, K^+ is the most abundant cation in plant cells, comprising up to 10% of plant dry weight that supports near-maximal growth rates.

The extraction of K^+ from soil and its distribution within the plant requires the presence of membrane-bound transport proteins. A large number of such transporters have now been identified at the molecular level, demonstrating the complex nature of K^+ transport. The physiological roles of these proteins in primary K^+ influx, efflux, compartmentation, and transport within the plant have been partially characterized (Lebaudy et al. 2007).

A variety of carriers also move cations into plant cells. The high-affinity transport system (HATS) is saturable that catalyzes thermodynamically active uptake of K^+ at low concentrations (<1 mM) (Cheeseman et al. 1980; Kochian et al. 1989; Maathuis and Sanders 1994; Briskin and Gawienowski 1996). Several genes have been identified that encode HATS transporters. They are grouped into four major families: HAK/KUP/KT (K^+ /H^+ symporters), HKT/TRK (K^+ /H^+ or K^+/Na^+ symporters), CPA (cation/H^+ antiporters), and Shaker channels. These transporters mediate active K^+ symport with H^+ (Maathuis and Sanders 1999; Rodríguez-Navarro 2000; Maser et al. 2001; Grabov 2007). The majority of HATS-mediated influx is catalyzed by members of the HAK/KUP/KT family, particularly under conditions of K^+ starvation (Gierth et al. 2005).

It has been found that HAK/KUP/KT function in the acquisition of K^+ at low $[K^+]$ ext., K^+ starvation to promote *HAK* gene transcription in a wide variety of plant systems, including barley, rice, *Arabidopsis thaliana*, *Capsicum annuum*, *Mesembryanthemum crystallinum*, *Solanum lycopersicum*, and *Phragmites australis* (Santa-María et al. 1997; Bañuelos et al. 2002; Su et al. 2002; Ahn et al. 2004; Armengaud et al. 2004; Martinez-Cordero et al. 2005; Shin and Schachtman 2004; Gierth et al. 2005; Nieves-Cordones et al. 2008; Takahashi et al. 2007).

According to the work of Muller et al. (1995) and Nakamura et al. (1995), GUS reporter system was used to localize the expression of at least two different K^+ channels, the Arabidopsis *KAT1* and its potato homolog *KST1*, both primarily expressed in guard cells. The expression and localization of another *Arabidopsis* K^+ uptake channel AKT1 was studied in *Brassica napus* and *Arabidopsis*. In those studies of the *AKT1* promoter, activity was not affected by nutrient depletion of K^+ and was restricted primarily to roots. Some promoter activity was detected in leaves, in specialized hydathode cells that are involved in the process of guttation. Cao et al. (1995) and Ketchum et al. (1989) individually found another K^+ uptake channel called AKT2, which appeared to be primarily expressed in the leaves of *Arabidopsis*.

A member of the high-affinity K^+ uptake system has been described in the wheat HKT1 protein (Rubio et al. 1995; Schachtman and Schroeder 1994). It is also a major route for the entry of Na^+ into roots, according to its specificity as a K^+/Na^+ symporter (Rubio et al. 1995).

OsHKT1, the homolog of HKT1, and its expression were analyzed by Golldack et al. (2002), in salt-tolerant rice Pokkali and salt-sensitive IR29 in response to external cation concentrations. By in situ hybridizations, the expression of *OsHKT1* was localized to the root epidermis and the vascular tissue inside the endodermis. In leaves, *OsHKT1* showed the strongest signals in cells surrounding the vasculature. The data suggest control over HKT expression as a factor that may distinguish salt stress-sensitive and salt stress-tolerant lines. *OsHKT1* includes an open reading frame encoding a protein of 530 amino acids. A comparison of amino acid sequence-deduced rice *OsHKT1* is 65% identical to that of wheat HKT1 (Schachtman and Schroeder 1994). OsHKT1 and the *Arabidopsis* homolog AtHKT1 (Uozumi et al. 2000) show 39% sequence identity and 56% sequence homology.

11.3 K^+ Channels and Transporters

In plants, the process of K^+ uptake by roots, translocation to aerial parts, and subsequent distribution and compartmentation constitute the K^+ trans-membrane support mediated chiefly by K^+ channels and transporters. Currently, such systems are best identified and characterized in *Arabidopsis thaliana*. Studies reveal that nonselective cation channels (NSCC) are responsible for K^+ uptake at high concentrations, that is, >10 mM, low-affinity inward-rectifying channel AtAKT1 regulates uptake at intermediate, that is, 1 mM concentrations, while under lower external K^+ concentrations like 100 μM, high-affinity K^+ transporter AtHAK5 along with AtAKT1 regulates K^+ uptake. AtHAK5 is then the only transporter with the capacity for K^+ uptake at very low concentrations like 10 μM (Nieves-Cordones et al. 2014).

HAK is the largest of the three families of K^+ transporters in plants where the members function as K^+/H^+ symporters. HAK genes have been reported in several plants like barley, *Arabidopsis,* and subsequently rice, wheat, and maize (Santa-María et al. 1997; Kim et al. 1998; Szczerba et al. 2009; Yang et al. 2014; Cheng et al. 2018; Zhang et al. 2012). Multiple roles are played by HAK transporters in uptake and translocation of K^+ as combating salinity and drought stress besides morphological development of root and shoot. AtHAK5 is the high-affinity K^+ transporter in *Arabidopsis* under severely low K^+ concentration and also maintains high K^+ acquisition and thereby growth under salinity stress (Nieves-Cordones et al. 2010), while AtKUP7 mediates K^+ transport through xylem sap, thus enabling root-shoot translocation of K^+ (Han et al. 2016).

Besides regulating the K^+ conductance of the plasmalemma, the shaker channels also control the tissue K^+ concentration, membrane, and osmotic potential. Shaker channels are found in microbes, plants, and animals, and several such channels have been cloned and characterized in higher plants. Pilot et al. (2003a, b) reported nine members in *Arabidopsis* each with its characteristic functional property, the pattern of expression, and position/location. The first two Shaker channels cloned in *Arabidopsis* plants in 1992 were AKT1 and KAT1 (Anderson et al. 1992). Ahmad et al. (2016) characterized them as inward-rectifying channels.

Generally, these channels permit massive influx or efflux, that is, exchange of K^+ between symplast and apoplast implying K^+ entrance for the inward or incoming and exit for the outward or outgoing channels while both entry and exit for low rectification channels. KAT, AKT1, and ATKC1, inward rectification, SKOR family-outward rectification, and AKT2 family, low rectification, are the three main functional types of Shaker channels. While AKT1 and ATKC1 channels in *Arabidopsis* mediate K^+ removal from the soil, SKOR and AKT2 play a role in long-distance K^+ transport in vascular tissue, while GORK, KAT1, and KAT2 channels function in the transport of K^+ in the guard cells controlling stomatal movements (Lebaudy et al. 2007; Saponaro et al. 2017).

From the studies of Pilot et al. (2003a, b) and Jegla et al. (2018), it was revealed that GORK1 and AKT1 are tetramers with sub-units displaying N-terminal trans-membrane along with a cytoplasmic domain made up of cyclic nucleotide-binding homologous domain followed by an ankyrin repeat domain with a characteristic plant K^+ channel motif, rich in hydrophobic and acidic residues at the C-terminal. While the trans-membrane region is responsible for ion transport, the cytosolic region carries out channel regulation. A common regulatory process for both GORK and AKT1 may be possible where a protein complex between a calcineurin B-like (CBL) and a CBL-interacting protein kinase (CIPK) can decode a specific calcium signature to enable phosphorylation and activate K^+ transport. CIPK5/CBL1 mediates GORK1 activation in case of wounding (Forster et al. 2019). CIPK23/CBL1 or CIPK23/CBL9 activates AKT1 on receiving signals for K^+ starvation (Xu et al. 2006). According to Lee et al. (2007) and Lefoulon et al. (2016), the K^+ status would determine whether the protein phosphatases AIP1 and ABI2 would reverse to dephosphorylate AKT1 and GORK1, respectively.

11.4 Adaptive Responses of Plants to Salinity Stress

Salinity stress with a predominance of NaCl is a major agricultural hurdle. The optimum cytosolic K^+-sodium (Na^+) ratio is the crucial feature controlling overall salinity tolerance in plants. The ability of different plants to retain K^+ under stress is considered a major step to combating salinity stress (Maathuis and Amtmann 1999; Chen et al. 2007).

K^+ functions as a charge balancing cation. Plants when subjected to high salinity, that is, Na^+ rapid membrane depolarization, occur with significant implications for nutrient and metabolite transport (Shabala et al. 2005). GORK, that is, guard cell outward-rectifying K-channel, and ROS-mediated nonselective cation channels (NSCC) initiate salinity stress-induced K^+ efflux from the cytosol to the apoplast (Jayakannan et al. 2013; Wu et al. 2015). Accumulation of ROS under different conditions of stress enhances both GORK and ROS-mediated NSCC channels causing more K-efflux. The rapid loss of cytosolic K^+ dismantles the Na^+/K^+ ratio, a characteristic of salt tolerance of plants (Hauser and Horie 2010).

For restoration of normal membrane potential, there are dual options, the first being to increase active H^+ pumping and the other to utilize K^+ efflux for charge balancing. As the second alternative involves lower energy, it appears to be a preferred short-term option. Chakraborty et al. (2016) recorded a rise in GORK transcript along with a simultaneous rise in HAK5 transcript in *Brassica*. The increase was observed to be the highest in *B. napus* (salt-tolerant) with a consistent K^+ content value in its tissues, thereby strongly indicating a compensation strategy. Thus, plants could utilize K^+ efflux as a safety valve to combat initial membrane depolarization due to salinity stress till the function is subsequently taken up by H^+-ATPases (Alvarez-Pizarro et al. 2009), and then K^+ is eventually regained by a high-affinity K^+ transport system.

As a determinant of cell fate (Shabala 2009), high cytosolic K^+ levels are essential for repressing caspase-like proteases and endonucleases. Under both salinity and oxidative stress, *Arabidopsis* mutants without GORK channels exhibited low programmed cell death events in comparison to wild type (Demidchik et al. 2010). Cell elimination through programmed cell death could be considered in certain circumstances, an obligatory part of acclimation (Shabala and Shabala 2011).

K^+ efflux could then function as a crucial metabolic switch preventing energy-expensive anabolic processes, thereby conserving energy for repair and subsequent adaptation (called metabolic hypothesis by Demidchik et al. (2014)). Potassium functions as an activator of more than 70 odd metabolic enzymes (Dreyer and Uozumi 2011; Anschütz et al. 2014). Plants under conditions of stress utilize a significant amount of ATP for defense, even though production of ATP drops severely (Sen and Mukherji 2004, 2007). The way out for avoiding this energy crisis is to severely restrict cell metabolism. Diminishing the cytosolic K^+ to below threshold amounts could inactivate cellular metabolism, thereby re-channeling the ATP toward defense processes.

Salinity stress induces OsHAK1 production and is downregulated in K^+ deprivation (Chen et al. 2015), thereby proving that OsHAK1 plays a crucial role in increasing salt tolerance in rice. OsHAK5 expressed greatly in root epidermis, stele, and vascular tissues are responsible for K^+ uptake under conditions of low external K^+ concentration as well as upward translocation (Yang et al. 2014). Overexpressed OsHAK1 lines greatly improve drought tolerance and grain yield and vice versa (Chen et al. 2017). AtHAK5 is the target of auxin response factor 2, and overexpression leads to longer primary roots even with low K^+ concentration (Zhao et al. 2016). AtKUP4/tiny root hair 1 is necessary for polar localization of the auxin efflux transporter PIN1 at the root tip for gravitropic movements and root hair formation (Daras et al. 2015; Rigas et al. 2013). Elumalai et al. (2002) reported AtKUP2/short hypocotyls 3 mutation affects cellular expansion leading to defects in development in shoots of Arabodopsis. Barley is well documented for its higher salt tolerance due to higher K^+ uptake in roots, efficient Na^+ exclusion, and vacuolar sequestration (Munns and Tester 2008; Wu 2018) and was researched by Cai et al. (2021) who reported 27 HAK genes with their phylogenetic relationships and adaptive responses to salt and osmotic stress and K deficiency.

11.5 Mechanism of Action of HAK and AKT

Excess Na^+ accumulation, particularly in the aerial regions of plants, causes salt stress, which is one of the most essential components of salinity (Peng et al. 2016; Zhao et al. 2021). Maintaining a balanced cytosolic Na^+/K^+ ratio has become a major salinity tolerance mechanism because Na^+ interferes with K^+ homeostasis, especially given its role in multiple metabolic activities. High amounts of Na^+ accumulate in plant cells under salt stress, eventually reaching hazardous levels and disrupting ion homeostasis (Wu 2018). Plants have devised methods for removing Na^+ from the cytoplasm to maintain low levels of the ion (Almeida et al. 2017). This is accomplished mostly by several transporters, which aid in the homeostasis of Na^+. At the cellular and whole-plant levels, K^+ transporters, along with voltage-gated channel proteins and their regulators, play critical roles in modulating K^+ uptake, release, and transportation (Zhao et al. 2021). Root epidermal and cortical cells absorb K^+ from the soil (Sun et al. 2015; Ragel et al. 2019). Once inside the root symplast, K^+ is either retained in vacuoles (Tang et al. 2020) or transferred to the shoot via xylem (Hu et al. 2020) to fulfill osmotic duties. In this transit from the soil to the different plant organs, K^+ crosses various cell membranes through K^+-specific transport systems (Ragel et al. 2019). The combined action of all the transporters results in the maintenance of K^+ homeostasis within the plant body. In this section, we would discuss the roles of the transporters in maintaining ionic balance with special reference to HAK and AKT 0 transporters.

In plant cells, several transporter proteins are involved in the uptake and distribution of K^+, which are divided into numerous families with different architectures and transport methods. They are Shaker-like voltage-dependent, the tandem-pore (TPK) (Dreyer and Uozumi 2011), the two-pore channels (TPC) (Isayenkov et al. 2011), the carrier-like families KT/HAK/KUP (Cai et al. 2021), HKT uniporters and symporters (Riedelsberger et al. 2021), and cation-proton antiporters (CPA) (Tsujii et al. 2020). The Shaker-like, voltage-gated, and K^+-selective channel AKT1 was the first K^+ transporter identified to have a role in nutrient absorption (Lebaudy et al. 2007; Sharma et al. 2013). The architecture of voltage-gated ion channels is conserved, with four subunits or homologous domains forming a core ion-conducting pore surrounded by four voltage sensors (Payandeh et al. 2011). Each subunit's voltage-sensing domain (VSD) is made up of segments S1 to S4; segment S4 contains many positively charged residues and is the major transmembrane voltage-sensing component. The ion permeation pore-gate domain is formed by the association of two extra transmembrane helices (S5 and S6) with the four subunits' intervening pore loops to produce a tetrameric structure that surrounds a central conduction channel and forms the pore domain (PD) (Barros et al. 2020). The S6 domain lines the pore and forms the activation gate at the pore's intracellular vestibule, whereas the S5 domain lines the PD's outer side facing the VSD (Jegla et al. 2018). The C terminus of plant channels contains a cytoplasmic region that is similar to the cyclic nucleotide-binding domain (CNBD) and is connected to the PD through a conserved helical linker (C-linker) (Jegla et al. 2018).

In terms of their responsiveness to membrane potential, voltage-gated K^+ channels in plants are split into three subfamilies. (1) Inward-rectifying (Kin) channels permit the uptake of K^+ because they activate upon membrane hyperpolarization and are closed when the driving force for K^+ is outwardly directed. Inward-rectifying (Kir) K^+ channels are part of a large superfamily of K^+ ion channels that also includes voltage-gated, two-pore, calcium-gated, and cyclic nucleotide-gated channels. Kir channels act as biological diodes because they have the unique capacity to mediate the inward flow of K^+ ions at hyperpolarizing membrane voltages more efficiently than the outward flow at depolarizing voltages (Chen and Swale 2018). (2) Outward-rectifying (Kout) channels act contrary wise and facilitate K^+ release. (3) Weak-rectifying (Kweak) channels can facilitate both, K^+ uptake and discharge (Sklodowski et al. 2017). Uptake of K^+ ion from the soil K^+ is taken up from the soil by a well-organized system of transport proteins, each of which contributes in its way (Hasanuzzaman et al. 2018). The identification of low-affinity and high-affinity transporters in different plant species such as barley (*Hordeum vulgare* L.) (Cai et al. 2021), rice (*Oryza sativa* L.) (Wang et al. 2021), tobacco (*Nicotiana tabacum*) (Song et al. 2019), and cassava (*Manihot esculenta*) (Ou et al. 2018) has been made possible by advances in molecular techniques and tools. In an experiment, a yeast mutant without the ability to take up K^+ was transformed with barley cDNA was the mutant able to thrive. This led to the identification of the high-affinity K transporter HvHAK1, which is similar to the HAK1 K^+ transporter in *Escherichia coli* and *Schwanniomyces occidentalis* (Santa-María et al. 1997). It is reported that in *Arabidopsis*, 75 genes code for proteins that aid in K^+ absorption and transport. Shaker-type K^+ channels (nine genes), two-pore K^+ channels (six genes), putative K^+/H^+ antiporters (six genes), KUP/HAK/KT transporters (13 genes), HKT transporters (one gene), cyclic-nucleotide gate channels (20 genes), and glutamate receptors are the seven groups of genes (Hasanuzzaman et al. 2018). In roots, the uptake of K^+ from the media is principally facilitated by two proteins, namely, AKT1 and HAK5, as these are expressed in the roots of *Arabidopsis* (Lara et al. 2020) and rice (Yang et al. 2014; Ahmad et al. 2016). In *Arabidopsis*, it is reported that hak5 or akt1 loss-of-function mutants were able to survive in a 100 M KCl solution, but the double hak5 akt1 mutant did not, demonstrating that AKT1 and HAK5 are high-affinity transporters that facilitate adequate K^+ absorption for plant development. In rice, Os-AKT1-mediated uptake of K^+ is controlled by a complex of two proteins, namely, calcineurin B-like protein1 (Os-CBL1) and CBL-interacting protein kinase23 (CIPK23) (Li et al. 2014). The proton-driven H^+/K^+ co-transporter AtHAK5 is the only mechanism responsible for K^+ uptake from the soil when external K^+ concentrations are less than 0.01 Mm. Meanwhile, AtHAK5 and AKT1 jointly contribute to K^+ absorption for K^+ values between 0.01 mM and 0.05 mM. AKT1, along with other low-affinity K^+ uptake systems, is responsible for K^+ uptake from the soil at greater external K^+ concentrations (Liu et al. 2020). Different environmental variables impact AKT1 and AtHAK5. Both transport proteins operate at various K^+ concentration spectra and have varying ion sensitivities. For example, in the presence of ammonium, AtHAK5 becomes sensitive (Rubio et al. 2008), whereas AKT1 stays unaffected (Sharma et al. 2013). On the other hand, barium ions inhibit AKT1, while AtHAK5 remains unaffected. It is also reported that the presence of Na^+ and hydrogen ions results in the stimulation of AtHAK5 (Sharma et al. 2013). AKT1 is a target of a regulatory network and contributes to high- and low-affinity K^+ uptake. It is reported that for activation of AKT1, CIPK23 (Sánchez-Barrena et al. 2020) and CBL1 or CBL9 (Saito and Uozumi 2020) is required. CBL1 and CBL9, two calcineurin B-like calcium sensors, bind to CIPK23, a CBL-interacting protein kinase that phosphorylates AKT1 (Xu et al. 2006). A 2C-type protein phosphatase (PP2C), AIP1, has been shown to bind and inactivate AKT1 in addition to numerous CIP kinases (Lee et al. 2007; Singh et al. 2018). PP2C phosphatases bind to the CIPK-CBL complex, inhibiting the kinase's phosphorylation activity and dephosphorylating AKT1 (Weinl and Kudla 2009). It is also reported that CBL10 binds to AKT1 directly and reduces its activity in a concentration-dependent and CIPK-independent way (Ren et al. 2013).

Aside from kinases and phosphatases, AtKC1 is another member of the Shaker-like family that affects AKT1's functionality. In a study, it was shown that under a low K^+ environment, the Shaker-like K^+ channel AtKC1, expressed by the AtLKT1 gene cloned from *Arabidopsis thaliana* low K^+-tolerant mutant Atlkt1, strongly regulates AKT1-mediated K^+ uptake. According to electrophysiological investigations, AtKC1 suppressed AKT1-mediated inward K^+ currents and changed the voltage dependency of AKT1 channels adversely. The findings show that the "silent" K^+ channel α-subunit AtKC1 inhibits AKT1-mediated K^+ uptake in *Arabidopsis* roots and, as a result, changes the root-to-shoot ratio under LK stress conditions (Wang et al. 2010). It is also described that the activation threshold of AKT1 is shifted by AtKC1 to more negative levels. Under unfavorable conditions, this would prevent K^+ efflux via AKT1. Under low K^+ concentrations, the reduction of potential outward currents protects the plant from K^+ loss (Geiger et al. 2009). AKT1-AtKC1 heteromers have changed not just the activation threshold but also the pore's K^+-dependent stability. The permeation pathway of K^+ channels becomes unstable and collapses when the external K^+ concentration declines. When K^+ concentrations are higher, the pore of AKT1-AtKC1 heteromers collapses than those of AKT1 homomers (Geiger et al. 2009). As a result, heteromers are a more efficient block of the unfavorable outward K^+ passage.

11.6 Conclusion

The information available from studies mainly on *Arabidopsis* distinctly characterizes a novel mechanism in plant stress response particularly in salinity stress. This lesser known mechanism can mediate Na^+ detoxification in leaves by recirculating it through the phloem to the roots. Multiple studies so far have focused on Na^+ toxicity in plants under saline/hypersaline conditions and mechanisms for combating this stress in terms of tolerance and/or adaptation. With rapid climate change, the inundation and submergence of low-lying coastal areas under saline water is inevitable accompanied by a massive loss of crop plants. So further work with transporters, their regulation, signaling model, target proteins, and their regulating complexes seem imperative.

HAK transporter present in the membranes of bacteria, fungi, and plants facilitates K^+ transport. Various functions of plants from mineral nutrition to cell growth and development regulation are controlled by the HAK transporters. AKT transporter helps in K^+ transport regulation through the tonoplast of the plant cell. Salinity stress induces OsHAK1 production and is downregulated in K^+ deprivation, thereby proving that OsHAK1 plays a crucial role in increasing salt tolerance in rice. OsHAK5 expressed greatly in root epidermis, stele, and vascular tissues are responsible for K^+ uptake under conditions of low external K^+ concentration as well as upward translocation. A common regulatory process for both GORK and AKT1 may be possible where a protein complex between a calcineurin B-like (CBL) and a CBL-interacting protein kinase (CIPK) can decode a specific calcium signature to enable phosphorylation and activate K^+ transport. These proteins are important in plants as it helps plant metabolism regulation during salinity stress. Detailed studies are required on these proteins in order to understand plant stress tolerance from close proximity. The CBL-CBIK signaling model is a significant way forward to enable better understanding plant abiotic stress response identifying molecular targets for producing genetically engineered stress-tolerant higher salinity-resistant crops. The functional characterization, expounding the details of synergistic actions in this intricate network and clarifying the molecular mechanisms of the target protein-regulating complexes, is suggested thrust areas for future research. Most of the information available is from the *Arabidopsis* plant, while work on other major crop plants is till sparse. More attention is needed on other crop plants as well. So the loss of crop yield due to environmental/abiotic stress may be strongly addressed, thereby successfully leading us to the SDG 2 of Zero Hunger.

References

Ahmad I, Mian A, Maathuis FJ (2016) Overexpression of the rice AKT1 potassium channel affects potassium nutrition and rice drought tolerance. J Exp Bot 67(9):2689–2698

Ahn SJ, Shin R, Schachtman DP (2004) Expression of KT/KUP genes in Arabidopsis and the role of root hairs in K+ uptake. Plant Physiol 134:1135–1145

Albrecht V, Ritz O, Linder S, Harter K, Kudla J (2001) The NAF domain defines a novel protein–protein interaction module conserved in Ca2+−regulated kinases. EMBO J 20:1051–1063

Almeida DM, Oliveira MM, Saibo NJM (2017) Regulation of Na+ and K+ homeostasis in plants: towards improved salt stress tolerance in crop plants. Genet Mol Biol 40(1 suppl 1):326–345

Alvarez-Pizarro JC, Gomes-Filho E, de Lacerda CF (2009) Salt-induced changes on H^{+}-ATPase activity, sterol and phospholipid content and lipid peroxidation of root plasma membrane from dwarf-cashew (*Anacardium occidentale* L.) seedlings. Plant Growth Regul 59:125–135

Anderson JA, Huprikar SS, Kochian LV, Lucas WJ, Gaber RF (1992) Functional expression of a probable *Arabidopsis thaliana* potassium channel in Saccharomyces cerevisiae. Proc Natl Acad Sci 89:3736–3740

Anschütz U, Becker D, Shabala S (2014) Going beyond nutrition: regulation of potassium homoeostasis as a common denominator of plant adaptive responses to environment. J Plant Physiol 171(9):670–687

Armengaud P, Breitling R, Amtmann A (2004) The potassium-dependent transcriptome of Arabidopsis reveals a prominent role of jasmonic acid in nutrient signalling. Plant Physiol 136:2556–2576

Ardie SW, Liu SK, Takano T (2010) Expression of the AKT1-type K+ channel gene from *Puccinellia tenuiflora*, PutAKT1, enhances salt tolerance in *Arabidopsis*. Plant Cell Rep 29: 865–874

Bañuelos MA, Garciadeblas B, Cubero B, Rodríguez-Navarro A (2002) Inventory and functional characterization of the HAK potassium transporters of rice. Plant Physiol 130(2):784–795

Barros F, de la Peña P, Domínguez P, Sierra LM, Pardo LA (2020) The EAG voltage-dependent K+ channel subfamily: similarities and differences in structural organization and gating. Front Pharmacol 11:411

Briskin DP, Gawienowski MC (1996) Role of the plasma membrane H+-ATPase in K+ transport. Plant Physiol 111:1199–1207

Busch W, Saier MH Jr (2002) The transporter classification (TC) system. Crit Rev Biochem Mol Biol 37(5):287–337

Buschmann PH, Vaidyanathan R, Gassmann W, Schroeder JI (2000) Enhancement of Na+ uptake currents, time-dependent inward-rectifying K+ channel currents, and K+ channel transcripts by K+ starvation in wheat root cells. Plant Physiol 122:1387–1398

Cai K, Zeng F, Wang J, Zhang G (2021) Identification and characterization of HAK/KUP/KT potassium transporter gene family in barley and their expression under abiotic stress. BMC Genomics 22(1):317

Cao Y, Ward JM, Kelly WB, Ichida AM, Gaber RF, Anderson JA, Uozumi N, Schroeder JI, Crawford NM (1995) Multiple genes, tissue specificity, and expression-dependent modulation contribute to the functional diversity of potassium channels in *Arabidopsis thaliana*. Plant Physiol 109:1093–1106

Chakraborty K, Bose J, Shabala L, Shabala S (2016) Difference in root K+ retention ability and reduced sensitivity of K+-permeable channels to reactive oxygen species confer differential salt tolerance in three *Brassica* species. J Exp Bot 67:4611–4625

Cheeseman JM, Lafayette PR, Gronewald JW, Hanson JB (1980) Effect of ATPase inhibitors on cell potential and K+ influx in corn roots. Plant Physiol 65:1139–1145

Chen R, Swale DR (2018) Inwardly rectifying potassium (Kir) channels represent a critical ion conductance pathway in the nervous systems of insects. Sci Rep 8(1):1–3

Chen Z, Zhou M, Newman IA, Mendham NJ, Zhang G, Shabala S (2007) Potassium and sodium relations in salinised barley tissues as a basis of differential salt tolerance. Funct Plant Biol 34(2):150–162

Chen G, Hu Q, Luo L, Yang T, Zhang S, Hu Y (2015) Rice potassium transporter OsHAK1 is essential for maintaining potassium-mediated growth and functions in salt tolerance over low and high potassium concentration ranges. Plant Cell Environ 38(12):2747–2765

Chen G, Liu C, Gao Z, Zhang Y, Jiang H, Zhu L (2017) OsHAK1, a high-affinity potassium transporter, positively regulates responses to drought stress in rice. Front Plant Sci 8:1885

Cheng X, Liu X, Mao W, Zhang X, Chen S, Zhan K et al (2018) Genome-wide identification and analysis of HAK/KUP/KT potassium transporters gene family in wheat (*Triticum aestivum* L.). Int J Mol Sci 19(12):3969

Daras G, Rigas S, Tsitsekian D, Iacovides TA, Hatzopoulos P (2015) Potassium transporter TRH1 subunits assemble regulating root-hair elongation autonomously from the cell fate determination pathway. Plant Sci 231:131–137

Deeken R, Geiger D, Fromm J, Koroleva O, Ache P, Langenfeld-Heyser R, Sauer N, May ST, Hedrich R (2002) Loss of the AKT2/3 potassium channel affects sugar loading into the phloem of Arabidopsis. Planta 216:334–344

Demidchik V, Tracey A, Cuin DS, Susan JS, Anthony JM, Sergey S (2010) Anatoliy Sokolik, Vladimir Yurin; *Arabidopsis* root K^+-efflux conductance activated by hydroxyl radicals: single-channel properties, genetic basis and involvement in stress-induced cell death. J Cell Sci 123(9): 1468–1479

Demidchik V, Straltsova D, Medvedev SS, Pozhvanov GA, Sokolik A, Yurin V (2014) Stress-induced electrolyte leakage: the role of K+-permeable channels and involvement in programmed cell death and metabolic adjustment. J Exp Bot 65:1259–1270

Dreyer I, Uozumi N (2011) Potassium channels in plant cells. FEBS J 278(22):4293–4303

Elumalai RP, Nagpal P, Reed JW (2002) A mutation in the Arabidopsis KT2/KUP2 potassium transporter gene affects shoot cell expansion. Plant Cell 14(1):119–131

Fernando A, Manuel NC, Vicente M, Francisco R (2011) Root K+ Acquisition in Plants: the *Arabidopsis thaliana* model. Plant Cell Physiol 52(9):1603–1612

Forster S, Schmidt LK, Kopic E, Anschutz U, Huang S, Schlucking K, Koster P, Waadt R, Larrieu A, Batistic O et al (2019) Wounding induced stomatal closure requires jasmonate-mediated activation of GORK K1 channels by a Ca21 sensor-kinase CBL1-CIPK5 complex. Dev Cell 48(1):87–99.e6

Geiger D, Becker D, Vosloh D, Gambale F, Palme K, Rehers M, Anschuetz U, Dreyer I, Kudla J, Hedrich R (2009) Heteromeric AtKC1{middle dot}AKT1 channels in *Arabidopsis* roots facilitate growth under K+-limiting conditions. J Biol Chem 284(32):21288–21295

Gierth M, Mäser P, Schroeder JI (2005) The potassium transporter AtHAK5 functions in K+ deprivation-induced high affinity K+ uptake and AKT1 K+ channel contribution to K+ uptake kinetics in Arabidopsis roots. Plant Physiol 137:1105–1114

Golldack D, Hua S, Francoise Q, Kamasani UR, Muñoz-Garay C, Enrique B, Olga VP, John B, Hans JB, Omar P (2002) Characterization of a HKT-type transporter in rice as a general alkali cation transporter. Plant J 31(4):529–542

Grabov A (2007) Plant KT/KUP/HAK potassium transporters: single family – multiple functions. Ann Bot 99:1035–1041

Grefen C, Blatt MR (2012) Do calcineurin B-like proteins interact independently of the serine threonine kinase CIPK23 with the K+ channel AKT1? Lessons learned from a menage a trois. Plant Physiol 159:915–919

Han M, Wu W, Wu WH, Wang Y (2016) Potassium transporter KUP7 is involved in K+ acquisition and translocation in Arabidopsis root under K+-limited conditions. Mol Plant 9(3):437–446

Hasanuzzaman M, Bhuyan MH, Nahar K, Hossain M, Mahmud JA, Hossen M, Masud AA, Fujita M (2018) Potassium: a vital regulator of plant responses and tolerance to abiotic stresses. Agronomy 8(3):31

Hashimoto K, Eckert C, Anschütz U, Scholz M, Held K, Waadt R, Reyer A, Hippler M, Becker D, Kudla J (2012) Phosphorylation of calcineurin B-like (CBL) calcium sensor proteins by their CBL-interacting protein kinases (CIPKs) is required for full activity of CBL-CIPK complexes toward their target proteins. J Biol Chem 287:7956–7968

Hauser F, Horie T (2010) A conserved primary salt tolerance mechanism mediated by HKT transporters: a mechanism for sodium exclusion and maintenance of high K^+/Na^+ ratio in leaves during salinity stress. Plant Cell Environ 33:552–565

Hirsch RE, Lewis BD, Spalding EP, Sussman MR (1998) A role for the AKT1 potassium channel in plant nutrition. Science 280:918–921

Hu W, Di Q, Zhang J, Liu J, Shi X (2020) Response of grafting tobacco to low potassium stress. BMC Plant Biol 20(1):1–1

Isayenkov S, Isner JC, Maathuis FJ (2011) Rice two-pore K+ channels are expressed in different types of vacuoles. Plant Cell 23(2):756–768

Jayakannan M, Bose J, Olga B, Zed R, Sergey S (2013) Salicylic acid improves salinity tolerance in *Arabidopsis* by restoring membrane potential and preventing salt-induced K^+ loss via a GORK channel. J Exp Bot 64(8):2255–2268

Jegla T, Busey G, Assmann SM (2018) Evolution and structural characteristics of plant voltage-gated K+ channels. Plant Cell 30(12):2898–2909

Kaddour R, Narsi N, Mïrah S, Berthomieu P, Lachaal M (2009) Comparative effect of potassium on K and Na uptake and transport in two accessions of Arabidopsis thaliana during salinity stress. C R Biol 332:784–794

Ketchum KA, Shrier A, Poole RJ (1989) Characterization of potassium-dependent currents in protoplasts of corn suspension cells. Plant Physiol 89:1184–1192

Kim EJ, Kwak JM, Uozumi N, Schroeder JI (1998) AtKUP1: an Arabidopsis gene encoding high-affinity potassium transport activity. Plant Cell 10(1):51–62

Kochian LV, Shaff JE, Lucas WJ (1989) High-affinity K+ uptake in maize roots – a lack of coupling with H+ efflux. Plant Physiol 91:1202–1211

Lara A, Ródenas R, Andrés Z, Martínez V, Quintero FJ, Nieves-Cordones M, Botella MA, Rubio F (2020) *Arabidopsis* K+ transporter HAK5-mediated high-affinity root K+ uptake is regulated by protein kinases CIPK1 and CIPK9. J Exp Bot 71(16):5053–5060

Lebaudy A, Véry AA, Sentenac H (2007) K+ channel activity in plants: genes, regulations and functions. FEBS Lett 581(12):2357–2366

Lee SC, Lan WZ, Kim BG, Li L, Cheong YH, Pandey GK, Lu G, Buchanan BB, Luan S (2007) A protein phosphorylation/dephosphorylation network regulates a plant potassium channel. Proc Natl Acad Sci U S A 104(40):15959–15964

Lefoulon C, Boeglin M, Moreau B, Véry AA, Szponarski W, Dauzat M, Michard E, Gaillard I, Chérel I (2016) The *Arabidopsis* AtPP2CA protein phosphatase inhibits the GORK K1 efflux channel and exerts a dominant suppressive effect on phosphomimetic-activating mutations. J Biol Chem 291:6521–6533

Li J, Long Y, Qi GN, Li J, Xu ZJ, Wu WH, Wang Y (2014) The Os-AKT1 channel is critical for K+ uptake in rice roots and is modulated by the rice CBL1-CIPK23 complex. Plant Cell 26(8): 3387–3402

Li W, Xu G, Alli A, Yu L (2018) Plant HAK/KUP/KT K+ transporters: function and regulation. Semin Cell Dev Biol 74:133–141

Liu WZ, Deng M, Li L, Yang B, Li H, Deng H, Jiang YQ (2015) Rapeseed calcineurin B-like protein CBL4, interacting with CBL-interacting protein kinase CIPK24, modulates salt tolerance in plants. Biochem Biophys Res Commun 467:467–471

Liu C, Tu B, Wang X, Li Y, Zhang Q, Liu X (2020) Transcript profile in vegetable soybean roots reveals potential gene patterns regulating K uptake efficiency. Agronomy 10(11):1796

Maathuis FJ (2006) The role of monovalent cation transporters in plant responses to salinity. J Exp Bot 57(5):1137–1147

Maathuis FJM, Amtmann A (1999) K+ nutrition and Na+ toxicity: the basis of cellular K+/Na+ ratios. Ann Bot 84:123–133

Maathuis FJM, Sanders D (1994) Mechanism of high-affinity potassium uptake in roots of Arabidopsis thaliana. Proc Natl Acad Sci U S A 91:9272–9276

Maathuis FJM, Sanders D (1999) Plasma membrane transport in context-making sense out of complexity. Curr Opin Plant Biol 2:236–243

Maathuis FJ, Ichida AM, Sanders D, Schroeder JI (1997) Roles of higher plant K^+ channels. Plant Physiol 114:1141–1149

Maathuis FJM, Filatov V, Herzyk P, Krijger GC, Axelsen KB, Chen SX (2003) Transcriptome analysis of root transporters reveals participation of multiple gene families in the response to cation stress. Plant J 35:675–692

Martinez-Cordero MA, Martinez V, Rubio F (2005) High-affinity K+ uptake in pepper plants. J Exp Bot 56:1553–1562
Maser P, Thomine S, Schroeder JI (2001) Phylogenetic relationships within cation transporter families of *Arabidopsis*. Plant Physiol 126:1646–1667
Muller RB, Ellenberg J, Provart N, Willmitzer L, Busch H, Becker D (1995) Cloning and electrophysiological analysis of KST1, an inward-rectifying K+ channel expressed in potato guard cells. EMBO J 14:2409–2416
Munns R, Tester M (2008) Mechanisms of salinity tolerance. Annu Rev Plant Biol 59(1):651–681
Nakamura RL, McKendree WL, Hirsch RE, Sedbrook JC, Gaber RF, Sussman MR (1995) Expression of an Arabidopsis potassium channel gene in guard-cells. Plant Physiol 109:371–374
Nieves-Cordones M, Miller AJ, Alemán F, Martínez V, Rubio F (2008) A putative role for the plasma membrane potential in the control of the expression of the gene encoding the tomato high-affinity potassium transporter HAK5. Plant Mol Biol 68:521–532
Nieves-Cordones M, Alemán F, Martínez V, Rubio F (2010) The *Arabidopsis thaliana* HAK5 K+ transporter is required for plant growth and K+ acquisition from low K+ solutions under saline conditions. Mol Plant 3(2):326–333
Nieves-Cordones M, Alemán F, Martínez V, Rubio F (2014) K+ uptake in plant roots. The systems involved their regulation and parallels in other organisms. J Plant Physiol 171:688–695
Ou W, Mao X, Huang C, Tie W, Yan Y, Ding Z, Wu C, Xia Z, Wang W, Zhou S, Li K (2018) Genome-wide identification and expression analysis of the KUP family under abiotic stress in cassava (*Manihot esculenta* Crantz). Front Physiol 9:17
Payandeh J, Scheuer T, Zheng N, Catterall WA (2011) The crystal structure of a voltage-gated sodium channel. Nature 475(7356):353–358
Peng Z, He S, Sun J, Pan Z, Gong W, Lu Y, Du X (2016) Na+ compartmentalization related to salinity stress tolerance in upland cotton (*Gossypium hirsutum*) seedlings. Sci Rep 6(1):1–4
Pilot G, Gaymard F, Mouline K, Cherel I, Sentenac H (2003a) Regulated expression of *Arabidopsis* Shaker K+ channel genes involved in K+ uptake and distribution in the plant. Plant Mol Biol 51: 773–787
Pilot G, Pratelli R, Gaymard F, Meyer Y, Sentenac H (2003b) Five-group distribution of the Shaker-like K+ channel family in higher plants. J Mol Evol 56:418–434
Pyo YJ, Gierth M, Schroeder JI, Cho MH (2010) High-affinity K+ transport in Arabidopsis: AtHAK5 and AKT1 are vital for seedling establishment and postgermination growth under low-potassium conditions. Plant Physiol 153(2):863–875
Qi Z, Hampton CR, Shin R, Barkla BJ, White PJ, Schachtman DP (2008) The high affinity K^+ transporter AtHAK5 plays a physiological role in planta at very low K^+ concentrations and provides a caesium uptake pathway in *Arabidopsis*. J Exp Bot 59:595–607
Ragel P, Raddatz N, Leidi EO, Quintero FJ, Pardo JM (2019) Regulation of K+ nutrition in plants. Front Plant Sci 10:281
Ren XL, Qi GN, Feng HQ, Zhao S, Zhao SS, Wang Y, Wu WH (2013) Calcineurin B-like protein CBL10 directly interacts with AKT1 and modulates K+ homeostasis in *Arabidopsis*. Plant J 74(2):258–266
Riedelsberger J, Miller JK, Valdebenito-Maturana B, Piñeros MA, González W, Dreyer I (2021) Plant HKT channels: an updated view on structure, function and gene regulation. Int J Mol Sci 22(4):1892
Rigas S, Ditengou FA, Ljung K, Daras G, Tietz O, Palme K et al (2013) Root gravitropism and root hair development constitute coupled developmental responses regulated by auxin homeostasis in the *Arabidopsis* root apex. New Phytol 197(4):1130–1141
Rodríguez-Navarro A (2000) Potassium transport in fungi and plants. Biochim Biophys Acta 1469 (1):1–30
Rodríguez-Navarro A, Rubio F (2006) High-affinity potassium and sodium transport systems in plants. J Exp Bot 57:1149–1160

Rubio F, Gassmann W, Schroeder JI (1995) Sodium-driven potassium uptake by the plant potassium transporter HKT1 and mutations conferring salt tolerance. Science 270:1660–1663

Rubio F, Nieves-Cordones M, Alemán F, Martínez V (2008) Relative contribution of AtHAK5 and AtAKT1 to K+ uptake in the high-affinity range of concentrations. Physiol Plant 134(4): 598–608

Saito S, Uozumi N (2020) Calcium-regulated phosphorylation systems controlling uptake and balance of plant nutrients. Front Plant Sci 11:44

Sánchez-Barrena MJ, Antonio CS, Natalia R, Imelda M, Álvaro C, Federico G, Juana MGR, Juan LB, Francisco JQ, José MP, Armando A (2020) Recognition and activation of the plant AKT1 potassium channel by the kinase CIPK23. Plant Physiol 182(4):2143–2153

Santa-María GE, Rubio F, Dubcovsky J, Rodríguez-Navarro A (1997) The HAK1 gene of barley is a member of a large gene family and encodes a high-affinity potassium transporter. Plant Cell 9(12):2281–2289

Saponaro A, Porro A, Chaves-Sanjuan A, Nardini M, Rauh O, Thiel G, Moroni A (2017) Fusicoccin activates KAT1 channels by stabilizing their interaction with 14-3-3 proteins. Plant Cell 29:2570–2580

Schachtman DP, Schroeder JI (1994) Structure and transport mechanism of a high-affinity potassium uptake transporter from higher plants. Nature 370:655–658

Sen S, Mukherji S (2004) Alterations in activities of acid phosphatase, alkaline phosphatase, ATPase and ATP content in response to seasonally varying Pi status in Okra (*Abelmoschus esculentus*). J Environ Biol 25(2):181–185

Sen S, Mukherji S (2007) Changes in phosphorus metabolism in *Lycopersicon esculentum* Mill.as affected by seasonal environmental conditions of West Bengal (India). Eco-Chronicle 2(2): 81–86

Shabala S (2009) Salinity and programmed cell death: unravelling mechanisms for ion specific signalling. J Exp Bot 60:709–711

Shabala S, Shabala L (2011) Ion transport and osmotic adjustment in plants and bacteria. Biomol Concepts 2(5):407–419

Shabala L, Cuin TA, Newman IA, Shabala S (2005) Salinity-induced ion flux patterns from the excised roots of *Arabidopsis* sos mutants. Planta 222:1041–1050

Shah WH, Rasool A, Saleem S, Mushtaq NU, Tahir I, Hakeem KR, Rehman RU (2021) Understanding the integrated pathways and mechanisms of transporters, protein kinases and transcription factors in plants under salt stress. Int J Genomics 12:5578727

Sharma T, Dreyer I, Riedelsberger J (2013) The role of K+ channels in uptake and redistribution of potassium in the model plant *Arabidopsis thaliana*. Front Plant Sci 4:224

Shi J, Kim KN, Ritz O, Albrecht V, Gupta R, Harter K, Luan S, Kudla J (1999) Novel protein kinases associated with calcineurin B-like calcium sensors in *Arabidopsis*. Plant Cell 11:2393–2405

Shin R, Schachtman DP (2004) Hydrogen peroxide mediates plant root cell response to nutrient deprivation. Proc Natl Acad Sci 101:8827–8832

Singh A, Yadav AK, Kaur K, Sanyal SK, Jha SK, Fernandes JL, Sharma P, Tokas I, Pandey A, Luan S, Pandey GK (2018) A protein phosphatase 2C, AP2C1, interacts with and negatively regulates the function of CIPK9 under potassium-deficient conditions in *Arabidopsis*. J Exp Bot 69(16):4003–4015

Sinha D, Tandon PK (2020) An overview of nitrogen, phosphorus and potassium: key players of nutrition process in plants. In: Mishra K, Tandon PK, Srivastava S (eds) Sustainable solutions for elemental deficiency and excess in crop plants. Springer, Singapore, pp 85–117

Sklodowski K, Riedelsberger J, Raddatz N, Riadi G, Caballero J, Chérel I, Schulze W, Graf A, Dreyer I (2017) The receptor-like pseudokinase MRH1 interacts with the voltage-gated potassium channel AKT2. Sci Rep 7(1):1–2

Song Z, Wu X, Gao Y, Cui X, Jiao F, Chen X, Li Y (2019) Genome-wide analysis of the HAK potassium transporter gene family reveals asymmetrical evolution in tobacco (*Nicotiana tabacum*). Genome 62(4):267–278

Su H, Golldack D, Zhao CS, Bohnert HJ (2002) The expression of HAK-type K+ transporters is regulated in response to salinity stress in common ice plant. Plant Physiol 129:1482–1493

Sun Y, Kong X, Li C, Liu Y, Ding Z (2015) Potassium retention under salt stress is associated with natural variation in salinity tolerance among *Arabidopsis* accessions. PLoS One 10(5):e0124032

Szczerba MW, Britto DT, Kronzucker HJ (2009) K+ transport in plants: physiology and molecular biology. J Plant Physiol 166(5):447–466

Takahashi R, Nishio T, Ichizen N, Takano T (2007) Cloning and functional analysis of the K+ transporter, PhaHAK2, from salt-sensitive and salt-tolerant reed plants. Biotechnol Lett 29:501–506

Tang RJ, Zhao FG, Yang Y, Wang C, Li K, Kleist TJ, Lemaux PG, Luan S (2020) A calcium signalling network activates vacuolar K+ remobilization to enable plant adaptation to low-K environments. Nat Plants 6(4):384–393

Tsujii M, Tanudjaja E, Uozumi N (2020) Diverse physiological functions of cation proton antiporters across bacteria and plant cells. Int J Mol Sci 21(12):4566

Uozumi N, Kim EJ, Rubio F, Yamaguchi T, Muto S, Tsuboi A, Bakker EP, Nakamura T, Schroeder JI (2000) The *Arabidopsis* HKT1 gene homolog mediates inward Na^+ currents in *Xenopus laevis* oocytes and Na^+ uptake in *Saccharomyces cerevisiae*. Plant Physiol 122:1249–1259

Véry AA, Sentenac H (2003) Molecular mechanisms and regulation of K^+ transport in higher plants. Annu Rev Plant Biol 54:575–603

Wang Y, He L, Li HD, Xu J, Wu WH (2010) Potassium channel α-subunit AtKC1 negatively regulates AKT1-mediated K+ uptake in Arabidopsis roots under low-K+ stress. Cell Res 20(7): 826–837

Wang X, Li J, Li F, Pan Y, Cai D, Mao D, Chen L, Luan S (2021) Rice potassium transporter OsHAK8 mediates K+ uptake and translocation in response to low K+ stress. Front Plant Sci 12: 730002

Weinl S, Kudla J (2009) The CBL–CIPK Ca^{2+}-decoding signaling network: function and perspectives. New Phytol 184(3):517–528

Wu H (2018) Plant salt tolerance and Na+ sensing and transport. Crop J 6(3):215–225

Wu H, Shabala L, Zhou M, Shabala S (2015) Chloroplast-generated ROS dominate NaCl-induced K^+ efflux in wheat leaf mesophyll. Plant Signal Behav 10:e1013793

Xu J, Li HD, Chen LQ, Wang Y, Liu LL, He L, Wu WH (2006) A protein kinase, interacting with two calcineurin B-like proteins, regulates K+ transporter AKT1 in *Arabidopsis*. Cell 125(7): 1347–1360

Yang T, Zhang S, Hu Y, Wu F, Hu Q, Chen G et al (2014) The role of a potassium transporter OsHAK5 in potassium acquisition and transport from roots to shoots in rice at low potassium supply levels. Plant Physiol 166(2):945–959

Zhang Z, Zhang J, Chen Y, Li R, Wang H, Wei J (2012) Genome-wide analysis and identification of HAK potassium transporter gene family in maize (*Zea mays* L.). Mol Biol Rep 39(8): 8465–8473

Zhao S, Zhang ML, Ma TL, Wang Y (2016) Phosphorylation of ARF2 relieves its repression of transcription of the K+ transporter gene HAK5 in response to low potassium stress. Plant Cell 28(12):3005–3019

Zhao S, Zhang Q, Liu M, Zhou H, Ma C, Wang P (2021) Regulation of plant responses to salt stress. Int J Mol Sci 22(9):4609

The Mechanism of Silicon Transport in Plants

12

Shafia Maryam and Alvina Gul

Abstract

Silicon is an abundant metalloid of Earth's crust, found in soil and rocks. Silicon is abundant as oxides or silicates. Silicon is an essential nutrient in diatoms, algae, and plants. Silicon demand in plants varies across species, as fertilizer. Silicon fertilizers are applied in combination with other essential nutrients. Common silicon fertilizers used are calcium silicate ($CaSiO_3$), potassium silicate (K_2SiO_3), and sodium silicate (Na_2SiO_3). The deficiency of silicon is detected by low stem strength and leaf freckling. Crops such as corn, wheat, oats, pumpkin, and cucumber were studied for the addition of silicon. Silicon absorption is increased by the supply of nitrogen and phosphorus. Silicon induces resistance to abiotic stresses such as saline stress along with drought, flooding, metal toxicity, nutrient deficiency along with nutrient inflation, UV radiation, and temperature inflation. Silicon is absorbed by roots and is deposited on different tissues. Silicon is transported in plants by two channels, Lsi1 and Lsi2 (low silicon 1 and 2). Lsi1 is a permeable channel from Nod26-like major intrinsic protein (NIP) III for uptake and distribution of silicon. This is an aquaporin membrane protein family. Lsi2 is part of an uncharacterized anion transporter family. Silicon has significance with agriculture, to increase crop yield for better harvest.

Keywords

Silicon · Lower silicon transporter 1 · Reactive oxygen species · Soil · Abiotic stress

S. Maryam · A. Gul (✉)
National University of Science and Technology, Islamabad, Pakistan
e-mail: Smaryam.phdabs18asab@student.nust.edu.pk; alvina_gul@asab.nust.edu.pk

K. Kumar, S. Srivastava (eds.), *Plant Metal and Metalloid Transporters*,
https://doi.org/10.1007/978-981-19-6103-8_12

Abbreviations

Si	Silicon
Lsi1	Lower silicon transporter 1
Lsi2	Lower silicon transporter2
ROS	Reactive oxygen species
RUBP-carboxylase	Ribulose bisphosphate carboxylase
SOD	Superoxide dismutase
POD	Peroxidase
CAT	Catalase
APX	Ascorbate peroxidase
GPOX	Guaiacol peroxidase
DHAR	Dehydroascorbate reductase
MDHAR	Monodehydroascorbate reductase

12.1 Silicon

Silicon is the second most abundant metalloid from Group 14 of the periodic table. Twenty-eight percent of Earth's crust is composed of silicon element. Silicon is found in soil and occurs in rocks as an amorphous brown powder or as gray crystals. It is an important component of sand, silt, and clay. Silicon is abundant as oxides or in the form of silicates. Silicon is rarely found in a free elemental form. Silica is the element, while the crystalline and amorphous compounds of silicon are calcium silicate ($CaSiO_3$), magnesium silicate ($MgSiO_3$) sodium silicate (Na_2SiO_3), or potassium silicate (K_2SiO_3). Silicic acid or mono silicic acid [Si $(OH)_4$, or H_4SiO_4] refers to the soluble forms of Si available and absorbed by plants.

12.2 Silicon in Plants

Silicon is an essential nutrient in diatoms, algae, and plants. Silicon is one of the most abundant trace element in humans after Fe and Zn. Silicon is helpful for plants for survival and growth (International Plant Nutrition Institute 2015). Silica or silicon dioxide (SiO_2) are glass-like compounds that are insoluble and not absorbed by plants. Silicic acid [Si $(OH)_4$] is absorbed by plant roots and transported to intercellular spaces for deposition (Epstein 2009). Grasses accumulate silicon as phytoliths to strengthen cell walls and sustain turgidity. In sugarcane, silica accumulation is required to hinder leaf freckling and overexposure to ultraviolet light. Leaves and stems deposited with silica are protected from insect attacks, diseases along with environmental stresses (Heckman 2013). Silicon in plants stimulates the production of organic compounds and enzymes responsible for defense response against fungus and insects. Silica stimulates the production of chitinase, peroxidase,

Table 12.1 Silicon accumulation percent of dry weight in major food crops

Sr. No.	Crops	% Dry weight	Reference
1	Rice	4.2	Hodson et al. (2005)
2	Wheat	2.5	Hodson et al. (2005)
3	Barley	1.8	Hodson et al. (2005)
4	Sugarcane	1.5	Hodson et al. (2005)
5	Soybean	1.4	Hodson et al. (2005)
6	Corn	0.8	Hodson et al. (2005)
7	Cassava	0.5	Hodson et al. (2005)

polyphenol oxidases, and flavonoid phytoalexins. These compounds act to protect the plant from fungal attacks. The absorption of silicon in plants varies by species, which indirectly influences immunity of plants.

Dicots especially legumes absorb the lowest concentration of silicon. Estimates lower than 0.5% are termed excluders. Some dicots such as sugarcane, cereals, or dryland grasses accumulate more silicon than excluders. The amount varies between 1% and 3%, due to which these plant species are termed intermediates. The highest silicon accumulation between 1% and 10% is in sedges and wetland grasses. These species are termed accumulators. Rice is one common silicon accumulator. Plants absorb uncharged silicic acid, $Si(OH)_4$. Plants store silicon as amorphous silica. Silicon with its imperative role in the development of plants to alleviate stress confers high vigor and resistance to growth constraints. Silicon is reported to provide higher yield and biomass in fiber crops. Silicon application in croplands enhances accumulation and improves yields of major agronomical crops. Silicon application to croplands with low availability is beneficial as fertilizer (Luyckx et al. 2017). Silicon reinforces cell walls through the deposition of silica. Silicic acid is transported by roots through the xylem and deposited under cuticles and other intercellular spaces, phytoliths, or plant opal. Silicon transport and uptake are different in all plant species due to a combination of proteins and genes (Table 12.1).

Silicon demand in plants varies across species, and as a result, Si fertilizer demand for all crops needs to be ascertained. Silicon fertilizers are applied in combination with other essential nutrients required for growth. Common silicon fertilizers used are calcium silicate ($CaSiO_3$) extracted from steel mills and mineral wollastonite. Calcium silicate provides an additional benefit as a limiting agent in low pH soil. Si is also applied as potassium silicate (K_2SiO_3) and sodium silicate (Na_2SiO_3) by means of drip irrigation. The deficiency of silicon is detected by low stem strength and leaf freckling. Crops such as corn, wheat, oats, pumpkin, and cucumber were studied for the addition of silicon. Positive results with better growth and stem strength were reported with exceptional results in rice.

12.3 Silicon in Soil

Soil is composed of various minerals and rocks. Thirty percent of soil is composed of silicon with a large concentration of soluble silicic acid. Soil solutions are composed of 0.1–0.6 mM of Si (Ma and Takahashi 2002). Sand is mostly composed of SiO_2, which contributes to its low solubility. In contrast, tropical soils along with peat, muck, and other soils with rich organic matter have low Si content. Soil with low water holding capacity has low Si absorption. Soluble Si is between 3.5 and 40 mg/l. In agricultural soil, average silicon concentration recorded is 14–20 mg/l. Silicon abundance in the soil makes it available for plant absorption. Studies designed to understand silicon absorption in soil report 1–4 mg SiO_2 per gram in control plant leaves. Silica persists in soil and resists degradation. Archeological and palaeocological research indicate that plant fossil obtained from soil has a significant amount of silica. Most plant species are tracked by silica content. Along with naturally soluble silica present in soil, plants accumulate supplementary silica required for extra strength. This phenomenon is common in grasses for mechanical strength, protection against biotic and abiotic stress along with structural strength and defense responses. Grasses are responsible for ecosystem stability, which indicates depletion of silicon from soil was devastating for the ecosystems. Silicon absorption is increased by the supply of nitrogen and phosphorus. This contributes to a systematic increase in yield (Jones and Handreck 1967; Lewin and Reimann 1969). An increase in silicon level lowers nitrogen, phosphorus, iron, and manganese in soil (Takahashi et al. 1990). Silicon applied on rice paddy decreases phosphorus content from stems and leaves and enhances nitrogen absorption in ears (Miyoshi and Ishii 1960). Various strategies have been applied to increase silicon absorption from the soil. NO_3 and NH_4 favor silicon absorption (Kono and Takahashi 1958). P22 is applied to clay mineral soil to enhance the release of silicon from soil and absorption by plants. Phosphate forms complex compounds such as aluminosilicates. These decrease phosphate fixation and phosphates from soil (Du Plessis and Burger 1966; Du Preez 1970; Vyas and Motiramani 1971). Application of lime on soil lowers silicon uptake capacity. This in turn lowers the transpiration rate (Jones and Handreck 1967). Silicon absorption from the soil is directly related to clay content in the soil while directly inverse to the pH of soil (Huang 1966; Jones and Handreck 1967; Obihara and Russell 1972). Soil acidity is neutralized by the formation of silicic acid to lower the solubility of manganese, iron, and aluminum (Ayres 1966; Garberg 1970; Nair and Aiyer 1968; Okuda and Takahashie 1963; Peaslee and Frink 1969). Hydrated soil absorbs Si more rapidly (Okuda and Takahashie 1963).

12.4 Silicon and Abiotic Stresses

Silicon induces resistance to abiotic stresses such as saline stress along with drought, flooding, metal toxicity, nutrient deficiency along with nutrient inflation, UV radiation, and temperature inflation (Etesami and Jeong 2018; Reynolds et al. 2016;

Coskun et al. 2016; Tripathi et al. 2017). Silicon supports root functioning and stimulates defense against water pressure (Struyf and Conley 2009).

12.4.1 Water-Deficit Stress

Drought or water deficiency reduces crop yield. It abruptly hinders growth with closing of stomatal apertures, low photosynthesis, transpiration, and water potential of cell (Yardanov et al. 2003). Silicon treatment of drought-encountered crops correlates to reducing water loss along with activating antioxidative defense response. Additionally, Si improves mineral uptake and osmotic homeostasis (Gunes et al. 2008). Si deposition in cell walls reduces transpiration rate to stimulate water storage (Agarie et al. 1998). Application of silicon dioxide on crops maintains water balance (Gong et al. 2005). Exogenous application of silicon enhances Ca levels to maintain cell stability along with photosynthetic efficiency by activating chlorophyll (Kaya et al. 2006). Potassium balances the osmotic balance in plants by maintaining flow in plants required for growth. High K^+ content in tissues due to water deprivation is also balanced by Si along with triggering action potential for H^+-ATPases within cell membrane (Liang et al. 2003). Si application stabilizes hemicellulose along with other solutes like osmolytes, proline, glycine betaine, sugars, and free amino acids to protect cells and membranes against negative impact of drought. Cell walls are strengthened by hemicellulose (Sabagh et al. 2020). Wang et al. (2015) reported improved root-to-shoot ratio with silicon supplied plants. Modulated root architecture supports water transport, with high uptake and translocation with root hydraulic conductance (Chen et al. 2011).

12.4.2 Temperature Stress

High or low temperature fluctuation is an abiotic stress that hinders growth and metabolism in plants. High temperature interferes with protein metabolism, plant pigment synthesis, photosynthesis, transpiration rate, and enzymatic processes (Gibson and Paulsen 1999). This alters balance between ROS and antioxidants to suppress plant defense. Si mitigates heat stress by binding to cell wall of vascular tissues and compressing excessive transpiration (Liang et al. 2007). Improved metabolism and reduction in transpiration loss helps plants to cope with damage by sunburn or wilting. Silicon increases ribulose bisphosphate carboxylase (RUBP-carboxylase) enzymes in plant tissues, leading to CO_2 metabolism and improved plant growth (Gunes et al. 2008). Si treatment on *Salvia splendens* modulates defense responses with production of SOD, POD, APX, and GPOX (Soundararajan et al. 2014).

Freezing by mechanical stress deposits ices on cell surfaces along with dehydration and saline deposition. This further instigates lipid peroxidation and production of ROS to agglomerate membranes (McKersie et al. 1993). Si elongate leaves and stem with stable deposition in cuticular regions. Stable cell wall rich in polysaccharides and lignin hinders plant injury from freezing (Hull 2004). Low temperature disturbs electron transport chain, rubisco enzyme activity, and carbon dioxide fixation along with pigment synthesis. RuBP carboxylase regulates metabolism of CO_2. Silicon relieves plant stress and improves growth by activating rubisco enzymes (Ashraf et al. 2010). Hydroponic treatment of Si on wheat, barley, maize, rice, sunflower, and cucumber is reported with tolerance against frost-induced wilt and nutrient deprivation (Zhu et al. 2004).

12.4.3 Ultraviolet Stress

Si application on wheat, barley, rice, soyabean, and maize activates tolerance to ultraviolet radiation stress by altered physiological and biochemical activity (Shen et al. 2014; Tripathi et al. 2017).

12.4.4 Mechanical Injury

Physical damage or wounds on plants by biotic or abiotic stress lead to cell apoptosis or plant death by mechanical injury. Mechanical injury triggers oxidative damage with membrane disruption and high ROS levels in plants (Tripathi et al. 2017). Si treatment on plants initiates expression of antioxidant enzymes to initiate defense in wheat, maize, barley, and many other species (Liang et al. 2007). Si further strengthens membranes and helps plants in recovering from injuries (Kim et al. 2014).

12.4.5 Heavy Metal Stress

Various anthropogenic activities along with natural disasters release excess heavy metals in environment detrimental for plant growth (Nagajyoti et al. 2010). Heavy metal accumulation in plants is visible by necrosis and leaf chlorosis. Heavy metals degrade cell membranes and impair chlorophyll biosynthesis. Si mediation ameliorates cadmium (Cd), copper (Cu), aluminum (Al), zinc (Zn), boron (B), and chromium (Cr) from plants (da Cunha and do Nascimento 2009; Gunes et al. 2007; Tripathi et al. 2012). Si is condensed in Casparian strips in endodermis of vascular tissues and deposited in intercellular spaces (da Cunha and do Nascimento 2009). Si enhances nutrient absorption and precipitates with harmful heavy metals reduction (Etesami and Jeong 2018; Liang et al. 2005). After immobilizing, metals are translocated and compartmentalized or distributed evenly (Adrees et al. 2015). Membranes are strengthened with lignification of Casparian strips, vascular tissues, and suberin lamellae to hinder heavy metal entry and distribution. Si stimulates release of root exudates to enzymatically reduce and stabilize metal ions below toxic levels (Imtiaz et al. 2016).

12.4.6 Excessive Mineral Nutrient Stress

Chemicals as fertilizers and pesticides applied to crops with beneficial impact have various side effects. Si crystals reduce nitrogen-based fertilizers. Si forms barrier in plants to prevent attack from pathogens, along with inhibition of insect feeding mechanisms by damaging germ tube required for epidermal penetration (Malhotra and Kapoor 2019). Si enhances nitrogen uptake to improve nodulation and improve nitrogen fixation in legumes (Pavlovic et al. 2013). Si further stimulates nitrogen and amino acid mobilization to modulate metabolism (Oliveira et al. 2019). Phosphorus is an essential mineral for plant growth required in soil, but with excess phosphorus, cells are damaged. Si triggers phosphorylation by rejecting uptake of ions such as Fe, Mn, and Al (Bityutskii et al. 2014). In acidic soil, supplementation of Si dwindles P absorption and enhances amount of total P in soil (Owino-Gerroh and Gascho 2005). Si balances potassium deficiency in soil by balancing availability between both soil and plants (Miao et al. 2010). Potassium scarcity causes membrane damage by lipid peroxidation, which is reverted by Si through modulation of antioxidant defense expression (Pei et al. 2010). Si fertilizers stimulate and contain Ca and Mg to promote Ca absorption and trigger proton movement across H^+-ATPases (Etesami and Jeong 2018). Si treatment functions to agglomerate nutrient content and diffuse adverse effects of micronutrient deficiencies (Savić and Marjanović-Jeromela 2013; Hernandez-Apaolaza 2014). Iron deficiency stimulates chlorosis. To stimulate Fe uptake and transport from roots to stem vascular tissues accumulate Fe in citrate and catechins from root to shoots along with Fe plaques on roots (Etesami and Jeong 2018). Si aids release of hydroxyl ions in roots, leading to oxidation of Fe^{2+} to Fe^{3+} (You-Qiang et al. 2012). Si mitigates Mn deprivation, with high Mn oxidation rhizosphere. Toxicity by Mn is reduced by oxidation of Mn leading to Mn absorption into cell walls (Li et al. 1999). Zn in roots and plant cell compartments activates synthesis of citrate compounds in roots and activates mechanistic pathways of Zn deficiency in plants (Pascual et al. 2016). Si diminishes Zn toxicity to impair transport and enhance absorption in silicate deposits. Si modulates optimal Cu and Zn concentrations in plants and deposits in cell wall (Frantz et al. 2011).

12.4.7 Saline Stress

Salinity is a common abiotic stress in environment with 20–30% agricultural land destroyed and damaged by salinity (Ashraf et al. 2010; Shrivastava and Kumar 2015). Saline soils generating ionic cytotoxicity are distributed by oxidation in plant cells to hinder growth (Khan et al. 2000). Sodium concentration is decreased by saline toxicity, along with high permeability of root cell membranes to recuperate stability of root cell (Luyckx et al. 2017). Silicon mitigates salinity stress in wheat, rice, barley, maize, soybean, cucumber, and tomato (Reezi et al. 2009). Silicon additionally mediates H^+-ATPase pump, along with transport of sodium from cells. Silicon is deposited in silica gel with cell wall to stimulate binding of salt and promote transportation to shoot (Lux et al. 2003). In leaves, silicon deposits minimize transpiration by polymerization to cells. This in turn dilutes salts to ameliorate stress (Malhotra and Kapoor 2019). Transpiration rate improves with high stomatal conductance. Supplementary silicon promotes gaseous exchange to support plant survival (Etesami and Jeong 2018). Exogenous application of silicon boosts water content in agricultural crops (Coskun et al. 2016). The defense systems in plants are associated with antioxidative enzymes such as superoxide dismutase (SOD), peroxidase (POD), catalase (CAT), APX (ascorbate peroxidase), guaiacol peroxidase (GPOX), dehydroascorbate reductase (DHAR), monodehydroascorbate reductase (MDHAR), etc. Many nonenzymatic antioxidants have also been reported to stimulate defenses. In barley and cucumber, glutathione, ascorbic acid, and tocopherol have been reported (Zhu et al. 2004). Silicon lowers oxidative stress by binding and polymerizing in plants. These compounds restrict electrolyte, reactive oxidation species, and malondialdehyde leakage to lower oxidative stress (Wang et al. 2010). Silicon uptake can be improved by upregulation of aquaporin genes. This improves ionic homeostasis in plants (Rios et al. 2017). Salinity in soil produces excessive sodium (Na), chloride (Cl), and sulfate (S) ionic levels with low amounts of potassium (K), calcium (Ca), and magnesium (Mg) ions in crops (Sibole et al. 2003). Silicon application to crops reduces sodium and upregulates potassium to regulate the Na-K pump and maintain osmotic balance in plants (Ashraf et al. 2010). Salinity also leads to agglomeration of hydroxyl radicals, superoxide radicals, hydrogen peroxide radicals, and other reactive oxygen species (ROS). These toxic ROS cause oxidation burst, disrupt lipid membrane, denature protein, and inhibit nucleic acid synthesis and photosynthesis (Gunes et al. 2007).

12.5 Silicon and Biotic Stress Mitigation

Literature states drought or herbivore attack is modulated by Si uptake (e.g., Reynolds et al. 2009; Yamaji and Ma 2011). Si combats pathogenic diseases in plants by mediating defense mechanisms to arrest pathogen. Monosilicic acid polymerized to polysialic acid forms amorphous silica complex in membranes (Hayasaka et al. 2008). These strengthen membranes with enzymes having enhanced tolerance against pests (Zargar et al. 2019). Si in association with phenolics, phytoalexins, peroxidases, β-glucanases, and PR1 proteins colonizes and damages fungal pathogens (Etesami and Jeong 2018; Rodrigues et al. 2003). Si influences the expression of pathogenesis-related genes or PR genes to activate defense response (Brunings et al. 2009). In tomato and sweet pepper, anthracnose disease is reduced by Si. Si further supports cuticle thickening and firmness of fruits (Somapala et al. 2016) (Fig. 12.1).

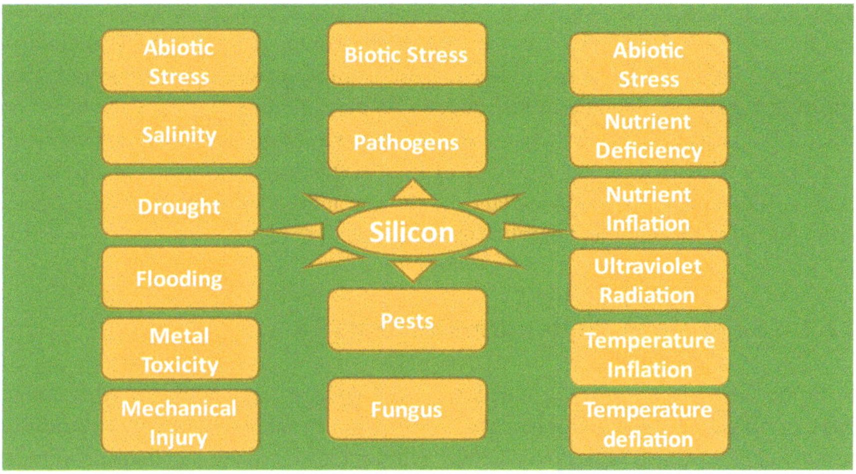

Fig. 12.1 Stress alleviation by silicon in plant species

12.6 Omics Studies on Silicon Application on Crops

Omics studies have been used to further understand the role of silicon in stress mediation in plants (Zargar et al. 2019). Transcriptomics and proteomics studies have revealed upregulation of silicon concentration in plants by upregulation of aquaporin genes, which consequently supports and stimulates water transport in saline-treated cucumber plants (Zhu et al. 2015). Si modulates expression level of *SbPIP* aquaporin genes in sorghum plants. Augmentation of silicon leads to activation of genes for salinity tolerance and antioxidative defense mechanism by *leDREB-1*, *leDREB-2*, *leDREB-*, *leLsi-1*, *leLsi-2*, *leLsi-3*, *leAPX*, *leSOD*, and *leCAT* in tomato plants (Muneer and Jeong 2015). In rice, resistance was induced against blast fungus. By 2D gel electrophoresis and liquid chromatography-mass spectroscopy, the impact of silicon against pathogenic fungi was studied. Si alters protein profile in plant to affect metabolism, photosynthesis, homeostasis, and cellular biochemistry to enhance defense responses to pathogens (Liu et al. 2014). Si ameliorates photosynthesis and saline stress (Muneer et al. 2014). Omics approach is a novel strategy to mediate Si productivity in crop.

12.7 Reactive Oxygen Species Regulation

Plants generate ROS such as superoxide anions, hydroxyl ions, and hydrogen peroxide in stress conditions in cellular compartments such as mitochondria, chloroplasts, and peroxisomes. These ROS degrade biomolecules, proteins, lipids, and nucleic acids (Tripathi et al. 2017). Antioxidative enzymes along with nonenzymatic antioxidants combat stress by ROS. Si with high antioxidative potential detoxifies free radicals, limits lipid peroxidation, and scavenges ROS by activating enzymes such as SOD, POD, CAT, and APX (Torabi et al. 2015). Si maintains membrane integrity and permeability for cell stability in stress condition (Etesami and Jeong 2018). Si stimulates tolerance to photorespiratory enzymes (Nwugo and Huerta 2011). Si triggers expression of antioxidant gene high glutathione level with low malondialdehyde, along with hydrogen peroxide levels (Liang et al. 2018; Ma et al. 2016a). Si augmentation in plants modulates antioxidation with enhanced plant immunity against biotic and abiotic stresses

12.8 Silicon and Phytohormone Cross Talk

Si further promotes root and shoot length and volume to promote biosynthesis of phytohormones (Hernandez-Apaolaza 2014). Plant hormones facilitate metabolism and assist in acclimatization in difficult environments by many mechanisms (Fahad et al. 2015). Si regulates the production of plant hormones ABA, GA, JA, BR, ET, SA, CK, and AUX along with developing immunity for biotic and abiotic stresses (Kim et al. 2014) Gibberellic acid promotes flowering, fruiting, and vegetative growth (Colebrook et al. 2014). Along with growth, gibberellic acid also promotes resistance to salinity and drought stress. Si exogenous treatment activates GA biosynthetic pathways to regulate development and stress tolerance yet delays senescence due to activation of cytokinin pathway. This is common in both plants that accumulate and do not accumulate silicon (Markovich et al. 2017). Si treatment on cucumber plants resulted in high gibberellic acid levels (Hamayun et al. 2010). Phytohormones by cross talk trigger stress resistance (Fahad et al. 2015) (Fig. 12.2).

12.9 Si Accumulation and Transporters in the Plant Kingdom

Variation of silicon concentration in aerial parts of plant from 0.1% to 10%. The variation is attributed to capacity of plant root to absorb silicon (Takahashi et al. 1990). To understand these differences in plants, silicon transporters are studied in detail. Genomic study of angiosperms (flowering plants) revealed the presence of genes encoding NIP III channel (*Lsi1*-like and *Lsi2*-like transporters). Similar results were reported for other monocots and dicots such as Gramineae, Arecaceae, banana (*Musa acuminata*), Solanaceae, Rosaceae, Cucurbitaceae, Leguminosae, grape (*Vitis vinifera*), citrus, and coffee (*Coffea canephora*). *Arabidopsis thaliana* does not include this list. No NIP III genes or *Lsi2* homologs are present in *Arabidopsis thaliana*. *Amborella trichopoda* contains *Lsi1*-like and *Lsi2*-like genes. In angiosperms, NIP III channels consist of conserved identical ar/R amino acid residues (G, S, G, and R) with some exceptions. These channels are permeable to silicon. Silicon transport is also influenced by other factors such as expression level, polarity, and cellular localization. An exception exists in tomato where the presence of NIP III channels and *Lsi2*-like genes do not promote silicon accumulation (Mitani and Ma 2005).

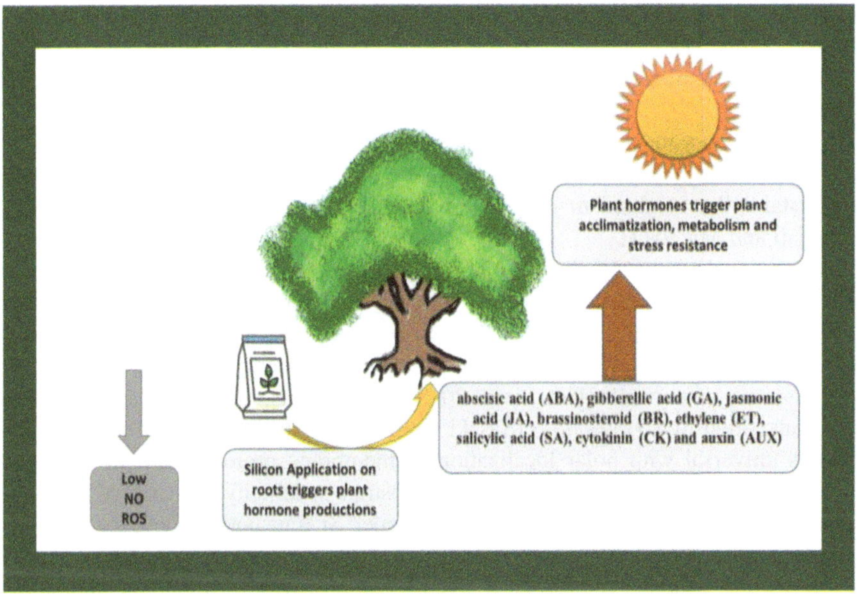

Fig. 12.2 Silicon application to roots triggering plant hormone production to support plant growth

Land plants such as Bryophyta, Lycopsida, and Equisetopsida of Pteridophyta accumulate silicon in high concentration. This is not common in other angiosperms (Ma and Takahashi 2002). EaNIP3s in Equisetum is permeable to silicon (Grégoire et al. 2012). Bryophyta and Pteridophyta contain *Lsi1*-like transporters, yet silicon transport by these transporters has not been reported by them. Differences between the transporters suggest that Si channels in angiosperms and Equisetum have evolved independent of each other. Presence on *Lsi2*-like genes in some Pteridophyta and Bryophyta does not stimulate Si accumulation. Gymnosperms have low Si levels, which is attributed to the fact that *Lsi1* and Lsi2 transporters are different in plants and diatoms (Hildebrand et al. 1997).

Si concentration differences have been recorded within species as well. In 38 varieties of *Oryza perennis*, Si concentration varies between 5.4% and 10.6% (Ma and Takahashi 2002). A total of 135 varieties of barley were studied to understand variation in Si content. In barley, the variation exists from 0% to 0.36%. The Si concentration decreases further in hull-less barley than in hulled barley. A correlation was obtained between Si concentrations of barley grains harvested over years. This concluded genetic control on Si concentration of barley grain. The data collected from studies impacts breeding Si-rich cultivars (Ma et al. 2002).

A positive correlation exists between Si uptake and expression of HvLsi2 gene in barley cultivars (Ma et al. 2007). Molecular mechanisms involved in rooting with Si transport are still under investigation. *Lsi1* and *Lsi2*-type transporters in databases of various plant species have still not been characterized. Molecular and physiological characterization of Si transporters is required to understand mechanism of Si uptake, transport, and accumulation in plant kingdom. Si itself is reported to regulate expression of transporter genes. In rice, Lsi1 is downregulated by Si and not affected in barley or maize (Ma et al. 2006; Chiba et al. 2009; Mitani et al. 2009). In case of Lsi2, the expression is reduced in rice, barley, and maize in the presence of Si (Ma et al. 2007; Mitani et al. 2009). Transcription factors have been investigated to understand the mechanism controlling gene expression of Si transporter genes (Yamaji and Ma 2011).

Polar localization of Si channels and efflux transporters in one direction strengthens root stele. Various other transporters such as boric acid channel NIP5;1 and the efflux transporter BOR1 in *A. thaliana* are localized in the roots (Takano et al. 2002, 2006). In maize roots, iron–phytosiderophore complex (*ZmYS1*) is transporter of minerals (Ueno et al. 2009). In rice roots, manganese and cadmium are transported by (*OsNramp5*) (Sasaki et al. 2012). Mechanisms for silicon transport is universal with polar localization of membrane proteins. Protein structural analysis of transporters requires study on substrate specificity. Benefits of Si in soil cannot be reaped by plants without any proper Si transport system. Arabidopsis plants modified with Lsi1 genes from rice and wheat accumulate silicon in shoots by expression of root-specific promoter (Montpetit et al. 2012). Silicon accumulation can be hindered by 35S promoter. Silicon uptake in various plant species requires a modified efficient system of silicon transporters.

12.10 Silicon Accumulation and Uptake

Silicon is absorbed by roots and is deposited on different tissues. Silicon deposits below cuticle prevent penetration of fungi and insects, along with reducing transpiration and increasing mechanical strength (Bélanger et al. 2003). Silicon is transported in plants by two channels Lsi1 and Lsi2 (low silicon 1 and 2). Lsi1 is a permeable channel from Nod26-like major intrinsic protein (NIP) III for uptake and distribution of silicon. This is an aquaporin membrane protein family. Lsi2 belongs to an uncharacterized anion transporter family. Although the transporters are reported to be in the plasma membrane, they may be found in other parts of the cell. The Lsi1 and Lsi2 are located on proximal and distal sides of roots. Silicon in the xylem is a monosilicic acid on Lsi6. Both Lsi1 andLsi6 are homologs of each other. Intervascular transport of Si is by Lsi6, which leads to preferential transfer to plant panicles (Feng et al. 2011). In every plant species, the expression of both transporters varies due to different expression pattern and Si accumulation requirement. Studies in rice concluded efficient transportation of Si in plants with corporation of both transporters. Studies indicate cereals possessing capacity to accumulate high amount of Si, which becomes a source for human and animal consumption. Si uptake, accumulation, and distribution in plants can be understood by understanding genetic variation. Genotypic variation is one cause of differential absorption of Si. Analysis of polygenes revealed differences over generations. Various traits were studied to understand these differences. Dai et al. (2005) identified ten QTLs for Si. Out of ten

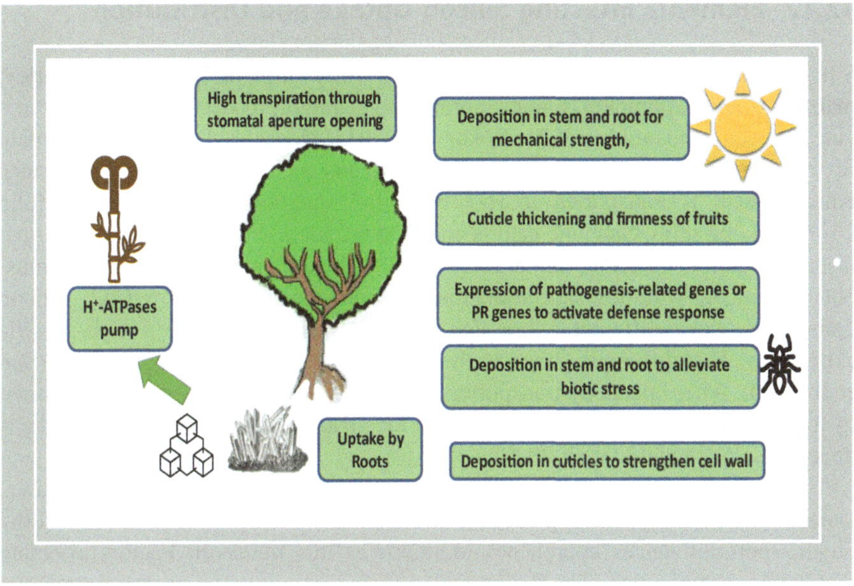

Fig. 12.3 Impact of silicon uptake and transport on plants

QTLs, four were specific to increasing Si concentrations in the hull, while the rest of four contributed to concentration in leaf and only two in stem. QTLs impact Si uptake. Fine mapping of these genomic regions results in identification of functional markers to develop transgenic varieties with better growth and defense genes. In wheat, Talsi1 gene from root-specific promoter was isolated for transfer in Arabidopsis (Montpetit et al. 2012) (Fig. 12.3).

12.11 Silicon Transport in Xylem

Roots take up minerals in grains to redistribute in plant nodes. Mineral distribution in plants is selective to accumulation in pinnacles, from lower nodes of vascular bundles to various large and small vascular bundles. Vascular bundles are arranged parallel and assembled in bundles connected to tissues (Kawahara et al. 1974; Chonan et al. 1985). Minerals are transferred from vascular bundles of roots to seeds. Transporters in xylem are located around the boundary of enlarged vascular bundles. Three transporters are involved in Si uptake and transfer to intervascular zone. Lsi6 is required for transferring and accumulating Si in vascular and pinnacles (Yamaji and Ma 2009). Si deposited in vascular bundles is as dumbbell-like shape with bulliform motor cells. Specific cell deposition is made possible by specific transporters.

12.12 Elements Effecting Silicon Uptake and Distribution

Silicon transporters have been characterized in roots of plants such as rice, barley, maize, soyabean, pumpkin, tomato, sorghum, cucumber, tobacco, plum, and grapevine (Mitani-Ueno and Ma 2021; Bokor et al. 2015; Noronha et al. 2020). The main Si transporters identified are *Lsi1* (channels) and *Lsi2* (anion-type transporter). Salicylic acid is transported in root by influx aquaporin channels of Nodulin 26-intrinsic proteins. Salicylic acid is converted to meta salicylic acid and exported from endodermis by efflux anion antiporter *Lsi2* loaded in xylem. High root transport of silicon with low accumulation in shoots is achieved by *Lsi2*. Silicon transporters have different expression patterns with different cell type-specific expression. This differently impacts the plant and influences plant processes (Sun et al. 2020). After uptake by transporters, the movement of Si in xylem is supported by transpiration. The final loading and polar localization in xylem and parenchyma is done by Lsi6 (Mandlik et al. 2020).

Uptake and transport along with distribution of Si in plants is still unknown. Detailed information on this process is still low. This lack of overall understanding on transport influences approaches to improves this transport. Factors affecting regulation are unknown. The downregulation and upregulation of transporters are regulated by Si in some plants, while it is differently regulated in other plants. Nutrient imbalance in plants impacts expression of Si transporters. This affects Si accumulation in shoots. Limited mineral supply induces accumulation in plant parts (Mitani-Ueno and Ma 2021; Wu et al. 2017; Chaiwong et al. 2018, 2020; De Tombeur et al. 2020; Minden et al. 2021). Mineral deficiency in plants increases silicate mobilization in plant (Gattullo et al. 2016; Wu et al. 2017; Chaiwong et al. 2020). Mineral mobilization in plants influences silicate minerals to liberate silicic acid. Potassium-deficient plants have upregulation of various transporter genes (Hosseini et al. 2017). Other minerals with nitrogen and phosphorus starvation stimulate uptake of Si distribution along with accumulation in aerial plant parts. Nitrogen deficiency stimulates accumulation in roots. Even distribution of silicon was observed under optimal growth conditions (Minden et al. 2021). Excessive mineral supply similarly supports expression of Si transporters. The upregulation or downregulation of transporters is dependent upon specific nutrients and nutrient concentration. In rice, elevated nitrogen and zinc concentration reduces silicon accumulation. With maize, silicon transport increases on response to metal toxicity (Bokor et al. 2015).

12.13 Silicon Uptake Mechanism: Influx and Efflux Transporters (Table 12.2)

Plants employ various mechanisms to transport Si from soil through membranes to plant cells. The to and from movement of molecules is an energy-dependent active process. Two main transporters are *Lsi1* (lower silicon transporter 1) and *Lsi2* (lower

Table 12.2 Lower silicon transporters reported in plant species

Sr No.	Transporters	Crops	References
1.	OsLsi1, OsLsi2, OsLsi6, OsNIP2	Rice *Oryza sativa*	Ma et al. (2007, 2008), Yamaji et al. (2008), Mitani et al. (2008)
2.	HvLsi1, HvLsi6	Barley *Hordeum vulgare*	Chiba et al. (2009), Yamaji et al. (2015)
3.	TaLsi1	Wheat	Montpetit et al. (2012)
4.	ZmLsi1, ZmLsi6	Maize (*Zea mays*)	Mitani et al. (2009)
5.	CSiT-1, CSiT-2	Cucumber	Wang et al. (2015)
6.	CmLsi1	Pumpkin *Cucurbita moschata*	Mitani et al. (2011)
7.	GmLsi1	Soybean *Glycine max*	Mitani et al. (2009)

silicon transporter 2). Both transporters belong to Nod26-like major intrinsic protein (NIP) III subgroup of the aquaporin membrane protein family (Ma and Yamaji 2015).

12.14 Silicon Transport

Silicon is transported from various transporters, some are channel-type transporter, while others are efflux transporters. Cell membranes with passive permeability to salicylic acid mediate lipid transport. This process is called lipid solution transport (Raven 2001). The movement of salicylic acid in plants requires facilitation by Si transporters (Ma and Yamaji 2015).

12.14.1 Channel-Type Transporters

These *Lsi1* transporters facilitate passive transport between extracellular environment and plasma membrane. The channel transporter are a subfamily of aquaporin-like proteins. It consists of a six transmembrane domains of conserved amino acid sequence with two Asn-Pro-Ala (NPA) motif. These NIP transporters facilitate passive transport of water, glycerol, selenite, ammonia, lactic acid, and boric acid along with other uncharged solutes (Wallace and Roberts 2005; Takano et al. 2006; Choi and Roberts 2007; Mitani et al. 2008; Zhao et al. 2010).

12.15 Silicon Uptake in Major Crops

Research on yield improvement is constantly conducted. Silicon application is assessed in various food crops. Some major crops have been discussed with mechanisms of silicon accumulation and transportation in plants.

12.15.1 Silicon Uptake in Rice

Rice is a siliceous crop with more than 10% dry mass constituting of silica and silicon compounds. High silicon accumulation assists sustainable growth of rice (Ma and Takahashi 2002; Tamai and Ma 2008). Silicon is transported from roots to all parts of the rice plants rapidly. Rice roots contain two transporters, passive transporter (*OsLsi1*) and active transporter *OsLsi2*. The active transporter through energy mediation effluxes Si, from cell to apoplast through proton antiport. The passive transporter by diffusion along concentration gradient transports silicon and belongs to the aquaporin family of transporters (Ma et al. 2007; Feng et al. 2011). The rice contains Lsi channel-type influx transporter belonging to NIP2 family of membrane proteins (Ma et al. 2006). The transport of silicon from root is facilitated by both *Lsi1* and *Lsi2* proteins (Cheng et al. 2007; Ma and Yamaji 2015). More influx transporters were discovered in rice. *Lsi6* along with efflux transporter *Lsi3* have been identified recently with evidence of assisting Si transport from vascular tissues to leaves (Yamaji and Ma 2009; Yamaji et al. 2015). Silicon is required by rice to complete its life cycle. Rice grown in lowlands with silicon deficiency causes wilting, followed by necrosis on mature leaves and decline in grain yield (Lewin and Reimann 1969). Silicon enhances resistance to brown spot, leaf scald, sheath blight, neck, and leaf blast in rice plants (Datnoff and Rodrigues 2005). Occurrence of powdery mildew is reduced (Fauteux et al. 2005, 2006). Si reduces lodging in rice along with helping alleviate drought impacts (Ma 2004).

Initial growth stages of growth in rice plants were not influenced by silicon uptake. 100 ppm silicon treatment caused enhanced leaf growth. Plant shoot length without silicon is delayed along with browning of leaves and eventually turned gray. Spots developed on leaves and heads sprouting were colored brown. Rice grains were smaller than normal, and hulled rice was dark brown (Mitsui and Takatoh 1963).

To identify Si transporter genes in rice plants, mutants tolerant to germanium were isolated. Geranium and silicon are analogues and are absorbed by the plant by similar mechanism. The study revealed two genes involved in the uptake of silicon. The genes were named Lsi*1* (lower silicon transporter 1) and *Lsi2* (lower silicon transporter 2) (Ma et al. 2002). Ten percent dry weight of rice shoot is composed of silicon. This amount is higher than other macronutrients such as potassium, phosphate, and nitrogen (Richmond and Sussman 2003). Silicon transporters are in plasma membrane between exo and endodermal cell in Casparian strips. Casparian strips are formed by lignin and suberin accumulation in cell wall (Schreiber 2010). Casparian strip performs physiological role by blocking backflow of water to

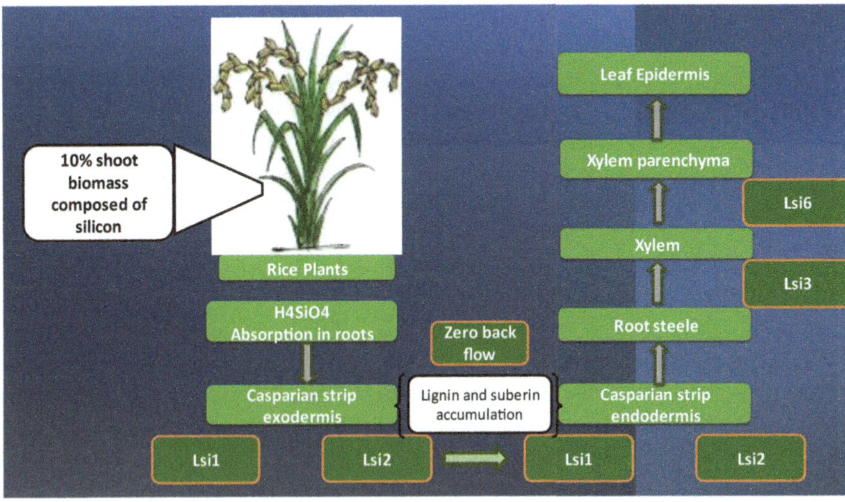

Fig. 12.4 Silicon accumulation in root is transported by *Lsi1* transporters from distal end toward *Lsi2* transporters on proximal end of Casparian strips. Silicon stimulates lignin and suberin accumulation in Casparian strips to prevent back flow of ions and water. Silicon strengthens leaf and shoot of plant to support plant growth and development

maintain root pressure to hinder water movement (Steudle 1994). Casparian strip functions to hinder apoplastic substance movement toward the interior tissues of the roots along with water and ions (Enstone et al. 2002). Lsi1 is found at distal end, while Lsi2 is present at proximal end. Cooperation between both transporters supports the root stele (Ma et al. 2006, 2007). The silicon transporters are in the Casparian strip. In silico study by Sakurai et al. (2015) established the presence of various silicon transporters in the Casparian strip promoting silicon absorption. Casparian strips devoid of silicon transporters absorb low amount of silicon, which effect growth and development of rice (Fig. 12.4).

Overexpression of *Lsi1* gene in rice triggers chilling in plants. Proteomic analysis by gene tags revealed high gene expression during injuries by cold. The identified proteins have multiple functions in plants from photosynthesis, metabolism, signal transduction, cellular homeostasis, protein synthesis, and other biochemical reactions (Azeem et al. 2016). Furthermore, Si reduces Cd toxicity by maintaining cell integrity and physiology. Silicon improves protein utilization in cell and decreases expression of glutathione S-transferases (GST). The research used iTRAQ (isobaric tags for relative and absolute quantification), ICP-MS (inductively coupled plasma mass spectrometry), fluorescent microscopy, etc. to understand these dynamics of silicon assimilation in rice crops (Ma et al. 2016b). With metal toxicity, Si transport genes upregulate production of mRNA to modulate morphology of root to tolerate metal toxicity in rice (Kim et al. 2014). Si deposition over epidermal wall reduces blast fungus disease in rice (Kim et al. 2002). Si enhances papillae formation to resist *B. graminis* (Bélanger et al. 2003). Rice infected with *Rhizoctonia solani* on treatment by silica developed resistance to sheath blights (Zhang et al. 2006).

Datnoff and Rodrigues (2005) studied rice infected with sheath blight, leaf blast, neck blast, leaf scald, discoloration, leaf blight, stem rot, etc. to analyze impact of silicon application. Si promotes inhibition of infection. In rice, Si transporters are classified as influx and efflux transporters called OsLsi1 and OsLsi2. Silicon present in soil downregulates transport by both these transporters (Mitani-Ueno and Ma 2021). Low supply of nitrogen with iron upregulates function of OsLsi1 and OsLsi2 (Wu et al. 2017; Chaiwong et al. 2020).

12.15.2 Silicon Uptake in Sugarcane

In sugarcane, Si treatment at root internodes modulates resistance and responsiveness to sugarcane borer *Eldana saccharina* analyzed by X-ray analysis (Keeping et al. 2009). Application of calcium silicate increases yield of sugar by 12 tons per hectare. The silicon used was trichloroacetic acid extractable and phosphate extractable. To increase silicon extraction from soil, acid solution of phosphate, along with sulfate, acetate, and water, was applied (Fox et al. 1967).

12.15.3 Silicon Uptake in Pepper

Silicon is required for growth development and metabolism in plants. *Lsi1* and *Lsi2* mediate cooperative absorption of Si. In silico genetic analysis of pepper (*Capsicum annuum*) reveals physiological processes for Si regulation. Si and phytohormones in coordination regulate stress tolerance, metabolism, and growth (Gómez-Merino et al. (2020).

12.15.4 Silicon Uptake in Tomato

Si treatment in tomato crop improved resistance capacity in the crop. The variation was recorded in photosynthetic characteristics along with modified proteome expression (Muneer et al. 2014). The expression pattern of cell type-specific expression leads to polar localization and functionality of transporters in plants. Tomato has low Si accumulation in shoot with high transport from shoot for high silicon absorption from soil by functional efflux transporter SILsi2 (Sun et al. 2020).

12.15.5 Silicon Uptake in Wheat

Silicon mediates localized defenses of plants against *B. graminis* f. sp. *tritici* attack. Si is accumulated in epidermal cells and activates papilla formation, along with callose production and release of glycosylated phenolics. The released phenolics accumulate with cell wall and are functionally like localized phytoalexins (Bélanger et al. 2003). Si absorption is different in monocots and dicots. While transpiration was considered a reason for Si uptake, evidence of active transportation by transporters proves otherwise. Silicon nanoparticles applied on wheat induced tolerance for ultraviolet radiation stress. Silicon nanoparticles are porous carriers with proteins (Tripathi et al. 2017). Roots accumulate relatively low Si, while shoots compose of 90% of all Si of plants. Mayland et al. (1991) reported rapid Si absorption in wheat plant. Transpiration was removed as reason for this uptake. Si absorption was recorded at 0.5 mM of elements in soil. This was same for pretreated wheat plants and nontreated wheat plants (Rafi et al. 1997). Various studies have been conducted to understand active silicon uptake in wheat. Silicon uptake in wheat is controlled by metabolic control of dinitrophenol and potassium cyanide. The Si absorption concentration was in accordance with Michaelis-Menten kinetics (Rains et al. 2006). Phosphate ions do not control uptake of silicon.

In wheat, Si transporter gene is called *Talsi*. Cloning and functional characterization of gene revealed it to be an ortholog of *OsLsi1* from rice. Both genes are part of Nod26-like intrinsic proteins (NIPs) III subgroup of the aquaporin membrane protein family (Montpetit et al. 2012). Si absorption by specific NIPs facilitated transport of monosilicic acid along with water and uncharged solutes. Proteins facilitate passive transport across plasma membrane. In the plant cell, silica oxide is moved by efflux transporters such as Lsi2. The Si by this route is deposited in the xylem by translocation from root to stem. Through stem, Si is deposited on aerial parts of plant. The Si permissible channels are characterized to further understand molecular dynamics of Si transporters. The transport capacity of all transporters varies. This difference is exceptionally potent to understand differences in species.

Si uptake is by roots through transpiration. Si is accumulated below cuticles to form multiple layers of Si deposit associated with cell wall. In wheat, the highest accumulation of silica was recorded in leaf blade followed by the lawn, leaf sheath, lemma, rachilla, and stem. The lowest Si concentration is present in root, which gradually increases with stomatal density. Silicified cells in wheat are correlated with concentration of Si in plant organs. By scanning electron microscope, continuous silica layer in cuticle along with epidermis and sclerenchyma cells was visualized. X-ray microanalysis of wheat reveled linear pattern of silica deposition. Si predominates in epidermis cells of leaves and cell walls. Si accumulation in straw of wheat plants is dependent upon geochemical cycle of Si. In comparison to natural ecosystem, concentration of amorphous silica is lower. This concentration further decreases overtime. The estimation of shoot uptake is based upon previous harvest records. The area of previous harvest along with production level and biomass portion of root and shoot with accumulated Si determine annual shoot Si uptake. The straw is removed from soil for increasing availability of Si. Crops with straw such as wheat in which straw is removed from the field additional Si in the form of fertilizers are required. Wheat crops deprived of silicon have various physical abnormalities. This significantly influences growth and development of plant. This concludes and characterizes silicon as a "quasi-essential" element (Rafi et al. 1997).

12.15.6 Silicon Uptake in Maize

In maize, silicon present in soil downregulates efflux transporters (Lsi2), while influx transporter Lsi1 is not affected by the presence of Si (Mitani et al. 2009; Chiba et al. 2009).

12.15.7 Silicon Uptake in Cucumber

Silicon treatment to cucumber increases stem rigidity along with rougher texture and high rigidity of mature leaves. Mature leaves with high Si levels were reported with shorter petioles and high chlorophyll, soluble protein, and RuBP carboxylase content. Silicate treatment also enhanced resistance to powdery mildew fungus *Sphaerotheca fuliginea* (Adatia and Besford 1986).

12.15.8 Silicon Uptake in Barley

Silicon uptake in barley increases root elongation with enhanced cell wall extensibility in meristem (Hattori et al. 2003). In barley, silicon present in soil downregulates efflux transporters (Lsi2), while influx transporter Lsi1 is not affected by the presence of Si (Mitani et al. 2009; Chiba et al. 2009).

Organic acids are exuded by roots of barley plants with iron deficiency. This indirectly increases Si availability in the rhizosphere (Gattullo et al. 2016). Potassium deficiency in plants upregulates expression of *HvLsi1*, *HvLsi2*, and *HvLsi6* to accumulate silicon in aerial plant parts (Hosseini et al. 2017).

12.15.9 Silicon Uptake in Arabidopsis

Silicon attributes to *Arabidopsis* defense priming followed by infection, and protection is conferred by altered priming. Silicon interferes with effector proteins released by pathogens to initiate defense responses (Vivancos et al. 2015).

12.15.10 Silicon Uptake in Cannabis

Cannabis is a crop with extensive stem structure. Cannabis stem is used for the production of industrial products. Silicon strengthens stems in cannabis.

12.16 Silicon Controversy

Silicon transport and impact in plants are not completely understood. Various details of mechanisms still need to be analyzed, understood, and processed. The benefits of silicon application are still a mystery as they provide limited biochemical advantages. Large-scale application of silicon on plants is not common practice, which questions over efficiency of silicon in plants. This is understood by considering the most important fact about silicon. All plant species have different absorption capacity for silicon. Plants are classified based upon this absorption capacity. Along with this classification, molecular tools were applied to characterize non-accumulators and accumulators. The classification is based upon the presence of functional Si transporters. The debate initiated by researchers today is that silicon has very limited nutritional role. Stress alleviation of plants assists in growth and development. Most advantages linked with silicon are considered from perspective of plant stress alleviation mechanisms. Plants have various mechanisms to fight stress. Prevention or deregulation of stress is controlled by various mechanisms in plants. Agricultural crops grow under stress. The negative impact is only upon the yield of crops, yet growth is still present. The effects of Si to limit stress while undebatable have only motivated agronomists to utilize silicon upon plants for stress alleviation. The mechanism involved in deregulation of this stress is not fully understood or reported (Liang et al. 2015). This ambiguity is creating doubt over capacity of silicon as a fertilizer. This debate cannot be sorted without detailed analysis on crop exposed to silicon. An apoplastic obstruction hypothesis was constructed to study beneficial effects of silicon. This model aims to stimulate critical thinking, along with understanding about properties of silicon.

12.17 Conclusion

Silicon with its essential role in plant growth is a requirement for healthy crops. Silicon stimulates growth in plants, which benefits to enhance crop yield. Si enhances defense responses in plants against a variety of stress conditions, while at the same case, it polymerizes to membranes, with minimum lodging and enhanced resistance to diseases and pests, along with protection against heavy metal toxicities. Silicon promotes leaf growth with maximum light interception, leading to photosynthesis, and affects fiber quality of plants. The yield in leafy plants and crop in greatly enhanced with silicon. This especially has significance with agriculture. To increase plant yields and biomass, silicon is applied, along with various mineral elements.

12.18 Future Recommendation

The complete transportation mechanism of silicon in various plant species still needs to be understood. The variation in transport reveals genomic differences. Genetic variation in crop varieties has revealed various reasons. Yet, the complete transcriptomic mechanism is not understood. Research to study mechanisms of Si-mediated resistance in plants will reveal the mechanisms behind isoformic antioxidative enzymes. Details of changed transcriptomic, proteomic, and metabolomic profiles along with cell imaging and functional genomics provide insight to plant immune responses dependent on Si.

References

Adatia MH, Besford RT (1986) The effects of silicon on cucumber plants grown in recirculating nutrient solution. Ann Bot 58(3):343–351

Adrees M, Ali S, Rizwan M, Zia-ur-Rehman M, Ibrahim M, Abbas F, Farid M, Qayyum MF, Irshad MK (2015) Mechanisms of silicon-mediated alleviation of heavy metal toxicity in plants: a review. Ecotoxicol Environ Saf 119:186–197

Agarie S, Uchida H, Agata W, Kubota F, Kaufman PB (1998) Effects of silicon on transpiration and leaf conductance in rice plants (Oryza sativa L.). Plant Prod Sci 1(2):89–95

Ashraf M, Afzal M, Ahmad R, Maqsood MA, Shahzad SM, Aziz A, Akhtar N (2010) Silicon management for mitigating abiotic stress effects in plants. Plant Stress 4(2):104–114

Ayres AS (1966) Calcium silicate slag as a growth stimulant for sugarcane on low-silicon soils. Soil Sci 101(3):216–227

Azeem S, Li Z, Zheng H, Lin W, Arafat Y, Zhang Z, Lin W (2016) Quantitative proteomics study on Lsi1 in regulation of rice (Oryza sativa L.) cold resistance. Plant Growth Regul 78(3): 307–323

Bélanger RR, Benhamou N, Menzies JG (2003) Cytological evidence of an active role of silicon in wheat resistance to powdery mildew (Blumeria graminis f. sp. tritici). Phytopathology 93(4): 402–412

Bityutskii N, Pavlovic J, Yakkonen K, Maksimović V, Nikolic M (2014) Contrasting effect of silicon on iron, zinc and manganese status and accumulation of metal-mobilizing compounds in micronutrient-deficient cucumber. Plant Physiol Biochem 74:205–211

Bokor B, Bokorová S, Ondoš S, Švubová R, Lukačová Z, Hýblová M, Lux A (2015) Ionome and expression level of Si transporter genes (Lsi1, Lsi2, and Lsi6) affected by Zn and Si interaction in maize. Environ Sci Pollut Res 22(9):6800–6811

Brunings AM, Datnoff LE, Ma JF, Mitani N, Nagamura Y, Rathinasabapathi B, Kirst M (2009) Differential gene expression of rice in response to silicon and rice blast fungus Magnaporthe oryzae. Ann Appl Biol 155(2):161–170

Chaiwong N, Prom-U-Thai C, Bouain N, Lacombe B, Rouached H (2018) Individual versus combinatorial effects of silicon, phosphate, and iron deficiency on the growth of lowland and upland rice varieties. Int J Mol Sci 19(3):899

Chaiwong N, Bouain N, Prom-U-Thai C, Rouached H (2020) Interplay between silicon and iron signaling pathways to regulate silicon transporter Lsi1 expression in rice. Front Plant Sci 11: 1065

Chen W, Yao X, Cai K, Chen J (2011) Silicon alleviates drought stress of rice plants by improving plant water status, photosynthesis, and mineral nutrient absorption. Biol Trace Elem Res 142(1): 67–76

Cheng L, Wang F, Shou H, Huang F, Zheng L, He F, Wu P (2007) Mutation in nicotianamine aminotransferase stimulated the Fe (II) acquisition system and led to iron accumulation in rice. Plant Physiol 145(4):1647–1657
Chiba Y, Mitani N, Yamaji N, Ma JF (2009) HvLsi1 is a silicon influx transporter in barley. Plant J 57(5):810–818
Choi WG, Roberts DM (2007) Arabidopsis NIP2; 1, a major intrinsic protein transporter of lactic acid induced by anoxic stress. J Biol Chem 282(33):24209–24218
Chonan N, Kawahara H, Matsuda T (1985) Ultrastructure of elliptical and diffuse bundles in the vegetative nodes of rice. Jpn J Crop Sci 54(4):393–402
Colebrook EH, Thomas SG, Phillips AL, Hedden P (2014) The role of gibberellin signalling in plant responses to abiotic stress. J Exp Biol 217(1):67–75
Coskun D, Britto DT, Huynh WQ, Kronzucker HJ (2016) The role of silicon in higher plants under salinity and drought stress. Front Plant Sci 7:1072
da Cunha KPV, do Nascimento CW (2009) Silicon effects on metal tolerance and structural changes in maize (Zea mays L.) grown on a cadmium and zinc enriched soil. Water Air Soil Pollut 197(1):323–330
Dai WM, Zhang KQ, Duan BW, Zheng KL, Zhuang JY, Cai R (2005) Genetic dissection of silicon content in different organs of rice. Crop Sci 45(4):1345–1352
Datnoff LE, Rodrigues FA (2005) The role of silicon in suppressing rice diseases. APSnet Features
De Tombeur F, Turner BL, Laliberté E, Lambers H, Mahy G, Faucon MP, Cornelis JT (2020) Plants sustain the terrestrial silicon cycle during ecosystem retrogression. Science 369(6508): 1245–1248
Du Plessis SF, Burger RD (1966) Silicate-phosphate interrelationships in soils. S Afr J Agric Sci 9(3):525–535
Du Preez P (1970) The effect of silicon on cane growth. In: Proceeding of the South African Sugar Technologist Association, 44th Annual Congress, pp 183–188
Enstone DE, Peterson CA, Ma F (2002) Root endodermis and exodermis: structure, function, and responses to the environment. J Plant Growth Regul 21(4):335–351
Epstein E (2009) Silicon: its manifold roles in plants. Ann Appl Biol 155(2):155–160
Etesami H, Jeong BR (2018) Silicon (Si): review and prospects on the action mechanisms in alleviating biotic and abiotic stresses in plants. Ecotoxicol Environ Saf 147:881–896
Fahad S, Hussain S, Matloob A, Khan FA, Khaliq A, Saud S, Huang J (2015) Phytohormones and plant responses to salinity stress: a review. Plant Growth Regul 75(2):391–404
Fauteux F, Rémus-Borel W, Menzies JG, Bélanger RR (2005) Silicon and plant disease resistance against pathogenic fungi. FEMS Microbiol Lett 249(1):1–6
Fauteux F, Chain F, Belzile F, Menzies JG, Bélanger RR (2006) The protective role of silicon in the Arabidopsis–powdery mildew pathosystem. Proc Natl Acad Sci 103(46):17554–17559
Feng J, Yamaji N, Mitani-Ueno N (2011) Transport of silicon from roots to panicles in plants. Proc Jpn Acad 87(7):377–385
Fox RL, Silva JA, Younge OR, Plucknett DL, Sherman GD (1967) Soil and plant silicon and silicate response by sugar cane. Soil Sci Soc Am J 31(6):775–779
Frantz JM, Khandekar S, Leisner S (2011) Silicon differentially influences copper toxicity response in silicon-accumulator and non-accumulator species. J Am Soc Hortic Sci 136(5):329–338
Garberg PK (1970) Soil silica and phosphorus response by maize in acid East African soils. East Afr Agric For J 35(4):396–408
Gattullo CE, Allegretta I, Medici L, Fijan R, Pii Y, Cesco S, Terzano R (2016) Silicon dynamics in the rhizosphere: connections with iron mobilization. J Plant Nutr Soil Sci 179(3):409–417
Gibson LR, Paulsen GM (1999) Yield components of wheat grown under high temperature stress during reproductive growth. Crop Sci 39(6):1841–1846
Gómez-Merino FC, Trejo-Téllez LI, García-Jiménez A, Escobar-Sepúlveda HF, Ramírez-Olvera SM (2020) Silicon flow from root to shoot in pepper: a comprehensive in silico analysis reveals a potential linkage between gene expression and hormone signaling that stimulates plant growth and metabolism. PeerJ 8:e10053

Gong H, Zhu X, Chen K, Wang S, Zhang C (2005) Silicon alleviates oxidative damage of wheat plants in pots under drought. Plant Sci 169(2):313–321

Grégoire C, Rémus-Borel W, Vivancos J, Labbé C, Belzile F, Bélanger RR (2012) Discovery of a multigene family of aquaporin silicon transporters in the primitive plant Equisetum arvense. Plant J 72(2):320–330

Gunes A, Inal A, Bagci EG, Pilbeam DJ (2007) Silicon-mediated changes of some physiological and enzymatic parameters symptomatic for oxidative stress in spinach and tomato grown in sodic-B toxic soil. Plant Soil 290(1):103–114

Gunes A, Pilbeam DJ, Inal A, Coban S (2008) Influence of silicon on sunflower cultivars under drought stress, I: growth, antioxidant mechanisms, and lipid peroxidation. Commun Soil Sci Plant Anal 39(13-14):1885–1903

Hamayun M, Sohn EY, Khan SA, Shinwari ZK, Khan AL, Lee IJ (2010) Silicon alleviates the adverse effects of salinity and drought stress on growth and endogenous plant growth hormones of soybean (Glycine max L.). Pak J Bot 42(3):1713–1722

Hattori T, Inanaga S, Tanimoto E, Lux A, Luxová M, Sugimoto Y (2003) Silicon-induced changes in viscoelastic properties of sorghum root cell walls. Plant Cell Physiol 44(7):743–749

Hayasaka T, Fujii H, Ishiguro K (2008) The role of silicon in preventing appressorial penetration by the rice blast fungus. Phytopathology 98(9):1038–1044

Heckman J (2013) Silicon: a beneficial substance. Better Crops 97(4):14–16

Hernandez-Apaolaza L (2014) Can silicon partially alleviate micronutrient deficiency in plants? Planta 240(3):447–458

Hodson MJ, White PJ, Mead A, Broadley MR (2005) Phylogenetic variation in the silicon composition of plants. Ann Bot 96(6):1027–1046

Hosseini SA, Maillard A, Hajirezaei MR, Ali N, Schwarzenberg A, Jamois F, Yvin JC (2017) Induction of Barley Silicon Transporter HvLsi1 and HvLsi2, increased silicon concentration in the shoot and regulated Starch and ABA homeostasis under osmotic stress and concomitant potassium deficiency. Front Plant Sci 8:1359

Huang PM (1966) Mechanism of neutral fluoride interaction with soil clay minerals and silica solubility scale for silicates common in soils. University of Wisconsin, Madison

Hull RJ (2004) Scientists start to recognize silicon's beneficial effects. Turfgrass Trends 8:69–73

Imtiaz M, Rizwan MS, Mushtaq MA, Ashraf M, Shahzad SM, Yousaf B, Tu S (2016) Silicon occurrence, uptake, transport and mechanisms of heavy metals, minerals, and salinity enhanced tolerance in plants with future prospects: a review. J Environ Manag 183:521–529

International Plant Nutrition Institute (2015). Nutri-facts. Silicon. No. 14. Available at http://www.ipni.net/publication/nutrifacts-na.nsf/0/A7B4AB4D35C153BF85257ECE006E0E34/$FILE/NutriFacts-NA-14.pdf. Accessed 18 October 2021

Jones LHP, Handreck KA (1967) Silica in soils, plants, and animals. In: Advances in agronomy, vol 19. Academic, New York, pp 107–149

Kawahara H, Chonan N, Matsuda T (1974) Studies on morphogenesis in rice plants: 7. The morphology of vascular bundles in the vegetative nodes of the culm. Jpn J Crop Sci 43(3): 389–401

Kaya C, Tuna L, Higgs D (2006) Effect of silicon on plant growth and mineral nutrition of maize grown under water-stress conditions. J Plant Nutr 29(8):1469–1480

Keeping MG, Kvedaras OL, Bruton AG (2009) Epidermal silicon in sugarcane: cultivar differences and role in resistance to sugarcane borer Eldana saccharina. Environ Exp Bot 66(1):54–60

Khan MA, Ungar IA, Showalter AM (2000) Effects of sodium chloride treatments on growth and ion accumulation of the halophyte Haloxylon recurvum. Commun Soil Sci Plant Anal 31(17-18):2763–2774

Kim SG, Kim KW, Park EW, Choi D (2002) Silicon-induced cell wall fortification of rice leaves: a possible cellular mechanism of enhanced host resistance to blast. Phytopathology 92(10): 1095–1103

Kim YH, Khan AL, Kim DH, Lee SY, Kim KM, Waqas M, Lee IJ (2014) Silicon mitigates heavy metal stress by regulating P-type heavy metal ATPases, Oryza sativa low silicon genes, and endogenous phytohormones. BMC Plant Biol 14(1):1–13

Kono M, Takahashi J (1958) Selective absorption of silica and calcium by rice and tomato plants. J Sci Soil 29:63–66

Lewin J, Reimann BE (1969) Silicon and plant growth. Annu Rev Plant Physiol 20(1):289–304

Li L, Yang S, Li X, Zhang F, Christie P (1999) Interspecific complementary and competitive interactions between intercropped maize and faba bean. Plant Soil 212(2):105–114

Liang Y, Chen QIN, Liu Q, Zhang W, Ding R (2003) Exogenous silicon (Si) increases antioxidant enzyme activity and reduces lipid peroxidation in roots of salt-stressed barley (Hordeum vulgare L.). J Plant Physiol 160(10):1157–1164

Liang Y, Si J, Römheld V (2005) Silicon uptake and transport is an active process in Cucumis sativus. New Phytol 167(3):797–804

Liang Y, Sun W, Zhu YG, Christie P (2007) Mechanisms of silicon-mediated alleviation of abiotic stresses in higher plants: a review. Environ Pollut 147(2):422–428

Liang Y, Nikolic M, Bélanger R, Gong H, Song A (2015) Silicon-mediated tolerance to salt stress. Silicon Agric 2015:123–142

Liang W, Ma X, Wan P, Liu L (2018) Plant salt-tolerance mechanism: a review. Biochem Biophys Res Commun 495(1):286–291

Liu M, Cai K, Chen Y, Luo S, Zhang Z, Lin W (2014) Proteomic analysis of silicon-mediated resistance to Magnaporthe oryzae in rice (Oryza sativa L.). Eur J Plant Pathol 139(3):579–592

Lux A, Luxová M, Abe J, Tanimoto E, Hattori T, Inanaga S (2003) The dynamics of silicon deposition in the sorghum root endodermis. New Phytol 158(3):437–441

Luyckx M, Hausman JF, Lutts S, Guerriero G (2017) Impact of silicon in plant biomass production: focus on bast fibres, hypotheses, and perspectives. Plan Theory 6(3):37

Ma JF (2004) Role of silicon in enhancing the resistance of plants to biotic and abiotic stresses. Soil Sci Plant Nutr 50(1):11–18

Ma JF, Takahashi E (2002) Soil, fertilizer, and plant silicon research in Japan. Elsevier, Amsterdam

Ma JF, Yamaji N (2015) A cooperative system of silicon transport in plants. Trends Plant Sci 20(7): 435–442

Ma JF, Tamai K, Ichii M, Wu GF (2002) A rice mutant defective in Si uptake. Plant Physiol 130(4): 2111–2117

Ma JF, Tamai K, Yamaji N, Mitani N, Konishi S, Katsuhara M, Ishiguro M, Murata Y, Yano M (2006) A silicon transporter in rice. Nature 440(7084):688–691

Ma JF, Yamaji N, Mitani N, Tamai K, Konishi S, Fujiwara T, Katsuhara M, Yano M (2007) An efflux transporter of silicon in rice. Nature 448(7150):209–212

Ma JF, Yamaji N, Mitani N, Xu XY, Su YH, McGrath SP, Zhao FJ (2008) Transporters of arsenite in rice and their role in arsenic accumulation in rice grain. Proc Natl Acad Sci 105(29): 9931–9935

Ma D, Sun D, Wang C, Qin H, Ding H, Li Y, Guo T (2016a) Silicon application alleviates drought stress in wheat through transcriptional regulation of multiple antioxidant defense pathways. J Plant Growth Regul 35(1):1–10

Ma J, Sheng H, Li X, Wang L (2016b) iTRAQ-based proteomic analysis reveals the mechanisms of silicon-mediated cadmium tolerance in rice (Oryza sativa) cells. Plant Physiol Biochem 104:71–80

Malhotra C, Kapoor RT (2019) Silicon: a sustainable tool in abiotic stress tolerance in plants. In: Plant abiotic stress tolerance. Springer, Cham, pp 333–356

Mandlik R, Thakral V, Raturi G, Shinde S, Nikolić M, Tripathi DK, Deshmukh R (2020) Significance of silicon uptake, transport, and deposition in plants. J Exp Bot 71(21):6703–6718

Markovich O, Steiner E, Kouřil Š, Tarkowski P, Aharoni A, Elbaum R (2017) Silicon promotes cytokinin biosynthesis and delays senescence in Arabidopsis and Sorghum. Plant Cell Environ 40(7):1189–1196

McKersie BD, Chen Y, de Beus M, Bowley SR, Bowler C, Inzé D, Botterman J (1993) Superoxide dismutase enhances tolerance of freezing stress in transgenic alfalfa (Medicago sativa L.). Plant Physiol 103(4):1155–1163

Miao BH, Han XG, Zhang WH (2010) The ameliorative effect of silicon on soybean seedlings grown in potassium-deficient medium. Ann Bot 105(6):967–973

Minden V, Schaller J, Venterink HO (2021) Plants increase silicon content as a response to nitrogen or phosphorus limitation: a case study with Holcus lanatus. Plant Soil 462(1):95–108

Mitani N, Ma JF (2005) Uptake system of silicon in different plant species. J Exp Bot 56(414): 1255–1261

Mitani N, Yamaji N, Ma JF (2008) Characterization of substrate specificity of a rice silicon transporter, Lsi1. Eur J Phys 456(4):679–686

Mitani N, Chiba Y, Yamaji N, Ma JF (2009) Identification and characterization of maize and barley Lsi2-like silicon efflux transporters reveals a distinct silicon uptake system from that in rice. Plant Cell 21(7):2133–2142

Mitani N, Yamaji N, Ago Y, Iwasaki K, Ma JF (2011) Isolation and functional characterization of an influx silicon transporter in two pumpkin cultivars contrasting in silicon accumulation. Plant J 66(2):231–240

Mitani-Ueno N, Ma JF (2021) Linking transport system of silicon with its accumulation in different plant species. Soil Sci Plant Nutr 67(1):10–17

Mitsui S, Takatoh H (1963) Nutritional study of silicon in graminaceous crops (part 1). Soil Sci Plant Nutr 9(2):7–11

Miyoshi H, Ishii H (1960) Effect of silicic acid and silicic slag on paddy rice. IV. Effect of silicic acid on the phosphate metabolism of paddy rice. J Sci Soil 31:146–148

Montpetit J, Vivancos J, Mitani-Ueno N, Yamaji N, Rémus-Borel W, Belzile F, Bélanger RR (2012) Cloning, functional characterization and heterologous expression of TaLsi1, a wheat silicon transporter gene. Plant Mol Biol 79(1-2):35–46

Muneer S, Jeong BR (2015) Proteomic analysis of salt-stress responsive proteins in roots of tomato (Lycopersicon esculentum L.) plants towards silicon efficiency. Plant Growth Regul 77(2): 133–146

Muneer S, Park YG, Manivannan A, Soundararajan P, Jeong BR (2014) Physiological and proteomic analysis in chloroplasts of Solanum lycopersicum L. under silicon efficiency and salinity stress. Int J Mol Sci 15(12):21803–21824

Nagajyoti PC, Lee KD, Sreekanth TVM (2010) Heavy metals, occurrence, and toxicity for plants: a review. Environ Chem Lett 8:199–216

Nair PK, Aiyer RS (1968) Status of available silica in the rice soils of Kerala state (India). Agric Res J 2013:20–26

Noronha H, Silva A, Mitani-Ueno N, Conde C, Sabir F, Prista C et al (2020) The grapevine NIP2;1 aquaporin is a silicon channel. J Exp Bot 71:6789–6798

Nwugo CC, Huerta AJ (2011) The effect of silicon on the leaf proteome of rice (Oryza sativa L.) plants under cadmium-stress. J Proteome Res 10(2):518–528

Obihara CH, Russell EW (1972) Specific adsorption of silicate and phosphate by soils. J Soil Sci 23(1):105–117

Okuda A, Takahashie E (1963) Physiological role of silicon in crops. 10. Effect of -some sugars and organic acids on the silicon uptake of rice. J Sci Soil 34:114–119

Oliveira LCP, Souza DCD, Silva LFL, Resende LV, Silva MLDS, Guerra TS (2019) Silicate fertilization in non-conventional vegetables in the southern region of Minas Gerais in Brazil. Rev Ceres 66:470–478

Owino-Gerroh C, Gascho GJ (2005) Effect of silicon on low pH soil phosphorus sorption and on uptake and growth of maize. Communications 35(15-16):2369–2378

Pascual MB, Echevarria V, Gonzalo MJ, Hernández-Apaolaza L (2016) Silicon addition to soybean (Glycine max L.) plants alleviate zinc deficiency. Plant Physiol Biochem 108:132–138

Pavlovic J, Samardzic J, Maksimović V, Timotijevic G, Stevic N, Laursen KH, Nikolic M (2013) Silicon alleviates iron deficiency in cucumber by promoting mobilization of iron in the root apoplast. New Phytol 198(4):1096–1107
Peaslee DE, Frink CR (1969) Influence of silicic acid on uptake of Mn, Al, Zn, and Cu by tomatoes (Lycopersicum esculentum) grown on an acid soil. Soil Sci Soc Am J 33(4):569–571
Pei ZF, Ming DF, Liu D, Wan GL, Geng XX, Gong HJ, Zhou WJ (2010) Silicon improves the tolerance to water-deficit stress induced by polyethylene glycol in wheat (Triticum aestivum L.) seedlings. J Plant Growth Regul 29(1):106–115
Rafi MM, Epstein E, Falk RH (1997) Silicon deprivation causes physical abnormalities in wheat (Triticum aestivum L). J Plant Physiol 151(4):497–501
Rains DW, Epstein E, Zasoski RJ, Aslam M (2006) Active silicon uptake by wheat. Plant Soil 280(1):223–228
Raven JA (2001) Silicon transport at the cell and tissue level. In: Studies in plant science, vol 8. Elsevier, Amsterdam, pp 41–55
Reezi S, Kalantari MBS, Okhovvat SM, Jeong BR (2009) Silicon alleviates salt stress, decreases malondialdehyde content and affects petal color of salt-stressed cut rose (Rosa x hybrida L.) hot lady. Afr J Biotechnol 8(8):1502–1508
Reynolds OL, Keeping MG, Meyer JH (2009) Silicon-augmented resistance of plants to herbivorous insects: a review. Ann Appl Biol 155:171–186
Reynolds OL, Padula MP, Zeng R, Gurr GM (2016) Silicon: potential to promote direct and indirect effects on plant defense against arthropod pests in agriculture. Front Plant Sci 7:744
Richmond KE, Sussman M (2003) Got silicon? The non-essential beneficial plant nutrient. Curr Opin Plant Biol 6(3):268–272
Rios JJ, Martínez-Ballesta MC, Ruiz JM, Blasco B, Carvajal M (2017) Silicon-mediated improvement in plant salinity tolerance: the role of aquaporins. Front Plant Sci 8:948
Rodrigues FÁ, Benhamou N, Datnoff LE, Jones JB, Bélanger RR (2003) Ultrastructural and cytochemical aspects of silicon-mediated rice blast resistance. Phytopathology 93(5):535–546
Sabagh AEL, Hossain A, Barutçular C, Islam MS, Ahmad Z, Wasaya A, Meena RS (2020) Adverse effect of drought on quality of major cereal crops: implications and their possible mitigation strategies. In: Agronomic crops. Springer, Singapore, pp 635–658
Sakurai G, Satake A, Yamaji N, Mitani-Ueno N, Yokozawa M, Feugier FG, Ma JF (2015) In silico simulation modeling reveals the importance of the Casparian strip for efficient silicon uptake in rice roots. Plant Cell Physiol 56(4):631–639
Sasaki A, Yamaji N, Yokosho K, Ma JF (2012) Nramp5 is a major transporter responsible for manganese and cadmium uptake in rice. Plant Cell 24(5):2155–2167
Savić J, Marjanović-Jeromela A (2013) Effect of silicon on sunflower growth and nutrient accumulation under low boron supply. Helia 36(58):61–68
Schreiber L (2010) Transport barriers made of cutin, suberin and associated waxes. Trends Plant Sci 15(10):546–553
Shen X, Li Z, Duan L, Eneji AE, Li J (2014) Silicon effects on the partitioning of mineral elements in soybean seedlings under drought and ultraviolet-B radiation. J Plant Nutr 37(6):828–836
Shrivastava P, Kumar R (2015) Soil salinity: a serious environmental issue and plant growth promoting bacteria as one of the tools for its alleviation. Saudi J Biol Sci 22(2):123–131
Somapala K, Weerahewa D, Thrikawala S (2016) Silicon rich rice hull amended soil enhances anthracnose resistance in tomato. Proc Food Sci 6:190–193
Soundararajan P, Sivanesan I, Jana S, Jeong BR (2014) Influence of silicon supplementation on the growth and tolerance to high temperature in Salvia splendens. Hortic Environ Biotechnol 55(4): 271–279
Steudle E (1994) Water transport across roots. Plant Soil 167(1):79–90
Struyf E, Conley DJ (2009) Silica: an essential nutrient in wetland biogeochemistry. Front Ecol Environ 7(2):88–94

Sun H, Duan Y, Mitani-Ueno N, Che J, Jia J, Liu J, Gong H (2020) Tomato roots have a functional silicon influx transporter but not a functional silicon efflux transporter. Plant Cell Environ 43(3): 732–744

Takahashi E, Ma JF, Miyake Y (1990) The possibility of silicon as an essential element for higher plants. Agric Food Chem 2(2):99–102

Takano J, Noguchi K, Yasumori M, Kobayashi M, Gajdos Z, Miwa K, Fujiwara T (2002) Arabidopsis boron transporter for xylem loading. Nature 420(6913):337–340

Takano J, Wada M, Ludewig U, Schaaf G, Von Wirén N, Fujiwara T (2006) The Arabidopsis major intrinsic protein NIP5; 1 is essential for efficient boron uptake and plant development under boron limitation. Plant Cell 18(6):1498–1509

Tamai K, Ma JF (2008) Reexamination of silicon effects on rice growth and production under field conditions using a low silicon mutant. Plant Soil 307(1):21–27

Torabi F, Majd A, Enteshari S (2015) The effect of silicon on alleviation of salt stress in borage (Borago officinalis L.). Soil Sci Plant Nutr 61(5):788–798

Tripathi DK, Singh VP, Kumar D, Chauhan DK (2012) Impact of exogenous silicon addition on chromium uptake, growth, mineral elements, oxidative stress, antioxidant capacity, and leaf and root structures in rice seedlings exposed to hexavalent chromium. Acta Physiol Plant 34(1): 279–289

Tripathi DK, Singh S, Singh VP, Prasad SM, Dubey NK, Chauhan DK (2017) Silicon nanoparticles more effectively alleviated UV-B stress than silicon in wheat (Triticum aestivum) seedlings. Plant Physiol Biochem 110:70–81

Ueno D, Yamaji N, Ma JF (2009) Further characterization of ferric-phytosiderophore transporters ZmYS1 and HvYS1 in maize and barley. J Exp Bot 60(12):3513–3520

Vivancos J, Labbé C, Menzies JG, Bélanger RR (2015) Silicon-mediated resistance of Arabidopsis against powdery mildew involves mechanisms other than the salicylic acid (SA)-dependent defence pathway. Mol Plant Pathol 16(6):572–582

Vyas MK, Motiramani DP (1971) Effect of organic matter, silicates, and moisture levels on availability of phosphate. J Indian Soc Soil Sci 19(1):39–43

Wallace IS, Roberts DM (2005) Distinct transport selectivity of two structural subclasses of the nodulin-like intrinsic protein family of plant aquaglyceroporin channels. Biochemistry 44(51): 16826–16834

Wang X, Ou-yang C, Fan ZR, Gao S, Chen F, Tang L (2010) Effects of exogenous silicon on seed germination and antioxidant enzyme activities of Momordica charantia under salt stress. J Anim Plant Sci 6:700–708

Wang HS, Yu C, Fan PP, Bao BF, Li T, Zhu ZJ (2015) Identification of two cucumber putative silicon transporter genes in Cucumis sativus. J Plant Growth Regul 34(2):332–338

Wu X, Yu Y, Baerson SR, Song Y, Liang G, Ding C, Zeng R (2017) Interactions between nitrogen and silicon in rice and their effects on resistance toward the brown planthopper Nilaparvata lugens. Front Plant Sci 8:28

Yamaji N, Ma JF (2009) A transporter at the node responsible for intervascular transfer of silicon in rice. Plant Cell 21(9):2878–2883

Yamaji N, Ma JF (2011) Further characterization of a rice silicon efflux transporter, Lsi2. Soil Sci Plant Nutr 57(2):259–264

Yamaji N, Mitatni N, Ma JF (2008) A transporter regulating silicon distribution in rice shoots. Plant Cell 20(5):1381–1389

Yamaji N, Sakurai G, Mitani-Ueno N, Ma JF (2015) Orchestration of three transporters and distinct vascular structures in node for intervascular transfer of silicon in rice. Proc Natl Acad Sci 112(36):11401–11406

Yardanov I, Velikova V, Tsonev T (2003) Plant responses to drought and stress tolerance. Bulg J Plant Physiol 2003:187–206

You-Qiang FU, Hong SHEN, Dao-Ming WU, Kun-Zheng CAI (2012) Silicon-mediated amelioration of Fe2+ toxicity in rice (Oryza sativa L.) roots. Pedosphere 22(6):795–802

Zargar SM, Mahajan R, Bhat JA, Nazir M, Deshmukh R (2019) Role of silicon in plant stress tolerance: opportunities to achieve a sustainable cropping system. 3Biotech 9(3):73

Zhang GL, Dai QG, Zhang HC (2006) Silicon application enhances resistance to sheath blight (Rhizoctonia solani) in rice. J Plant Physiol Mol Biol 32(5):600–606

Zhao XQ, Mitani N, Yamaji N, Shen RF, Ma JF (2010) Involvement of silicon influx transporter OsNIP2; 1 in selenite uptake in rice. Plant Physiol 153(4):1871–1877

Zhu Z, Wei G, Li J, Qian Q, Yu J (2004) Silicon alleviates salt stress and increases antioxidant enzymes activity in leaves of salt-stressed cucumber (Cucumis sativus L.). Plant Sci 167(3): 527–533

Zhu YX, Xu XB, Hu YH, Han WH, Yin JL, Li HL, Gong HJ (2015) Silicon improves salt tolerance by increasing root water uptake in Cucumis sativus L. Plant Cell Rep 34(9):1629–1646

The Copper Transport Mechanism in Plants 13

Alvina Gul, Namra Haq, and Khola Rafique

Abstract

Heavy metals are required by plants in trace amounts for adequate growth and development, and their absence may lead to several detrimental effects during plant growth and development. Among them, one of the imperative trace elements required by plants for normal growth and development is copper. Copper (Cu) also serves as an important cofactor of many proteins. However, the details regarding these many Cu proteins are certainly limited. Nonetheless, the role played by these Cu proteins is of paramount significance. Cu holds an indispensable position in this regard; nevertheless, if its amount surpasses the required limit, it can lead to serious repercussions. Therefore, in order to maintain such a delicate balance, there exists an innate system within plants, which controls its absorption, distribution, and excretion within plants. There exists a unique set of proteins within plants termed as transport proteins, which regulate this delicate balance within plants. In the upcoming discussion, three of the most significant transport proteins also known as transporters are brought to light. These transport proteins include P type ATPases, that are responsible for the transport of Cu ions across the cell membrane, COPT proteins, that are responsible for the transport of Cu ions to various different cellular compartments, and chaperones that do not actually contain Cu but work like others, which are possessed with Cu. NRamp family gene analysis in soyabean seedlings also

A. Gul (✉) · N. Haq
Atta-ur-Rahman School of Applied Biosciences, National University of Sciences and Technology, Islamabad, Pakistan
e-mail: alvina_gul@asab.nust.edu.pk

K. Rafique
Pest Warning and Quality Control of Pesticides, Agriculture Department, Rawalpindi, Punjab, Pakistan

K. Kumar, S. Srivastava (eds.), *Plant Metal and Metalloid Transporters*,
https://doi.org/10.1007/978-981-19-6103-8_13

revealed their role during Cu and other heavy metal strain. The expression of this gene family also gets altered during heavy metal toxicity. The role of SPL7 transcription factor, in Cu homeostasis, has also been highlighted. In addition to it, the related role of Cu transport systems in biosynthesis and homeostasis has also been discussed.

Keywords

Copper · Copper response regulator · Homeostasis · COPT · NRAMP

Abbreviations

ATP	Adenosine triphosphate
Cu	Copper
CR	Copper response regulator
N ramp	Natural resistance-associated macrophage protein
ROS	Reactive oxygen species

13.1 Introduction

Plants, in addition to light and water, also require certain metal elements, in small amounts, which ensure apposite growth and development. Such elements are obtained either from the soil or from foliar applications (Yruela 2009). To draw these elements from soil in highly calculated way, plants have undergone evolution and, as a consequence of it, have evolved structures and ways to get to these mineral elements and also to ensure its efficient distribution. Presently, 17 elements are regarded by biologists as essential. Depending upon their required concentration, they are termed as micronutrients or macronutrients. If the required concentration of the mineral is below 100 mg/kg DW, the required minerals will be categorized as micronutrients, and if the required concentration of the mineral is above 1000 mg/kg DW, they are termed as macronutrients (Printz et al. 2016). To make it possible, plants have successfully evolved an in-built set of transporter proteins, channels, and pumps. These innate devices within plants help plant to not only absorb but also transfer and distribute minerals as per requirements. In reduced suitability of Fe during various physiological developmental stages, Cu serves as a strong alternative (Burkhead et al. 2009). In general, Cu occurrence is 60 mg/kg, and in European environment, its range is between 11.4 and 17 mg/kg (Alloway 2013).

Copper (Cu) is a vital element—a micronutrient, which is required by plants in minuscule amounts, for its proper growth and development. It is an important transition metal active in redox reaction, which carries many plants physiological processes as it can present in several oxidation states (Yruela 2005). The major functions performed by copper include its contribution in electron transport chain vital for both linked with photosynthesis and respiration. It also aids plants in sensing ethylene, metabolism occurring in cell wall, protection against oxidative stress, and synthesis of molybdenum cofactor (Yruela 2009). The significance of copper in plants' life cannot be overlooked.

Nevertheless, if the amount of Cu exceeds compared to the bare minimum of it being required, it can lead to plant damage and deterioration. It not only hampers the regular plant growth but also negatively affects the regular cellular processes occurring in the cells. Several studies have mentioned the negative effects as a result of excessive amounts of Cu on plant germination and growth, photosynthetic activity, and antioxidant response in case of several agricultural crops. The inhibition of mineral nutrition, biosynthesis of chlorophyll, and activity of antioxidant enzymes have been proved (Mir et al. 2021). Therefore, it can cause both damage and deterioration simultaneously. Such an increased concentration of Cu in the soil leading to higher levels of toxicity may occur when Cu becomes rich in parental materials, and pH of soil promotes metal availability, or soil pollution occurs by mining activities and waste deposits, or through intensive usage of plant disease control Cu-containing chemicals in agricultural, or rigorous use of manure or sewage sludge (Rehman et al. 2019; Kumar et al. 2021). For most of the crops, the serious toxicity level is over 20–30 mg/kg leaf dry weight, whereas in metallophytes tolerant to Cu, leaves may possess around 1000 μg/g leaf dry weight (Kupper et al. 2009; Monni et al. 2000).

However, as mentioned earlier, its absence can also have negative effects on growth and development of plants. The dearth of copper causes alterations in the expression of genes and also exhibit several deficiency symptoms in plant parts such as leave structure that gets distorted. The leaves first turn yellow, which can even lead to tissue death (Marschner 1995). Hence, its appropriate amount is required by plants to ensure proper growth and development. This indicates that, in plants, there exists a well-established mechanism of metal uptake by the roots from the soil and also for its translocation and distribution in all the required parts of plants. The maintenance of their concentration in the cytosol is indispensable for normal plant growth and development. To regulate these complicated and intricate processes, different transport proteins present within plants play their role.

Biochemical and molecular techniques have already helped scientists to understand these processes and will help further to explore ways to grow plants in heavy metal-contaminated zones, by developing varieties of plants, which will not absorb Cu heavy metals beyond requirement or will be able to get hold on them by developing in them both efficient and effective efflux, compartmentalization, or detoxification mechanism in response to its absorption besides requirement. Phytoremediation gained momentum to deal with the issue of soil contaminated with heavy metal (Salt et al. 1998). In the upcoming section, the transport mechanism of this indispensable metal in plants will be elucidated.

13.2 Mechanism of Copper (Cu) Transport in Plants

The importance of copper (Cu) in plants cannot be denied, however its excess can have repercussions is an established fact too. This suggests that there exists an in-built machinery that regulates the adequate transport of copper in plants. This further suggests that the device present in it is both delicate and sophisticated, which ensures regulated transport of copper in plants. However, the research on it was really limited, until research was conducted in which the transport processes in yeast and various other eukaryotic organisms were published (Nevo and Nelson 2006).

Under physiological circumstances, Cu exists in two forms, the reduced and oxidized Cu states as Cu I and Cu II, respectively. This dual nature helps it to bind with a variety of substrates (Cohu and Pilon 2010). This dual nature facilitates it to make bonds with diverse group of molecules prominently proteins, not only to run biochemical reactions but also to maintain structural ensembles (Festa and Thiele 2011). Nevertheless, redox Cu has the potential to produce reactive oxygen species (ROS) via a popular method called Fenton reaction, thereby damaging the proteins, DNA, and other biological molecules (Hänsch and Mendel 2009).

Cu homeostasis in plants is mediated by SPL7, Cu-responsive transcription factor. This SPL7 is regarded as a functional homolog of Cu response regulator 1 (CR1), which has semblance with SBP domain transcription factor that has been found in signaling of Cu in *Chlamydomonas reinhardtii* (Yamasaki et al. 2004; Kropat et al. 2005; Sommer et al. 2010). These SBP domains appear to be highly conserved domains for DNA binding and able to identify the TNCGTACAA site and particularly the GTAC core sequence (Printz et al. 2016). In plants, direct interaction between Cu and SPL7 is still not evident. However, it has been hypothesized that under sufficient Cu conditions, SPL7 may attach with Cu via specific Cu-complexes interactions, and this results in the lack of SPL7 ability to join the GTAC motif in the promotor of target genes (Garcia-Molina et al. 2014). This significant aspect of SPL7 regulation of Cu homeostasis needs further research. The research studies exhibited that there are several components, which are responsible for both efficient and effective Cu transport machinery. The different components of Cu transport machinery will be discussed in this chapter.

13.3 P-Type ATPase Copper Transporters

P-type ATPases have been reported in various different organisms that are meant to transport heavy metals across the plasma membrane. These heavy metals are both beneficial and harmful for plants, and this is basically dependent on the amount of it being present in the plant cell. P-type ATPases are a large superfamily that has subgroups in it. Subgroups make use of ATP to pump a large number of substrates carrying charge across various biological membranes and are differentiated on the basis of phosphorylated intermediate during the reaction process. Based on the type of substrate, P-type ATPase transport, they are sorted into five groups.

CPx ATPases belong to P-type ATPases, which have been associated with heavy metal transportation across the cell membrane (Solioz and Vulpe 1996). CPx ATPases are not only responsible for the absorption of heavy metal in the plants but also prevent the accumulation of heavy metals to deleterious extent. In human Menkes disorder, the gene encodes a defected copper pump, which results in the accumulation of copper to toxic levels. CPx ATPases are generally linked with Cu and Cd translocation; nevertheless, in *E. coli* and *Synechocystis* PCC 6803, they have been linked with Zn translocation too (Beard et al. 1997).

Copper (Cu) is a metal that holds a unique position, as far as its role in plants is considered. It is both a blessing and a menace for a cell, and the only thing that matters is its amount. Therefore, to maintain such a delicate balance, several constituents play their part. P-type ATPase copper transporters, which are homologous to human and yeast genes, have been reported to regulate the transport of copper across the endomembrane system in Arabidopsis (Hirayama et al. 1999; Woeste and Kieber 2000). Burkhead et al. (2009) stated that heavy metal transferring P-type ATPases (HMAs) five to eight are found to be related to Cu homeostasis. Nevertheless, HMA 5 among all four of them has been greatly linked with outflow of Cu and vascular translocation. It has been found in inordinate amounts in both plant roots and flowers and are significantly increased, when plethora of Cu gets accumulated in these plant parts (Andrés-Colás et al. 2006).

According to Axelsen and Palmgren (2001), three other putative Cu-translocating genes have been recognized; however, their mode of action has not been eloquently described. Cu transfer to chloroplast is also vital as it serves as a cofactor for stromal enzyme copper/zinc superoxide dismutase (Cu/Zn SOD) and for thylakoid lumen protein, which plays role in ETC initiating in cytochrome B_6f complex and culminating in photosystem I. In this transfer of Cu to chloroplast, the role of P-type ATPase Cu transporter cannot be discounted (Shikanai et al. 2003).

13.4 COPT Copper Transporters

The dual nature of Cu has led the plants and other organisms to develop an advanced homeostatic network in order to control uptake, transfer, utilization, and detoxification/export of Cu (Himelblau and Amasino 2000; Clemens 2001). The Cu homeostasis involves a main step regarding the uptake of Cu via cell membrane, and for mediation of its uptake, different forms of transporter proteins have been reported. Among these, the main group is the COPT (COPper transporter)/Ctr (Copper transporter) proteins belonging to several protein families in diverse organisms (Puig and Thiele 2002). Another transporter type for Cu movement from cytosol into organelles in plants and humans is P-type adenosine triphosphate pump (Williams et al. 2000; Williams and Mills 2005). In few studies, it has also been documented that other metal transporters can also carry Cu into the cells. For example, in Arabidopsis, two transporters of OPT/YSL Fe transporter family, namely, YSL1 and YSL3, can carry Cu from plant leaves to seeds (Waters et al.

2006). Similarly, ZIP2 and ZIP4 belonging to ZIP Zn transporter family transport Cu (Puig et al. 2007a, b).

The role of COPT/Ctr proteins in Cu uptake has been described principally in yeast, *Saccharomyces cerevisiae* (Dancis et al. 1994). Later, COPT/Ctr proteins transporting Cu were characterized in diverse organisms, for instance, in the case of plants: AtCOPT1, AtCOPT2, AtCOPT3, AtCOPT4, and AtCOPT5 in Arabidopsis (Sancenon et al. 2004) and OsCOPT1 and OsCOPT5 in rice (Yuan et al. 2010). In rice, the COPT family is composed of seven members, COPT1 to COPT7. Among these, COPT1 and COPT5 are able to develop homodimers or a heterodimer. Both of these COPTs can bind to multiple sites of XA13 protein in rice, which is considered as a susceptible protein to plant pathogenic bacterium *Xanthomonas oryzae* pv. *oryzae* (Xoo) (Yuan et al. 2010). Except the two COPTs, namely, COPT1 and COPT5, the rest of COPTs have been described to function individually or jointly to carry Cu transport in distinctive rice tissues (Yuan et al. 2011).

Copper ($Cu^{+}2$) gets reduced to Cu+, in order to be carried by COPT transporters. The COPT family comprises of six constituents, among which COPT1, COPT2, and COPT6 exist on plasma membrane and COPT3 and COPT5 in internal membranes. Cu uptake by COPT proteins is an important phenomenon and has been studied by comparable yeast mutants (Sanz et al. 2019). Seven member COPT-type gene families are found in rice plant, one of the major crop plants found in the world (Yuan et al. 2011). COPT proteins have been reported in transfer of Cu in many major parts of plants. Puig (2014) reviewed and stated that COPT1 plays role in absorption of Cu in plant roots, COPT6 plays role in the distribution of Cu in plant shoots, and COPT5 activates and organizes Cu from organelles meant for storage. Therefore, in the light of aforementioned functions of COPT proteins, it can be stated that the COPT proteins play an indispensable role in Cu homeostasis in plant. COPT regulates Cu, which has a role in Arabidopsis circadian clock. Sancenon et al. (2004) reported that during the period of Cu shortage, COPT1 in SPL7-dependent fashion gets activated, which ensures efficient absorption of Cu from a culture medium. Perea-García et al. (2013) reported, that in response to Cu scarcity, expression enhances manifold in SPL7-dependent manner. Furthermore, the expression of two transport proteins of ZIP family, ZIP2 and ZIP4, that mediate the transport of divalent cations is regulated by the presence of Cu (Wintz et al. 2003; Del Pozo et al. 2010).

13.5 Copper Chaperones

Copper chaperone found in plants is similar to the one found in all eukaryotic organisms, which has been revealed by complementation studies conducted in *S. cerevisiae* (Koch et al. 1997; Peña et al. 1999). These are proteins that do not possess Cu but perform job similar to the proteins containing Cu (Andrés-Colás et al. 2006). It is vital to maintain the levels of Cu within plant cells. These are actually a set of soluble proteins, which possess a special domain that is meant to bind

Cu. Thus, their Cu-chelating potential helps them to both efficiently and effectively regulate Cu homeostasis within plant cell (Shin et al. 2012) to evade copper-induced harmful effects. Moreover, Cu chaperones carry out delivery of Cu to particular Cu proteins and compartments. Brewer (2010) highlighted the significance of maintaining Cu within plant cells, and if the levels of free Cu go unchecked, it generated super oxide and hydrogen peroxide reactive oxygen species, and hydroxyl radicals negatively affect proteins, lipids, and DNA of the cells. In order to avoid the levels of free Cu within a cell, it needs to be chelated within it. This guarantees efficient and effective transfer and homeostasis.

In the case of Arabidopsis, its genome encodes seven Cu chaperones, namely, Cu chaperone for superoxide dismutase (CCS), antioxidant protein1 (ATX1), ATX1-like Cu chaperone (CCH), cytochrome c oxidase 11 (COX11), COX17, and two homologs of the yeast Cu chaperone (HCC1 and HCC2) (Puig et al. 2007a, b; Burkhead et al. 2009; Attallah et al. 2011). CCS carries Cu to Cu/Zn superoxide dismutases (SODs) in the chloroplast, cytoplasm, and peroxisome (Burkhead et al. 2009). ATX1 and CCH exhibit maximum sequential homology with the yeast protein (ATX1), and both of these can complement the yeast ATX1 mutant; however, they possess diverse properties and roles in Cu homeostasis (Shin et al. 2012). ATX1 enhances tolerance against Cu excess as well as deficiency through its Cu-binding MXCXXC motif (Shin and Yeh 2012; Shin et al. 2012. Additionally, Cu chaperones COX11, COX17, HCC1, and HCC2 function in mitochondrial respiration (Attallah et al. 2011). While extensive studies have been conducted on Cu chaperones, considerable information is still missing, specifically involving whether Cu chaperones and Cu are transported into the nucleus and induce plant defense responses.

13.6 Natural Resistance-Associated Macrophage Protein (NRAMP)

The NRAMP genes have been widely reported in organisms ranging from bacteria to yeast, including plants, mice, and human beings. This gene family has been widely found in the transport of heavy metal divalent ions across the plasma membrane (Nevo and Nelson 2006). In plants, numerous members of this gene family have also been reported, and their functions have been characterized. For example, in the case of Arabidopsis, six NRAMP proteins have been demonstrated (Mäser et al. 2001). Among these proteins, AtNRAMP1 is responsible for regulating Fe homoeostasis (Curie et al. 2000), and as a high-affinity transporter, it is involved in the uptake of Mn (Cailliatte et al. 2010). The two proteins, namely, AtNRAMP3 and AtNRAMP4, exist on the vacuolar membrane, and during the phase of seed germination, both perform mobilization of vacuolar Fe (Lanquar et al. 2005). AtNRAMP6 is directed to endomembrane compartment, which is vesicular-shaped, and this protein works as a metal transporter intracellularly with known association with Cd tolerance (Cailliatte et al. 2009).

In rice, it has been reported that three NRAMP proteins take part in Fe, Mn, and Cd uptake (Takahashi et al. 2011; Sasaki et al. 2012; Yang et al. 2014), whereas OsNrat1 participates in the uptake of Al from tip cell walls of roots into the cell, which creates Al tolerance in rice (Li et al. 2014). Likewise, in legumes, many NRAMP genes have been detected. For example, AhNRAMP1, a NRAMP gene from peanut, has been shown that it is considerably induced by Fe deficiency in leaves and roots, and this gene, when heterologously expressed in tobacco, results in accumulation of Fe in young plant leaves and Fe deprivation tolerance (Xiong et al. 2012). Additionally, gene MtNRAMP1 is specifically restricted to the plasma membrane in case of a model legume named *Medicago truncatula*, and this gene shows highest expression levels in roots and nodules, depicting its major involvement as a transporter in apoplastic uptake of Fe in rhizobia-infected cells (Tejada-Jiménez et al. 2015).

In soyabean studies, it has been revealed that the gene regulation gets affected by the shortage of N, P, K, Fe, and S. Additionally, the regulation gets affected by the buildup of Fe, Cu, Cd, and Mn. This suggests that Gm NRAMP genes play role in various different stress-related pathways and perhaps are involved in cross talk in nutrient stress pathways (Illing et al. 2012). In order to study the Gm NRAMP responses during heavy metal stresses, expression of these genes was calculated, by exposing soyabean seedlings to plethora of Fe, Cu, Cd, and Mn. In this study, only 10NRAMP gene expression was noticeable. Under excess Cu, GmNRAMP5a expression was enhanced in both leaves and roots, and expression of GmNRAMP1a was amplified in roots; however, the expression of GmNRAMP2a was diminished in both leaves and roots, respectively. Nevertheless, two NRAMP genes exhibited a unique, rather conflicting drift in soyabean leaves and roots, in response to inordinate concentration of Cu (Qin et al. 2017).

13.7 Relating the Biosynthetic and Homeostatic Roles of Cu Transport Systems

Cu serves as a cofactor, and thus, it can be contended that all the Cu transport proteins have some role in biosynthesis of different products (Burkhead et al. 2009). One of the biosynthetic functions can be viewed in the context of three ATP-driven pumps, namely, HMA6, HMA7, and HMA8. On the other hand, homeostatic function of the transport protein can be viewed in the context of regulating apposite concentrations of Cu in different compartments both locally and widely in different plants, which can be noted over a period of time. This appears to be one of the main roles played by the members of COPT family. COPT1 and COPT5 phenotypes are found where Cu concentrations are comparatively low and are evident in tissues, where transport proteins are generally expressed; however, this cannot be explained by the absence of Cu enzyme function, and this can only be explained in terms of Cu/Zn superoxide dismutase, which stops its function at once Cu shortfall happens. This decrease in the concentration of Cu/Zn superoxide dismutase follows the increase in miR398 through SPL7 (Yamasaki et al. 2007, 2009), which is among one of the four Cu-associated RNAs (Burkhead et al. 2009). Besides other functions, it was documented that Cu micro RNAs regulate the plethora of Cu in order for it to be available as cofactor where and when needed by the Cu proteins (Burkhead et al. 2009).

Cu homeostasis in plants is controlled by the SPL7 (squamosa promoter binding protein-like) transcription factor, which is active during Cu deficiency (Yamasaki et al. 2007; Bernal et al. 2012). In case of Arabidopsis, major targets of SPL7 include the genes COPT1, COPT5, and COPT6. COPT1 is engaged in encoding high-affinity Cu transporter of the roots, which is involved in primary Cu uptake (Sancenon et al. 2004). COPT6 gene has been shown to express in shoots and is found in the plasma membrane (Jung et al. 2012). Both genes are upregulated in plants during Cu deficiency so as to enhance the absorption capability at a systemic level (COPT1) and much precisely in photosynthetic plant cells (COPT6) (Sancenon et al. 2004; Jung et al. 2012). COPT5 gene is also expressed under Cu deficiency conditions, and it carries Cu efflux from the vacuole, demonstrating its role in Cu remobilization (Klaumann et al. 2011). It has been shown that Cu binds very firmly to its targets (Lippard and Berg 1994), and consequently, any competing Cu-utilizing proteins must be removed when Cu becomes deficient to permit the favored delivery of Cu to plastocyanin. This mechanism in plants regarding "copper economy" encompasses the posttranscriptional regulation of dispensable Cu enzymes by several microRNAs, which are in turn controlled by SPL7 (Yamasaki et al. 2007). The transcripts that encode the vital Cu proteins like plastocyanin are not directed for degradation by the microRNAs (Abdel-Ghany and Pilon 2008), which suggests that such proteins are important targets for deficient Cu.

13.8 Conclusion

Copper (Cu) is a vital element, and its requirement by plants as a micronutrient, as a transporter, and as a cofactor has not been elucidated thoroughly. Nonetheless, research has been conducted in the past and is still being continued on the functions of Cu in plants. As a result of these research efforts, it has been revealed that as a micronutrient, it is required by the plants in miniscule amounts; nevertheless, it is imperative for plant growth and development. In addition to it, the role of Cu as a cofactor cannot be discounted. The role of copper as Cu proteins is sine qua non for the normal functioning of plant proteins. In the light of discussion, it can be concluded that it is sine qua non for plant's survival; however, it should always be understood that it is required by the plants in very low amounts, and if it exceeds the limit, it can prove detrimental to plant growth and development and can even threaten its very existence. Therefore, in order to maintain such a delicate balance, plant has developed an efficient as well as an effective metal transport system, in which several players play their role to regulate the concentrations of different metals like copper in them.

Copper (Cu) homeostasis in plants is mediated by SPL7 regulator, which is the functional homolog of copper response regulator (CRR1), which has some semblance with the one reported in *Chlamydomonas reinhardtii.* To maintain the levels of Cu in plants, different transport proteins play their part. The major and the most important proteins among them include P-type ATPase copper transport proteins, which regulate movement of copper across the plasma membrane. P-type ATPase is not only responsible for the uptake of Cu in plants but also prevents its inordinate accumulation that can lead to deleterious consequences. Additionally, COPT transporters exist in plants, a six-member family, in which COPT1, COPT2, and COPT6 are located on cell membrane and COPT 3 and COPT5 are located on internal membranes. These are responsible for transport of Cu to various different parts of plants. The role of Cu chaperones, one being without Cu, is similar to the one with Cu. These too play role in the regulation of Cu within plant cells. The expression of NRamp genes also gets altered during heavy metal toxicity in plants, when studied in soyabean seedlings.

Cu transport systems have a linked function in two most significant activities occurring within a cell, that is, biosynthesis and homeostasis. Biosynthetic role can be explained in terms of three ATP-driven pumps, namely, HMA6, HMA7, and HMA8. Homeostatic role can be explained in terms of regulating apposite concentration of Cu to be made available to proteins as per requirement. Nevertheless, this hypothetical statement needs to be proven in future. To sum up, more extensive research studies are suggested to enhance our understanding of Cu homeostasis within plants. Nonetheless, as yet, we can say that Cu transport is a complex phenomenon, which is overseen by an intricate machinery built within plants.

References

Abdel-Ghany SE, Pilon M (2008) MicroRNA-mediated systemic down-regulation of copper protein expression in response to low copper availability in Arabidopsis. J Biol Chem 283: 15932–15945

Alloway BJ (2013) Sources of heavy metals and metalloids in soils. In: Alloway BJ (ed) Heavy metals in soils environmental pollution. Springer, Dordrecht, pp 11–50

Andrés-Colás N, Sancenón V, Rodríguez-Navarro S, Mayo S, Thiele DJ, Ecker JR et al (2006) The Arabidopsis heavy metal P-type ATPase HMA5 interacts with metallochaperones and functions in copper detoxification of roots. Plant J 45:225–236

Attallah CV, Welchen E, Martin AP, Spinelli SV, Bonnard G, Palatnik JF, Gonzalez DH (2011) Plants contain two SCO proteins that are differentially involved in cytochrome c oxidase function and copper and redox homeostasis. J Exp Bot 62:4281–4294

Axelsen KB, Palmgren MG (2001) Inventory of the superfamily of P-type ion pumps in Arabidopsis. Plant Physiol 126:696–706

Beard SJ, Hashim R, Membrillo-Hernandez J, Hughes MW, Poole RK (1997) Zinc (II) tolerance in Escherichia coli K-12: evidence that the zntA gene (o732) encodes a cation transport ATPase. Mol Microbiol 25:883–891

Bernal M, Casero D, Singh V, Wilson GT, Grande A, Yang H, Dodani SC, Pellegrini M, Huijser P, Connolly EL (2012) Transcriptome sequencing identifies SPL7-regulated copper acquisition genes FRO4/FRO5 and the copper dependence of iron homeostasis in Arabidopsis. Plant Cell 24:738–761

Brewer GJ (2010) Copper toxicity in the general population. Clin Neurophysiol 121:459–460

Burkhead JL, Reynolds KA, Abdel-Ghany SE, Cohu CM, Pilon M (2009) Copper homeostasis. New Phytol 182:799–816

Cailliatte R, Lapeyre B, Briat JF, Mari S, Curie C (2009) The NRAMP6 metal transporter contributes to cadmium toxicity. Biochem J 422:217–228. https://doi.org/10.1042/BJ20090655

Cailliatte R, Schikora A, Briat J-F, Mari S, Curie C (2010) High-affinity manganese uptake by the metal transporter NRAMP1 is essential for arabidopsis growth in low manganese conditions. Plant Cell 22:904–917. https://doi.org/10.1105/tpc.109.073023

Clemens S (2001) Molecular mechanisms of plant metal tolerance and homeostasis. Planta 212: 475–486. https://doi.org/10.1007/s004250000458

Cohu CM, Pilon M (2010) Cell biology of copper. In: Rüdiger H, Mendel R-R (eds) Cell biology of metals and nutrients plant cell monographs. Springer, Heidelberg, pp 55–74

Curie C, Alonso JM, Le Jean M, Ecker JR, Briat JF (2000) Involvement of NRAMP1 from Arabidopsis thaliana in iron transport. Biochem J 347:749–755. https://doi.org/10.1042/bj3470749

Dancis A, Yuan DS, Haile D, Askwith C, Eide D, Moehle C, Kaplan J, Klausner RD (1994) Molecular characterization of a copper transport protein in S. cerevisiae: an unexpected role for copper in iron transport. Cell 76:393–402. https://doi.org/10.1016/0092-8674(94)90345-X

Del Pozo T, Cambiazo V, González M (2010) Gene expression profiling analysis of copper homeostasis in Arabidopsis thaliana. Biochem Biophys Res Commun 393:248–252

Festa RA, Thiele DJ (2011) Copper: an essential metal in biology. Curr Biol 21:R877–R883

Garcia-Molina A, Xing S, Huijser P (2014) Functional characterisation of Arabidopsis SPL7 conserved protein domains suggests novel regulatory mechanisms in the Cu deficiency response. BMC Plant Biol 14:231. https://doi.org/10.1186/s12870-014-0231-5

Hänsch R, Mendel RR (2009) Physiological functions of mineral micronutrients (Cu, Zn, Mn, Fe, Ni, Mo, B, Cl). Curr Opin Plant Biol 12:259–266

Himelblau E, Amasino RM (2000) Delivering copper within plant cells. Curr Opin Plant Biol 3: 205–210

Hirayama T, Kieber JJ, Hirayama N, Kogan M, Guzman P, Nourizadeh S, Alonso JM, Dailey WP, Dancis A, Ecker JR (1999) Responsive-to-antagonist1, a Menkes/Wilson disease-related copper transporter, is required for ethylene signaling in Arabidopsis. Cell 97:383–393

Illing AC, Shawki A, Cunningham CL, Mackenzie B (2012) Substrate profile and metal-ion selectivity of human divalent metal-ion transporter-1. J Biol Chem 287:30485–30496

Jung HI, Gayomba SR, Rutzke MA, Craft E, Kochian LV, Vatamaniuk OK (2012) COPT6 is a plasma membrane transporter that functions in copper homeostasis in arabidopsis and is a novel target of SQUAMOSA promoter binding protein-like 7. J Biol Chem 287:33252–33267

Klaumann S, Nickolaus SD, Fürst SH, Starck S, Schneider S, Ekkehard NH, Trentmann O (2011) The tonoplast copper transporter COPT5 acts as an exporter and is required for interorgan allocation of copper in Arabidopsis thaliana. New Phytol 192:393–404

Koch KA, Peña MMO, Thiele DJ (1997) Copper-binding motifs in catalysis, transport, detoxification and signaling. Chem. & . Biol 4:549–560

Kropat J, Tottey S, Birkenbihl RP, Depège N, Huijser P, Merchant S (2005) A regulator of nutritional copper signaling in Chlamydomonas is an SBP domain protein that recognizes the GTAC core of copper response element. Proc Natl Acad Sci U S A 102:18730–18735

Kumar V, Pandita S, Sidhu GPS, Sharma A, Khanna K, Kaur P, Bali AS, Setia R (2021) Copper bioavailability, uptake, toxicity and tolerance in plants: a comprehensive review. Chemosphere 262:127810

Kupper H, Götz B, Mijovilovich A, Küpper FC, Meyer-Klaucke WE (2009) Complexation and toxicity of copper in higher plants I. Characterization of copper accumulation, speciation, and toxicity in Crassula helmsii as a new copper accumulator. Plant Physiol 151:702–714

Lanquar V, Lelièvre F, Bolte S, Hamès C, Alcon C, Neumann D et al (2005) Mobilization of vacuolar iron by AtNRAMP3 and AtNRAMP4 is essential for seed germination on low iron. EMBO J 24:4041–4051. https://doi.org/10.1038/sj.emboj.7600864

Li JY, Liu J, Dong D, Jia X, McCouch SR, Kochian LV (2014) Natural variation underlies alterations in Nramp aluminum transporter (NRAT1) expression and function that play a key role in rice aluminum tolerance. Proc Natl Acad Sci U S A 111:6503. https://doi.org/10.1073/pnas.1318975111

Lippard SJ, Berg JM (1994) Principles of bioinorganic chemistry. University Science Books, New York

Marschner H (1995) Mineral nutrition of higher plants. Academic Press, London

Mäser P, Thomine S, Schroeder JI, Ward JM, Hirschi K, Sze H et al (2001) Phylogenetic relationships within cation transporter families of Arabidopsis. Plant Physiol 126:1646–1667. https://doi.org/10.1104/pp.126.4.1646

Mir AR, Pichtel J, Hayat S (2021) Copper: uptake, toxicity and tolerance in plants and management of Cu-contaminated soil. Biometals 34(4):737–759. https://doi.org/10.1007/s10534-021-00306-z

Monni S, Salemaa M, White C, Tuittila E (2000) Copper resistance of Calluna vulgaris originating from the pollution gradient of a Cu-Ni smelter, in southwest Finland. Environ Pollut 109:211–219

Nevo Y, Nelson N (2006) The NRAMP family of metal-ion transporters. Mol Cell Res 1763(7): 609–620

Peña MMO, Lee J, Thiele DJ (1999) A delicate balance: homeostatic control of copper uptake and distribution. J Nutr 129:1251–1260

Perea-García A, Garcia-Molina A, Andrés-Colás N, Vera-Sirera F, Pérez-Amador MA, Puig S et al (2013) Arabidopsis copper transport protein COPT2 participates in the cross talk between iron deficiency responses and low-phosphate signaling. Plant Physiol 162:180–194

Printz B, Lutts S, Hausman JF, Sergeant K (2016) Copper trafficking in plants and its implication on cell wall dynamics. Front Plant Sci 7:601

Puig S (2014) Function and regulation of the plant COPT family of high-affinity copper transport proteins. Adv Bot 2014:476917

Puig S, Thiele DJ (2002) Molecular mechanisms of copper uptake and distribution. Curr Opin Chem Biol 6:171–180. https://doi.org/10.1016/S1367-5931(02)00298-3

Puig S, Andres-Colas N, Garcia-Molina A, Penarrubia L (2007a) Copper and iron homeostasis in Arabidopsis: response to metal deficiencies, interactions and biotechnological applications. Plant Cell Environ 30:271–290. https://doi.org/10.1111/j.1365-3040.2007.01642.x

Puig S, Mira H, Dorcey E et al (2007b) Higher plants possess two different types of ATX1-like copper chaperones. Biochem Biophys Res Commun 354:385–390

Qin L, Han P, Chen L, Walk TC, Li Y, Hu X, Xie L, Liao H, Liao X (2017) Genome-wide identification and expression analysis of NRAMP family genes in soybean (Glycine Max L.). Front Plant Sci 8:1436

Rehman N, Maqbool Z, Peng D, Liu L (2019) Morpho-physiological traits, antioxidant capacity and phytoextraction of copper by ramie (Boehmeria nivea L.) grown as fodder in copper-contaminated soil. Environ Sci Pollut Res 26:5851–5861

Salt DE, Smith RD, Raskin I, Annu. (1998) Phytoremediation. Annu Rev Plant Physiol 49:643–668

Sancenon V, Puig S, Mateu-Andres I, Dorcey E, Thiele DJ, Penarrubia L (2004) The Arabidopsis copper transporter COPT1 functions in root elongation and pollen development. J Biol Chem 279:15348–15355. https://doi.org/10.1074/jbc.M313321200

Sanz A, Pike S, Khan MA et al (2019) Copper uptake mechanism of Arabidopsis thaliana high-affinity COPT transporters. Protoplasma 256:161–170

Sasaki A, Yamaji N, Yokosho K, Ma JF (2012) Nramp5 Is a major transporter responsible for manganese and cadmium uptake in rice. Plant Cell 24:2155–2167. https://doi.org/10.1105/tpc.112.096925

Shikanai T, Müller-Moulé P, Munekage Y et al (2003) PAA1, a P-Type ATPase of Arabidopsis, functions in copper transport in chloroplasts. Plant Cell 15:1333–1346

Shin LJ, Yeh KC (2012) Overexpression of Arabidopsis ATX1 retards plant growth under severe copper deficiency. Plant Signal Behav 7:1082–1083

Shin L-J, Lo J-C, Ye K-C (2012) Copper chaperone antioxidant protein1 is essential for copper homeostasis. Plant Physiol 159(3):1099–1110
Solioz M, Vulpe C (1996) CPx-type ATPases: a class of P-type ATPases that pump heavy metals. Trends Biochem Sci 21(7):237–241
Sommer F, Kropat J, Malasarn D, Grossoehme NE, Chen X, Giedroc DP et al (2010) The CRR1 nutritional copper sensor in Chlamydomonas contains two distinct metal-responsive domains. Plant Cell 22:4098–4113
Takahashi R, Ishimaru Y, Nakanishi H, Nishizawa NK (2011) Role of the iron transporter OsNRAMP1 in cadmium uptake and accumulation in rice. Plant Signal Behav 6:1813–1816. https://doi.org/10.4161/psb.6.11.17587
Tejada-Jiménez M, Castro-Rodríguez R, Kryvoruchko I, Lucas MM, Udvardi M, Imperial J et al (2015) Medicago truncatula natural resistance-associated macrophage protein1 is required for iron uptake by rhizobia-infected nodule cells. Plant Physiol 168:258–272. https://doi.org/10.1104/pp.114.254672
Waters BM, Chu HH, Didonato RJ, Roberts LA, Eisley RB, Lahner B, Salt DE, Walker EL (2006) Mutations in Arabidopsis yellow stripe-like 1 and yellow stripe-like 3 reveal their roles in metal ion homeostasis and loading of metal ions in seeds. Plant Physiol 141:1446–1458. https://doi.org/10.1104/pp.106.082586
Williams LE, Mills RF (2005) P1B-ATPases-an ancient family of transition metal pumps with diverse function in plants. Trends Plant Sci 10:491–502. https://doi.org/10.1016/j.tplants.2005.08.008
Williams LE, Pittman JK, Hall JL (2000) Emerging mechanisms for heavy metal transport in plants. Biochim Biophys Acta 1465:104–126. https://doi.org/10.1016/S0005-2736(00)00133-4
Wintz H, Fox T, Wu Y-Y, Feng V, Chen W, Chang H-S et al (2003) Expression profiles of Arabidopsis thaliana in mineral deficiencies reveal novel transporters involved in metal homeostasis. J Biol Chem 278:47644–47653
Woeste KE, Kieber JJ (2000) A strong loss-of-function mutation in RAN1 results in constitutive activation of the ethylene response pathway as well as a rosette-lethal phenotype. Plant Cell 12: 443–455
Xiong H, Kobayashi T, Kakei Y, Senoura T, Nakazono M, Takahashi H et al (2012) AhNRAMP1 iron transporter is involved in iron acquisition in peanut. J Exp Bot 63:4437–4446. https://doi.org/10.1093/jxb/ers117
Yamasaki K, Kigawa T, Inoue M, Tateno M, Yamasaki T, Yabuki T et al (2004) A novel zinc-binding motif revealed by solution structures of DNA-binding domains of Arabidopsis SBP-family transcription factors. J Mol Biol 337:49–63
Yamasaki H, Abdel-Ghany SE, Cohu CM, Kobayashi Y, Shikanai T, Pilon M (2007) Regulation of copper homeostasis by micro-RNA in Arabidopsis. J Biol Chem 282:16369–16378
Yamasaki H, Hayashi M, Fukazawa M, Kobayashi Y, Shikanai T (2009) SQUAMOSA promoter binding protein-like7 is a central regulator for copper homeostasis in Arabidopsis. Plant Cell 21: 347–361
Yang M, Zhang W, Dong H, Zhang Y, Lv K, Wang D et al (2014) OsNRAMP3 is a vascular bundles-specific manganese transporter that is responsible for manganese distribution in rice. PLoS ONE 8:e83990. https://doi.org/10.1371/journal.pone.0083990
Yruela I (2005) Copper in plants. Braz J Plant Physiol 17:145–156
Yruela I (2009) Copper in plants: acquisition, transport and interactions. Funct Plant Biol 36(5): 409–430
Yuan M, Chu Z, Li X, Xu C, Wang S (2010) The bacterial pathogen Xanthomonas oryzae overcomes rice defenses by regulating host copper redistribution. Plant Cell 22:3164–3176. https://doi.org/10.1105/tpc.110.078022
Yuan M, Li X, Xiao J, Wang S (2011) Molecular and functional analyses of COPT/Ctr-type copper transporter-like gene family in rice. BMC Plant Biol 11(1):1–12

Plant Metal Tolerance Proteins: Insight into Their Roles in Metal Transport and Homeostasis for Future Biotechnological Applications

14

Natasha Das, Praveenya Tirunagari, and Mrinal K. Maiti

Abstract

Similar to all living organisms, plants require appropriate supplies of metal micronutrients for their normal metabolism, growth, and development. Plants being sessile in nature respond to external variations (both types and concentrations) of these metal and metalloid elements by employing a complex network of membrane transport system for efficient uptake, translocation, and compartmentalization of metal(loid)s in order to maintain the ion homeostasis. Among the different gene families involved in metal(loid) transport in plants, the ubiquitous cation diffusion facilitator (CDF) is a family of transmembrane transporters that efflux divalent cations from the cytoplasm to either subcellular locations or outside the cell. In plants, CDFs are called metal tolerance proteins (MTPs) and have shown specificity in transporting Zn^{2+}, Mn^{2+}, and Fe^{2+} but can also transport Cd^{2+}, Co^{2+}, and Ni^{2+}, and some of these are also toxic to plants. Thus, the MTPs are presumed to carry out important and essential roles in mineral nutrition maintenance, stress tolerance, and homeostasis of metal(loid)s in plants. This chapter summarizes the recent developments through both in silico genome-wide analyses and functional characterization studies of MTP transporters in both dicot and monocot model plants, such as Arabidopsis and rice, respectively, together with other plant species of known genome sequence. A comprehensive understanding of the MTP family of proteins will help us to grasp clearly their roles in plant metal(loid) tolerance and cellular homeostasis at the physiological and biochemical levels. Further, it is anticipated that an in-depth study of the transcriptional and posttranscriptional regulation of the *MTP* genes in different plant species during metal(loid) stress will help to identify the candidate genes,

N. Das · P. Tirunagari · M. K. Maiti (✉)
Department of Biotechnology, Indian Institute of Technology Kharagpur, Kharagpur, India
e-mail: maitimk@bt.iitkgp.ac.in

K. Kumar, S. Srivastava (eds.), *Plant Metal and Metalloid Transporters*,
https://doi.org/10.1007/978-981-19-6103-8_14

which could be employed for crop biofortification and environmental bioremediation in future.

Keywords

Cellular homeostasis · Crop biofortification · Environmental bioremediation · Metal micronutrients · Metal stress tolerance · Metal transporter protein

14.1 Introduction

In order to better comprehend the cellular homeostasis within plant cell, knowledge on the different transport pathways for metal ion influx, efflux, and subcellular partitioning is of utmost necessity. Essential elements such as zinc (Zn), manganese (Mn), and iron (Fe) are required for many biochemical functions as enzyme co-factors or structural components of proteins and pigments essential for plant metabolism and development. Along with these essential elements, plants also transport several nonessential elements such as cadmium (Cd), cobalt (Co), and nickel (Ni). The fact that both essential and nonessential elements are toxic at elevated concentrations (i.e., above certain threshold level) for plant cells, plants have evolved different techniques such as detoxification, extrusion, and exclusion for avoiding these metalloid stress effects (Emamverdian et al. 2015). As a part of plant resistance or tolerance system, plants have also developed an array of transporters localized at the plasma membrane, envelope membranes of cell organelles such as the vacuole, mitochondria, and plastids, and these transporters play key roles in protecting the cells against toxicity caused by these harmful elements. Most recently, scientists have also discovered a sensing system in which transporter proteins also work as receptors, that is, transceptors to maintain metal homeostasis within plant cells (Cointry and Vert 2019; Narayan et al. 2020; Yadav et al. 2021). As mentioned earlier, metal ions are inevitable in biological processes for plant growth and development, and for that, cell organelles have compartmentalized cellular spaces for carrying out specific functions such as photosynthesis, respiration, biosynthesis of metabolites, macromolecules and phytohormones, and detoxification of metal(loid)s. However, for compartmentalizing and accumulating metal(loid)s as a detoxification process, the vacuole plays the critical role in plant's mechanism to reduce metal(loid) stress-related conditions at the cellular level. Moreover, for the plant transport machineries of micronutrients, Zn, Mn, and Fe serve as substrates for a range of membrane transporters in plants, such as heavy metal ATPase (HMA), zinc- or iron-regulated transporter-like protein (ZIP) and cation diffusion facilitator (CDF) family, which are referred to as metal tolerance proteins (MTPs) in higher plants. MTPs play the pivotal role in cellular homeostasis and detoxification of metals.

The CDF/MTP family members, virtually found in all organisms spanning the Eukaryota, Eubacteria, and Archaea kingdoms, were first reported by Nies and Silver (1995). MTP family proteins have shown specificity in transporting Zn^{2+},

Mn^{2+}, and Fe^{2+} but can also transport Cd^{2+}, Co^{2+}, and Ni^{2+} and thus play a pivotal role in cellular homeostasis of metal ions in plants. Along with the cell membrane, MTP proteins are reportedly found in membranes of various intracellular compartments, such as vacuole, mitochondria, endoplasmic reticulum (ER), Golgi/ trans-Golgi, and pre-vacuolar compartment, where they are involved in either the efflux of metals from the cytoplasm into these organelles (Ricachenevsky et al. 2013) or efflux from the plasma membrane or sequestration into the vacuole (Delhaize et al. 2003) or Golgi (Pedas et al. 2014). CDF/MTP members seem to act as $metal^{2+}$ (Me^{2+})/H^+ antiporters and contain amino and carboxy cytoplasmic termini having six transmembrane domains (TMD), with a few exceptions (Guffanti et al. 2002; Chao and Fu 2004; Grass et al. 2005; Kawachi et al. 2008). Table 14.1 lists the MTP transporters described in the texts with the information on the corresponding substrate(s) of that particular MTP protein and their subcellular localization, which are available on characterized plant MTP proteins. Based on their substrate specificities, the CDF/MTP family is classified into three subfamilies, namely, Mn-CDF, Zn-CDF, and Zn/Fe-CDF (Montanini et al. 2007). The first CDF gene in *Arabidopsis thaliana* was characterized as the zinc transporter 1 (*ZAT1*) and later renamed as metal tolerance protein 1 gene (*AtMTP1*) (van der Zaal et al. 1999; Delhaize et al. 2003). Phylogenetic tree of MTP family members comprising of different groups for *Oryza sativa*, *Arabidopsis thaliana*, and *Glycine max* (Fig. 14.1) is prepared as a representative one to understand the evolutionary relationships of the MTP members belonging to various groups and families. Different plant species contain numerous members, indicating diverse functions in vivo. Closely related isoform exists within MTP gene family, indicating possible redundancies and specialized functions. In this review, the role of different MTPs in uptake, distribution, and detoxification of ions as a part of metal ion homeostasis for maintaining optimal metal ion concentration in cytoplasm is discussed to get a comprehensive idea about the importance of this transporter family for its future application in biotechnology.

14.2 Regulation of Cellular Metal Homeostasis

14.2.1 Role of MTPs in Vacuolar Compartmentalization for Metal Homeostasis

Several MTP transporters localized to the vacuolar membrane have been reported to play a key role in protecting the cell against metal toxicity by means of metal efflux from the cell or intracellular metal sequestration into the vacuole that serves as a major reservoir to regulate cellular metal homeostasis. In *Arabidopsis*, Zn-CDF proteins have been studied from the first identified metal tolerance protein 1, AtMTP1 (van der Zaal et al. 1999), and are reported to play a critical role in sequestration of excess Zn into the vacuole (Desbrosses-Fonrouge et al. 2005). It has been reported the *AtMTP1* gene, which is normally expressed constitutively in both roots as well as in shoot, when overexpressed was also able to enhance Zn tolerance

Table 14.1 Functionally characterized plant MTP family members showing their subcellular localization and the metal substrates for transport across biological membranes

Plant species	Protein name	Subcellular location	Metals as substrate	Reference
Arabidopsis. halleri	*AhMTP1*	Tonoplast	Zn	Shahzad et al. (2010)
Arabidopsis thaliana	AtMTP1	Tonoplast	Zn	Kobae et al. (2004), Desbrosses-Fonrouge et al. (2005)
	AtMTP3	Tonoplast	Co, Zn	Arrivault et al. (2006)
	AtMTP8	Tonoplast	Mn	Eroglu et al. (2016), Chu et al. (2017), Eroglu et al. (2017)
	AtMTP11	Pre-vacuolar compartment/ trans Golgi network (TGN)	Mn	Delhaize et al. (2007), Peiter et al. (2007)
Soybean (*Glycine max*)	GmMTP8.1	Endoplasmic reticulum (ER)	Mn	Li et al. (2021a, b)
Barley (*Hordeum vulgare*)	HvMTP1	Tonoplast	Co, Zn	Podar et al. (2012)
	HvMTP8.1/ HvMTP8.2	Golgi compartments	Mn	Pedas et al. (2014)
Rice (*Oryza sativa*)	OsMTP1	Tonoplast	Cd, Co, Fe, Ni, Zn	Lan et al. (2012), Yuan et al. (2012), Menguer et al. (2013)
	OsMTP8.1/ OsMTP8.2	Tonoplast	Mn	Delhaize et al. (2003), Chen et al. (2013), Eroglu et al. (2016), Takemoto et al. (2017)
	OsMTP9	Plasma membrane	Mn	Ueno et al. (2015)
	OsMTP11	Trans Golgi network (TGN)	Mn	Ma et al. (2018)
Stylosanthes hamate	ShMTP8	Tonoplast	Cu, Mn	Delhaize et al. (2003, 2007)
Pyrus bretschneideri Rehd	PbMTP8.1	Pre-vacuolar compartment (PVC)	Mn	Li et al. (2021a, b)
Camellia sinensis L.	CsMTP8.1	Plasma membrane	Mn	Li et al. (2017)
	CsMTP8.2	Plasma membrane	Mn	Zhang et al. (2020)
Poplar (*Populus trichocarpa*)	PtMTP11.1/ PtMTP11.2	Golgi compartments	Mn	Peiter et al. (2007)
Populus trichocarpa	PtrMTP6	Pre-vacuolar compartment (PVC)	Co, Mn	Yang et al. (2021)

(continued)

Table 14.1 (continued)

Plant species	Protein name	Subcellular location	Metals as substrate	Reference
B. vulgaris spp. *maritima*	BmMTP10	Golgi compartments	Mn	Erbasol et al. (2013)
	BmMTP11			
Tobacco (*Nicotiana tabacum*)	NtMTP2	Tonoplast	Co, Ni	Papierniak-Wygladala et al. (2020)

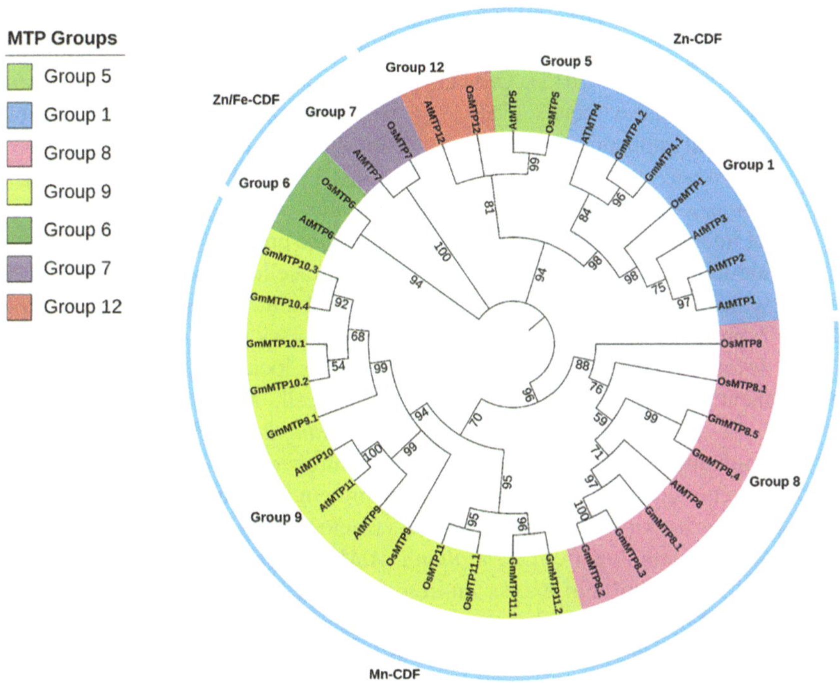

Fig. 14.1 Phylogenetic analysis of metal tolerance proteins in *Oryza sativa*, *Arabidopsis thaliana*, and *Glycine max*. The phylogenetic tree was constructed by the maximum likelihood method based on the Jones-Taylor-Thornton (JTT) matrix-base model with a bootstrap of 1000 replicates using MEGA10.2.4 software. MTP proteins are distributed into seven primary groups, Group 1, Group 5, Group 6, Group 7, Group 8, Group 9, and Group 12 (which are highlighted in different colors) with the bootstrap values mentioned, and are clustered into three major substrate-specific groups, Zn-CDF (Groups 1, 5, 12), Zn/Fe-CDF (Group 6, 7), and Mn-CDF (Groups 8, 9), and *AtMTP*, *OsMTP*, and *GmMTP* indicate the metal tolerance proteins of *Arabidopsis thaliana*, *Oryza sativa*, and *Glycine max*, respectively

in *Arabidopsis* (van der Zaal et al. 1999). However, RNA interference (RNAi)-mediated silencing (Desbrosses-Fonrouge et al. 2005) or T-DNA insertion mutation (Kobae et al. 2004) of this gene increases Zn sensitivity, indicating its important role in regulation of Zn homeostasis. Similarly, in Zn hyperaccumulator plant *Arabidopsis halleri*, metal tolerance protein 1 (AhMTP1) gene is considered to have a role in Zn hypertolerance (Shahzad et al. 2010).

An ortholog of MTPs in rice, the tonoplast localized *OsMTP1* has been found to have a considerably broader substrate specificity than that of the other highly Zn-specific member like *AtMTP1* (Kobae et al. 2004; Desbrosses-Fonrouge et al. 2005) and hybrid poplar (*Populus trichocarpa* x *Populus deltoids*) *PtdMTP1* (Blaudez et al. 2003). The *OsMTP1* primarily transport Zn; additionally, it can also transport Co, Fe, Cd, and Ni upon heterologous expression in yeast (Menguer et al. 2013; Yuan et al. 2012). OsMTP1 function as a detoxification system for excess levels of these metals made this gene a perfect candidate for possible biotechnological applications, such as biofortification and phytoremediation.

Previous report on *Medicago truncatula* metal tolerance protein (MtMTP1) described it as a Zn transporter involved in Zn efflux from the cytosol (Chen et al. 2009). But interestingly, *M. truncatula* metal tolerance protein2 (MtMTP2) found to be a nodule-induced Zn-efflux protein that help in intracellular compartmentalization of Zn, which is a prerequisite for effective symbiotic nitrogen fixation. This MtMTP2 protein localized in nodule cell and expression is upregulated during nodule development. Apart from nodule, it is also localized to an intracellular compartment in root epidermal and endodermal cell, and putative localization would be the endoplasmic reticulum. Studies on *Mtmtp2* mutants exhibited abnormal accumulation of Zn in nodules and impaired nitrogen fixation as well as growth. It has been reported that loss of function of this protein also leads to severe reduction of nitrogenase activity and altered nodule development (León-Mediavilla et al. 2018). Another ortholog from *A. thaliana*, *AtMTP3*, plays a crucial role in the storage and sequestration of excess Zn in to the vacuole (Arrivault et al. 2006). In contrast to AtMTP1, AtMTP3 is expressed primarily in roots and reported to be engaged in maintenance of metal homeostasis by excluding Zn from shoot under Zn oversupply as well as in under Fe deficiency (Arrivault et al. 2006). The *AtMTP3* expression-related study shows that expression of *MTP3* is positively regulated by Fe deficiency and also by excess Zn and Co (Arrivault et al. 2006). Although *AtMTP3* function similarly to that of *AtMTP1* and *AhMTP1* at cellular level, however, AtMTP3 is reported to confer higher Zn as well as Co tolerance.

In CDF subfamilies, the first characterized member of the Mn-CDF subgroup was from Mn-accumulating tropical legume *Stylosanthes hamata*, and the tonoplast-localized Mn transporter *ShMTP8* has been shown to be involved in Mn internalization into vacuole to avoid metal toxicity in yeast and *A. thaliana* (Delhaize et al. 2003). Members of the Mn-CDF subclade from other plants like pear and tobacco described as a Mn transporter (Hou et al. 2019; Liu et al. 2019). A very recent studies in yeast showed the putative Mn-CDF transporter in *Pyrus bretschneideri* Rehd, PbMTP8.1, localizes to pre-vacuolar compartment (PVC) and plays a major role in Mn tolerance, accumulation, and overall in maintaining Mn homeostasis (Li et al.

2021a, b). In *A. thaliana*, MTP8, MTP9, MTP10, and MTP11 are the four members of Mn-CDF subfamily. So far, only *AtMTP8* and *AtMTP11* have been characterized in detail. Among them, the tonoplast-localized transporter, AtMTP8, functions as Mn and Fe transporter (Eroglu et al. 2016, 2017; Chu et al. 2017) and sequester Mn into vacuoles. Ortholog of AtMTP8 found in rice *Oryza sativa* (Chen et al. 2013) shows conserve nature of MTP8 among different plant species, with the possible exception of barley (*Hordeum vulgare*), where homologs of MTP8 have been localized to the Golgi apparatus (Pedas et al. 2014). Under excess Mn supply, the *AtMTP8* expression was also found to be increased (Eroglu et al. 2016, 2017; Chu et al. 2017). Consistent with the function of AtMTP8 in Mn detoxification, high Mn has been found to damage the growth of *Atmtp8* mutants (Eroglu et al. 2016, 2017; Chu et al. 2017).

It has been known in plants that the Fe nutrition is antagonistically affected by Mn, and Fe deficiency symptoms are induced by increased Mn availability. Thus, *AtMTP8* plays a critical role in the Fe deficiency response that was confirmed through the studies on *mtp8* mutants. The mutants on media with limited Fe availability in the presence of Mn showed chlorosis and significantly low levels of Fe in shoots (Eroglu et al. 2016). Moreover *AtMTP8* expression is induced by Fe deficiency and expressed particularly to cells of the epidermis and the cortex in roots (Eroglu et al. 2016). Iron-regulated transporter 1 (IRT1) is a central iron transporter responsible for the uptake of iron from the rhizosphere to root epidermal cells in *Arabidopsis*, and this AtIRT1 transporter also takes up Mn during Fe-limiting condition. During Fe limitation, AtMTP8 showed Mn-specific role by demonstrating Mn detoxification that was taken up by nonspecific Fe transporter AtIRT1 (Eroglu et al. 2016). Functionality of MTP8 in dicot plants such as *Arabidopsis* and cucumber is in stark contrast with that of graminaceous species rice. This difference may be due to the availability of Mn in their natural environment or their Fe acquisition strategies. Unlike the orthologs in dicot species such as *Arabidopsis* and cucumber, the *OsMTP8.1* is expressed solely in the rice shoot. Interestingly as opposed to what happened in Arabidopsis, in rice, *Osmtp8.1* mutant contained less Mn content (Chen et al. 2013). Under the reducing soil condition in lowland, rice, which generally results in excess of both Mn and Fe concomitantly (Marschner and Rengel 2012), often leads to translocation of excess Mn to shoot. In this situation, additional expression of *OsMTP8.1* in leaves apart from shoot helps the whole plant capacity of Mn compartmentalization in vacuoles. Another MTP8 in rice *OsMTP8.2* was expressed in both root and shoot. The expression level of *OsMTP8.2* is low compared to OsMTP8.1 (Takemoto et al. 2017) in different rice tissues. The phenotype of the double mutant *mtp8.1mtp8.2* in rice indicated that insufficient Mn sequestration caused the necrotic leaf blades. However, contrary to *Arabidopsis*, so far no confirmatory evidence has been observed that suggests involvement of *OsMTP8.1* and *OsMTP8.2* in Fe deficiency and transport or elevation in transcript level during Fe deficiency.

Another member belonging to Mn-CDF cluster in rice, that is, OsMTP9, is responsible for radial transport of Mn into the root stele out of exodermis and endodermis cells of roots. Members of the Mn-CDF subclade in pear and tobacco have also been described as Mn transporters, but their importance in Mn homeostasis and subcellular localization is still unknown (Hou et al. 2019; Liu et al. 2019).

14.2.2 Plasma Membrane-Localized MTP Transporter Responsible for Distal Transport of Mn

Contrary to most other MTP members, OsMTP9 is a plasma membrane-localized Mn efflux transporter that together with the Mn influx transporter natural resistance-associated macrophage protein 5 (Nramp5) plays a critical role in distal Mn translocation in rice plant. Both the Nramp5 and OsMTP9 are localized in the plasma membrane of the exodermis and endodermis cells of the rice roots. However, the relative positions of these two transporters are polar opposite of one another. The influx transporter Nramp5 is localized toward the distal side, whereas the efflux transporter OsMTP9 is localized at the proximal end (Ueno et al. 2015). It is established that, following the uptake of Mn from the external environment by the distally located influx transporter Nramp5, the proximally located efflux transporter OsMTP9 then releases this uptake Mn toward the apoplast of root stele including the xylem vessels (Ueno et al. 2015). Moreover, increased Mn accumulation in roots but decreased uptake in shoots due to knockout of OsMTP9 clearly demonstrates the significance of OsMTP9 in distal transport of Mn in rice plant (Ueno et al. 2015). Apart from the model plants, MTP transporters belonging to the Mn-CDF subgroup have also been identified in nonmodel plants such as *Brassica rapa* (Li et al. 2018), *Pyrus bretschneideri* Rehd. (Hou et al. 2019), and tea (*Camellia sinensis* L.). The functional homologue of MTP8 in *C. sinensis*, *CsMTP8*, localizes to the plasma membrane, and its expression in the leaves is upregulated by excess Mn and is involved in Mn efflux from shoot cells (Li et al. 2017). However, overexpression studies in yeast and *Arabidopsis* probably confer its role in Mn homeostasis via excess Mn efflux from the leaf cells (Li et al. 2017). Similar functional analysis study confirmed that another MTP member of tea plant, CsMTP8.2, is also a Mn-specific transporter and participates in tolerance to Mn toxicity by enhancing the efflux of Mn from cells (Zhang et al. 2020).

14.2.3 MTP Transporter as Manganese Transport Proteins in Endomembranes

Even though most of the MTP8-like proteins are similar, still functional variations are observed among different homologues, as, for example, MTP8 members ShMTP8, AtMTP8, and OsMTP8.1/8.2 are all tonoplast localized and detoxify Mn by sequestration into the vacuoles (Delhaize et al. 2003; Chen et al. 2013; Eroglu et al. 2016; Takemoto et al. 2017), whereas in barley (*Hordeum vulgare*), MTP8 proteins, HvMTP8.1 and HvMTP8.2, modulate Mn homeostasis by delivering Mn to the Golgi apparatus (Pedas et al. 2014) but not in vacuole.

The characteristics and functions of soybean (*Glycine max*) MTP, GmMTPs, in response to Mn toxicity have been documented recently. Surprisingly, GmMTP8.1 was found to be localized to the ER in tobacco leaf epidermis cells. Transgenic *Arabidopsis* overexpressing *GmMTP8.1* indicates its role in maintaining Mn homeostasis by transporting Mn into ER that helps in reduction of cytosolic Mn (Li et al. 2021a, b). In some earlier study, an endomembrane-type Ca-ATPase from *A. thaliana*, AtECA1, was found to be ER localized and demonstrated elevated growth of Arabidopsis under Mn toxicity conditions by pumping Mn into the ER by reducing cytosolic Mn (Wu et al. 2002). Recent report on transgenic Arabidopsis overexpressing *GmMTP8.1* showed less Mn accumulation in shoots but not in roots, compared to non-transgenic or wild type under excess Mn conditions, indicating participation of GmMTP8.1 in the extrusion of Mn from leaf cells (Li et al. 2021a, b). Figure 14.2 depicts localization and mode of action of some MTP members involved in metal(loid) transport and homeostasis in plant cells.

In Arabidopsis, another Golgi-localized transporter, the P2A-type ATPase *AtECA3* (*Arabidopsis thaliana* ENDOPLASMIC RETICULUM-TYPE CALCIUM-TRANSPORTING ATPase) is also responsible for maintaining supply of Mn and Ca ions by importing them into the Golgi vesicles (Peiter et al. 2007). Similar to ECA3, AtMTP11 is localized in both Golgi (Peiter et al. 2007) and the PVC (Delhaize et al. 2007; Peiter et al. 2007). Contrary to AtECA3, which is required for growth under low-Mn conditions, AtMTP11 is needed to tolerate under high-Mn condition (Mills et al. 2008). Both the AtMTP11and AtECA3 are believed to play a critical role in Mn homeostasis by transporting Mn into the Golgi apparatus that finally travel via exocytosis of secretory vesicles and get released into extracellular space maintaining the Mn ion homeostasis (Peiter et al. 2007; Mills et al. 2008). Importantly, mutation in AtMTP11 in Arabidopsis results in hypersensitivity to elevated Mn levels, whereas overexpression of AtMTP11 showed hypertolerance to Mn. Interestingly, increased tolerance was not detected in the presence of other metals such as Zn, Co, or Ni, further highlighting the importance of AtMTP11 in Mn homeostasis in Arabidopsis (Delhaize et al. 2007; Peiter et al. 2007).

Moreover, two poplar homologs, that is, PtMTP11.1 and PtMTP11.2, are assumed to function similarly since they complement the Mn-sensitive yeast mutant *pmr1* by targeting to Golgi compartments, similar to AtMTP11 (Peiter et al. 2007). Likewise, the BmMTP10 and BmMTP11 proteins from *Beta vulgaris* spp. *maritima* demonstrated similar *in planta* localization and Mn detoxification activity to that of Arabidopsis homolog AtMTP11, indicating evolutionary significance of these proteins in heavy metal homeostasis among plant species. Similarly, another group 11 member from rice, OsMTP11, is responsible for Mn homeostasis and tolerance in rice. However, the localization of OsMTP11 is yet unclear, as some studies on rice protoplasts and tobacco epidermal cells suggest it to be localized to trans-Golgi network (TGN), whereas during high extracellular Mn concentration, OsMTP11 has been reported to be relocalized to the plasma membrane (Ma et al. 2018). Although expression of OsMTP11 is induced by high Mn, epigenetic factors such as DNA methylation have also been found to play a significant role in regulating the

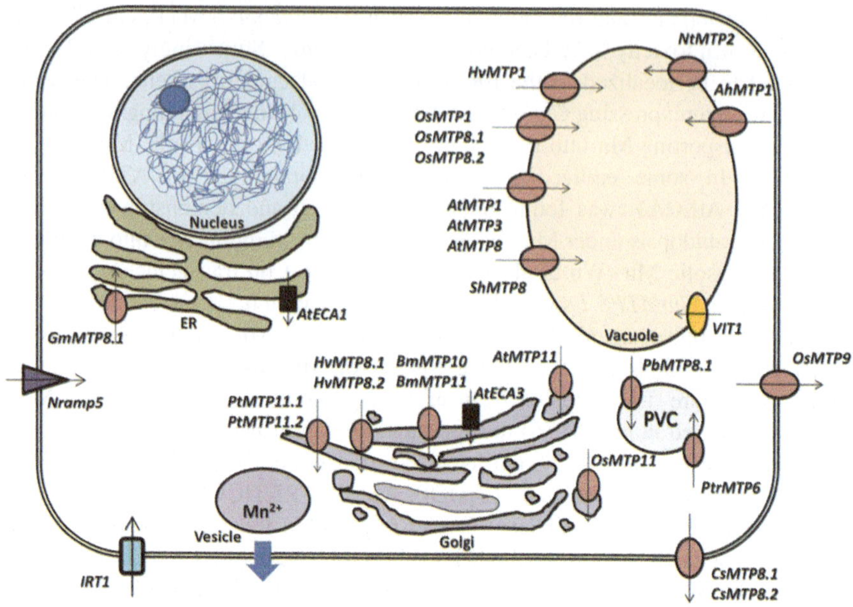

Fig. 14.2 Diagrammatic representation of localization and mode of action of different MTP members involved in metal(loid) transport and homeostasis in plant cells. The tonoplast localized MTP members (AhMTP1, AtMTP1, AtMTP3, AtMTP8, HvMTP1, OsMTP1, OsMTP8.1/OsMTP8.2, ShMTP8, NtMTP2) play a key role by means of intracellular metal sequestration into the vacuole that serves as a major reservoir to regulate cellular metal homeostasis. Furthermore, cytosolic metal(loid)s are sequestered into Golgi compartments or Trans-Golgi network (TGN) by specific MTP transporters localized in Golgi (HvMTP8.1/HvMTP8.2, PtMTP11.1/PtMTP11.2, BmMTP10, BmMTP11) and TGN (AtMTP11, OsMTP11). Similarly, endoplasmic reticulum (ER)-localized (GmMTP8.1) and pre-vacuolar compartment (PVC)-localized (PbMTP8.1, PtrMTP6) also contribute in metal(loid) homeostasis by reducing the concentration of cytosolic metal(loid)s. Contrary to these endomembrane-bound MTP transporters, the plasma membrane-bound members (OsMTP9, CsMTP8.1, CsMTP8.2) efflux the cytosolic metal(loid)s out of the cells to reduce toxicity and maintain homeostasis. A few other transporters (Nramp5, AtECA3, IRT1, VIT1) playing significant role in conjunction with MTP members in cellular metal homeostasis are depicted in this schematic diagram

OsMTP11 expression (Zhang and Liu 2017). Recently, tobacco MTP localized in tonoplast, NtMTP2, has been characterized as the housekeeping protein that controls the basal availability of micronutrients and when expressed in yeast shows tolerance to Co and Ni by sequestering excess metal into vacuole specifically in leaves (Papierniak-Wygladala et al. 2020). When *NtMTP2* expressed in yeast, it conferred transport and tolerance exclusively for Co and Ni but not for other metals such as Zn, Mn, Cu, Fe, or Cd.

14.2.4 MTP Member Assures Mn Homeostasis During Seed Development and Germination

Besides its role in Mn detoxification, *AtMTP8* plays a critical role in seed vigor as this protein expressed in embryos of *Arabidopsis* seeds (Eroglu et al. 2016). In *Arabidopsis*, vacuolar iron transporter (VIT1) transfers Fe from the cytosol to the vacuole for intracellular Fe storage. The embryos lacking the expression of AtVIT1 were still found to build up Fe hotspots in MTP8-expressing cell types, suggesting the role of MTP8 in Fe transport as well. It has been found that *AtMTP8* expression is specific in root epidermis and the cortex cells and strongly induced by Fe deficiency (Eroglu et al. 2016). The early phase of seed germination, that is, during imbibition, seeds take up Mn from the environment, and MTP8 sequesters excess Mn preventing the seeds from Mn overflows, thereby protecting the embryos from cytotoxicity (Eroglu et al. 2017). Consequently, AtMTP8, helps in establishment of plants by facilitating the cell type-specific vacuolar loading of Mn and Fe in developing embryos during seed imbibition (Eroglu et al. 2017).

14.3 Potential of MTP in Biotechnological Application

One of the promising strategies for enhancement of micronutrient contents in seeds or grains for human consumption is biofortification through transgenic expression of different MTPs along with other genes. To function optimally, the human body needs several micronutrients such as Fe, Zn, iodine (I), and selenium (Se) under both normal and disease conditions. Among different essential micronutrients, one of the most important one is Zn, which is found in small amounts in some food items, especially in cereals. Inadequate intake of which could cause deficiency syndromes. Since Zn cannot be stored in the human body, a regular dietary intake of this micronutrient is the only way for maintaining the recommend amounts in the human body. Thus, people following only the plant-based diets may suffer from zinc deficiencies in higher incidence. The root to shoot barrier in plants and the grain filling process in cereals seem to be the major hindrance in the biofortification. Several new findings reveal that MTP family members play a major role in the root to shoot distribution of Zn, and thus endosperm-specific overexpression of *MTP* has been proposed as a suitable candidate for biofortification of rice with Zn (Ricachenevsky et al. 2013). Interestingly, the $H_{90}D$ mutation in rice OsMTP1 abolished Zn transport and enhanced Fe tolerance (Menguer et al. 2013) allowing OsMTP1 to transport Fe along with Zn. This observation clearly establishes OsMTP1 as a suitable candidate for biofortification approaches (Kawachi et al. 2012; Podar et al. 2012) as simultaneous enhancement of Fe and Zn concentrations would be highly desired strategy to combat deficiency of these minerals in humans. Even from QTL localization study, the *OsMTP1* was considered as a high priority candidate gene for enrichment of Fe and Zn concentrations in seeds (Anuradha et al. 2012).

In *Arabidopsis*, MTP8 not only sequesters Mn but is also identified as critical determinant for Fe deficiency tolerance, which makes it an attractive candidate gene for biofortification purpose to enhance the micronutrient content in seeds (White and Broadley 2009; Bouis et al. 2011; Ricachenevsky et al. 2013; Vatansever et al. 2017). Although recent studies on overexpression of *MTP8* did not cause a consistent increase in Mn and Fe accumulation but in the bulk samples, Zn concentrations were consistently increased (Höller et al. 2022).

Transgenic or gene editing techniques are thought to be the most powerful path for phytoremediation, as several scientific efforts carried out in recent years indicate that generation of hyperaccumulators by these advanced tools using suitable transporter proteins and regulators might help toward remediation of heavy metal contaminated soil. Long-distance metal transport by phloem tissue is an important strategy of plants toward shoot protection. Previous studies in *Arabidopsis* supported this concept as heavy metal, like Cd, is reported to be retranslocated to other tissue via the phloem, and this distribution is considered to be a part of the excluder strategy (Van Belleghem et al. 2007). Recent studies in *Populus trichocarpa* showed that the *PtrMTP6* was mainly expressed in the phloem tissue of stems and roots and overexpressing *PtrMTP6* increased Mn and Co accumulation in young tissues (Yang et al. 2021). Results from this study suggest that poplar might be suitable plant species for bioremediation of ecosystems affected by the industries like electric vehicle and battery production. Plants overexpressing certain MTP genes may lead to the production of elevated amounts of thiol compounds and chelating compounds like phytochelatins (PC), therefore forming complex with metals, which enable efficient sequestration of toxic metal ions in the vacuoles. This strategy helps plants to combat metal stress, and this pathway is exploited for hyperaccumulation of toxic ions within plant biomass. It was reported that upon external stress with cadmium (Cd) and arsenic (As), overexpression of OsMTP1 gene from indica rice (*Oryza sativa* L. cv. IR64) in tobacco accelerated hyperaccumulation of Cd and enhanced tolerance as well as accumulation of As (Das et al. 2016). Since OsMTP1 study suggests for its broad substrate specificity, this gene can be employed in suitable host plants for phytoremediation purpose in multi-contaminated sites.

14.4 Future Prospects

The plant MTPs comprise a family of membrane proteins with the C-terminal cation efflux domain that transports divalent cations such as Mn^{2+}, Zn^{2+}, Fe^{2+}, Cd^{2+} Co^{2+}, Ni^{2+}, Cu^{2+}, etc. across the biological membrane. Although the MTP transporters are postulated to play a critical role in maintenance of plant mineral nutrition and heavy metal stress resistance, but the biological functions of several MTP members in diverse plant species still remain elusive. In the last few decades, bioinformatics technique has emerged as one of the convenient and useful ways to investigate different interactions and function of specific gene families. Nowadays, due to the advent of modern sequencing technology, sequencing of entire genome of important plants can be completed enabling classification and comparative genomics.

However, genome-wide identification study on different plant species including crop plants can be carried out to identify the MTP gene family members and interpret their in silico study derived results. Due to the availability of genome sequences, recently a number of MTP proteins have been identified through genome-wide analysis in some important plants, like *Triticum aestivum* and *Citrus sinensis* to name a few (Vatansever et al. 2017; Fu et al. 2017). Through bioinformatics analyses, information on phylogenetic classification, gene ontology, gene structure, and transcriptomic data of MTP family members from different plant species have been assembled, which are also providing a basis for the analysis of the mechanism and function of uncharacterized MTP proteins. With respect to possible posttranscriptional control mechanisms, potential targets of miRNAs have also been investigated through bioinformatics that corresponds to several MTPs. The results accumulated through in silico studies will provide important hints and set down a theoretical foundation in clarifying the roles of MTPs in heavy metal(loid) tolerance and characterization for future biotechnological applications.

References

Anuradha K, Agarwal S, Rao YV, Rao KV, Viraktamath BC, Sarla N (2012) Mapping QTLs and candidate genes for iron and zinc concentrations in unpolished rice of Madhukar×Swarna RILs. Gene 508(2):233–240. https://doi.org/10.1016/j.gene.2012.07.054

Arrivault S, Senger T, Krämer U (2006) The Arabidopsis metal tolerance protein AtMTP3 maintains metal homeostasis by mediating Zn exclusion from the shoot under Fe deficiency and Zn oversupply. Plant J 46(5):861–879. https://doi.org/10.1111/j.1365-313X.2006.02746.x

Blaudez D, Kohler A, Martin F, Sanders D, Chalot M (2003) Poplar metal tolerance protein 1 confers zinc tolerance and is an oligomeric vacuolar zinc transporter with an essential leucine zipper motif. Plant Cell 15(12):2911–2928. https://doi.org/10.1105/tpc.017541

Bouis HE, Hotz C, McClafferty B, Meenakshi JV, Pfeiffer WH (2011) Biofortification: a new tool to reduce micronutrient malnutrition. Food Nutr Bull 32(1):31–40. https://doi.org/10.1177/15648265110321S105

Chao Y, Fu D (2004) Kinetic study of the antiport mechanism of an Escherichia coli zinc transporter. J Biol Chem 279(13):12043–12050. https://doi.org/10.1074/jbc.M313510200

Chen M, Shen X, Li D, Ma L, Dong J, Wang T (2009) Identification and characterization of MtMTP1, a Zn transporter of CDF family, in the Medicago truncatula. Plant Physiol Biochem 47(11–12):1089–1094. https://doi.org/10.1016/j.plaphy.2009.08.006

Chen Z, Fujii Y, Yamaji N, Masuda S, Takemoto Y, Kamiya T, Yusuyin Y, Iwasaki K, Kato S, Maeshima M, Ma JF, Ueno D (2013) Mn tolerance in rice is mediated by MTP8.1, a member of the cation diffusion facilitator family. J Exp Bot 64(14):4375–4387. https://doi.org/10.1093/jxb/ert243

Chu HH, Car S, Socha AL, Hindt MN, Punshon T, Guerinot ML (2017) The Arabidopsis MTP8 transporter determines the localization of manganese and iron in seeds. Sci Rep 7(1):11024. https://doi.org/10.1038/s41598-017-11250-9

Cointry V, Vert G (2019) The bifunctional transporter-receptor IRT 1 at the heart of metal sensing and signalling. New Phytol 223(3):1173–1178. https://doi.org/10.1111/nph.15826

Das N, Bhattacharya S, Maiti MK (2016) Enhanced cadmium accumulation and tolerance in transgenic tobacco overexpressing rice metal tolerance protein gene OsMTP1 is promising for phytoremediation. Plant Physiol Biochem 105:297–309. https://doi.org/10.1016/j.plaphy.2016.04.049

Delhaize E, Kataoka T, Hebb DM, White RG, Ryan PR (2003) Genes encoding proteins of the cation diffusion facilitator family that confer manganese tolerance. Plant Cell 15(5):1131–1142. https://doi.org/10.1105/tpc.009134

Delhaize E, Gruber BD, Pittman JK, White RG, Leung H, Miao Y, Jiang L, Ryan PR, Richardson AE (2007) A role for the AtMTP11 gene of Arabidopsis in manganese transport and tolerance. Plant J 51(2):198–210. https://doi.org/10.1111/j.1365-313X.2007.03138.x

Desbrosses-Fonrouge AG, Voigt K, Schröder A, Arrivault S, Thomine S, Krämer U (2005) Arabidopsis thaliana MTP1 is a Zn transporter in the vacuolar membrane which mediates Zn detoxification and drives leaf Zn accumulation. FEBS Lett 579(19):4165–4174. https://doi.org/10.1016/j.febslet.2005.06.046

Emamverdian A, Ding Y, Mokhberdoran F, Xie Y (2015) Heavy metal stress and some mechanisms of plant defense response. Sci World J 2015:756120. https://doi.org/10.1155/2015/756120

Erbasol I, Bozdag GO, Koc A, Pedas P, Karakaya HC (2013) Characterization of two genes encoding metal tolerance proteins from Beta vulgaris subspecies maritima that confers manganese tolerance in yeast. Biometals 26(5):795–804. https://doi.org/10.1007/s10534-013-9658-7

Eroglu S, Meier B, von Wirén N, Peiter E (2016) The vacuolar manganese transporter MTP8 determines tolerance to iron deficiency-induced chlorosis in Arabidopsis. Plant Physiol 170(2): 1030–1045. https://doi.org/10.1104/pp.15.01194

Eroglu S, Giehl RFH, Meier B, Takahashi M, Terada Y, Ignatyev K, Andresen E, Küpper H, Peiter E, von Wirén N (2017) Metal tolerance protein 8 mediates manganese homeostasis and iron reallocation during seed development and germination. Plant Physiol 174(3):1633–1647. https://doi.org/10.1104/pp.16.01646

Fu XZ, Tong YH, Zhou X, Ling LL, Chun CP, Cao L et al (2017) Genome-wide identification of sweet orange (Citrus sinensis) metal tolerance proteins and analysis of their expression patterns under zinc, manganese, copper, and cadmium toxicity. Gene 629:1–8. https://doi.org/10.1016/j.gene.2017.07.072

Grass G, Otto M, Fricke B, Haney CJ, Rensing C, Nies DH, Munkelt D (2005) FieF (YiiP) from Escherichia coli mediates decreased cellular accumulation of iron and relieves iron stress. Arch Microbiol 183(1):9–18. https://doi.org/10.1007/s00203-004-0739-4

Guffanti AA, Wei Y, Rood SV, Krulwich TA (2002) An antiport mechanism for a member of the cation diffusion facilitator family: divalent cations efflux in exchange for K+ and H+. Mol Microbiol 45(1):145–153. https://doi.org/10.1046/j.1365-2958.2002.02998.x

Höller S, Küpper H, Brückner D, Garrevoet J, Spiers K, Falkenberg G, Andresen E, Peiter E (2022) Overexpression of metal tolerance protein8 reveals new aspects of metal transport in Arabidopsis thaliana seeds. Plant Biol 24(1):23–29. https://doi.org/10.1111/plb.13342

Hou L, Gu D, Li Y, Li J, Li J, Chen X, Zhang W (2019) Phylogenetic and expression analysis of Mn-CDF transporters in pear (Pyrus bretschneideri Rehd.). Plant Mol Biol Report 37(1–2): 98–110. https://doi.org/10.1007/s11105-019-01142-9

Kawachi M, Kobae Y, Mimura T, Maeshima M (2008) Deletion of a histidine-rich loop of AtMTP1, a vacuolar Zn(2+)/H(+) antiporter of Arabidopsis thaliana, stimulates the transport activity. J Biol Chem 283:8374–8383. https://doi.org/10.1074/jbc.M707646200

Kawachi M, Kobae Y, Kogawa S, Mimura T, Krämer U, Maeshima M (2012) Amino acid screening based on structural modeling identifies critical residues for the function, ion selectivity and structure of Arabidopsis MTP1. FEBS J 279(13):2339–2356. https://doi.org/10.1111/j.1742-4658.2012.08613.x

Kobae Y, Uemura T, Sato MH, Ohnishi M, Mimura T, Nakagawa T, Maeshima M (2004) Zinc transporter of Arabidopsis thaliana AtMTP1 is localized to vacuolar membranes and implicated in zinc homeostasis. Plant Cell Physiol 45(12):1749–1758. https://doi.org/10.1093/pcp/pci015

Lan HX, Wang ZF, Wang QH, Wang MM, Bao YM, Huang J et al (2012) Characterization of a vacuolar zinc transporter OZT1 in rice (Oryza sativa L.). Mol Biol Rep 40:1201–1210. https://doi.org/10.1007/s11033-012-2162-2

León-Mediavilla J, Senovilla M, Montiel J, Gil-Díez P, Saez Á, Kryvoruchko IS, Reguera M, Udvardi MK, Imperial J, González-Guerrero M (2018) MtMTP2-facilitated zinc transport into intracellular compartments is essential for nodule development in Medicago truncatula. Front Plant Sci 9:990. https://doi.org/10.3389/fpls.2018.00990

Li Q, Li Y, Wu X, Zhou L, Zhu X, Fang W (2017) Metal transport Protein 8 in Camellia sinensis confers superior manganese tolerance when expressed in yeast and Arabidopsis thaliana. Sci Rep 7(1):39915. https://doi.org/10.1038/srep39915

Li X, Wu Y, Li B, He W, Yang Y, Yang Y (2018) Genome-wide identification and expression analysis of the cation diffusion facilitator gene family in turnip under diverse metal ion stresses. Front Genet 9:103. https://doi.org/10.3389/fgene.2018.00103

Li J, Dong R, Jia Y, Huang J, Zou X, An N, Song J, Chen Z (2021a) Characterization of metal tolerance proteins and functional analysis of GmMTP8.1 involved in manganese tolerance in soybean. Front Plant Sci 12:683813. https://doi.org/10.3389/fpls.2021.683813

Li J, Zheng L, Fan Y, Wang Y, Ma Y, Gu D, Lu Y, Zhang S, Chen X, Zhang W (2021b) Pear metal transport protein PbMTP8.1 confers manganese tolerance when expressed in yeast and Arabidopsis thaliana. Ecotoxicol Environ Saf 208:111687. https://doi.org/10.1016/j.ecoenv.2020.111687

Liu J, Gao Y, Tang Y, Wang D, Chen X, Yao Y, Guo Y (2019) Genome-wide identification, comprehensive gene feature, evolution, and expression analysis of plant metal tolerance proteins in tobacco under heavy metal toxicity. Front Genet 10:345. https://doi.org/10.3389/fgene.2019.00345

Ma G, Li J, Li J, Li Y, Gu D, Chen C, Cui J, Chen X, Zhang W (2018) OsMTP11, a trans-Golgi network localized transporter, is involved in manganese tolerance in rice. Plant Sci 274:59–69. https://doi.org/10.1016/j.plantsci.2018.05.011

Marschner P, Rengel Z (2012) Nutrient availability in soils. In: Marschner's mineral nutrition of higher plants. Academic, New York, pp 315–330

Menguer PK, Farthing E, Peaston KA, Ricachenevsky FK, Fett JP, Williams LE (2013) Functional analysis of the rice vacuolar zinc transporter OsMTP1. J Exp Bot 64:2871–2883. https://doi.org/10.1093/jxb/ert136

Mills RF, Doherty ML, López-Marqués RL, Weimar T, Dupree P, Palmgren MG, Pittman JK, Williams LE (2008) ECA3, a Golgi-localized P2A-type ATPase, plays a crucial role in manganese nutrition in Arabidopsis. Plant Physiol 146(1):116–128. https://doi.org/10.1104/pp.107.110817

Montanini B, Blaudez D, Jeandroz S, Sanders D, Chalot M (2007) Phylogenetic and functional analysis of the cation diffusion facilitator (CDF) family: improved signature and prediction of substrate specificity. BMC Genomics 8(1):107. https://doi.org/10.1186/1471-2164-8-107

Narayan OP, Verma N, Jogawat A, Dua M, Johri AK (2020) Role of sulphate transporter (PiSulT) of endophytic fungus Serendipita indica in plant growth and development. bioRxiv. https://doi.org/10.1101/2020.01.07.897710

Nies DH, Silver S (1995) Ion efflux systems involved in bacterial metal resistances. J Ind Microbiol Biotechnol 14:186–199. https://doi.org/10.1007/BF01569902

Papierniak-Wygladala A, Kozak K, Barabasz A, Palusińska M, Całka M, Maślińska K, Antosiewicz DM (2020) Identification and characterization of a tobacco metal tolerance protein, NtMTP2. Metallomics 12(12):2049–2064. https://doi.org/10.1039/d0mt00210k

Pedas P, Schiller Stokholm MS, Hegelund JN, Ladegård AH, Schjoerring JK, Husted S (2014) Golgi localized barley MTP8 proteins facilitate Mn transport. PLoS ONE 9(12):e113759. https://doi.org/10.1371/journal.pone.0113759

Peiter E, Montanini B, Gobert A, Pedas P, Husted S, Maathuis FJ, Blaudez D, Chalot M, Sanders D (2007) A secretory pathway-localized cation diffusion facilitator confers plant manganese tolerance. Proc Natl Acad Sci U S A 104(20):8532–8537. https://doi.org/10.1073/pnas.0609507104

Podar D, Scherer J, Noordally Z, Herzyk P, Nies D, Sanders D (2012) Metal selectivity determinants in a family of transition metal transporters. J Biol Chem 287(5):3185–3196. https://doi.org/10.1074/jbc.M111.305649

Ricachenevsky FK, Menguer PK, Sperotto RA, Williams LE, Fett JP (2013) Roles of plant metal tolerance proteins (MTP) in metal storage and potential use in biofortification strategies. Front Plant Sci 4:144. https://doi.org/10.3389/fpls.2013.00144

Shahzad Z, Gosti F, Frérot H, Lacombe E, Roosens N, Saumitou-Laprade P, Berthomieu P (2010) The five AhMTP1 zinc transporters undergo different evolutionary fates towards adaptive evolution to zinc tolerance in Arabidopsis halleri. PLoS Genet 6(4):e1000911. https://doi.org/10.1371/journal.pgen.1000911

Takemoto Y, Tsunemitsu Y, Fujii-Kashino M, Mitani-Ueno N, Yamaji N, Ma JF, Kato SI, Iwasaki K, Ueno D (2017) The tonoplast-localized transporter MTP8.2 contributes to manganese detoxification in the shoots and roots of Oryza sativa L. Plant Cell Physiol 58(9): 1573–1582. https://doi.org/10.1093/pcp/pcx082

Ueno D, Sasaki A, Yamaji N, Miyaji T, Fujii Y, Takemoto Y, Moriyama S, Che J, Moriyama Y, Iwasaki K, Ma JF (2015) A polarly localized transporter for efficient manganese uptake in rice. Nat Plants 1:15170. https://doi.org/10.1038/nplants.2015.170

Van Belleghem F, Cuypers A, Semane B, Smeets K, Vangronsveld J, d'Haen J, Valcke R (2007) Subcellular localization of cadmium in roots and leaves of Arabidopsis thaliana. New Phytol 173(3):495–508. https://doi.org/10.1111/j.1469-8137.2006.01940.x

van der Zaal BJ, Neuteboom LW, Pinas JE, Chardonnens AN, Schat H, Verkleij JA, Hooykaas PJ (1999) Overexpression of a novel Arabidopsis gene related to putative zinc-transporter genes from animals can lead to enhanced zinc resistance and accumulation. Plant Physiol 119(3): 1047–1055. https://doi.org/10.1104/pp.119.3.1047

Vatansever R, Filiz E, Froglu S (2017) Genome-wide exploration of metal tolerance protein (MTP) genes in common wheat (Triticum aestivum): insights into metal homeostasis and biofortification. Biometals 30:217–235. https://doi.org/10.1007/s10534-017-9997-x

White PJ, Broadley MR (2009) Biofortification of crops with seven mineral elements often lacking in human diets—iron, zinc, copper, calcium, magnesium, selenium and iodine. New Phytol 182(1):49–84. https://doi.org/10.1111/j.1469-8137.2008.02738.x

Wu Z, Liang F, Hong B, Young JC, Sussman MR, Harper JF, Sze H (2002) An endoplasmic reticulum-bound Ca(2+)/Mn(2+) pump, ECA1, supports plant growth and confers tolerance to Mn(2+) stress. Plant Physiol 130(1):128–137. https://doi.org/10.1104/pp.004440

Yadav B, Jogawat A, Lal SK, Lakra N, Mehta S, Shabek N, Narayan OP (2021) Plant mineral transport systems and the potential for crop improvement. Planta 253(2):45. https://doi.org/10.1007/s00425-020-03551-7

Yang F, Gao Y, Liu J, Chen Z, de Dios VR, Gao Q, Zhang M, Peng Z, Yao Y (2021) Metal tolerance protein MTP6 is involved in Mn and Co distribution in poplar. Ecotoxicol Environ Saf 226:112868. https://doi.org/10.1016/j.ecoenv.2021.112868

Yuan L, Yang S, Liu B, Zhang M, Wu K (2012) Molecular characterization of a rice metal tolerance protein, OsMTP1. Plant Cell Rep 31(1):67–79. https://doi.org/10.1007/s00299-011-1140-9

Zhang M, Liu B (2017) Identification of a rice metal tolerance protein OsMTP11 as a manganese transporter. PLoS ONE 12(4):e0174987. https://doi.org/10.1371/journal.pone.0174987

Zhang X, Li Q, Xu W, Zhao H, Guo F, Wang P, Wang Y, Ni D, Wang M, Wei C (2020) Identification of MTP gene family in tea plant (Camellia sinensis L.) and characterization of CsMTP8.2 in manganese toxicity. Ecotoxicol Environ Saf 202:110904. https://doi.org/10.1016/j.ecoenv.2020.110904

Co-Transport Mechanism in Plants for Metals and Metalloids

15

Varun Kumar and Indraneel Sanyal

Abstract

Metals and metalloids are important ingredients for the regular growth and development of plants. The uptake of various ions from surroundings take place with the aid of different types of transporters. The presence of excess levels of certain metals and metalloids are viewed to be major environmental contaminants. Such pollutants cause imbalances in the redox metabolism, which leads to the generation of reactive oxygen species (ROS) in cellular compartments and cause oxidative damage to cell biomolecules, metabolic activities, and limits plant development. To avoid such negative repercussions of ROS, plants have developed several defensive systems, which shield the internal cellular components of plants and maintain ion homeostasis. Controlling the absorption, transport, and translocation of these metals and metalloids in plants cells take place with the aid of several transporters, such as CDF, Lsi, YSL, HMAs, ZIP, NRAMP, ABC, and aquaglyceroporin. These transporters are distributed in different parts of plant cells and organelles including plasma membrane, vacuoles, mitochondria, tonoplast, chloroplasts and peroxisomes. These transporters perform essential roles in entrance, distribution, and homeostasis of various metal ions inside the cells and organelles of the plant. Many metals and metalloid transport themselves from more than one transporters, due

V. Kumar
Plant Transgenic Laboratory, Molecular Biology and Biotechnology Division, CSIR-National Botanical Research Institute, Lucknow, India

I. Sanyal (✉)
Plant Transgenic Laboratory, Molecular Biology and Biotechnology Division, CSIR-National Botanical Research Institute, Lucknow, India

Academy of Scientific and Innovative Research (AcSIR), Ghaziabad, India
e-mail: i.sanyal@nbri.res.in

K. Kumar, S. Srivastava (eds.), *Plant Metal and Metalloid Transporters*, https://doi.org/10.1007/978-981-19-6103-8_15

to their similar structural properties like aquaglyceroporins, phosphate transporters, and hexose transporters. In this chapter, the present state of knowledge regarding these metals and metalloid transporters function, molecular mechanisms, biofortification, phytoremediation, and phytomining activities have been discussed. The purpose of this chapter is to present an overview of roles and relevance of such transporters in plant growth and development.

Keywords

Co-transport · ABC transporter · ZIP · HMA · NRAMP

15.1 Introduction

Plants not only need light but also water and inorganic elements to support their growth and development. During evolution, plants have developed specialized structures and functions that provide enough access for each essential chemical element for their survival, breeding, and equal distribution. For all vascular plants, 17 elements are classified into two main groups depending on whether they are required in low amounts (usually below 100 mg/kg DW), the micronutrients, or if the concentration required at higher strengths (more than 1000 mg/kg DW), the macronutrients. Every live cell uses metals in enzymes that catalyze cellular metabolism, respiration, and transcription factors for different chemical reactions. Typically, these transition element cations are classified as manganese (Mn), iron (Fe), cobalt (Co), nickel (Ni), copper (Cu), zinc (Zn), molybdenum (Mo), and tungsten (W) and form semi-covalent and covalent bonds with a specific electron-donating functional group in protein or small organic molecules, thus stabilizing certain folds and imparting chemical properties that are absent from the nature's amino acid repertoire (Bozzi and Gaudet 2021). The most widely used transition metal in a biological system is iron which efficiently cycles between Fe^{2+} and Fe^{3+} oxidation state, making it a perfect co-factor for the redox (electron transfer) reaction and is therefore vital for all of the known organisms. Furthermore, iron-sulfur clusters and iron-containing cytochromes provide the means for electron transport to mitochondria that enable efficient ATP synthesis; for example, globin proteins are used to transport and store oxygen. Copper and manganese are also widely used for biological redox processes, with manganese co-factors well known for supporting water oxidation by photosystem II, to diatomic oxygen. On the other hand, redox-inactive zinc often has a structural function in many proteins and functions as Lewis acid in enzyme active sites (Archibald 1983; Posey and Gherardini 2000; Cassat and Skaar 2013). Like other species, to maintain optimum growth, plants must precisely regulate the intake, allocation, and storage of these essential metals. On the one hand, the shortage of these metals results in growth delays and developmental problems throughout the plant's life cycle, and also overaccumulation of these heavy metals would lead to toxicity. Therefore cells have to control the homeostasis of metal ions to prevent such toxicity, and this is carried out for closely monitored methods of absorption, storage, and secretion. Non-essential toxic heavy metals exist in natural soils, and discharge of toxic heavy metals generally accompanies volcanic eruptions (Song et al. 2014). The anthropogenic discharge of heavy metals has considerably increased during the industrialized phase of the late nineteenth and early twentieth centuries. Our environment has been damaged by mining, garbage incinerators, pesticides, and fertilizers. Besides, water wells first created in Bangladesh and India to supply clean water for the population were subsequently found to contain a deadly amount of arsenic, which lead to the pollution of rice fields (Zhao et al. 2010; Meharg and Rahman 2003; Williams et al. 2006). Non-essential heavy metals may grab the transporters required for the uptake of essential heavy metals because of chemical similarity between the two metal species, and consequently they enter and accumulate in the plant cell. The iron and zinc transporter IRT1 has been demonstrated to be the main entry point of hazardous cadmium, while high-affinity phosphate transporters and members of the MIP family import arsenate (V) and arsenite (III), respectively (Vert et al. 2001; Zhao et al. 2010).

The low solubility of metal ions in oxygenated water makes them limitable to most biological systems and creates a thermodynamic barrier for organisms, as it is entropically unfavorable to concentrate a few chemical resources within a small membrane-bound compartment (Kolber et al. 1994). To avoid this barrier, evolution has provided two separate mechanisms, identified as primary and secondary active transport. In primary active transport, membrane-bound proteins bind and translocate the requested substrates across the lipid bilayer by joining or "connecting" the transport process to an unrelated energetically beneficial chemical activity such as ATP hydrolysis. ATP is also consumed by primary active transport in order to determine common ion gradients like Na^+, K^+, H^+, and Cl^-. When coupled with selective membrane permeability by ion-specific channels, these gradients lead to a net charge separation or membrane potential, typically negative intracellularly (Bozzi and Gaudet 2021).

The term "metalloid" is usually used to indicate a chemical element with physical and chemical characteristics, a transition between metals and non-metals (Coyle et al. 2002). Common characteristics of metalloids include lustrous, brittle solids with intermediate to excellent electrical conductivity. They mainly form amphoteric to low acid oxides and have nearly 2.0% electronegativity and about 200 kcal/mol ionization energy. Arsenic (As), antimony (Sb), boron (B), germanium (Ge), silicon (Si), and tellurium (Te) are the six elements that are commonly identified as metalloids. Excessive levels of metalloids in soil, water, or air environment results in their entry into the food chain (Dhanker et al. 2012). The natural surplus of these metalloids is significantly different. Silicon is the second largest element behind oxygen in the Earth's crust and constitutes approximately 28% by weight of the crust; instead, tellurium is found in trace amounts of $1\times10^{-7}\%$ (Mukhopadhyay et al. 2014). In the biological system, the role of metalloids ranges from essential to toxic and influences biological systems, and they must move across the cellular membrane and accumulate in cells (De Carvalho 2008).

In recent years the factors regulating metal and metalloid transportation across cell membranes, intracellular homeostasis, and cell regulatory responses to the

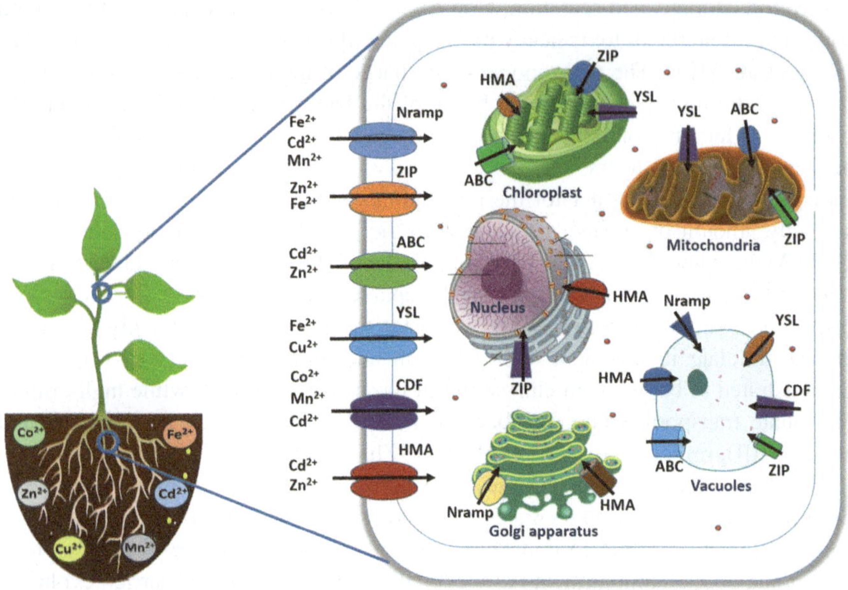

Fig. 15.1 Diagrammatic illustration of the uptake and movement of metals and metalloids in plants through transporters

changed environmental supply of metal ions such as Cu^{2+}, Mn^{2+}, Fe^{2+}, and Zn^{2+} have been the subject of many studies of particular interest, due to their vital role in cell metabolism as co-factors of many enzymes. Its intracellular concentration is usually kept at a modest physiological level, which is slightly constant. Toxic metals such as Cd^{2+}, Co^{2+}, and Ni^{2+} could offer a significant health concern to animals and could check plant growth, mainly because they prevent the transport, homeostasis, or function of the essential metals (Hediger 1997; Eide 1998; Nelson 1999; Radisky and Kaplan 1999). In order to overcome their deleterious effects, cells have evolved various strategies, among which transport or sequestration into organelles and binding by thiols are the most prominent.

A panel of specific but relatively well-conserved transporters, channels, and pumps have evolved, allowing plants to adjust their mineral uptake, translocation, and distribution capacities in response to environmental fluctuations. These mineral transportation activities involve a variety of protein-reliant mechanisms, including natural resistance-associated macrophage protein (NRAMP) transporter, ABC transporters, zinc-regulated transporter (ZRT), IRT-like (ZIP) family of proteins, P1B-like ATPase heavy metal (HMA) proteins, and cation diffusion facilitator (CDF) family of transporters, as well as yellow stripe-like (YSL) transporters (Fig. 15.1).

15.2 Cation Diffusion Facilitators (CDF) Transporter

The CDF family, also known as the cation efflux family (Lang et al. 2011), was described for the first time by Nies and Silver in Nies and Silver 1995. The protein members of the CDF family have been identified in all six kingdoms of life (Migocka et al. 2015a) and are involved in trace element transport and tolerance (Kim et al. 2004). Members of the CDF family contain three distinguishing features: an N-terminal signature sequence, a cation efflux domain, and around six predicted transmembrane domains (Paulsen and Saier 1997). Between transmembrane domains 4 and 5, most eukaryotic CDFs have a histidine (His)-rich cytoplasmic loop (Paulsen and Saier 1997). Plant CDF family members have been recognized as metal tolerance proteins, MTPs (Migocka et al. 2015b). In *Arabidopsis* and rice, the MTP family includes 12 and 10 members, respectively. Unfortunately, their roles are still poorly understood (Migocka et al. 2015b). MTP proteins from various plant species have recently been classified into seven groups based on the results of phylogenetic analysis and annotation of *Arabidopsis* MTPs (Gustin et al. 2011). Groups 1, 5, and 12 belong to the Zn-CDFs, groups 6 and 7 to the Fe/Zn-CDFs, and groups 8 and 9 to the Mn-CDF cluster (Gustin et al. 2011). MTP1–MTP4 members of group 1, MTP8 proteins of group 8, and MTP9–MTP11 members of group 9 are the most researched MTP proteins in plants so far (Migocka et al. 2015b). The majority of these MTPs have been functionally described in a variety of plant species. Although group 1 of plant MTPs are related to Zn-CDFs, its members are capable of transporting a variety of metals into the vacuole of plant cells, including Zn, Cd, Co, Ni, or Fe (Kim et al. 2004; Xu et al. 2009; Lang et al. 2011; Menguer et al. 2013; Migocka et al. 2015a). MTP5 has recently been revealed to form a functional compound with MTP12 to transport Zn into the Golgi apparatus in *Arabidopsis* (Fujiwara et al. 2015). Group 8, which consists solely of MTP8 proteins, is essential in Mn homeostasis, necessitating a proton gradient while transferring Mn^{2+} across membranes (Migocka et al. 2015b; Eroglu et al. 2017; Li et al. 2017a). They are required for active Mn sequestration in intracellular organelles or Mn delivery to Mn-dependent enzymes (Migocka et al. 2015b). Meanwhile, MTP9–MTP11 proteins are found in group 9 (Montanini et al. 2007; Gustin et al. 2011), and MTP 9 and MTP 10 are more closely linked and form a distinct internal clade. This association suggests that MTP9/10 and MTP11 proteins have different functions (Gustin et al. 2011). Several MTP11 transporters from different species confer Mn^{2+} tolerance (Delhaize et al. 2007; Zhang and Liu 2017), whereas there is currently insufficient evidence of the involvement of MTP9/10 proteins in Mn transport or tolerance to Mn (Migocka et al. 2015b), even though they are also classified as Mn-CDFs. According to current research, MTP proteins contribute to heavy metal detoxification and encourage their buildup (Kim et al. 2004; Delhaize et al. 2007; Xu et al. 2009; Lang et al. 2011; Menguer et al. 2013; Migocka et al. 2015a, 2015b; Eroglu et al. 2017; Li et al. 2017a; Zhang and Liu 2017). MTPs or other metal transporters' actions in heavy metal hyperaccumulation may boost phytoremediation efficiency (Kim et al. 2004; Gustin et al. 2011). However, understanding about the functional attribution of plant MTPs is still lacking and the functional features of groups 6 and 7 are yet to be determined.

The CDF family has a vital role in Mn, Zn, and Fe homeostasis (Shao et al. 2017; Socha and Guerinot 2014). The CDF family is divided into three subfamilies: Mn-CDF, Zn-CDF, and Zn/Fe-CDF, which differ in substrate specificity (Montanini et al. 2007). CDF family members are also known as metal tolerance proteins (MTPs) in plants. *Arabidopsis thaliana* and *Oryza sativa* MTP families have 12 and 10 members each. These MTPs are mostly involved in Mn, Zn, and Fe transport, but they can also transport Cd, Co, Cu, and Ni (Ricachenevsky et al. 2013). MTP8, which contributes to Mn detoxification in *A. thaliana* (Eroglu et al. 2017), *O. sativa* (Chen et al. 2013; Takemoto et al. 2017), and cucumber (Migocka et al. 2014), is the most fully functionally described MTP to date. MTP8.1 and MTP8.2 in barley (*Hordeum vulgare*) aid in loading Mn into the Golgi apparatus (Pedas et al. 2014). Researchers recently verified the significance of *A. thaliana* AtMTP8 in iron reallocation and Mn homeostasis during seed development and germination (Eroglu et al. 2017).

Although many MTP family genes have been identified in non-model plants, including *Triticum aestivum* (Vatansever et al. 2017), *Citrus sinensis* (Fu et al. 2017), *Brassica rapa* (Li et al. 2017b), and *Pyrus bretschneider* Rehd. (Hou et al. 2019), it is unknown whether similar MTP genes in other plant species, particularly Mn-CDF members in tea plants (*Camellia sinensis*), are categorized as Mn hyperaccumulators, with Mn concentrations in tea leaf tissues on the order of 1000 mg kg^{-1}. (Yemane et al. 2008). The ideal soil pH range for tea plants is 4.5–5.5 (Yan et al. 2020). The soil pH decreases when the tea saplings are planted, resulting in an increase in exchangeable Mn in the soil (Fung and Wong 2002). Despite the presence of high Mn in tea gardens, tea plants may thrive in acidic soils. However, the particular explanation for this is still unknown. CsMTP8, termed CsMTP8.1 in this study, was the first identified Mn-specific transporter from *C. sinensis*, and it played an important role in Mn homeostasis by enhancing Mn efflux from yeast and *A. thaliana* (Li et al. 2017a). Furthermore, heterologous expression of *C. sinensis* MTP11 in transgenic yeast results in improved tolerance to Mn and Co. (Yuan et al. 2017). They focused on the MTP family, which is notably implicated in Mn homeostasis, to carefully investigate the molecular mechanisms of Mn buildup in the tea plant. They found 13 CsMTP genes from tea plants, examined the sequence and structural properties of probable CsMTP genes, and determined the expression profiles of CsMTPs. Furthermore, using transcriptome analysis, they discovered that CsMTP8.2 expression (Log_2 fold-change) in shoots was downregulated approximately 2.32-fold when subjected to high Mn^{2+} treatments (50 M for 15 days) and finally functionally characterized CsMTP8.2 in terms of its role in plant Mn poisoning response.

15.3 Lsi Transporter

The role of silicon is characterized by its ability to protect the plant from various biotic and abiotic stresses (Ma and Takahashi 2002; Ma 2004; Ma and Yamaji 2006, 2008). Silicon magnifies the resistance capacity of plants to diseases caused by both fungi and bacteria in a variety of plant species, including rice blast, powdery mildew, sheath blight, ringspot, rust, leaf spot, and gray leaf spot (Fauteux et al. 2005). Insect pests such as stem borer, brown plant-hopper, rice green leafhopper, and white-backed plant-hopper are also suppressed by silicon (Savant et al. 1996). Furthermore, Si enhances plant resistance to abiotic stresses such as salt, metal toxicity, nutritional imbalance, lodging, drought, radiation, high temperature, freezing, and UV irradiation (Ma and Takahashi 2002; Ma 2004; Ma and Yamaji 2006, 2008). Silicon deposited beneath the cuticular layer works as a physical barrier, preventing fungus and insects from penetrating inside, increasing the mechanical strength, and decreasing transpiration.

The identification of silicon transporters in rice reported the discovery of two types of Si transporter genes in the plasma membrane of root cells: *Oryza sativa* Low Silicon 1 (OsLsi1) influx and 2 (OsLsi2) efflux. Lsi1 is defined as a Nod26-like intrinsic protein (NIP) subfamily of the aquaporin family and Lsi2 subfamily of anion channel transporters (Ma et al. 2007); both of the transporters were identified to be largely silicon transporters in rice and perform silicon uptake (Chen et al. 2011). Lsi1 and Lsi2 homologous genes have been identified in several plant species, including barley (Chiba et al. 2009), maize (Mitan et al. 2009), pumpkin (Mitani et al. 2011), horsetail (Grégoire et al. 2012), and wheat (Sun et al. 2017). Specific Si transporters are found in the plasma membranes of exodermis and endodermis cells in the roots, although they are polarized differently. Lsi1 is located on the distal side, while Lsi2 is confined on the proximal side. For example, HvLsi1 in barley is restricted to epidermal and cortical cells in the seminal roots and hypodermal cells in lateral roots, whereas ZmLsi1 in maize is restricted to epidermal and hypodermal cells in seminal and crown roots and cortical cells in the lateral roots (Ma et al. 2006; Chiba et al. 2009; Mitan et al. 2009). On the other hand, CmLsi1 in pumpkin did not display polar localization but was found in all root cells (Mitani-Ueno et al. 2011). Very little research has been done on Lsi2 homologs, but it has been observed that Lsi2 in barley, maize, and cucumber is localized in the endodermis without polarity (Mitan et al. 2009; Sun et al. 2018). These Si transporter location and polarity changes have been considered important variables for efficient Si absorption (Ma and Yamaji 2015). The most potent combination for active Si uptake is the location of OsLsi1 at the distal side and OsLsi2 at the proximal side of both the endodermis and exodermis, according to mathematical modeling (Sakurai et al. 2015). However, the precise mechanisms behind the interspecific and intraspecific variations in Si accumulation remain unknown.

Arsenic (As) is extensively present in rice fields as arsenite; due to submerged anaerobic conditions, it is taken up by two silicon (Si) transporters (Song et al. 2014). Rice root histology is defined by two Casparian strips in both the exo- and endodermis regions. As travels radially/laterally through the epidermis, exodermis, cortex, endodermis, and stele shortly after uptake. At this point, pericycle cells will select whether or not to translocate As(III) to the shoots. Both exo- and endodermis cells have Lsi influx and efflux transporters that work in opposite directions, with the former assisting in the arsenic influx into the cell and the latter participating in arsenic efflux toward the stele. As a result, Lsi2 is in charge of As(III) translocation in the xylem sap, whereas the mutant plant is unable to translocate arsenic into the shoot, leaves, straw, and grains due to a loss-of-function mutation, and as a result, there is low arsenic accumulation in the rice grain (Ma and Yamaji 2008), whereas rice Lsi1 (OsLsi1) allows the uptake of methylated arsenic species such as MMA and DMA (Li et al. 2009). For high Si uptake in rice, a cooperative transport mediated by Lsi1 and Lsi2 is necessary (Ma and Yamaji 2015).

15.4 Yellow Stripe-Like Proteins (YSL) Transporter

YSL transporters (yellow stripe-like proteins) are a phenotype recognized in maize, where interveinal chlorotic (yellow) zones are observed (Beadle 1929). This chlorosis is caused by insufficient iron uptake, resulting from a mutation in a root epidermal transporter (YS1; Beadle 1929; Curie et al. 2009). Members of the YSL family are only found in plants, while YSL is a subfamily of the larger OPT (oligopeptide transporter) family, which is also found in fungi (Lubkowitz 2011). Metal complexes with nicotianamine (NA) or its derivatives are accepted as substrates by YSL transporters rather than free metals (Curie et al. 2009). NA is a non-proteogenic amino acid that is produced by the enzyme NA synthase from S-adenosyl-methionine (Higuchi et al. 1999). H^+-symport stimulates transport by YSL proteins (Schaaf et al. 2004). Furthermore, some plant OPT transporters have been associated with metal transfer (Lubkowitz 2011; Zhai et al. 2014; Bashir et al. 2015). There is little information known about the structure of these proteins, with various models offering a range of 11–16 transmembrane regions (Lubkowitz 2011).

YSL transporters are generally associated with metal uptake from the soil in monocots and long-distance metal distribution in both monocots and dicots (Conte and Walker 2011). In *Arabidopsis*, two FRO proteins, FRO4 and FRO5, are strong candidates for converting Cu^{2+} to Cu^+, which would be introduced into the plant via COPT transporters (Bernal et al. 2012; Gayomba et al. 2013). Furthermore, Cu^{2+} can bind to MA precursor NA and be transported by YSL proteins indicating the presence of a Strategy II copper absorption strategy (DiDonato Jr et al. 2004). Furthermore, as evidenced by research in the tomato mutant “chloronerva,” YSL transporters would be implicated in metal unloading from the xylem (Conte and Walker 2011). This mutant possesses interveinal chlorosis due to a mutation in the gene encoding a NA synthase (Ling et al. 1999). Notably, the *Arabidopsis* small ubiquitin-like modifier (SUMO) E3 ligase SIZ1, which regulates the expression of AtYSL1 and AtYSL3, plays a role in excess Cu tolerance in plants (Chen et al. 2011). The Cu re-oxidation from Cu^+ to Cu^{2+} is anticipated to occur once inside the xylem for proper root-to-shoot transfer (Ryan et al. 2013). Cu must be chelated, and histidine and nicotianamine are the best candidates for xylem Cu-binding partners

(Printz et al. 2016). However, in rice, deoxymugineic acid, which is generated from nicotianamine, may also be responsible for Cu chelation. Furthermore, Cu availability may influence which molecule is engaged in Cu translocation: NA may chelate Cu under starvation but not under high concentrations (Irtelli et al. 2009). OsYSL16, a Cu-nicotianamine transporter related to Cu transport, has been found in rice to translocate the phloem. OsYSL16 transports Cu-nicotianamine complexes into the phloem, allowing Cu to be transported to young tissues and seeds, but it is not involved in primary Cu absorption (Zheng et al. 2012). Significantly, OsYSL16 is involved in Fe homeostasis, implying that Fe and Cu levels are regulated by partially overlapping processes (Lee et al. 2012; Printz et al. 2016). Furthermore, rice YSL16 participates in the supply of Cu-NA to growing young tissues and seeds via phloem transport. In the roots, Zn and Fe shortage increased OsYSL16 expression but not Mn and Cu deficiency. OsYSL16 knockout significantly reduced Cu-NA translocation from older to younger leaves and the flag leaf to the panicle (Zheng et al. 2012). It was found that *Arabidopsis* YSL2, when complexed with NA, transports both iron and copper, as does ZmYS1. The pattern of YSL2 expression suggests that metal-NA transport occurs in various types of cells in both roots and shoots, implying that metals are regularly transported into cells as NA complexes. The principal function of YSL2 appears to be the lateral transport of metals through the vasculature, based on its expression pattern. This movement appears to be particularly important when the iron is abundant, as YSL2 expression is reduced under iron-deficient development circumstances (DiDonato Jr et al. 2004). Furthermore, YSL (yellow strip 1-like) proteins, whose genes are constitutively overexpressed in the roots and shoots of *N. caerulescens*, are implicated in the translocation of nicotianamine-metal (particularly nicotianamine-Ni) complexes from root to shoot in hyperaccumulator plants (Gendre et al. 2007). Se persists in the form of selenite ions in the root cells after uptake by its particular hyperaccumulators, and root-to-shoot translocation of Se occurs via sulfate transport mechanisms (Sors et al. 2005).

YS1 from maize is the most well-studied member of this family (Roberts et al. 2004; Schaaf et al. 2004). ZmYS1 protein accumulates in the roots and leaves of Fe-deficient plants and acts as a proton-coupled symporter to transport Fe-PS. Fe solubilization by mugineic acid results in the formation of an MA-Fe^{3+} complex, which is carried out by yellow stripe-like (YSL) transporters (Conte and Walker 2012). It may also be involved in the homeostasis of Cu, Zn, or Ni as mugineic acid (MA) complexes. *A. thaliana* has eight putative YSL proteins based on sequence similarity to ZmYS1. The expression of AtYSL2 in metal uptake-defective yeast strains promoted Fe-NA and Cu-NA uptake. The presence of *Arabidopsis* YSL2 in the root endodermis and pericycle cells facing the metaxylem tubes suggests that it is involved in the lateral migration of Fe and/or Cu within the veins (Schaaf et al. 2005). These proteins appear to be involved in metal-NA unloading from the vasculature into growing tissues, metal-NA immobilization from senescent leaves, and metal-NA efficient loading into seeds. When exposed to high Fe levels, the transcript levels of *AtYSL1* increase (Jean et al. 2005). *Arabidopsis* YSL1-deficient mutants have reduced amounts of Fe-NA in their seeds and have a temporary germination deficiency that can be recovered by supplementation of iron. During leaf senescence, AtYSL1 and AtYSL3 were increased. The ysl1ysl3 double knockout mutants, which have interveinal chlorosis in leaves due to low Fe levels and reduced fertility due to faulty anther and embryo development, were less efficient in mobilizing metals, primarily Cu, from senescent leaves. Under metal deficiency, AtYSL2 and AtYSL3 are differentially expressed, and heterologous production of AtOPT3 in yeast reveals that it can transport Cu^{2+}, Mn^{2+}, and Fe^{2+} (Wintz et al. 2003). There are 19 potential OsYSL genes in the rice genome. Fe^{2+}-NA and Mn^{2+}-NA complexes are transported by OsYSL2, but not Fe^{2+}-NA complexes (Koike et al. 2004; Colangelo and Guerinot 2006).

15.5 Heavy Metal ATPases (HMAs) Transporters

P1B-type ATPases, also known as heavy metal ATPases (HMAs), play an important role in transporting metal ions toward their electrochemical gradient in plants by utilizing the energy given by ATP hydrolysis (Axelsen and Palmgren 1998). According to the functional properties of HMAs, these transporters can be divided into two classes depending on their metal-substrate specificity: The monovalent cations transport Cu/Ag group and the second group (Zn/Co/Cd/Pb) transport divalent cations (Verbruggen et al. 2009). It has also been reported that the genes encoding bivalent cation transporters such as HMAs are upregulated in the roots and shoots of Zn/Cd hyperaccumulators such as *N. caerulescens* and *A. halleri* and downregulated in their non-hyperaccumulator relatives (Papoyan and Kochian 2004; Talke et al. 2006; Hanikenne et al. 2008). HMAs are unique transporter in terms of tissue distribution, subcellular localization, and metal selectivity. *A. thaliana* contains eight HMAs, with AtHMA1–AtHMA4 referring to the Zn/Co/Cd/Pb subgroup. Certain transporters are found in different plant areas and play a vital role in heavy metal detoxification, accumulation, vacuolar sequestration, and efflux. In *Arabidopsis*, the ATPases HMA2 and HMA4 can also transport and accumulate Zn from roots to shoots. HMA5 ATPase ensures Cu transfer, and the non-essential heavy metals are transported using the same transporters. Using genome sequence analysis, nine HMA genes in rice (*Oryza sativa* L.) were identified, with OsHMA1–OsHMA3 referring to the Zn/Co/Cd/Pb subgroup. Furthermore, OsHMA9 may transport Zn, Cd, and Pb, despite being phylogenetically related to the Cu/Ag subgroup. Whereas OsHMA3 transports only Cd and participates in the sequestration of Cd into vacuoles in root cells. In RiceXPro database (http://ricexpro.dna.affrc.go.jp/), OsHMA1 is highly expressed in the leaf edge and the root, inflorescence, anther, pistil, lemma, palea, ovary, embryo, and endosperm. OsHMA2 is found at the plasma membrane and carries Zn and Cd. Zn deficiency reduced the expression of OsHMA2 in the roots. OsHMA2 expression was found mostly in rice roots, where OsHMA2 transcripts were plentiful in vascular bundles. Furthermore, the Zn and Cd concentrations in OsHMA2-suppressed rice leaves dropped, but the Zn concentration increased in the roots associated with the wild-type (WT).

Modifying HMA gene expression is an effective method for lowering Cd levels in rice grains. The Cd concentration in the grains of OsHMA3 overexpressing rice was lower than in WT rice. For example, the ATPases OsHMA2 and OsHMA3 are engaged in the transport and accumulation of Cd in rice: OsHMA2 is involved in Cd distribution to growing tissues, while OsHMA3 is important in compartmentalization in root cell vacuoles (Li et al. 2017c). OsHMA3 has been identified as a responsive gene for quantitative trait loci of Cd concentration in the rice cultivars 'Anjana Dhan' and 'Cho-kokoku', and their loss of function mutation of this protein results in excessive Cd accumulation in the shoots. A null mutant of AtHMA4 collected sufficient Zn in the roots but not in the shoots, implying that AtHMA4 is involved in Zn loading to the xylem. The disruption of root-to-shoot Cd translocation in a hma2hma4 double mutant demonstrates that AtHMA2 and AtHMA4 are important participants in Cd translocation in *A. thaliana*. Tandem gene duplication and HMA4 deregulation lead to Zn and Cd hyperaccumulation in *A. halleri*, which has similarly been seen in *Noccaea caerulescens*. In comparison to dicots, there are minimal reports of HMAs from monocots. The HMA4 (a type of HMAs) proteins are found in the membranes of xylem parenchyma cells, and their role in the Zn/Co/Cd/Pb group is enhanced by overexpression of the HMA4 gene, which sequentially generates a high efflux of Cd and Zn from the cortical cells of roots (via symplast) for uploading in xylem vessels for shoot uptake (Mills et al. 2003). It has also been discovered that the action of HMA4 positively influences other genes involved in metal or metalloid hyperaccumulation; as a result, this root-to-shoot translocation functions as a driving factor for the hyperaccumulation by creating a form of feedback loop. Roots respond to metal scarcity (Hanikenne et al. 2008; Rascio and Navari-Izzo 2011). According to Mikkelsen et al. (2012), HvHMA1 is a broad-specificity metal exporter from barley chloroplasts that acts as a scavenging mechanism for mobilizing plastid Zn and Cu when cells become low in these elements. HvHMA1 may be involved in the mobilization of Zn and Cu from aleurone cells during grain filling and germination in grains. AtHMA6/PAA1 (located in the chloroplast envelope) and AtHMA8/PAA2 (located in thylakoids) are two more transporters that can alter HM transport, and mutations of AtHMA6 and AtHMA8 significantly lower chloroplast copper concentration. The outer membrane of the chloroplast envelope is not a selective barrier due to the presence of pores with broad specificity, and AtHMA6 is likely to be localized in the inner membrane of the chloroplast envelope (Catty et al. 2011).

15.6 ZIP Transporter

Zinc (Zn) is a micronutrient that is required by both plants and humans. Zn deficiency affects over half of the world's agricultural soils. The insufficient availability of Zn decreases agricultural output and quality. The zinc-regulated, iron-regulated transporter (ZRT/IRT)-like protein (ZIP) family has been identified and characterized in prokaryotes, eukaryotes, and archaeotes and has been validated to be involved in intracellular trafficking, detoxification, and balancing the uptake, distribution, and utilization of Zn and Fe (Grotz and Guerinot 2006; Kavitha et al. 2015). There is very little information known on the structural characteristics and mechanism of Zn transport, by plant ZIP family transporters. Plant ZIP family transporters are expected to have 6–9 transmembrane (TM) domains (helices), with 8 being the most common (Guerinot 2000). The Zn transporters have molecular weights ranging from 33.1 to 51.4 kDa and protein sequences ranging from 322 to 478 amino acids (Vatansever et al. 2016). The high-resolution crystal structures of ZIP transporters are required for understanding their mechanism. The plant ZIP protein's crystal structure is not yet known. However, a high-resolution crystal structure of a prokaryotic ZIP protein from the bacterium *B. bronchiseptica* was recently determined, and its metal (Cd^{2+} and Zn^{2+}) transport mechanism was also predicted (Zhang et al. 2017).

The BbZIP transporter protein crystal structure revealed eight TM domains (TM1–TM8). The structure of BbZIP was determined using an inward-open confirmation and an occluded extracellular side with binuclear metal core in the center. The eight TMs created a helix bundle that was tightly related. Between rotating 180 degrees, the first three TMs (TM1–TM3) can be overlaid on the last three TMs (TM6–TM8), and TM4 and TM5 are symmetrically related and sandwiched by the two 3 TM repeats (Zhang et al. 2017). TM4 and TM5 interact with many conserved amino acid residues (His177, Asn178, Pro180, Glu181, and Gly182) and metal-binding motifs (Gln207, Asp/Asn208, Pro210, Glu211, and Gly212) to create the binuclear metal center. The BbZIP structure revealed four Cd^{2+} and seven Zn^{2+} metal-binding sites (Zhang et al. 2017). Both the N and C termini of BbZIP are accessible to the extracellular area, and the potential metal transport channel discovered on the extracellular side is inhibited by hydrophobic residues of the TM2 (Met99 and Ala102), TM5 (Leu200 and Iso204), and TM7 (Met269) (Zhang et al. 2017). The invariant Ser106 on TM2 is located at the bottom of the shallow and negatively charged entrance cavity and is critical for directing metals into the transport channel. The entry cavity contains two invariant metal-chelating residues, Asp113 and Asp305, which are important for recruiting metal substrates. Multiple conserved metal-binding sites around the metal outlet chamber are visible in the BbZIP structure, indicating that these represent a route of metal release to the cytoplasm. Metal-chelating residues such as His177, Glu276, His275, Pro180, Pro210, and Asp144 release the bound metal into the cytoplasm (Zhang et al. 2017). The binding of Zn^{2+} is coordinated by Glu181, Gln207, and Glu211 and two molecules of water (Guerinot 2000; Gaither and Eide 2001; Eide 2006; Zhang et al. 2017).

The formalized structural features of ZIP proteins enable the transport of divalent metal cations such as Zn, Fe, Mn, Ni, Cd, Co, and Cu (Milner et al. 2013; Potocki et al. 2014; Ivanov and Bauer 2016). Pedas and Husted (2009) define formalized formal ZIP proteins are involved in a variety of biological processes such as biofortification of grains with Zn, metal homeostasis, enzyme activity, secondary metabolite formation, and stress responses because these metals play catalytic, cocatalytic, and structural roles in organisms (Gaither and Eide 2001; Rutherford and Bird 2004; Eide 2006). As a result, ZIP family proteins are regarded to be essential transporters. The major Fe^{2+} transporter in *Arabidopsis* has been identified as AtIRT1. The transporters AtIRT1 (Eide et al. 1996), AtIRT2 (Vert et al. 2001), and AtIRT3 (Lin et al. 2009) appear to operate in *Arabidopsis* for Fe^{2+} absorption and transport. *Arabidopsis* IRTs can transport divalent metal ions such as Zn^{2+}, Mn^{2+}, and Cd^{2+} (Eide et al. 1996; Vert et al. 2001; Chiang et al. 2006; Lee and An 2009; Lin et al. 2009). The IRT genes are reported to be expressed under various metal stress conditions, with higher expression levels under Fe shortage. Under Zn stress circumstances, ZIP transporters balance Zn^{2+} absorption, use, and storage (Ramesh et al. 2003; Palmgren et al. 2008). ZIP transporters are found in various cell organelles and play an important role in Zn homeostasis and plant adaptation to low and high Zn soils (Tiong et al. 2015). Most ZIP transporters in crops are poorly understood, with only a few ZIPs functionally defined in plants (Kavitha et al. 2015).

15.7 NRAMP (Natural Resistance-Associated Macrophage Protein) Transporters

NRAMP transporters (natural resistance-associated macrophage protein) were recognized as Fe transporter in macrophages of rats (Vidal et al. 1993). This transporter family can be found in all the three domains of life (Nevo and Nelson 2006). Members of the NRAMP family have a remarkable protein sequence identity with mammalian proteins of 28% (yeast), 40% (plant), and 55% (fly) (46%, 58%, and 73% similarity, respectively). NRAMPs, which are generally membrane-spanning monomeric proteins with approximately 12 extremely hydrophobic trans-membrane domains, are now recognized as a widespread family of metal transporters with homologs in fungi, mammals, plants, and bacteria (Cellier et al. 1995; Ehrnstorfer et al. 2014). Nramp transporters have various metal substrates, most notably Fe^{2+}, Mn^{2+}, Co^{2+}, Cu^{2+}, Cd^{2+}, Ni^{2+}, and Zn^{2+}, driven by an H^+ symport (Nevo and Nelson 2006; Gunshin et al. 1997). Metal-binding sites are integrated by a carbonyl from a peptide bond in TM6, a Met in the same domain, and two Asp from TM1, in planar geometry (Ehrnstorfer et al. 2014). Some Nramp transporters have been implicated in iron and manganese uptake by the root epidermis (Cailliatte et al. 2010). Later, various divalent cation transporters, both similar and dissimilar, were discovered. The discovered NRAMPs from various organisms originally represent Fe transport behavior in yeast (Pinner et al. 1997; Eide 1998). Some *Arabidopsis* NRAMPs (AtNRAMP1, AtNRAMP3, and AtNRAMP4) are high-affinity Fe transporters that rescue yeast mutant fet3fet4's low Fe sensitivity phenotype (Curie et al. 2000; Thomine et al. 2000). Related to Fe uptake, Mn transportability of AtNRAMP1 and AtNRAMP3 and AtNRAMP4 has been shown (Cailliatte et al. 2010; Lanquar et al. 2010). Furthermore, numerous investigations have shown that NRAMPs retain the ability to transport heavy metals like Ni and Cd (Mizuno et al. 2005; Oomen et al. 2009). Nrat1, an NRAMP-like protein, has been shown to mediate and significantly contribute to Al^{3+} transport and absorption in rice (Xia et al. 2010). However, various studies have been conducted to understand the role of *Arabidopsis* NRAMPs in Fe and another metal uptake in rice NRAMPs

(Takahashi et al. 2011; Sasaki et al. 2012; Ishimaru et al. 2012). According to Chakrabarty et al. (2009), OsNRAMP1 expression is upregulated following As exposure in rice, despite its function being unclear in As transport previously. They present evidence for the involvement of OsNRAMP1 in As(III), as well as Cd and Fe, transport and accumulation in *Arabidopsis* and yeast (fet3fet4) mutants. Later, it was discovered that OsNRAMP1 localization inside the pericycle region might allow for the rapid mobilization of toxic metals in aerial tissues by facilitating xylem-mediated transport and limiting the level of toxicity utilized by these metals. OsNRAMP, which is found near Lsi2, performs the same function, namely, As(III) efflux within the root (Kumar et al. 2015). NRAMP is a highly conserved family of integral membrane-bound protein channels involved in ion transport in various organisms, including humans, plants, bacteria, and others. NRAMP (OsNRAMP1, OsNRAMP2, and OsNRAMP3) were initially discovered in rice (Belouchi et al. 1997).

15.8 ABC Transporter

ABC transporters are one of the most abundant protein families present in all living species. ABC transporters are powered by ATP hydrolysis and can function as both exporters and importers. The plant genome encodes for more than 100 ABC transporters, exceeding the number encoded by other organisms. Only 22 of the 130 *Arabidopsis* genes have been functionally investigated. They are found in most plant cell membranes, including the plasma membrane, tonoplasts, chloroplasts, mitochondria, and peroxisomes and serve a variety of roles. Initially identified as detoxification transporters, they have since been essential for organ growth, plant nutrition, plant development, response to abiotic stressors, pathogen resistance, and plant interaction with its environment. Membrane-bound ABC proteins are made up of four primary subunits, two transmembrane domains (TMD) and two nucleotide-binding domains (NBD), which work together to drive active transport during ATP hydrolysis (Higgins 1992). Individual genes encode these subunits (ABCI subfamily), two genes encoding one NBD and one TMD (half-size ABCs) that form heterodimers, one gene encoding one NBD and one TMD (half-size) that forms homodimers, or a single gene encoding one NBD and one TMD (half-size) that forms homodimers (full-size ABCs). ABCA through ABCD protein subunits contains a forward TMD-NBD domain arrangement, whereas the ABCG subfamily has a reverse NBD-TMD domain organization. The soluble ABCE and ABCF protein subfamilies are made up of only two NBDs, but the ABCI subfamily is made up of several genes that encode only one domain, such as an NBD, TMD, or accessory domain.

Some of the various domains encoded by ABCIs have been found to assemble into multi-subunit ABC transporters, similar to prokaryotic ABC proteins (Verrier et al. 2008). Many plant ABC transporters that have been identified so far are involved in hazardous metal transport, protecting plants against the detrimental effects of toxic heavy metals. It will be interesting to examine why higher creatures evolved ABC transporters that are primarily exporters, whereas bacteria evolved to have more nutrient importers. Plants create glutathione-derived heavy metal chelators, phytochelatins, in reaction to heavy metals such as cadmium, lead, and arsenic (Grill et al. 1989; Cobbett 2000). SpHMT1 was named after the first putative vacuolar phytochelatin transporter and was discovered in *Schizosaccharomyces pombe*. SpHMT1 is an ABCB subclass half-size ABC transporter. The equivalent mutant, hmt1, was hypersensitive to cadmium, which was attributed to a significantly reduced number of high-molecular-weight phytochelatin-cadmium (HMWPC-Cd-S-2) complexes (Ortiz et al. 1992).

15.9 Aquaglyceroporin Transporter

Aquaglyceroporin channels play a role in diffusing uncharged substances like water and glycerol (Agre and Kozono 2003). Trivalent metalloids such as arsenic(III) and antimony(III) share structural similarities and charge distribution with glycerol at physiological pH, allowing them to be readily absorbed by plants via aquaglyceroporins (Alegre 2004). According to reports, the glpF and Fps1 genes encode glycerol transporters that belong to the major intrinsic protein (MIP) family. MIP family members are related to the transportation of water, glycerol, urea, and tiny charged solutes, and the transport pathway does not require energy (Arima et al. 2003). Fps1 is involved in maintaining osmolarity by regulating the level of glycerol within the cell (Carbrey et al. 2003). Metalloid transport is controlled by the transcription of the Fps gene, which is inhibited by the activity of the mitogen-activated protein kinase Hog1 (Caperna et al. 2007). The Hog1 protein kinase phosphorylates the Fps gene at the N-terminus, at threonine 231. Metalloid uptake was increased when Hog1 was deleted, and Fps1 was expressed (Carbrey et al. 2003). Plant aquaporins are classified into four subcategories based on their location: small basic intrinsic proteins (SIP) found in the endoplasmic reticulum, nodulin 26-like intrinsic membrane protein (NIP), plasma membrane intrinsic protein (PIP), and tonoplast intrinsic protein (TIP) found in the plasma membrane (Chen et al. 2011). NIP transporters mediate $As(OH)_3$, $B(OH)_3$, $Si(OH)_3$, and $Sb(OH)_3$ transfer. *S. cerevisiae* missing the Fps1 gene and expressing NIP genes shown enhanced susceptibility to metalloids in an experiment (Chen et al. 2016). Because aquaglyceroporins are bidirectional, they not only mediate absorption but also participate in metalloid efflux, which contributes to tolerance (Chen et al. 2016). Because of the existence of the aquaglyceroporin gene aqpS, which works to efflux arsenic, the legume *Sinorhizobium meliloti* demonstrated arsenic tolerance, in contrast to the arsenite transporter gene arsB (Cheong et al. 2007).

Ji et al. (2017) discovered a calcium-dependent protein kinase (CPK31) protein in *Arabidopsis thaliana* that interacts with the aquaporin (NIP1;1), which is important in As(III) uptake. These researchers hypothesized that because CPK31 and NIP1;1 have an overlapping expression in *A. thaliana*, they might work cooperatively in the roots. The NIP 5;1, NIP 6;1, and NIP1;3 channels have been identified as boric acid channels and are essential for effective boron uptake. The NIP3;1 gene is encoded by the tassel-less (tls1) gene, and tls1 mutants displayed less B in fluorescence assays. The BOR1 is an important B exporter in B uptake via the xylem (Carbrey et al. 2003). Plants suffer from boron shortage when BOR1 is removed. BOR1 expression is controlled at the post-translational level. To avoid over-accumulation of B inside the plant system at a high-level of B, the BOR1 is degraded (Caperna et al. 2007). BOR1 is also thought to control the distribution of B to different areas of the shoot. As is found mainly in rice fields due to flooded, anaerobic circumstances, and it is taken up by two silicon (Si) transporters, namely, Lsi1 (low silicon 1), a Si-influx transporter, and Lsi2 (low silicon 2), a Si-efflux transporter (Song et al. 2014). Because they are highly expressed in rice roots, Lsi1 and Lsi2 are regarded as significant arsenite transporters in rice. Lsi1 and Lsi2 of rice (*Oryza sativa* L.) are aquaporins and anion channel transporters, respectively, that are known to be predominantly silicon transporters in rice and contribute to silicon uptake (Chen et al. 2011).

The OsLsi1 and OsLsi2 homologs in barley, maize, and wheat have been identified as being important for Si absorption in the plants (Caperna et al. 2007). Silicon influx transporter OsNIP2;1 has been found in rice (*Oryza sativa* L.) that mediates Si absorption, and this transporter also mediates the uptake of selenium (Zhao et al. 2010).

15.10 Conclusions

The presence of metals and metalloids in the environment exists in different forms and plays important roles in plant growth and development. The uptake of various metals and metalloids from the soil and water takes place with the aid of several types of transporters such as CDF, Lsi, YSL, HMAs, ZIP, NRAMP, ABC, and aquaglyceroporin. These transporters are found in different parts of plants from root to organelles such as mitochondria, nucleus, chloroplasts, vacuoles, Golgi apparatus, and endoplasmic reticulum. The uptake of these metal and metalloids has an essential role in a range of enzymes, co-factors, transcription factors, and a variety of proteins. The maintenance of ion homeostasis is particularly crucial for cellular and physiological normal functioning. Many ions display similar properties like chemical species, size, and partially transported by more than one transporters, such as As metalloids influx into the cells by NIPs and on the one hand effluxing from the cells by Lsi transporter. Therefore, plants have adopted sophisticated transporting systems for influx and efflux of metals with combinations of transporters and controlling the ion transportation for their physiological roles and requirements. In nature, not every metal and metalloid have a beneficial role in plants' growth and development; for example, numerous heavy metals and metalloids including As, Pb, Cd, Cr, Ni, and Hg show toxicity in the plant. Even translocation of excess amount of essential elements also might induce harmful consequences in plant cells. The presence of such non-essential metals in environments by nature and due to excessive anthropogenic activity leads to a major decrease in plant productivity. The presence of such high concentrations of metals and metalloids in soil, believed to be key pollutants, persists in the environment and exert a detrimental influence on plant growth. Transportation of such non-essential elements in plants induces reactive oxidative stress-mediated consequences. This leads to damage the internal integrity of cells by altering loss of enzyme function, membrane permeability to ion transport, membrane lipid peroxidation, destruction of nucleic acids, and other physiological activities that finally cause cell death. During the course of evolution, plants have also equipped themselves with diverse defensive mechanisms, such as excess metal sequestration in a separate compartment like vacuoles and several chelators within cell organelles which utilize or store excess metals into the vacuole. Thus, such metal and metalloid transporters play an important role in plant development and productivity. A key issue for researchers is to identify how these transporters selectively control metals in plants. Therefore, the action mechanisms of the different transporters become important to study the structure and function of these transporters, with the use of the yeast two-hybrid method which detects protein-protein interaction with certain domains of the transporters. Having knowledge about the structure and function of these transporters will aid in the construction of genetically engineered plants, which shall boost the accumulation of the metals for nutritional purposes (biofortification) and show enhanced phytoremediation and phytomining activities. Thus these metals and metalloid transporters have a highly hotheaded topic and developing area in current plant biology. This chapter presents an insight into the discovery of a new range of ion transporters that will change our concepts about transporters in plants.

References

Agre P, Kozono D (2003) Aquaporin water channels: molecular mechanisms for human diseases. FEBS Lett 555(1):72–78. https://doi.org/10.1016/s0014-5793(03)01083-4

Alegre L (2004) Die and let live: leaf senescence contributes to plant survival under drought stress. Funct Plant Biol 31:203–216. https://doi.org/10.1071/FP03236

Archibald F (1983) *Lactobacillus plantarum*, an organism not requiring iron. FEMS Microbiol Lett 19(1):29–32. https://doi.org/10.1111/j.1574-6968.1983.tb00504.x

Arima H, Yamamoto N, Sobue K, Umenishi F, Tada T, Katsuya H, Asai K (2003) Hyperosmolar mannitol stimulates expression of aquaporins 4 and 9 through a p38 mitogen-activated protein kinase-dependent pathway in rat astrocytes. J Biol Chem 278(45):44525–44534. https://doi.org/10.1074/jbc.M304368200

Axelsen KB, Palmgren MG (1998) Evolution of substrate specificities in the P-type ATPase superfamily. J Mol Evol 46(1):84–101. https://doi.org/10.1007/pl00006286

Bashir K, Ishimaru Y, Itai RN, Senoura T, Takahashi M, An G, Oikawa T, Ueda M, Sato A, Uozumi N, Nakanishi H (2015) Iron deficiency regulated OsOPT7 is essential for iron homeostasis in rice. Plant Mol Biol 88(1-2):165–176. https://doi.org/10.1007/s11103-015-0315-0

Beadle GW (1929) Yellow stripe-a factor for chlorophyll deficiency in maize located in the pr pr chromosome. Am Nat 63(685):189–192

Belouchi A, Kwan T, Gros P (1997) Cloning and characterization of the OsNramp family from *Oryza sativa*, a new family of membrane proteins possibly implicated in the transport of metal ions. Plant Mol Biol 33(6):1085–1092. https://doi.org/10.1023/a:1005723304911

Bernal M, Casero D, Singh V, Wilson GT, Grande A, Yang H, Dodani SC, Pellegrini M, Huijser P, Connolly EL, Merchant SS (2012) Transcriptome sequencing identifies SPL7-regulated copper acquisition genes FRO4/FRO5 and the copper dependence of iron homeostasis in *Arabidopsis*. Plant Cell 24(2):738–761. https://doi.org/10.1105/tpc.111.090431

Bozzi AT, Gaudet R (2021) Molecular mechanism of Nramp-family transition metal transport. J Mol Biol 433(16):166991. https://doi.org/10.1016/j.jmb.2021.166991

Cailliatte R, Schikora A, Briat JF, Mari S, Curie C (2010) High-affinity manganese uptake by the metal transporter NRAMP1 is essential for *Arabidopsis* growth in low manganese conditions. Plant Cell 22(3):904–917. https://doi.org/10.1105/tpc.109.073023

Caperna TJ, Shannon AE, Richards MP, Garrett WM, Talbot NC (2007) Identification and characterization of aquaporin-9 (AQP9) in porcine hepatic tissue and hepatocytes in monolayer culture. Domest Anim Endocrinol 32(4):273–286. https://doi.org/10.1016/j.domaniend.2006.03.008

Carbrey JM, Gorelick-Feldman DA, Kozono D, Praetorius J, Nielsen S, Agre P (2003) Aquaglyceroporin AQP9: solute permeation and metabolic control of expression in liver. Proc Natl Acad Sci 100(5):2945–2950. https://doi.org/10.1073/pnas.0437994100

Cassat JE, Skaar EP (2013) Iron in infection and immunity. Cell Host Microbe 13(5):509–519. https://doi.org/10.1016/j.chom.2013.04.010

Catty P, Boutigny S, Miras R, Joyard J, Rolland N, Seigneurin-Berny D (2011) Biochemical characterization of AtHMA6/PAA1, a chloroplast envelope Cu (I)-ATPase. J Biol Chem 286(42):36188–36197. https://doi.org/10.1074/jbc.M111.241034

Cellier M, Prive G, Belouchi A, Kwan T, Rodrigues V, Chia W, Gros P (1995) Nramp defines a family of membrane proteins. Proc Natl Acad Sci 92(22):10089–10093. https://doi.org/10.1073/pnas.92.22.10089

Chakrabarty D, Trivedi PK, Misra P, Tiwari M, Shri M, Shukla D, Kumar S, Rai A, Pandey A, Nigam D, Tripathi RD (2009) Comparative transcriptome analysis of arsenate and arsenite stresses in rice seedlings. Chemosphere 74(5):688–702. https://doi.org/10.1016/j.chemosphere.2008.09.082

Chen CC, Chen YY, Tang IC, Liang HM, Lai CC, Chiou JM, Yeh KC (2011) *Arabidopsis* SUMO E3 ligase SIZ1 is involved in excess copper tolerance. Plant Physiol 156(4):2225–2234. https://doi.org/10.1104/pp.111.178996

Chen D, Cao B, Qi L, Yin L, Wang S, Deng X (2016) Silicon-moderated K-deficiencyinduced leaf chlorosis by decreasing putrescine accumulation in sorghum. Ann Bot 118:305–315. https://doi.org/10.1093/aob/mcw111

Chen Z, Fujii Y, Yamaji N, Masuda S, Takemoto Y, Kamiya T, Yusuyin Y, Iwasaki K, Kato SI, Maeshima M, Ma JF (2013) Mn tolerance in rice is mediated by MTP8. 1, a member of the cation diffusion facilitator family. J Exp Bot 64(14):4375–4387. https://doi.org/10.1093/jxb/ert243

Cheong YH, Pandey GK, Grant JJ, Batistic O, Li L, Kim BG, Lee SC, Kudla J, Luan S (2007) Two calcineurin B-like calcium sensors, interacting with protein kinase CIPK23, regulate leaf transpiration and root potassium uptake in *Arabidopsis*. Plant J 52(2):223–239. https://doi.org/10.1111/j.1365-313X.2007.03236.x

Chiang HC, Lo JC, Yeh KC (2006) Genes associated with heavy metal tolerance and accumulation in Zn/Cd hyperaccumulator *Arabidopsis halleri*: a genomic survey with cDNA microarray. Environ Sci Technol 40(21):6792–6798. https://doi.org/10.1021/es061432y

Chiba Y, Mitani N, Yamaji N, Ma JF (2009) HvLsi1 is a silicon influx transporter in barley. Plant J 57(5):810–818. https://doi.org/10.1111/j.1365-313X.2008.03728.x

Cobbett CS (2000) Phytochelatins and their roles in heavy metal detoxification. Plant Physiol 123(3):825–832. https://doi.org/10.1104/pp.123.3.825

Colangelo EP, Guerinot ML (2006) Put the metal to the petal: metal uptake and transport throughout plants. Curr Opin Plant Biol 9(3):322–330. https://doi.org/10.1016/j.pbi.2006.03.015

Conte SS, Walker EL (2011) Transporters contributing to iron trafficking in plants. Mol Plant 4(3): 464–476. https://doi.org/10.1093/mp/ssr015

Conte SS, Walker EL (2012) Genetic and biochemical approaches for studying the yellow stripe-like transporter family in plants. Curr Top Membr 69:295–322. https://doi.org/10.1016/B978-0-12-394390-3.00011-2

Coyle P, Philcox JC, Carey LC, Rofe AM (2002) Metallothionein: the multipurpose protein. Cell Mol Life Sci 59(4):627–647. https://doi.org/10.1007/s00018-002-8454-2

Curie C, Alonso JM, Jean ML, Ecker JR, Briat JF (2000) Involvement of NRAMP1 from *Arabidopsis thaliana* in iron transport. Biochem J 347(3):749–755

Curie C, Cassin G, Couch D, Divol F, Higuchi K, Le Jean M, Misson J, Schikora A, Czernic P, Mari S (2009) Metal movement within the plant: contribution of nicotianamine and yellow stripe 1-like transporters. Ann Bot 103(1):1–11. https://doi.org/10.1093/aob/mcn207

De Carvalho MHC (2008) Drought stress and reactive oxygen species: Production, scavenging and signaling. Plant Signal Behav 3:156–165. https://doi.org/10.4161/psb.3.3.5536

Delhaize E, Gruber BD, Pittman JK, White RG, Leung H, Miao Y, Jiang L, Ryan PR, Richardson AE (2007) A role for the AtMTP11 gene of *Arabidopsis* in manganese transport and tolerance. Plant J 51(2):198–210. https://doi.org/10.1111/j.1365-313X.2007.03138.x

Dhanker R, Kumar R, Hwang JS (2012) Predation by *Pseudodiaptomus annandalei* (Copepoda: Calanoida) on rotifer prey: size selection, egg predation and effect of algal diet. J Exp Mar Biol Ecol 414:44–53. https://doi.org/10.1016/j.jembe.2012.01.011

DiDonato RJ Jr, Roberts LA, Sanderson T, Eisley RB, Walker EL (2004) Arabidopsis yellow stripe-like 2 (YSL2): a metal-regulated gene encoding a plasma membrane transporter of nicotianamine–metal complexes. Plant J 39(3):403–414. https://doi.org/10.1111/j.1365-313X.2004.02128.x

Ehrnstorfer IA, Geertsma ER, Pardon E, Steyaert J, Dutzler R (2014) Crystal structure of a SLC11 (NRAMP) transporter reveals the basis for transition-metal ion transport. Nat Struct Mol Biol 21(11):990–996. https://doi.org/10.1038/nsmb.2904

Eide D, Broderius M, Fett J, Guerinot ML (1996) A novel iron-regulated metal transporter from plants identified by functional expression in yeast. Proc Natl Acad Sci 93(11):5624–5628. https://doi.org/10.1073/pnas.93.11.5624

Eide DJ (1998) The molecular biology of metal ion transport in *Saccharomyces cerevisiae*. Annu Rev Nutr 18(1):441–469. https://doi.org/10.1146/annurev.nutr.18.1.441

Eide DJ (2006) Zinc transporters and the cellular trafficking of zinc. Biochim Biophys Acta 1763(7):711–722. https://doi.org/10.1016/j.bbamcr.2006.03.005

Eroglu S, Giehl RFH, Meier B, Takahashi M, Terada Y, Ignatyev K et al (2017) Metal tolerance protein 8 mediates manganese homeostasis and iron reallocation during seed development and germination. Plant Physiol 174:1633–1647. https://doi.org/10.1104/pp.16.01646

Fauteux F, Rémus-Borel W, Menzies JG, Bélanger RR (2005) Silicon and plant disease resistance against pathogenic fungi. FEMS Microbiol Lett 249(1):1–6. https://doi.org/10.1016/j.femsle.2005.06.034

Fu XZ, Tong YH, Zhou X, Ling LL, Chun CP, Cao L, Zeng M, Peng LZ (2017) Genome-wide identification of sweet orange (*Citrus sinensis*) metal tolerance proteins and analysis of their expression patterns under zinc, manganese, copper, and cadmium toxicity. Gene 629:1–8. https://doi.org/10.1016/j.gene.2017.07.072

Fujiwara T, Kawachi M, Sato Y, Mori H, Kutsuna N, Hasezawa S, Maeshima M (2015) A high molecular mass zinc transporter MTP 12 forms a functional heteromeric complex with MTP 5 in the Golgi in *Arabidopsis thaliana*. FEBS J 282(10):1965–1979. https://doi.org/10.1111/febs.13252

Fung KF, Wong MH (2002) Effects of soil pH on the uptake of Al, F and other elements by tea plants. J Sci Food Agric 82(1):146–152. https://doi.org/10.1002/jsfa.1007

Gaither LA, Eide DJ (2001) Eukaryotic zinc transporters and their regulation. Biometals 14(3-4): 251–270. https://doi.org/10.1023/a:1012988914300

Gayomba SR, Jung HI, Yan J, Danku J, Rutzke MA, Bernal M, Krämer U, Kochian LV, Salt DE, Vatamaniuk OK (2013) The CTR/COPT-dependent copper uptake and SPL7-dependent copper deficiency responses are required for basal cadmium tolerance in *A. thaliana*. Metallomics 5(9):1262–1275. https://doi.org/10.1039/c3mt00111c

Gendre D, Czernic P, Conéjéro G, Pianelli K, Briat JF, Lebrun M, Mari S (2007) TcYSL3, a member of the YSL gene family from the hyper-accumulator *Thlaspi caerulescens*, encodes a nicotianamine-Ni/Fe transporter. Plant J 49(1):1–15. https://doi.org/10.1111/j.1365-313X.2006.02937.x

Grégoire C, Rémus-Borel W, Vivancos J, Labbé C, Belzile F, Bélanger RR (2012) Discovery of a multigene family of aquaporin silicon transporters in the primitive plant *Equisetum arvense*. Plant J 72(2):320–333. https://doi.org/10.1111/j.1365-313X.2012.05082.x

Grill E, Löffler S, Winnacker EL, Zenk MH (1989) Phytochelatins, the heavy-metal-binding peptides of plants, are synthesized from glutathione by a specific γ-glutamylcysteine dipeptidyl transpeptidase (phytochelatin synthase). Proc Natl Acad Sci U S A 86(18):6838–6842. https://doi.org/10.1073/pnas.86.18.6838

Grotz N, Guerinot ML (2006) Molecular aspects of Cu, Fe and Zn homeostasis in plants. Biochim Biophys Acta 1763(7):595–608. https://doi.org/10.1016/j.bbamcr.2006.05.014

Guerinot ML (2000) The ZIP family of metal transporters. Biochim Biophys Acta 1465(1–2): 190–198. https://doi.org/10.1016/s0005-2736(00)00138-3

Gunshin H, Mackenzie B, Berger UV, Gunshin Y, Romero MF, Boron WF et al (1997) Cloning and characterization of a mammalian proton-coupled metalion transporter. Nature 388:482–488. https://doi.org/10.1038/41343

Gustin JL, Zanis MJ, Salt DE (2011) Structure and evolution of the plant cation diffusion facilitator family of ion transporters. BMC Evol Biol 11:76. https://doi.org/10.1186/1471-2148-11-76

Hanikenne M, Talke IN, Haydon MJ, Lanz C, Nolte A, Motte P, Kroymann J, Weigel D, Krämer U (2008) Evolution of metal hyperaccumulation required cis-regulatory changes and triplication of HMA4. Nature 453(7193):391–395. https://doi.org/10.1038/nature06877

Hediger MA (1997) Membrane permeability. The diversity of transmembrane transport processes. Curr Opin Cell Biol 9(4):543–546

Higgins CF (1992) ABC transporters: from microorganisms to man. Annu Rev Cell Biol 8(1): 67–113. https://doi.org/10.1146/annurev.cb.08.110192.000435

Higuchi K, Suzuki K, Nakanishi H, Yamaguchi H, Nishizawa NK, Mori S (1999) Cloning of nicotianamine synthase genes, novel genes involved in the biosynthesis of phytosiderophores. Plant Physiol 119(2):471–480. https://doi.org/10.1104/pp.119.2.471

Hou L, Gu D, Li Y, Li J, Li J, Chen X, Zhang W (2019) Phylogenetic and expression analysis of Mn-CDF transporters in pear (*Pyrus bretschneideri* Rehd.). Plant Mol Biol Report 37(1): 98–110. https://doi.org/10.3389/fgene.2018.00103

Irtelli B, Petrucci WA, Navari-Izzo F (2009) Nicotianamine and histidine/proline are, respectively, the most important copper chelators in xylem sap of *Brassica carinata* under conditions of copper deficiency and excess. J Exp Bot 60(1):269–277. https://doi.org/10.1093/jxb/ern286

Ishimaru Y, Takahashi R, Bashir K, Shimo H, Senoura T, Sugimoto K, Ono K, Yano M, Ishikawa S, Arao T, Nakanishi H (2012) Characterizing the role of rice NRAMP5 in manganese, iron and cadmium transport. Sci Rep 2(1):1–8. https://doi.org/10.1038/srep00286

Ivanov R, Bauer P (2016) Sequence and coexpression analysis of iron-regulated ZIP transporter genes reveals crossing points between iron acquisition strategies in green algae and land plants. Plant Soil 418:61–73. https://doi.org/10.1007/s11104-016-3128-2

Jean ML, Schikora A, Mari S, Briat JF, Curie C (2005) A loss-of-function mutation in AtYSL1 reveals its role in iron and nicotianamine seed loading. Plant J 44(5):769–782. https://doi.org/10.1111/j.1365-313X.2005.02569.x

Ji R, Zhou L, Liu J, Wang Y, Yang L, Zheng Q, Zhang C, Zhang B, Ge H, Yang Y, Zhao F, Luan S, Lan W (2017) Calcium-dependent protein kinase CPK31 interacts with arsenic transporter AtNIP1;1 and regulates arsenite uptake in *Arabidopsis thaliana*. PLoS One 12(3):e0173681. https://doi.org/10.1371/journal.pone.0173681

Kavitha PG, Kuruvilla S, Mathew MK (2015) Functional characterization of a transition metal ion transporter, OsZIP6 from rice (*Oryza sativa* L.). Plant Physiol Biochem 97:165–174. https://doi.org/10.1016/j.plaphy.2015.10.005

Kim D, Gustin JL, Lahner B, Persans MW, Baek D, Yun DJ et al (2004) The plant CDF family member TgMTP1 from the Ni/Zn hyperaccumulator *Thlaspi goesingense* acts to enhance efflux of Zn at the plasma membrane when expressed in *Saccharomyces cerevisiae*. Plant J 39:237–251. https://doi.org/10.1111/j.1365-313X.2004.02126.x

Koike S, Inoue H, Mizuno D, Takahashi M, Nakanishi H, Mori S, Nishizawa NK (2004) OsYSL2 is a rice metal-nicotianamine transporter that is regulated by iron and expressed in the phloem. Plant J 39(3):415–424. https://doi.org/10.1111/j.1365-313X.2004.02146.x

Kolber ZS, Barber RT, Coale KH, Fitzwateri SE, Greene RM, Johnson KS, Lindley S, Falkowski PG (1994) Iron limitation of phytoplankton photosynthesis in the equatorial Pacific Ocean. Nature 371(6493):145–149

Kumar S, Dubey RS, Tripathi RD, Chakrabarty D, Trivedi PK (2015) Omics and biotechnology of arsenic stress and detoxification in plants: current updates and prospective. Environ Int 74:221–230. https://doi.org/10.1016/j.envint.2014.10.019

Lang M, Hao M, Fan Q, Wang W, Mo S, Zhao W, Zhou J (2011) Functional characterization of BjCET3 and BjCET4, two new cation-efflux transporters from *Brassica juncea* L. J Exp Bot 62(13):4467–4480. https://doi.org/10.1093/jxb/err137

Lanquar V, Ramos MS, Lelièvre F, Barbier-Brygoo H, Krieger-Liszkay A, Krämer U, Thomine S (2010) Export of vacuolar manganese by AtNRAMP3 and AtNRAMP4 is required for optimal photosynthesis and growth under manganese deficiency. Plant Physiol 152(4):1986–1999

Lee S, An G (2009) Over-expression of OsIRT1 leads to increased iron and zinc accumulations in rice. Plant Cell Environ 32(4):408–416. https://doi.org/10.1111/j.1365-3040.2009.01935.x

Lee S, Ryoo N, Jeon JS, Guerinot ML, An G (2012) Activation of rice yellow Stripe1-like 16 (OsYSL16) enhances iron efficiency. Mol Cells 33(2):117–126. https://doi.org/10.1007/s10059-012-2165-9

Li Q, Li Y, Wu X, Zhou L, Zhu X, Fang W (2017a) Metal transport protein 8 in *Camellia sinensis* confers superior manganese tolerance when expressed in yeast and *Arabidopsis thaliana*. Sci Rep 7(1):1–11. https://doi.org/10.1038/srep39915

Li LZ, Tu C, Wu LH, Peijnenburg WJ, Ebbs S, Luo YM (2017b) Pathways of root uptake and membrane transport of Cd in the zinc/cadmium hyperaccumulating plant *Sedum plumbizincicola*. Environ Toxicol Chem 36(4):1038–1046. https://doi.org/10.1186/s12993-021-00179-9

Li H, Luo N, Li YW, Cai QY, Li HY, Mo CH, Wong MH (2017c) Cadmium in rice: transport mechanisms, influencing factors, and minimizing measures. Environ Pollut 224:622–630. https://doi.org/10.1016/j.envpol.2017.01.087

Li RY, Ago Y, Liu WJ, Mitani N, Feldmann J, McGrath SP, Ma JF, Zhao FJ (2009) The rice aquaporin Lsi1 mediates uptake of methylated arsenic species. Plant Physiol 150(4):2071–2080. https://doi.org/10.1104/pp.109.140350

Lin YF, Liang HM, Yang SY, Boch A, Clemens S, Chen CC et al (2009) *Arabidopsis* IRT3 is a zinc-regulated and plasma membrane localized zinc/iron transporter. New Phytol 182:392–404. https://doi.org/10.1111/j.1469-8137.2009.02766.x

Ling HQ, Koch G, Bäumlein H, Ganal MW (1999) Map-based cloning of chloronerva, a gene involved in iron uptake of higher plants encoding nicotianamine synthase. Proc Natl Acad Sci U S A 96(12):7098–7103. https://doi.org/10.1073/pnas.96.12.7098

Lubkowitz M (2011) The oligopeptide transporters: a small gene family with a diverse group of substrates and functions? Mol Plant 4(3):407–415. https://doi.org/10.1093/mp/ssr004

Ma JF, Yamaji N (2015) A cooperative system of silicon transport in plants. Trends Plant Sci 20(7):435–442. https://doi.org/10.1016/j.tplants.2015.04.007

Ma JF, Takahashi E (2002) Soil, fertilizer, and plant silicon research in Japan. Elsevier, Amsterdam

Ma JF, Yamaji N (2006) Silicon uptake and accumulation in higher plants. Trends Plant Sci 11(8):392–397. https://doi.org/10.1016/j.tplants.2006.06.007

Ma JF, Yamaji N (2008) Functions and transport of silicon in plants. Cell Mol Life Sci 65(19): 3049–3057. https://doi.org/10.1007/s00018-008-7580-x

Ma JF (2004) Role of silicon in enhancing the resistance of plants to biotic and abiotic stresses. Soil Sci Plant Nutr 50(1):11–18. https://doi.org/10.1080/00380768.2004.10408447

Ma JF, Tamai K, Yamaji N, Mitani N, Konishi S, Katsuhara M, Ishiguro M, Murata Y, Yano M (2006) A silicon transporter in rice. Nature 440(7084):688–691. https://doi.org/10.1038/nature04590

Ma JF, Yamaji N, Mitani N, Tamai K, Konishi S, Fujiwara T, Katsuhara M, Yano M (2007) An efflux transporter of silicon in rice. Nature 448(7150):209–212. https://doi.org/10.1038/nature05964

Meharg AA, Rahman MM (2003) Arsenic contamination of Bangladesh paddy field soils: implications for rice contribution to arsenic consumption. Environ Sci Technol 37(2): 229–234. https://doi.org/10.1021/es0259842

Menguer PK, Farthing E, Peaston KA, Ricachenevsky FK, Fett JP, Williams LE (2013) Functional analysis of the rice vacuolar zinc transporter OsMTP1. J Exp Bot 64(10):2871–2883. https://doi.org/10.1093/jxb/ert136

Migocka M, Kosieradzka A, Papierniak A, Maciaszczyk-Dziubinska E, Posyniak E, Garbiec A, Filleur S (2015a) Retracted: two metal-tolerance proteins, MTP1 and MTP4, are involved in Zn homeostasis and Cd sequestration in cucumber cells. J Exp Bot 66(3):1001–1015. https://doi.org/10.1093/jxb/ery342

Migocka M, Papierniak A, Kosieradzka A, Posyniak E, Maciaszczyk-Dziubinska E, Biskup R, Garbiec A, Marchewka T (2015b) Retracted: cucumber metal tolerance protein cs MTP 9 is a plasma membrane H^+-coupled antiporter involved in the Mn^{2+} and Cd^{2+} efflux from root cells. Plant J 84(6):1045–1058. https://doi.org/10.1111/tpj.13056

Migocka M, Papierniak A, Maciaszczyk-Dziubińska E, Poździk P, Posyniak E, Garbiec A, Filleur S (2014) Retracted: cucumber metal transport protein MTP8 confers increased tolerance to manganese when expressed in yeast and *Arabidopsis thaliana*. J Exp Bot 65(18):5367–5384. https://doi.org/10.1093/jxb/eru295

Mikkelsen MD, Pedas P, Schiller M, Vincze E, Mills RF, Borg S, Møller A, Schjoerring JK, Williams LE, Baekgaard L, Holm PB (2012) Barley HvHMA1 is a heavy metal pump involved in mobilizing organellar Zn and Cu and plays a role in metal loading into grains. PLoS One 7(11):e49027. https://doi.org/10.1371/journal.pone.0049027

Mills RF, Krijger GC, Baccarini PJ, Hall JL, Williams LE (2003) Functional expression of AtHMA4, a P1B-type ATPase of the Zn/Co/Cd/Pb subclass. Plant J 35(2):164–176. https://doi.org/10.1046/j.1365-313x.2003.01790.x

Milner MJ, Seamon J, Craft E, Kochian LV (2013) Transport properties of members of the ZIP family in plants and their role in Zn and Mn homeostasis. J Exp Bot 64:369–381. https://doi.org/10.1093/jxb/ers315

Mitan N, Yamaji N, Ma JF (2009) Identification of maize silicon influx transporters. Plant Cell Physiol 50(1):5–12. https://doi.org/10.1093/pcp/pcn110

Mitani N, Yamaji N, Ago Y, Iwasaki K, Ma JF (2011) Isolation and functional characterization of an influx silicon transporter in two pumpkin cultivars contrasting in silicon accumulation. Plant J 66(2):231–240. https://doi.org/10.1111/j.1365-313X.2011.04483.x

Mitani-Ueno N, Yamaji N, Ma JF (2011) Silicon efflux transporters isolated from two pumpkin cultivars contrasting in Si uptake. Plant Signal Behav 6(7):991–994. https://doi.org/10.4161/psb.6.7.15462

Mizuno T, Usui K, Horie K, Nosaka S, Mizuno N, Obata H (2005) Cloning of three ZIP/Nramp transporter genes from a Ni hyperaccumulator plant *Thlaspi japonicum* and their Ni^{2+}-transport abilities. Plant Physiol Biochem 43(8):793–801. https://doi.org/10.1016/j.plaphy.2005.07.006

Montanini B, Blaudez D, Jeandroz S, Sanders D, Chalot M (2007) Phylogenetic and functional analysis of the cation diffusion facilitator (CDF) family: improved signature and prediction of substrate specificity. BMC Genomics 8(1):1–16. https://doi.org/10.1186/1471-2164-8-107

Mukhopadhyay R, Bhattacharjee H, Rosen BP (2014) Aquaglyceroporins: generalized metalloid channels. Biochim Biophys Acta 1840(5):1583–1591. https://doi.org/10.1016/j.bbagen.2013.11.021

Nelson N (1999) Metal ion transporters and homeostasis. EMBO J 18(16):4361–4371. https://doi.org/10.1093/emboj/18.16.4361

Nevo Y, Nelson N (2006) The NRAMP family of metal-ion transporters. Biochimica et Biophysica acta (BBA)-molecular. Cell Res 1763(7):609–620. https://doi.org/10.1016/j.bbamcr.2006.05.007

Nies DH, Silver S (1995) Ion efflux systems involved in bacterial metal resistances. J Ind Microbiol 14(2):186–199

Oomen RJ, Wu J, Lelièvre F, Blanchet S, Richaud P, Barbier-Brygoo H, Aarts MG, Thomine S (2009) Functional characterization of NRAMP3 and NRAMP4 from the metal hyperaccumulator *Thlaspi caerulescens*. New Phytol 181(3):637–650. https://doi.org/10.1111/j.1469-8137.2008.02694.x

Ortiz DF, Kreppel L, Speiser DM, Scheel G, McDonald G, Ow DW (1992) Heavy metal tolerance in the fission yeast requires an ATP-binding cassette-type vacuolar membrane transporter. EMBO J 11(10):3491–3499

Palmgren MG, Clemens S, Williams LE, Krämer U, Borg S, Schjørring JK, Sanders D (2008) Zinc biofortification of cereals: problems and solutions. Trends Plant Sci 13(9):464–473. https://doi.org/10.1016/j.tplants.2008.06.005

Papoyan A, Kochian LV (2004) Identification of *Thlaspi caerulescens* genes that may be involved in heavy metal hyperaccumulation and tolerance. Characterization of a novel heavy metal transporting ATPase. Plant Physiol 136(3):3814–3823. https://doi.org/10.1104/pp.104.044503

Paulsen IT, Saier MH (1997) A novel family of ubiquitous heavy metal ion transport proteins. J Membr Biol 156:99–103. https://doi.org/10.1007/s002329900192

Pedas P, Husted S (2009) Zinc transport mediated by barley ZIP proteins are induced by low pH. Plant Signal Behav 4(9):842–845. https://doi.org/10.4161/psb.4.9.9375

Pedas P, Schiller Stokholm M, Hegelund JN, Ladegård AH, Schjoerring JK, Husted S (2014) Golgi localized barley MTP8 proteins facilitate Mn transport. PLoS One 9(12):e113759. https://doi.org/10.1371/journal.pone.0113759

Pinner E, Gruenheid S, Raymond M, Gros P (1997) Functional complementation of the yeast divalent cation transporter family SMF by NRAMP2, a member of the mammalian natural resistance-associated macrophage protein family. J Biol Chem 272(46):28933–28938. https://doi.org/10.1074/jbc.272.46.28933

Posey JE, Gherardini FC (2000) Lack of a role for iron in the Lyme disease pathogen. Science 288(5471):1651–1653. https://doi.org/10.1126/science.288.5471.1651

Potocki S, Valensin D, Kozlowski H (2014) The specificity of interaction of Zn^{2+}, Ni^{2+} and Cu^{2+} ions with the histidine-rich domain of the TjZNT1 ZIP family transporter. Dalton Trans 43(26):10215–10223. https://doi.org/10.1039/c4dt00903g

Printz B, Lutts S, Hausman JF, Sergeant K (2016) Copper trafficking in plants and its implication on Cell Wall dynamics. Front Plant Sci 7:601. https://doi.org/10.3389/fpls.2016.00601

Radisky D, Kaplan J (1999) Regulation of transition metal transport across the yeast plasma membrane. J Biol Chem 274(8):4481–4484. https://doi.org/10.1074/jbc.274.8.4481

Ramesh SA, Shin R, Eide DJ, Schachtman DP (2003) Differential metal selectivity and gene expression of two zinc transporters from rice. Plant Physiol 133(1):126–134. https://doi.org/10.1104/pp.103.026815

Rascio N, Navari-Izzo F (2011) Heavy metal hyperaccumulating plants: how and why do they do it? And what makes them so interesting? Plant Sci 180(2):169–181. https://doi.org/10.1016/j.plantsci.2010.08.016

Ricachenevsky FK, Menguer PK, Sperotto RA, Williams LE, Fett JP (2013) Roles of plant metal tolerance proteins (MTP) in metal storage and potential use in biofortification strategies. Front Plant Sci 4:144. https://doi.org/10.3389/fpls.2013.00144

Roberts LA, Pierson AJ, Panaviene Z, Walker EL (2004) Yellow stripe 1. Expanded roles for the maize iron-phytosiderophore transporter. Plant Physiol 135(1):112–120. https://doi.org/10.1104/pp.103.037572

Rutherford JC, Bird AJ (2004) Metal-responsive transcription factors that regulate iron, zinc, and copper homeostasis in eukaryotic cells. Eukaryot Cell 3(1):1–13. https://doi.org/10.1128/EC.3.1.1-13.2004

Ryan BM, Kirby JK, Degryse F, Harris H, McLaughlin MJ, Scheiderich K (2013) Copper speciation and isotopic fractionation in plants: uptake and translocation mechanisms. New Phytol 199(2):367–378. https://doi.org/10.1111/nph.12276

Sakurai G, Satake A, Yamaji N, Mitani-Ueno N, Yokozawa M, Feugier FG, Ma JF (2015) In silico simulation modeling reveals the importance of the Casparian strip for efficient silicon uptake in rice roots. Plant Cell Physiol 56(4):631–639. https://doi.org/10.1093/pcp/pcv017

Sasaki A, Yamaji N, Yokosho K, Ma JF (2012) Nramp5 is a major transporter responsible for manganese and cadmium uptake in rice. Plant Cell 24(5):2155–2167. https://doi.org/10.1105/tpc.112.096925

Savant NK, Snyder GH, Datnoff LE (1996) Silicon management and sustainable rice production. Adv Agron 58:151–199. https://doi.org/10.1016/S0065-2113(08)60255-2

Schaaf G, Ludewig U, Erenoglu BE, Mori S, Kitahara T, Von Wirén N (2004) ZmYS1 functions as a proton-coupled symporter for phytosiderophore-and nicotianamine-chelated metals. J Biol Chem 279(10):9091–9096. https://doi.org/10.1074/jbc.M311799200

Schaaf G, Schikora A, Häberle J, Vert G, Ludewig U, Briat JF, Curie C, von Wirén N (2005) A putative function for the *Arabidopsis* Fe–phytosiderophore transporter homolog AtYSL2 in Fe and Zn homeostasis. Plant Cell Physiol 46(5):762–774. https://doi.org/10.1093/pcp/pci081

Shao JF, Yamaji N, Shen RF, Ma JF (2017) The key to Mn homeostasis in plants: regulation of Mn transporters. Trends Plant Sci 22(3):215–224. https://doi.org/10.1016/j.tplants.2016.12.005

Socha AL, Guerinot ML (2014) Mn-euvering manganese: the role of transporter gene family members in manganese uptake and mobilization in plants. Front Plant Sci 5:106. https://doi.org/10.3389/fpls.2014.00106

Song WY, Park J, Eisenach C, Maeshima M, Lee Y, Martinoia E (2014) ABC transporters and heavy metals. In: Plant ABC transporters. Springer, Cham, pp 1–17. https://doi.org/10.1199/tab.0153

Sors TG, Ellis DR, Salt DE (2005) Selenium uptake, translocation, assimilation and metabolic fate in plants. Photosynth Res 86:373–389. https://doi.org/10.1007/s11120-005-5222-9

Sun H, Duan Y, Qi X, Zhang L, Huo H, Gong H (2018) Isolation and functional characterization of CsLsi2, a cucumber silicon efflux transporter gene. Ann Bot 122:641–648. https://doi.org/10.1093/aob/mcy103

Sun H, Guo J, Duan Y, Zhang T, Huo H, Gong H (2017) Isolation and functional characterization of CsLsi1, a silicon transporter gene in *Cucumis sativus*. Physiol Plant 159(2):201–214. https://doi.org/10.1111/ppl.12515

Takahashi R, Ishimaru Y, Nakanishi H, Nishizawa NK (2011) Role of the iron transporter OsNRAMP1 in cadmium uptake and accumulation in rice. Plant Signal Behav 6(11): 1813–1816. https://doi.org/10.4161/psb.6.11.17587

Takemoto Y, Tsunemitsu Y, Fujii-Kashino M, Mitani-Ueno N, Yamaji N, Ma JF, Kato SI, Iwasaki K, Ueno D (2017) The tonoplast-localized transporter MTP8. 2 contributes to manganese detoxification in the shoots and roots of *Oryza sativa* L. Plant Cell Physiol 58(9): 1573–1582. https://doi.org/10.1080/15592324.2017.1422466

Talke IN, Hanikenne M, Krämer U (2006) Zinc-dependent global transcriptional control, transcriptional deregulation, and higher gene copy number for genes in metal homeostasis of the hyperaccumulator *Arabidopsis halleri*. Plant Physiol 142(1):148–167. https://doi.org/10.1104/pp.105.076232

Thomine S, Wang R, Ward JM, Crawford NM, Schroeder JI (2000) Cadmium and iron transport by members of a plant metal transporter family in *Arabidopsis* with homology to Nramp genes. Proc Natl Acad Sci U S A 97:4991–4996. https://doi.org/10.1073/pnas.97.9.4991

Tiong J, McDonald G, Genc Y, Shirley N, Langridge P, Huang CY (2015) Increased expression of six ZIP family genes by zinc (Zn) deficiency is associated with enhanced uptake and root-to-shoot translocation of Zn in barley (*Hordeum vulgare*). New Phytol 207(4):1097–1109. https://doi.org/10.1111/nph.13413

Vatansever R, Filiz E, Eroglu S (2017) Genome-wide exploration of metal tolerance protein (MTP) genes in common wheat (*Triticum aestivum*): insights into metal homeostasis and biofortification. Biometals 30(2):217–235. https://doi.org/10.1007/s10534-017-9997-x

Vatansever R, Özyigit İİ, Filiz E (2016) Comparative and phylogenetic analysis of zinc transporter genes/proteins in plants. Turk J Biol 40(3):600–611. https://doi.org/10.3906/biy-1501-91

Verbruggen N, Hermans C, Schat H (2009) Molecular mechanisms of metal hyperaccumulation in plants. New Phytol 181(4):759–776. https://doi.org/10.1111/j.1469-8137.2008.02748.x

Verrier PJ, Bird D, Burla B, Dassa E, Forestier C, Geisler M, Klein M, Kolukisaoglu Ü, Lee Y, Martinoia E, Murphy A (2008) Plant ABC proteins–a unified nomenclature and updated inventory. Trends Plant Sci 13(4):151–159. https://doi.org/10.1016/j.tplants.2008.02.001

Vert G, Briat JF, Curie C (2001) *Arabidopsis* IRT2 gene encodes a root-periphery iron transporter. Plant J 26(2):181–189. https://doi.org/10.1046/j.1365-313x.2001.01018.x

Vidal SM, Malo D, Vogan K, Skamene E, Gros P (1993) Natural resistance to infection with intracellular parasites: isolation of a candidate for Bcg. Cell 73(3):469–485. https://doi.org/10.1016/0092-8674(93)90135-d

Williams PN, Islam MR, Adomako EE, Raab A, Hossain SA, Zhu YG, Feldmann J, Meharg AA (2006) Increase in rice grain arsenic for regions of Bangladesh irrigating paddies with elevated arsenic in groundwaters. Environ Sci Technol 40(16):4903–4908. https://doi.org/10.1021/es060222i

Wintz H, Fox T, Wu YY, Feng V, Chen W, Chang HS, Zhu T, Vulpe C (2003) Expression profiles of *Arabidopsis thaliana* in mineral deficiencies reveal novel transporters involved in metal homeostasis. J Biol Chem 278(48):47644–47653. https://doi.org/10.1074/jbc.M309338200

Xia J, Yamaji N, Kasai T, Ma JF (2010) Plasma membrane-localized transporter for aluminum in rice. Proc Natl Acad Sci 107(43):18381–18385. https://doi.org/10.1073/pnas.1004949107

Xu J, Chai T, Zhang Y, Lang M, Han L (2009) The cation-efflux transporter BjCET2 mediates zinc and cadmium accumulation in *Brassica juncea* L. leaves. Plant Cell Rep 28(8):1235–1242. https://doi.org/10.1007/s00299-009-0723-1

Yan P, Wu L, Wang D, Fu J, Shen C, Li X, Zhang L, Zhang L, Fan L, Wenyan H (2020) Soil acidification in Chinese tea plantations. Sci Total Environ 715:136963. https://doi.org/10.1016/j.scitotenv.2020.136963

Yemane M, Chandravanshi BS, Wondimu T (2008) Levels of essential and non-essential metals in leaves of the tea plant (*Camellia sinensis* L.) and soil of Wushwush farms, Ethiopia. Food Chem 107(3):1236–1243. https://doi.org/10.1016/j.foodchem.2007.09.058

Yuan L, Chen Y, Wei X, Tong H (2017) Cloning and function analysis of metal tolerance gene (CsMTP11) in tea plant (*Camellia sinensis* LO Kuntze). Acta Agron Sin 43(5):708–717

Zhai Z, Gayomba SR, Jung HI, Vimalakumari NK, Piñeros M, Craft E, Rutzke MA, Danku J, Lahner B, Punshon T, Guerinot ML (2014) OPT3 is a phloem-specific iron transporter that is essential for systemic iron signaling and redistribution of iron and cadmium in *Arabidopsis*. Plant Cell 26(5):2249–2264. https://doi.org/10.1105/tpc.114.123737

Zhang T, Liu J, Fellner M, Zhang C, Sui D, Hu J (2017) Crystal structures of a ZIP zinc transporter reveal a binuclear metal center in the transport pathway. Sci Adv 3(8):e1700344. https://doi.org/10.1126/sciadv.1700344

Zhang M, Liu B (2017) Identification of a rice metal tolerance protein OsMTP11 as a manganese transporter. PLoS One 12(4):e0174987. https://doi.org/10.1371/journal.pone.0174987

Zhao K, Liu X, Xu J, Selim HM (2010) Heavy metal contaminations in a soil–rice system: identification of spatial dependence in relation to soil properties of paddy fields. J Hazard Mater 181(1-3):778–787. https://doi.org/10.1016/j.jhazmat.2010.05.081

Zheng L, Yamaji N, Yokosho K, Ma JF (2012) YSL16 is a phloem-localized transporter of the copper-nicotianamine complex that is responsible for copper distribution in rice. Plant Cell 24(9):3767–3782. https://doi.org/10.1105/tpc.112.103820

Metal Nanoparticle Implication, Transport, and Detection in Plants

16

Kareem A. Mosa, Aya Awad, Rand Alhaj Yahya, Shatha N. Alameeri, Kalidoss Ramamoorthy, and Muna A. Ali

Abstract

Widespread dispersion of metal nanoparticles into the environment is mainly due to their applications in various industries. Due to the excessive load of metal nanoparticles, their interaction with the surrounding plants and water bodies in the environment is an alarming concern. Several publications have reported metal nanoparticles' phytotoxicity on various plants highlighting its negative impact on metal nanoparticles. Hence, there is high risk of metal nanoparticles entering into the food chain and becoming cautious to humans. It is inevitable and mandatory to measure the impact of metal nanoparticles, transport, and the possible detection methods. In this book chapter, we summarize both positive and negative impacts of metal nanoparticles, their transport in different plants, their detection methods, and the possible molecular interactions of plant-metal nanoparticles.

Keywords

Metal nanoparticles · Phytotoxicity · Plant growth · Genomics · Proteomics

K. A. Mosa (✉)
Department of Applied Biology, College of Sciences, University of Sharjah, Sharjah, United Arab Emirates

Department of Biotechnology, Faculty of Agriculture, Al-Azhar University, Cairo, Egypt
e-mail: kmosa@sharjah.ac.ae

A. Awad · R. A. Yahya · S. N. Alameeri · K. Ramamoorthy · M. A. Ali
Department of Applied Biology, College of Sciences, University of Sharjah, Sharjah, United Arab Emirates

K. Kumar, S. Srivastava (eds.), *Plant Metal and Metalloid Transporters*,
https://doi.org/10.1007/978-981-19-6103-8_16

Abbreviations

AFM	Atomic force microscope
APX	Ascorbate peroxidase
CA	Chromosomal aberrations
CAD	Cinnamyl alcohol dehydrogenase
CAT	Catalase
GA	Gibberellic acid
GR	Glutathione reductase
HMA	Heavy metal ATPases
IAA	Indole-3-acetic acid
ICP	Inductively coupled plasma
ICP-MS	Inductively coupled plasma mass spectrometry
LA-ICP-Ms	Laser ablation inductively coupled plasma mass spectrometry
m/z	Mass to charge ratio
MI	Mitotic index
miRNA	MicroRNA
MTP	Metal tolerance protein
NGS	Next-generation sequencing
NP	Nanoparticle
PAL	Phenylalanine ammonia lyase
PVP	Polyvinylpyrrolidone
qRT-PCR	Quantitative real-time PCR
ROS	Reactive oxygen species
SAR	Systemic acquired resistance
SEM	Scanning electron microscopy
SEM-EDX	Scanning electron microscopy with energy dispersive X-ray
SOD	Superoxide dismutase
SWCNH	Single-walled carbon nano-horn
TEM	Transmission electron microscopy
t-ZR	Trans-zeatin riboside
XAS	X-ray absorption spectrometry
XRF	X-ray fluorescence spectrometry
ZIP	ZRT, IRT-like protein
μ-XAS	Micro-X-ray absorption spectroscopy
μ-XRF	Micro-X-ray fluorescence mapping

16.1 Introduction

The manufacturing of engineered nanoparticles (NPs) has been quite dominant in recent years and is extensively used in research and other fields. The abundance of NPs and their wrong dispersal in soil and water cause the continuous uptake by plants (Bakshi & Kumar 2021; Maurer-Jones et al. 2013). According to the environmental conditions, the NPs' properties tend to change their aggregation, precipitation, and oxidation states (Amde et al. 2017; Levard et al. 2012). In other terms, the physical and chemical parameters of the environment determine the stability of the NPs. Plants can take NPs through leaves and roots and transport them to the entire system (Ma & Yan 2018; Wang et al. 2013; Raliya et al. 2016). Exposure to high concentration of NPs has been proved to be toxic for the plants, causing damage to the cell wall, plasma membrane, and interfering with various plant metabolic processes. In contrast, minute quantities are required for normal plant growth and metabolism (Zhao et al. 2020; Mazumdar and Ahmed 2011; Mirzajani et al. 2013).

Increased usage of NPs raised major concerns about their impact on several aspects of the human life and the environment. Therefore, to thoroughly comprehend the plant-NP interaction and mechanisms involved in phytotoxicity, highly developed and specialized analytical techniques have been used to identify and specify the plant transport, translocation, cellular internalization, and biotransformation of NPs (Mahdi et al. 2017; Yan et al. 2020). Engineered NPs vary in size and morphological features; they display definite chemical and physical features with contrasting ecological practices and toxicity when compared to their bulk equivalents (Auffan et al. 2009; Chen 2018; Rastogi et al. 2017). Different behavior and translocation characteristics are observed for distinct NPs. Therefore, the major toxic effect caused by NPs does not only depend on the concentration, elemental composition, or initial properties (including particle size, structure, shape, mass concentration, and state of aggregation), but also the physicochemical properties of the NPs in different matrices within the sample play a major role requiring its analysis (Zhang et al. 2012). This chapter summarizes the positive and negative effects of metal NPs, their transport in the plant system, and their detection methods.

16.2 Metal NPs Implications on Plants

The role of NPs in the agriculture field is to enhance plant's growth (Rossi et al. 2019) and seed germination, provide sustained delivery of micronutrients (Gao et al. 2021), increase crop productivity (Kolenčík et al. 2020; Tripathi et al. 2017a; Khot et al. 2012; Reddy et al. 2016), and decrease the effect of toxic metals (Bidi et al. 2021). However, recent studies showed that NPs can influence the development and growth of different plant species either positively or negatively based on the properties of the used NPs in different morphological, physiological, biochemical, and genotoxic levels (Table 16.1 and Fig. 16.1). Currently, NPs are used in many industrial sectors leading to increased risk on the environment due to the extensive exposure to these NPs. Therefore, metal NPs including copper, silver, zinc, titanium, cerium, nickel, iron, aluminum, and other NPs' impacts on different plants are under study as they can result in reduced root growth and germination (Wang et al. 2020), alteration of chlorophyll synthesis (Malandrakis et al. 2021), as well as reduction in the growth and development of several plant species (Dykman & Shchyogolev 2018; Joner et al. 2008). Alternatively, increasing the rate of seed germination and

Table 16.1 Examples of positive and negative effects of various NPs on different plant species

Nanoparticle	Concentration[a]	Plant species	Effects	References
Positive effects				
Fe	50 mg/L	*Oryza sativa* L.	Reduction of arsenic phytotoxicity by strengthening antioxidant defense and down regulating *Lsi1* and *Lsi2* genes involved in arsenic uptake	(Bidi et al. 2021)
ZnO	10 mg/L	*Coffea arabica* L.	Increased fresh and dry weights, and photosynthetic rate	(Rossi et al. 2019)
ZnO	2.6 mg/L	*Helianthus annuus* L.	Increased grain yield, and oil content	(Kolenčík et al. 2020)
TiO_2	2.6 mg/L	*Helianthus annuus* L.	Increased grain yield, and oil content	(Kolenčík et al. 2020)
Negative effects				
Ag	1000 μM	*Cucumis sativus* L.	Reduction in shoot and root lengths, fresh shoot and root biomass, dry shoot and root biomass, and chlorophyll content. Increased oxidative damage	(Tripathi et al. 2017b)
CuO	2000 mg/L	*Oryza sativa*	Inhibited seed germination and early seedling growth	(Wang et al. 2020)
Fe_2O_3	180 μg per plant	*Triticum aestivum*	Caused oxidative stress, degradation of chlorophyll, and plant growth inhibition	(Lu et al. 2020)
CuO	250 mg per plant	*Lactuca sativa* L.	Dry weight reduction, and photosynthetic activity reduction	(Xiong et al. 2017)
CuO	250 mg per plant	*Brassica oleracea* L.	Dry weight reduction, and photosynthetic activity reduction	(Xiong et al. 2017)

[a]Highest concentration only is mentioned when more than one concentration was studied

improving growth and development of different parts for various plant species like tomato, rice, soybean, and corn were observed as a result of being exposed to the single-walled carbon nano-horns (SWCNHs) (Lahiani et al. 2015). In addition, the development of nanotechnology products in the agriculture field, including nano-fertilizers, nano-herbicides, and nano-pesticides, enhances crop productivity (Mali et al. 2020; Fraceto et al. 2016). Recently, NPs can be used as nano-sensors, nano-zeolites, and hydrogels for controlling the strength of the plant and soil and improving their quality (Fraceto et al. 2016). Mesoporous silicon NPS have been developed and used as metalloid NPs to deliver DNA, protein, and several chemicals to

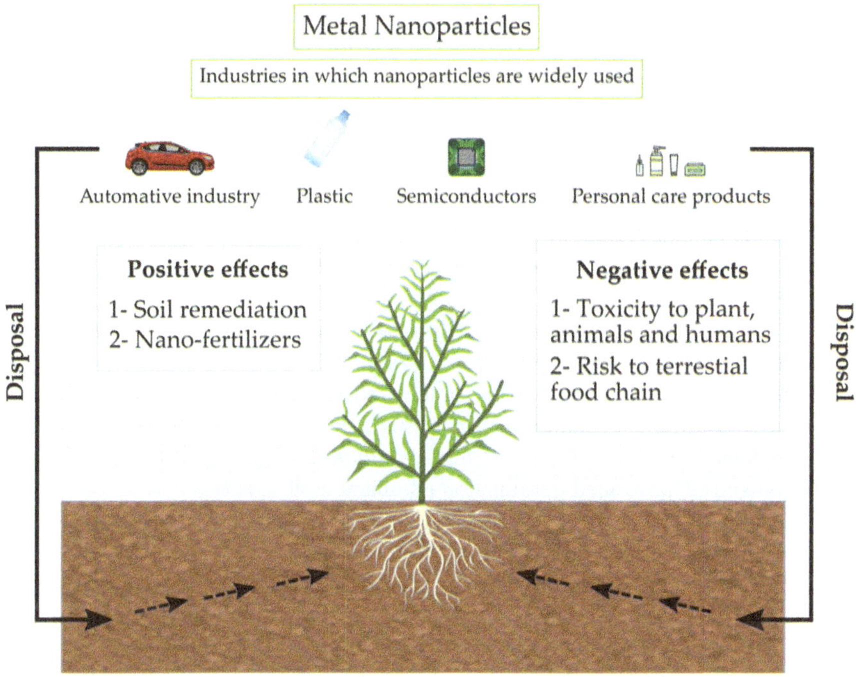

Fig. 16.1 Positive and negative impacts of metal NPs on plants. Metal NPs are widely used in several industries, including automotive industry, plastic, semiconductors, and personal care products. This will lead to the release of metal NPs into the environment. Depending on NP type, properties, and other environmental conditions, NPs could have positive or negative effects on plant growth (Liu et al. 2020)

different plant parts (Torney et al. 2007). Silica NPs showed a nontoxic impact on the plant; however, some research works detected a toxic impact on the plant after they are added due to the decrees in the pH value of the growth media (Slomberg & Schoenfisch 2012). Therefore, the impact of metal NPs on the atmosphere, agriculture, plant, and society can be both progressive and destructive.

16.2.1 Metal NPs Implications on Seed Germination

Based on many factors including the NP type, size, concentration, and exposure time, the effect of NPs can be either positive or negative on different plant species (Acharya et al. 2020; Kasote et al. 2019; Pariona et al. 2017; Almutairi and Alharbi 2015; Mahakham et al. 2017; Savithramma et al. 2012; Vannini et al. 2014; Arora et al. 2012; Parveen and Rao 2015; Yasur and Rani 2013; Pokhrel 2013; Zari et al. 2015; Sundaria et al. 2019).

As several studies were conducted, the observed results showed that the size of some metal NPs has an impact on the phytotoxicity level in which the smaller the size used, the more toxic impact it has on the plants. For example, a more toxic effect of 25 nm copper oxide NPs treated soybean plant was observed in terms of seed yield reduction and oxidative stress induction, when compared to 50 nm and 250 nm NPs (Yusefi-Tanha et al. 2020).

As TiO_2 NPs used for *Brassica napus* plant treatment at 2 and 10 mg/L concentration, improvement in the growth of shoots and seedlings was observed in the plant along with increasing the germination rate and plumule growth (Feizi et al. 2012). On the other hand, different metal NPs have an opposite impact on seed germination and growth such as the exposure of *Brassica nigra* to CuO NPs that resulted in the inhibition of seed germination and growth process (Zafar et al. 2017). After exposing the seeds of both corn and ryegrass to 2000 mg/L and before their incubation, Zn and ZnO NPs resulted in the prevention of seed germination. While in *Arachis hypogaea*, Zn NPs induce an increase in the seed germination at low concentration, it decreases at high concentration (Irmak et al. 2015). Ag NPs treated black-eyed (*Vigna unguiculata* L.), garbanzo (*Cicer arietinum* L.), and lentil (*Lens culinaris*) beans showed reduced seed germination (Budhani et al. 2019).

16.2.2 Metal NPs Implications on Plant Growth and Root Elongation

As metal NPs can induce changes in several plant parts on different levels; physiological, genetic characteristics, and biochemical levels, the inhibition of growth is related to the phytotoxic level, and it is shown as a decrease in the root elongation, leaf growth, flowering delay, root biomass, and yield reduction (Rajput et al. 2021). As stated earlier, in addition to the adverse effects on plant growth, NPs can also be used as nano-fertilizers, nano-pesticides, and growth stimulators (Rastogi et al. 2017). For example, the treatment of *Arabidopsis thaliana* with a low concentration of Ag NPs lead to an increase in the plant growth, while higher concentration resulted in a significant decrease in plant growth (Wang et al. 2019; Kaveh et al. 2013). Moreover, the exposure of *Vigna radiata* and *Sorghum bicolor* plants to Ag NPs promotes the inhibition of root length at high concentration and increases the length at lower concentration (Lee et al. 2012; Bahri et al. 2016). As well, in *Triticum aestivum*, Ag NPs maximize the biomass and reduce the shoot's weight at the concentrations of 50 and 75 mg/L, respectively under heat stress (Iqbal et al. 2019).

The abovementioned NPs ZnO NPs, Ag NPs, and TiO_2 NPs are reported to increase plant growth by enhancing the intake of nutrients and other critical components. However, when *Nicotiana tabacum* plants were exposed to TiO_2 NPs (<25 nm), inhibition of plant growth was detected as it is significantly affected by the microRNAs (miRNAs) expression profile that was recently discovered for the small endogenous non-coding RNAs ($\sim$20–22 bases). Additionally, it is considered a significant gene regulator playing a critical role in the plant development and tolerance to different abiotic stress types including drought, salinity, cold, and heavy metal. Tobacco plants react by regulating gene expression when exposed to several heavy metals and NPs (Frazier et al. 2014). Furthermore, roots are considered the primary target of some NPs as it is the most susceptible zone especially the root apex. Root growth decrease in both *Cicer arietinum* and *Glycine max* plants was detected as a result of being exposed to CuO NPs (Dimkpa et al. 2012). Besides, ZnO NPs showed no effect on the root growth and biomass of *Zea mays* plants at high concentrations, whereas higher concentrations of ZnO NPs (2000 mg/L) prevented root elongation in *Zea mays and Oryza sativa* (Yanık and Vardar 2015).

16.2.3 Metal NP Implications on Photosynthetic Pigments

Photosynthesis is a process used to produce carbohydrates from inorganic substrates using certain pigments to capture the light at a specific wavelength; a number of studies exhibited that metal NPs could have an impact on the pigment content and photosynthetic activity especially on terrestrial plants (Shaw et al. 2014). At the productive stage of *Zea mays* plants, TiO_2 NPs indicated that it could enhance the chloroplast protection from intensive light by improving the pigment content and photosynthesis through regulating the activity of antioxidant enzymes, superoxide dismutase (SOD), catalase (CAT), and peroxidase. Besides, it plays a role as a photo-substance used in the production of pigments to improve the rate of photosynthesis (Hong et al. 2005). Using CuO NPs, ZnO NPs, and Ag NPs was found to have a negative impact on photosynthetic activity and pigments level leading to a decrease in the photosynthetic activity and chlorophyll content of *Vigna radiata* L. when it was exposed to 100 mg/L concentration (Nair and Chung 2014). ZnO NPs reduced the activity when (24 $\pm$ 3 nm) at 800 mg/kg was used, but using 400 mg/kg concentration showed no effect on the *Zea mays* plants (Zhao et al. 2013), and Ag NPs resulted in a reduction in the chlorophyll production and photosynthesis in *Thalassiosira weissflogii* plants (Miao et al. 2009). However, using MnNPs and Fe_2O_3 (6 nm) resulted in enhancing the photosynthetic rate and improving the growth in *Glycine max* and *Vigna radiata*, respectively (Pradhan et al. 2013). Different metallic NPs such as Ag, TiO_2, CeO_2, and Al_2O_3 showed positive effects in different plant species which have been discussed recently (Landa 2021). In addition, the chlorophyll and carotenoid contents were not significantly altered in rice plants exposed to 50 mg/L Fe NPs (Bidi et al. 2021).

16.2.4 Metal NP Implications on Oxidative Stress and Plant Hormones

The accumulation of reactive oxygen species (ROS) can be induced by environmental stress especially when plants are exposed to metal NPs, leading to producing an additional amount of ROS thus, oxidation of proteins, lipids, and DNA in terrestrial plants (Nair and Chung 2014). Depending on the plant species, ROS showed either an increase or decrease in their levels, whereas when *Lolium perenne* and *Cucurbita mixta* plants were exposed to Fe_3O_4 NPs, the elevated levels of ROS showed a significant increase and induced changes in the membrane stability (Wang et al. 2011). Enhanced production of ROS was observed using CeO_2 NPs and Zn NPs as

Brassica rapa and *Allium cepa* resulted in high oxidative stress when *Brassica rapa* was exposed to CeO_2 NPs using 1000 mg/L concentration and *Allium cepa* treated with different concentrations of 0.2, 0.4, and 0.8 g/L of Zn NPs (Ghosh et al. 2016). TiO_2 consists of two crystal phases including rutile and anatase that showed, based on some studies, an increase in the ROS activity in plants. The TiO_2 crystals play a critical role in industrial activities as both phases can be activated from exposure to ultraviolet light. The anatase phase is being used abundantly, and it is considered more toxic than the rutile phase causing a significant increase in the oxidative stresses on the plant that can induce irreversible DNA damage and cell death.

Furthermore, plant hormones which are organic signal molecules that occur at extremely low concentrations and are produced within plants to control the growth, development, and reproduction process can be affected by several NPs (Santner et al. 2009). Auxin and cytokinin of *Capsicum annuum* L. treated with Ag NPs were significantly increased in total in the plant leaves. Studies showed that exposure to Ag NPs at 0.05 mg/L concentrations or lower resulted in reducing the plant biomass, and it was observed that their biomass decreased by 29–42%, where roots decreased by 18–38% in comparison to control plants, which concludes that leaves showed more sensitivity when exposed to NPs compared to the roots (Ward et al. 2019). Alternatively, CeO_2 NPs did not show any influence on the hormones of cotton, since it showed no effect on the indole-3-acetic acid (IAA) or the gibberellic acid (GA) in the leaves of transgenic cotton; however, a decrease up to 25% in the trans-zeatin riboside (t-ZR) was observed when conventional cotton was exposed to a concentration of 500 mg/L (Le et al. 2014).

16.2.5 Metal NP Implications on Plant Enzymes

There are different positive responses that have been reported in response to metal NPs with plants. Watermelon plants sprayed with different metal NPs (B, CuO, MnO, SiO, TiO, and ZnO) showed increased biomass and yield protection from *Fusarium* wilt (Elmer et al. 2018). As plants and agriculture fields are continuously exposed to high amounts of NPs due to industrial activities, a number of changes can induce an impact on the specificity, activity, and stability of microbial enzymes in plants. Thus, studies and measurements about enzyme activities can be used to detect significant changes and effects on the plant and the environment. The effects of different concentrations of ZnO NPs and TiO_2 NPs treatments on *Linum usitatissimum* L. flax plant were studied to detect changes in the enzyme activities and production of secondary metabolites in the phase of cell suspension cultures of the flax plant. Results showed a significant effect on the activity of phenylalanine ammonia lyase (PAL) which is considered the key enzyme of the phenol synthesis process as it is one of the important enzymes for plant's response against a number of biotic and abiotic stresses, changes were detected when the plant was treated with ZnO NPs using 30 mg/L concentration, whereas no significant changes were observed on the PAL enzyme when treated with TiO_2 NPs (Sotelo-Boyás et al. 2016). Similarly, cinnamyl alcohol dehydrogenase (CAD) is a specialized enzyme in the reduction of cinnamaldehydes into cinnamyl alcohols for the polymerization of the cell wall, showing high activity when exposed to both ZnO NPs at 60 mg/L and 150 mg/L and TiO NPs. Phenol recorded its maximum content when treated with 150 mg/L TiO_2 NPs, and lignan showed an increase in the amount of overall intervals when exposed to TiO NPs.

The same results were observed when the plant was treated with ZnO NPs. As s a result, this will enhance the cell wall rigidity and hydrophobic properties and promotes mineral transport cross the vascular pores (Karimzadeh et al. 2019). Rice plants treated with Fe NPs showed no significant difference in the activity of CAT, SOD, ascorbate peroxidase (APX), and glutathione reductase (GR) enzymes. However, under arsenic stress, Fe NP treatment caused a significant increase in the level of the enzymes compared to rice plants untreated with Fe NPs (Bidi et al. 2021).

16.2.6 Metal NP Implications on Plant Morphology

The morphological research has been carried out on some metal NPs to study their effects on different plant species including the changes in biomass, leaf length, roots and shoots length, and the xylem content. Exposure of rice plants to 50 mg/L Fe NPs showed no significant difference in plant dry weight and height compared to control plants (Bidi et al. 2021). Furthermore, studies showed that exposure to ZnO NPs induced increases in the biomass, length of shoots and roots, and the root area when compared to the control treatments conducted on *Cyamopsis tetragonoloba* (López-Moreno et al. 2010). An increase in the uptake of *Glycine max* was observed when treated with 500 mg/L of ZnO NPs which resulted in damage in the leaves (Clapa et al. 2020), while a decrease was shown at higher concentration (1000–4000 mg/L) (López-Moreno et al. 2010). Researchers concluded changes in the content of the meta-xylem in the *Zea mays* when treated with Ag NPs, $AgNO_3$, and $ZnSO_4$ coated with citrate, but ZnO NPs showed no significant effect. Reports have clearly demonstrated that the influence of NPs on plants differs according to various morphological characteristics (Pokhrel 2013).

16.2.7 Metal NP Implications on Plant Physiology

The use of many metal NPs has played a significant role in plant physiology. It may result in physiological changes either directly or indirectly by altering the development of reactive oxygen species and a number of enzymes including dismutase, CAT, and peroxidase, in addition to the production of chlorophyll, phenol, and leaf protein. *Bacopa monnieri* plants treated with Ag NPs showed higher protein and carbohydrate content, lower phenol content, peroxidase, and CAT activities. Additionally, TiO_2 NPs showed an increase in the chlorophyll content, malondialdehyde, CAT, peroxidase, and SOD activities at lower concentrations (200 mg/mL) when *Lemna minor* was exposed to it. On the contrary, TiO_2 NPs resulted in a critical disruption of the cell membrane at higher concentrations (500 mg/mL) due to the elimination of reactive oxygen species (Li et al. 2013). When the *Cucurbita* plant was exposed to Fe_2O_3, no evidence was found on the absorption of NPs which hypothesized that if the size of the NPs added is high, it directly affects the translocation of the NPs into the plasma membrane and the absorption of NPs into cell walls. However, the reported impact of suspended Fe_2O_3 NPs on watermelon in the liquid medium is that suspended NPs had been directly absorbed and translocated into different plant sections, which could contribute to increased seedling germination and improved physiological activity to certain degrees. The positive results of NPs grew steadily and then slowed down with an increase in the dosage of the treatments (Asli and Neumann 2009). The TiO_2 nano-conjugates with Alizarin Red S were instantly ingested and transferred to the *A. thaliana* seedlings. In conclusion, mucilaginous compounds could form a pectin hydrogel capsule around the roots of the plants and play an important role in inhibiting or promoting TiO_2 complexes with Alizarin Red S when exposed to TiO_2 NPs.

16.3 Transport of Metal NPs in Plants

There are two main routes for metal NP uptake by plants: foliar uptake through leaves and root uptake (Fig. 16.2). After uptake, the NPs are translocated through plant tissues. The uptake and translocation routes will be discussed in this section.

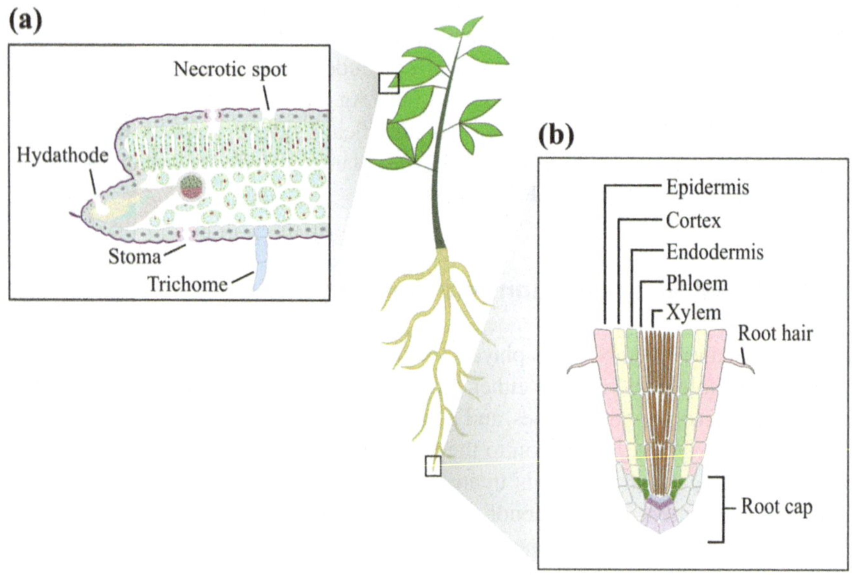

Fig. 16.2 The features that control NP uptake by plants through (**a**) foliar and (**b**) root uptake (adapted from Avellan et al. 2021)

16.3.1 Foliar Uptake and Translocation of Metal NPs

The leaf surface has several types of features and apertures that serve as potential entry points for metal NPs. Polar apertures include stomata, hydathodes, trichomes, and necrosis spots (Fig. 16.2a). In addition, metal NPs can enter through the non-polar leaf cuticles and their pores (Avellan et al. 2021). Most of the studies investigated the foliar uptake of metal NPs through stomata. The stomatal openings are in the range of 10–100 μm (Avellan et al. 2019). Scanning electron microscopy with energy dispersive X-ray (SEM-EDX) analysis of copper oxide NPs after foliar uptake by lettuce and cabbage showed the presence of NP aggregates around stomatal openings (Xiong et al. 2017). Similarly, iron oxide NPs were detected in the stomata by SEM-EDX analysis of wheat seedlings upon foliar treatment with iron oxide NPs (Lu et al. 2020). Additionally, dark field microscopy of wheat leaves after foliar exposure of gold NPs coated with citrate showed the absence of NPs where the stomata are located, possibly indicating their uptake through the stomatal pathway (Avellan et al. 2019).

Hydathodes are small apertures at the leaf tips with a size up to several microns (Bombo et al. 2019; Schwab et al. 2016). They lack a cuticle layer and are directly connected to the vascular system. They secrete the excess water (guttation) from the leaves of many herbaceous plants when transpiration is reduced when the stomata are closed or when the humidity is high (Schwab et al. 2016; Huang 1986). The secreted water droplets may be drawn back in the leaf when the stomata open and transpiration increases (Huang 1986). Thus, NPs suspended in the water droplet can gain access to the leaf and enter the vascular system. There are few studies that investigated the foliar uptake of NP by hydathodes. The uptake of cerium dioxide NPs after foliar application on cucumber leaves was suggested to be through hydathodes and stomata (Hong et al. 2014).

Trichomes are appendages of the epidermis on most of the plant surfaces and play several protective roles (Schwab et al. 2016). In addition, trichomes play important roles in metal absorption and uptake. After the foliar application of $ZnSO_4$ to the leaves of the sunflower plant, trichomes were particularly important for Zn uptake (Li et al. 2019). Furthermore, trichomes of plants with heavy metal tolerance can accumulate heavy metals as part of the detoxification strategy (Koul et al. 2021). Until now, there is no study that investigated the uptake of metal NPs by trichomes after foliar application. Wounds and necrotic zones are other potential routes for metal NPs entry to leaves. After the foliar exposure of lettuce plants to lead, micro X-ray fluorescence mapping (μ-XRF) showed lead-rich spots around necrotic zones of the leaves (Uzu, et al. 2010).

The cuticle layer covers the majority of the leaf surface and has pores that are expected to allow only small hydrophobic NPs (<0.5–2 nm) to enter through this route (Avellan et al. 2021). However, many studies reported the entry of larger or less hydrophobic NPs through the cuticle. The uptake of silver NPs by lettuce after foliar exposure was suggested to be through the cuticle and the stomata. Although uncoated silver NPs were hydrophilic, they might be coated by the cuticular waxes increasing their lipophilicity and allowing their diffusion through the cuticle (Larue et al. 2014). Gold NPs with the size of 12 nm and coated with polyvinylpyrrolidone (PVP) were detected in the leaf mesophyll in areas devoid of stomata. Thus, it was suggested that PVP-coated gold NPs crossed and/or disrupted the cuticle layer. However, in the same study, NPs of similar size coated with citrate remained on the cuticle surface with uptake suggested through the stomata (Avellan et al. 2019). This indicates that the coating affects the uptake route of NPs.

After metallic NPs cross the cuticle and epidermis layers, NPs either follow the symplastic or the apoplastic pathways in the mesophyll before being translocated to the bundle sheath and the phloem (Avellan et al. 2021). The apoplastic pathway occurs when NPs penetrate the cell wall pores and spread into the area falling between the cell wall and the plasma membrane without passing through the cell membrane (Ullah et al. 2020). On the other hand, the symplastic pathway occurs when NPs penetrate the cell membrane and reach the cytoplasm. After that, NPs can translocate to the other plant compartments. After the foliar exposure of arsenic NPs, the NPs were detected in the leaves and roots of *Spinacia oleracea* (Shahid et al. 2019). Similarly, copper oxide NPs were detected in the leaves and roots of lettuce (*Lactuca sativa* L.) and cabbage (*Brassica oleracea* L.) after foliar exposure. Copper concentration increased in the leaves and roots of both species with the increase in the applied dose of copper oxide NPs. Additionally, the values of translocation factors from leaves to roots were significantly higher in cabbage than in lettuce (Xiong et al. 2017). Thus, the plant species is an important factor controlling the uptake and translocation of NPs. Another important factor that affects NP translocation is the coating. Zinc oxide NPs coated with SiO_2 were detected in the dosed leaf, upper leaf, and the stem of the tomato plant after foliar exposure. In contrast, uncoated zinc oxide NPs were only detected in the dosed leaf (Gao et al. 2021). This was suggested to be because there are natural pathways for Si translocation in plants (Wang et al. 2017). In addition, NPs can translocate from the leaves and be released in the rhizosphere. After the foliar exposure of wheat plants with gold NPs coated with citrate or PVP, the NPs were not only detected in the shoots and roots, but also in the soil surrounding the roots (Avellan et al. 2019).

16.3.2 Root Uptake and Translocation of Metal NPs

The other route of metal NPs entry to plants is through the root (Fig. 16.2b). This occurs when metal NPs are released in the soil, landfill, or water. The barriers facing metal NPs after approaching the root include the cuticle, epidermis, cortex, endodermis, and Casparian strip before they enter the xylem (Lv et al. 2019). Plant roots have small root hairs functioning in secreting negatively charged mucilage or small molecules such as organic acids. The negatively charged root surface attracts the positively charged NPs leading to strong adsorption and accumulation of NPs on the root surface (Avellan et al. 2017). Exposing *Triticum aestivum* seedlings to positive, negative, or neutral CeO_2 NPs showed variable accumulation and translocation. Positively charged CeO_2 NPs adhered strongly to plant roots, while negatively

Table 16.2 Examples of NPs detection in different plant tissues

Plant	Nanoparticle (size)	Detected in	References
Triticum aestivum L.	Ag (15 nm)	cytoplasm of apical meristem cells	(Zhang et al. 2019)
	Au (13 nm & 33 nm)		(Zhang et al. 2019)
Triticum aestivum L.	Ag (Non-specified size)	Epidermis, root hairs	(Pradas del Real et al. 2017)
Zea mays	Zno & CuO (Non-specified size)	Periderm, Vascular tissues	(Ahmed et al. 2021)
Arabidopsis thaliana	Au (12 nm)	Detaching border-like cells	(Avellan et al. 2017)
Arabidopsis thaliana	Ag (20 nm, 40 nm, 80 nm)	Border cells, columella cells, columella initials, root cap, lateral root cap, epidermis	(Geisler-Lee et al. 2012)
Zea mays	PbS (15)	Inside the cell wall, intercellular space and cytoplasm of the cortical cell	(Ullah et al. 2020)
Allium cepa L.	NiO (<50 nm)	Epidermal cells, apical meristem cells, cortex, vascular tissues	(Manna et al. 2021)

charged or neutral forms translocated more efficiently to the leaves after 34 h (Spielman-Sun et al. 2017).

Similar to foliar uptake of NPs, metal NPs exposed to roots can be translocated via apoplastic or symplastic pathways. Upon root uptake of PbS NPs by *Zea mays* plant, NPs accumulated inside the cell wall in intercellular space and in the cytoplasm of the cortical cell of roots. This indicated that PbS NPs translocated in maize tissues by apoplastic and symplastic pathways (Ullah et al. 2020). Likewise, Ag NPs were suggested to be transported in wheat roots via apoplastic and symplastic pathways (Pradas del Real et al. 2017). The translocation of Ag NPs by *Medicago sativa* was suggested to be through the apoplastic pathway (Stegemeier et al. 2015). This indicates that metal NPs could follow either apoplastic or symplastic or even both pathways when translocating within plant tissues after uptake by roots. After NPs uptake by roots, NPs were detected in different plant tissues (Table 16.2).

16.4 Detection Methods of Metal NPs

Various advanced analytical techniques have been used in recent years to detect the metal NPs in the plant tissues and subcellular organelles (Castillo-Michel et al. 2017; Shrivastava et al. 2019; Sanzari et al. 2019). Analytical methods must be sensitive enough to measure low concentrations. The methods should also minimize sample disruption to guarantee that laboratory studies represent the unbothered environmental circumstance (Shrivastava et al. 2019; Borm et al. 2006). Since studying NPs during plant-NP interactions is becoming progressively important in research, a variety of analytical tools have been proposed to provide the required details of plant-NP interaction, uptake mechanisms of NP, and their translocation pathway. These approaches include imaging and quantitative analysis techniques. Imaging analysis techniques mainly focus on visualizing the morphology and distribution of NPs under exposure as well as their chemical composition. Meanwhile, quantitative analysis experiments are applied to detect size distribution, particle number, and mass accumulation. Both techniques are usually coupled to provide thorough data on plant-NP interaction in complex biological samples. The following literature will provide a detailed summary of the most used and appropriate techniques available for both imaging to characterize NPs and quantitative analysis to study plant-NP interaction discussing their benefits and limitations. Additionally, this section will propose the information generated from metal NP application in complex environmental media from previous studies (Yan and Chen 2018).

16.4.1 Imaging Analysis

Electron microscopy acts on giving precise information on the predicted field of the particles observing the true dimensions of the particle size, as well as the morphology and structure. Among the popular methods to visualize NPs transmission electron microscopy (TEM) and scanning electron microscopy (SEM), a concentrated electron beam is passed through a specimen during the TEM procedure. Due to the interaction of the sample with electrons, an image is created. It then magnifies the image and focuses on an image sensor. In contrary to light microscopes, TEM is able to image at a high resolution (down to the sub-nanometer), visualizing as little as a single atom column, which is thousands of times smaller than what a light microscope would see. However for SEM procedure, instead of the streak of electrons passing through the sample, the beam is centered on the sample's surface. Due to the interaction of electrons and atoms present on the surface of the sample, signals are produced resulting in the identification of the sample's surface composition and topography (Shrivastava et al. 2019; Stokes 2008). According to one of the researches, analysis of cucumber plant roots treated with CeO_2 NPs by TEM after 21 days showed images of needle-like clusters. These clusters were clearly present in the epidermis of roots (Zhang et al. 2012). Zinc oxide NPs coated with SiO_2 were detected by TEM in the intracellular leaf space of tomato plants (Gao et al. 2021).

Other imaging approaches used in visualizing plant-metal NP interactions and localizations are X-ray fluorescence spectrometry (XRF) and X-ray absorption spectrometry (XAS). These are mainly the two techniques in X-ray spectroscopy. Both act on measuring the absorption of X-radiation or its spectra. The photoelectric effect is the process where the element is controlled during the absorption of X-ray photons. Briefly, once the X-ray radiation is directed on the sample, photons of certain energy will shine. Atoms will only start X-ray photons absorption when the energy of the incident X-ray equals the binding energy of core electrons (Lin et al. 2010). X-ray photons are emitted in a fluorescence form and with certain energy obtained from the difference between the two electronic levels binding energies; this will lead to element specificity characterization based on absorption spectroscopy (Zhang et al. 2012). In XRF, elemental maps are generated based on fluorescence signals recorded at separate positions. This technique is advantageous to localize and identify elemental concentrations in biological samples since it's a non-destructive manner. XRF analysis confirmed the accumulation of copper NPs in cucumber after root exposure to copper NPs, with significantly higher levels of copper detected in roots (Mosa et al. 2018).

On the other hand, XAS focuses on increasing the beam of incident X-ray on samples beyond their binding energy to create a characteristic spectrum of X-ray absorption named the absorption edge (Ma et al. 2011). For bulk analysis, XAS spectroscopy can be used; however, minor elements will not be detected in a complex heterogeneous sample. To overcome this obstacle, microbeam X-ray absorption mapping (μ-XAS) analysis is used which acts on decreasing the size range of the beam to μm instead of nm to detect minor elements (Lv et al. 2019). Recently, a new imaging technique named XANES has been introduced to characterize the entire lateral segregation of elements in a wide area of a sample. An elemental map is created to detect the area of interest and then with the continuous increase of incident X-ray on the area of interest; μ-X-ray fluorescence signals are collected. Through this technique, both major and minor elements can be detected by the XANES spectra obtained by aligning and stacking the elemental maps (Zhang et al. 2012). XANES analysis detected ZnO NPs in the roots of corn seedlings after root exposure to ZnO NPs (López-Moreno et al. 2010). Atomic force microscope (AFM) can be used to characterize various metal NP properties either prior or after interaction with plants (Dimkpa et al. 2012). This technique gives a three-dimensional surface contour image (Shrivastava et al. 2019).

16.4.2 Quantitative Analysis

According to recent studies, it is illustrated that precise quantitative knowledge of metal or rare earth-based NPs still rely on inductively coupled plasma mass spectrometry methods (ICP-MS) to analyze their samples' metallic element concentration for varying sample types involving solid, liquid, and suspensions. ICP-MS instruments contain the sample ion source, inductively coupled plasma (ICP), ion optics, mass analyzer, and detector. For metal analysis, ICP sources are primarily used. It's a perfect source of ionization for mass spectrometry, as it can ionize up to 90% of a variety of elements. Different mass-to-charge classes are covered by the mass spectrophotometer (e.g., ion trap, quadruple or time-of-flight); they vary in mass precision and resolution. First, the liquid sample is converted to fine spray in the sample introduction system that is transferred directly to the argon plasma. Then the sample is ionized through high-temperature plasma to create ions that would be extracted and set into an electrostatic lens known as ion optics. The ion optics would focus the ion beam into a mass spectrophotometer which would filter the ions based on their mass-charge ratio (m/z) to be identified by the detector (Wilschefski and

Baxter 2019). In a study performed on AuNPs treated watermelons, ICP-MS confirmed the accumulation of AuNPs within *C. lanatus* with a concentration of 58.6%, 28.0%, and 13.4% in the stem, leaves, and roots, respectively (Raliya et al. 2016). In addition, ICP-MS was used to detect the accumulation of cerium in radish plants after treatment with CeO_2 NPs and showed the highest accumulation in roots (Gui et al. 2017).

One of the most widely used methods is laser ablation inductively coupled plasma mass spectrometry (LA-ICP-MS). The laser ablation procedure directs a laser beam to the surface of the sample generating fine particles. Ablated particles are then transferred to the ICP-MS instrument to detect elemental analysis (Wen et al. 2007). LA-ICP-MS has a lower lateral resolution than that of electron or X-ray microanalysis. Nonetheless, it allows for in situ determination of metal distribution in biological materials (Pozebon et al. 2017). Research was performed on rice plants that were subjected to AuNPs that were positively, neutrally, or negatively charged had analyzed the interaction through LA-ICP-MS to detect level of dissemination of differently charged elemental AuNPs in rice tissue. Surface charges had a major effect on AuNP absorption and translocation in the roots, according to the results (Koelmel et al. 2013). LA-ICP-MS was utilized as an in situ monitoring tool to detect the behavior of SiNPs in sweet basil after root exposure. It revealed the accumulation of SiNPs in specific regions such as main veins, leaf margins, and tips (Ko et al. 2019).

ICP-MS can be used to classify individual particles in the single-particle mode, in a process called single-particle inductively coupled plasma mass-spectrometry (SP-ICP-MS). SP-ICP-MS has been commonly used in biological samples to quantify particle size, distribution, concentration, and elemental composition of NPs, making it an important instrument in NP quantification within matrices (Pace et al. 2012). The mass spectrophotometer used during the process focused on the detection of pulse signals generated by each NP that goes through the plasma. Signal intensity and frequency are affected by the particles' size and concentration, making SP-ICP-MS a responsive technique for low concentration analysis of NPs (Fleischer et al. 1999). A strong enzymatic digestion technique is applied to let out the NPs from the matrix in order to deal with biological tissues. However, this threatens the final analysis data due to the possible dissolution of the metal NPs. To clear this dilemma, a study carried out by Dan et al. showed high recovery of AuNPs from tomato plants without causing any dissolution by using a Macerozyme R-10 enzyme (Dan et al. 2015). However, so far, the implementation of SP-ICP-MS has been mainly limited to dormant NPs such as Au, Ag, CeO_2, CuO, and TiO_2. In recent studies, SP-ICP-MS quantitative analysis was conducted on *Arabidopsis thaliana* treated with 10 nm Ag NPs to detect the size of the distribution. After successfully releasing Ag NPs from plant matrices by enzymatic digestion, SP-ICP-MS was utilized and characteristics of Ag NPs obtained showed distribution in both shoots and roots agreeing with TEM micrograph results. This indicates that the enzymatic digestion combined with SP-ICP-MS generates a detailed quantitative analysis (Bao et al. 2016).

16.5 Molecular Analysis of Metal NPs

Either the chemical or mechanical interactions can happen between plants and metal NPs based on the size, surface area, and catalytic reactivity of various metal NPs (Dietz and Herth 2011). The positive and negative impacts of the metal NPs completely depend on such interactions; hence, the understanding of the interactions should be thoroughly understood. The United States Environmental Protection Agency advised researchers to conduct a detailed analysis of a phytotoxic study which focuses on seed germination and root elongation upon NP treatment (EPA 1996). These morphological and physiological measurements may not be sensitive enough when assessing the toxicity of NPs. Several biochemical indicators, such as metabolite structure, membrane integrity, and enzyme activity, have been used to assess the effect of environmental stress on plants to date. However, there are numerous, complex, and still inadequately known modes of action of NPs on the cellular and molecular levels (Jha and Pudake 2016). Table 16.3 is showing examples of the effects of different NPs on different plant species at the molecular levels. Details will be discussed below.

16.5.1 Plant-NP Interactions: Genomic and Transcriptomic Approaches

In general, gene expression analysis is conducted along with morphological and/or physiological research to understand the molecular insights of the complex pathways (Babajani et al. 2019). Gene expression tests are commonly used to complement morphological and/or physiological research in order to gain valuable insight into complex pathways of toxicity and modes of action that cannot be achieved by other means (Ankley et al. 2006). To analyze modifications in gene expression, high-throughput genomic tools are used such as cDNA microarrays and quantitative real-time PCR (qRT-PCR). While transcriptional studies in a variety of species, including bacteria, humans, and mammalian cell lines, have been widely used to investigate the molecular mechanism of NP toxic effect (Asharani et al. 2009), only narrow studies have been conducted out to evaluate the specific basis of molecular interactions between metal NPs and plant phytotoxicity. Xu et al. used high-throughput methods such as cDNA microarrays and qRT-PCR analysis to understand the molecular responses in plants exposed to metal NPs (Xu et al. 2011). A thorough evaluation of gene expression studies in plants exposed to various forms of metal NPs will be discussed to showcase a connection between the phenotypic effects of plants under NP treatment and the associated genes.

Wheat plants treated with Ag NPs showed cellular transcriptome and proteome levels were altered due to NPs (Dimkpa et al. 2012). By metal ion sequestration, Ag NPs induced oxidative stress as shown by an increase in oxidized glutathione and high expression of a metallothionein gene involved in detoxification. Moreover, Ag NPs caused plant defense in *Arabidopsis*, which was identified by the upregulation of genes related to pathogenesis known as PR1, PR2, and PR5 involved in systemic

Table 16.3 Examples of the effects of NPs on different plant species at the genomic, transcriptomic, and proteomic levels

Plant species	Plant part	Nanoparticle	Molecular analysis technique	Effect[a]	References
Genomic level					
Arabidopsis thaliana	Rosette leaves	Ag	qRT-PCR	↑: *PR1*, *PR2*, *PR5*	(Chu et al. 2012)
Arabidopsis thaliana	Whole plant	Ag	Microarray	↑: 286 genes mainly linked to metal and oxidative stress such as SOD, proton exchanger and peroxidase	(Kaveh et al. 2013)
				↓: 81 genes responsible for plant defense mechanisms and hormonal stimulants such as auxin-regulated gene involved in organ size-ARGOS, ethylene signaling pathway, and SAR against pathogens	
Nicotiana tabacum	Whole plant	TiO_2	qRT-PCR	↑: miR395, miR399, *APX*, *ADH*	(Frazier et al. 2014)
Arabidopsis thaliana	Root	Au	Microarray, qRT-PCR	↓: *IRT1*, *IRT2*, *COPT2*, *ZIP*	(Taylor et al. 2014)
Proteomic level					
Eruca sativa	Root	Ag	2D-IEF/ SDS-PAGE[b], CHIP-q-TOF MS/MS	↑: Proteins involved in metabolism, defence/ stress related proteins	(Vannini et al. 2013)
				↓: Transport proteins, cell cycle proteins	
Nicotiana tabacum)	Root	Ag	2-DE/SDS-PAGE, mass spectrometry	↑↓: Proteins involved in response to abiotic and biotic stimuli, response to oxidative stress	(Peharec Štefanić et al. 2019)
				↓: Proteins involved in protein biosynthesis	
	Leaf			↑↓: Proteins involved in photosynthesis,	

(continued)

Table 16.3 (continued)

Plant species	Plant part	Nanoparticle	Molecular analysis technique	Effect[a]	References
				response to abiotic and biotic stimuli	
				↓: Proteins involved in glycolytic processes, mRNA processing, translation, protein folding	
				↑: Proteins involved in proteolysis	

[a]Effect on gene expression at transcriptomic level, or proteins at proteomic level. ↑: upregulation, ↓: downregulation, ↑↓: differentially expressed

[b]2D-IEF/SDS-PAGE: Two-dimensional isoelectrofocusing/sodium dodecyl sulphate-polyacrylamide gel electrophoresis

acquired resistance (SAR) (Chu et al. 2012). Interestingly, on a transcriptional level in *Arabidopsis* plants treated with Ag NPs, whole-genome cDNA expression microarray analysis was performed and marking an upregulation in 286 genes mainly linked to metal and oxidative stress such as SOD, proton exchanger, and peroxidase. On the other hand, the exposure led to downregulation of 81 genes responsible for plant defense mechanisms and hormonal stimulants such as auxin-regulated gene involved in organ size-ARGOS, ethylene signalling pathway, and SAR against pathogens (Kaveh et al. 2013). In further studies conducted on Ag NPs to examine toxicological endpoints such as mitotic index (MI), chromosomal aberrations (CA), and micronucleus induction, *V. faba* root tip cells were treated with different concentrations of Ag NPs. Results confirm that Ag NP exposure induced the number of CA was directly proportional with the Ag NP concentration, whereas MI reduction was observed in the treated plants compared to control. A significant decrease was demonstrated with the increase of Ag NPs dose. This study indicates that Ag NPs could have entered the plant system and damaged mitosis leading to CA and MN (Patlolla et al. 2012).

Terrestrial plants treated with CuO NPs showed induced accumulation of oxidative mutagenic DNA lesions compared to control plants (Atha et al. 2012). CuO NP disrupted the photosynthesis process of the *Lemna gibba* plant, resulting in the suppression of photosystem II reaction centers, a decline in electron transport, and a reduction in photosynthetic pigments (Perreault et al. 2014). TiO_2, which has a negative effect on plant growth and production, is another metal NP that has a negative impact. The miRNAs are one of the post-transcriptional gene regulators belonging to a limited endogenous class of noncoding RNAs. Mainly, miRNAs act by degrading targeted mRNAs or inhibiting their translation to alter gene expression (Zhang et al. 2006). MiRNAs, which regulate gene expression, have been shown to play an important role in plant response to NPs. When tobacco plants were exposed to TiO_2, NPs affected miRNA expression levels (Frazier et al. 2014). Tobacco seedlings went through growth reduction upon exposure to a small amount of TiO_2, significantly activating miR395 and miR399 expression by 285-fold and 143-fold, respectively.

Microarray studies are usually conducted to determine genes responsible for the uptake of metal NPs and transport throughout the plant. In regard to gold NPs, no previous analysis was conducted. However, microarray analysis of *Cupriavidus metallidurans* bacterium has detected induced expression levels of metal homeostasis genes, when exposed to gold NPs (Reith et al. 2009). The orthologs of these genes in the *Arabidopsis* plant were *HMA5* (At1g63440), *mtLPD1* (At1g48030), and *mtHSC1* (At4g37910). They played a role in the response and absorption of a variety of metals. For analysis, qPCR was done using cDNA obtained from root tissue of the *Arabidopsis* plant. Results showed that three genes encoding channel proteins involved in metal transport were down regulated. Iron-regulated transporter *IRT1* was downregulated by 132-folds as well as *IRT2* expression by 38.4-folds. Furthermore, the copper transporter (*COPT2*) and nickel transporter (*ZIP*) were also downregulated by 22.8- and 23.6-fold, respectively. Since gold NPs were able to activate these operating mechanisms in response to toxicity, by downregulating cation transporters and minimizing the absorbance of gold NPs, it indicates that gold exists in the environment as a cation form and that these transporters are sensitive to gold (Taylor et al. 2014). Zn homeostasis in plants is controlled through various transporter proteins (Clemens et al. 2002). ZIP (ZRT, IRT-like proteins), HMA (heavy metal ATPases), and MTP (metal tolerance protein) transporter protein are some of the most known families. For Zn acquisition in *Arabidopsis thaliana*, *AtZIP4*, *AtZIP9*, and *AtZIP12* are involved to transport Zn from the rhizosphere to the aerial tissues (Jain et al. 2013). Root-to-shoot Zn translocation is allowed by *AtHMA3* and *AtHMA4* (Hussain et al. 2004). *AtMTP3* is important in the process of detoxifying excessive Zn (Arrivault et al. 2006). Therefore, to prevent surplus Zn toxicity, a group of molecular working systems acts together to handle Zn toxicity (Lin and Aarts 2012). *Arabidopsis thaliana* treated with ZnO NPs showed a decrease in the expression of ZIP family genes in roots and shoots. The excess presence of Zn has proved to downregulate the ZIP gene. Therefore, the act of suppression works to sustain Zn stability by forbidding the consumption and accumulation of excess Zn. However, in regard to *AtHMA3* and *AtHMA4*, high expression was noticed in shoots after exposure to ZnO NPs, which could prevent the entrance of excess Zn into the vascular system, followed by a decrease in expression of *AtMTP1* and *AtMTP3* in shoots and roots. This indicates that exposure to ZnO NPs induced the transcription of genes responsible for metal ion homeostasis (Sturikova et al. 2018). Moreover, a significant increase was noticed in Fe-uptake genes (*AtIRT1*, *AtIRT2*, and *AtFRO2*). Previous research claims that Fe accumulation is decreased in the presence of Zn toxicity, causing the upregulation of *AtIRT1*, *AtIRT2*, and *AtFRO2* genes (Fukao et al. 2011).

16.5.2 Plant-NP Interactions: Proteomic Approach

Similar to genome analysis of the phytotoxic study, proteomic approaches also shed a light on understanding the effects of metal NPs by proteome levels of the treated plants. Usually, observed phenotypic traits of the NP-exposed plants are strongly linked to the changes in protein build-up, since alteration in transcript level doesn't always go along with the change in protein expression. It is therefore extremely necessary to investigate changes in plant proteome because proteins are direct effectors of a plant stress response. Proteomic technologies play a great role in analyzing post-translational modifications, protein-protein/toxicant interactions, function, and subcellular localization analysis. Techniques include the major approach differential protein expression profiling, as well as two-dimensional polyacrylamide gel electrophoresis. A study on *Eruca sativa* roots treated with Ag NPs and $AgNO_3$ analyzed the plant proteome levels. Both Ag NPs and $AgNO_3$ shared common responses and actions such as the accumulation of proteins in charge of sulfur metabolism, proteins involved in transport, cell cycle, protein folding, stress/defense response, and activation of ROS detoxification pathways (Vannini et al. 2013). Therefore, focusing on proteomic research will aid in the identification and detection of main proteins when exposed to metal NPs acting as biomarkers for NP phytotoxicity. The comparative proteomic approach reveals that Ag and $AgNO_3$ NPs treated with *Nicotiana tabacum* plants showed biotic, abiotic stimuli, and oxidative stress proteins were differentially regulated in roots and photosynthetic pigment proteins were downregulated in leaves (Peharec Štefanić et al. 2019). A positive response of aluminum oxide NPs exposed to soybean plants stressed with flood stress plants improved survival percentage, and biomass increase was shown using quantitative proteomic study (Yasmeen et al. 2016).

16.6 Conclusion

Now, we summarized the positive and negative effects of various metal NPs on plants. Based on the type and properties of the used NPs, different morphological, physiological, biochemical, and genotoxic effects are observed. Metal NPs are transported from the environment to plants via foliar and root uptake. A wide variety of analytical techniques based on imaging and quantitative analysis were employed to illustrate plant-NP interaction. After plant uptake, metal NPs can alter seed germination, plant growth, plant morphology, phytohormones, and photosynthesis and can induce oxidative stress. Hence, it's important to understand the molecular mechanism of plant-NP interaction in order to understand and study the fate of NPs in our environment. NP phytotoxicity assessment is a crucial step to synthesize NPs with minimum phytotoxic effects on plants. Along with various advanced technologies discussed here to elucidate the molecular mechanisms, still the latest technique like next-generation sequencing (NGS) to study the genome of many plants, which its genome is unknown, can be implicated.

References

Acharya P, Jayaprakasha GK, Crosby KM et al (2020) Nanoparticle-mediated seed priming improves germination, growth, yield, and quality of watermelons (Citrullus lanatus) at multi-locations in Texas. Sci Rep 10(1):1–16. https://doi.org/10.1038/s41598-020-61696-7

Ahmed B, Rizvi A, Syed A, Elgorban AM, Khan MS, HA AL-S, Musarrat J, Lee J (2021) Differential responses of maize (Zea mays) at the physiological, biomolecular, and nutrient levels when cultivated in the presence of nano or bulk ZnO or CuO or Zn^{2+} or Cu^{2+} ions. J Hazard Mater 419:126493. https://doi.org/10.1016/j.jhazmat.2021.126493

Almutairi Z, Alharbi A (2015) Effect of silver nanoparticles on seed germination of crop plants. J Adv Agric 4(1):280–285. https://doi.org/10.24297/jaa.v4i1.4295

Amde M, Liu J, Tan Z et al (2017) Transformation and bioavailability of metal oxide nanoparticles in aquatic and terrestrial environments. A Rev Environ Pollut 230(1987):250–267. https://doi.org/10.1016/j.envpol.2017.06.064

Ankley GT, Daston GP, Degitz SJ, Denslow ND, Hoke RA, Kennedy SW, Miracle AL, Perkins EJ, Snape J, Tillitt DE, Tyler CR, Versteeg D (2006) Toxicogenomics in regulatory ecotoxicology. Environ Sci Technol 40(13):4055–4065. https://doi.org/10.1021/es0630184

Arora S, Sharma P, Kumar S et al (2012) Gold-nanoparticle induced enhancement in growth and seed yield of Brassica juncea. Plant Growth Regul 66(3):303–310. https://doi.org/10.1007/s10725-011-9649-z

Arrivault S, Senger T, Krämer U (2006) The Arabidopsis metal tolerance protein AtMTP3 maintains metal homeostasis by mediating Zn exclusion from the shoot under Fe deficiency and Zn oversupply. Plant J 46(5):861–879. https://doi.org/10.1111/j.1365-313X.2006.02746.x

AshaRani PV, Low Kah Mun G, Hande MP, Valiyaveettil S (2009) Cytotoxicity and genotoxicity of silver nanoparticles in human cells. ACS Nano 3(2):279–290. https://doi.org/10.1021/nn800596w

Asli S, Neumann PM (2009) Colloidal suspensions of clay or titanium dioxide nanoparticles can inhibit leaf growth and transpiration via physical effects on root water transport. Plant Cell Environ 32(5):577–584. https://doi.org/10.1111/j.1365-3040.2009.01952.x

Atha DH, Wang H, Petersen EJ, Cleveland D, Holbrook RD, Jaruga P, Dizdaroglu M, Xing B, Nelson BC (2012) Copper oxide nanoparticle mediated DNA damage in terrestrial plant models. Environ Sci Technol 46(3):1819–1827. https://doi.org/10.1021/es202660k

Auffan M, Rose J, Bottero J et al (2009) Towards a definition of inorganic nanoparticles from an environmental, health and safety perspective. Nature Nanotech 4(10):634–641. https://doi.org/10.1038/nnano.2009.242

Avellan A, Yun J, Morais BP, Clement ET, Rodrigues SM, Lowry GV (2021) Critical review: role of inorganic nanoparticle properties on their foliar uptake and in planta translocation. Environ Sci Technol 55(20):13417–13431. https://doi.org/10.1021/acs.est.1c00178

Avellan A, Yun J, Zhang Y, Spielman-Sun E, Unrine JM, Thieme J, Li J, Lombi E, Bland G, Lowry GV (2019) Nanoparticle size and coating chemistry control foliar uptake pathways, translocation, and leaf-to-rhizosphere transport in wheat. ACS Nano 13:5291–5305. https://doi.org/10.1021/acsnano.8b09781

Avellan A, Schwab F, Masion A, Chaurand P, Borschneck D, Vidal V, Rose J, Santaella C, Levard C (2017) Nanoparticle uptake in plants: gold nanomaterial localized in roots of Arabidopsis thaliana by X-ray computed Nanotomography and hyperspectral imaging. Environ Sci Technol 51(15):8682–8691. https://doi.org/10.1021/acs.est.7b01133

Babajani A, Iranbakhsh A, Oraghi Ardebili Z, Eslami B (2019) Differential growth, nutrition, physiology, and gene expression in Melissa officinalis mediated by zinc oxide and elemental selenium nanoparticles. Environ Sci Pollut Res Int 26(24):24430–24444. https://doi.org/10.1007/s11356-019-05676-z

Bahri S, Bhatia SS, Moitra S et al (2016) Influence of silver nanoparticles on seedlings of Vigna radiata (L.) R. Wilczek. Delhi Univ J Undergrad Res Innov 2:142–148

Bakshi M, Kumar A (2021) Copper based nanoparticles in the soil-plant environment: assessing their applications, interactions, fate, and toxicity. Chemosphere:130940. https://doi.org/10.1016/j.chemosphere.2021.130940

Bao D, Oh ZG, Chen Z (2016) Characterization of silver nanoparticles internalized by arabidopsis plants using single particle ICP-MS analysis. Front Plant Sci 7:32. https://doi.org/10.3389/fpls.2016.00032

Bidi H, Fallah H, Niknejad Y et al (2021) Iron oxide nanoparticles alleviate arsenic phytotoxicity in rice by improving iron uptake, oxidative stress tolerance and diminishing arsenic accumulation. Plant Physiol Biochem 163:348–357. https://doi.org/10.1016/j.plaphy.2021.04.020

Bombo AB, Pereira AES, Lusa MG et al (2019) A mechanistic view of Interactions of a nanoherbicide with target organism. J Agric Food Chem 67(16):4453–4462. https://doi.org/10.1021/acs.jafc.9b00806

Borm PJA, Robbins D, Haubold S, Kuhlbusch T, Fissan H, Donaldson K, Schins R, Stone V, Kreyling W, Lademann J, Krutmann J, Warheit D, Oberdorster E (2006) The potential risks of nanomaterials: a review carried out for ECETOC. Part Fibre Toxicol 3(1):11. https://doi.org/10.1186/1743-8977-3-11

Budhani S, Egboluche NP, Arslan Z et al (2019) Phytotoxic effect of silver nanoparticles on seed germination and growth of terrestrial plants. J Environ Sci Health C 37(4):330–355. https://doi.org/10.1080/10590501.2019.1676600

Castillo-Michel HA, Larue C, Pradas del Real AE, Cotte M, Sarret G (2017) Practical review on the use of synchrotron based micro- and nano- X-ray fluorescence mapping and X-ray absorption spectroscopy to investigate the interactions between plants and engineered nanomaterials. Plant Physiol Biochem 110:13–32. https://doi.org/10.1016/j.plaphy.2016.07.018

Chen H (2018) Metal based nanoparticles in agricultural system: behavior, transport, and interaction with plants. Chem Speciat Bioavailab 30(1):123–134. https://doi.org/10.1080/09542299.2018.1520050

Chu H, Kim H-J, Su Kim J, Kim M-S, Yoon B-D, Park H-J, Kim CY (2012) A nanosized ag–silica hybrid complex prepared by γ-irradiation activates the defense response in Arabidopsis. Radiat Phys Chem 81(2):180–184. https://doi.org/10.1016/j.radphyschem.2011.10.004

Clapa D, Borsai O, Leopold L et al (2020) The effect of TiO2 and ZnO2 nanoparticles upon some biometrical characteristics in soybean (Glycine max L. Merril) in vitro cultures. Sci Bull Ser F Biotechnol 24:31–36

Clemens S, Palmgren MG, Krämer U (2002) A long way ahead: understanding and engineering plant metal accumulation, vol 7. Elsevier Ltd., Oxford, pp 309–315

Dan Y, Zhang W, Xue R, Ma X, Stephan C, Shi H (2015) Characterization of gold nanoparticle uptake by tomato plants using enzymatic extraction followed by single-particle inductively coupled plasma–mass spectrometry analysis. Environ Sci Technol 49(5):3007–3014. https://doi.org/10.1021/es506179e

Dietz K-J, Herth S (2011) Plant nanotoxicology. Trends Plant Sci 16(11):582–589. https://doi.org/10.1016/j.tplants.2011.08.003

Dimkpa C, McLean J, Latta D et al (2012) CuO and ZnO nanoparticles: phytotoxicity, metal speciation, and induction of oxidative stress in sand-grown wheat. J Nanopart Res 14(9):1–15. https://doi.org/10.1007/s11051-012-1125-9

Dykman L, Shchyogolev S (2018) The effect of gold and silver nanoparticles on plant growth and development. In: Saylor Y, Irby V (eds) Metal nanoparticles. Nova Science Publishers, New York, pp 263–300

Elmer W, De La Torre-Roche R, Pagano L et al (2018) Effect of metalloid and metal oxide nanoparticles on fusarium wilt of watermelon. Plant Dis 102(7):1394–1401. https://doi.org/10.1094/PDIS-10-17-1621-RE

EPA U (1996) US environmental protection agency: ecological effects test guidelines (OPPTS 850.4200): seed/germination/root elongation toxicity test. US environmental protection agency, Washington, DC

Feizi H, Rezvani Moghaddam P, Shahtahmassebi N et al (2012) Impact of bulk and nanosized titanium dioxide (TiO2) on wheat seed germination and seedling Growth. Biol Trace Elem Res 146(1):101–106. https://doi.org/10.1007/s12011-011-9222-7

Fleischer A, O'Neill MA, Ehwald R (1999) The pore size of non-graminaceous plant cell walls is rapidly decreased by borate ester cross-linking of the pectic polysaccharide rhamnogalacturonan II1. Plant Physiol 121(3):829–838. https://doi.org/10.1104/pp.121.3.829

Fraceto LF, Grillo R, de Medeiros GA et al (2016) Nanotechnology in agriculture: which innovation potential does it have? Front Environ Sci 4. https://doi.org/10.3389/fenvs.2016.00020

Frazier T, Burklew C, Zhang B (2014) Titanium dioxide nanoparticles affect the growth and microRNA expression of tobacco (Nicotiana tabacum). Funct Integr Genomics 14(1):75–83. https://doi.org/10.1007/s10142-013-0341-4

Fukao Y, Ferjani A, Tomioka R, Nagasaki N, Kurata R, Nishimori Y, Fujiwara M, Maeshima M (2011) iTRAQ analysis reveals mechanisms of growth defects due to excess zinc in Arabidopsis. Plant Physiol 155(4):1893–1907. https://doi.org/10.1104/pp.110.169730

Gao X, Kundu A, Bueno V, Rahim AA, Ghoshal S (2021) Uptake and translocation of mesoporous SiO_2-coated ZnO nanoparticles to Solanum lycopersicum following foliar application. Environ Sci Technol 55(20):13551–13560. https://doi.org/10.1021/acs.est.1c00447

Geisler-Lee J, Wang Q, Yao Y, Zhang W, Geisler M, Li K, Huang Y, Chen Y, Kolmakov A, Ma X (2012) Phytotoxicity, accumulation and transport of silver nanoparticles by Arabidopsis thaliana. Nanotoxicology 7(3):323–337. https://doi.org/10.3109/17435390.2012.658094

Ghosh M, Jana A, Sinha S et al (2016) Effects of ZnO nanoparticles in plants: Cytotoxicity, genotoxicity, deregulation of antioxidant defenses, and cell-cycle arrest. Mutat Res Genet Toxicol Env Mutagen 807:25–32. https://doi.org/10.1016/j.mrgentox.2016.07.006

Gui X, Rui M, Song Y et al (2017) Phytotoxicity of CeO2 nanoparticles on radish plant (Raphanus sativus). Environ Sci Pollut Res Int 24(15):13775–13781. https://doi.org/10.1007/s11356-017-8880-1

Hong J, Peralta-Videa JR, Rico C et al (2014) Evidence of translocation and physiological impacts of foliar applied CeO2 nanoparticles on cucumber (Cucumis sativus) Plants. Environ Sci Technol 48(8):4376–4385. https://doi.org/10.1021/es404931g

Hong F, Zhou J, Liu C et al (2005) Effect of nano-TiO2 on photochemical reaction of chloroplasts of spinach. Biol Trace Elem Res 105(1):269–279. https://doi.org/10.1385/BTER:105:1-3:269

Huang J (1986) Ultrastructure of bacterial penetration in plants. Annu Rev Phytopathol 24(1): 141–157. https://doi.org/10.1146/annurev.py.24.090186.001041

Hussain D, Haydon MJ, Wang Y, Wong E, Sherson SM, Young J, Camakaris J, Harper JF, Cobbett CS (2004) P-type ATPase heavy metal transporters with roles in essential zinc homeostasis in Arabidopsis. Plant Cell 16(5):1327–1339. https://doi.org/10.1105/tpc.020487

Iqbal M, Raja NI, Mashwani Z et al (2019) Assessment of AgNPs exposure on physiological and biochemical changes and antioxidative defence system in wheat (Triticum aestivum L) under heat stress. IET Nanobiotechnol 13(2):230–236. https://doi.org/10.1049/iet-nbt.2018.5041

Irmak S, Cil AN, Yucel H et al (2015) Effects of zinc application on yield and some yield components in peanut arachis hypogaea in the easthern mediterranean region. J Agric Sci 22(1):109–116. https://doi.org/10.1501/Tarimbil_0000001373

Jain A, Sinilal B, Dhandapani G, Meagher RB, Sahi SV (2013) Effects of deficiency and excess of zinc on morphophysiological traits and spatiotemporal regulation of zinc-responsive genes reveal incidence of cross talk between micro- and macronutrients. Environ Sci Technol 47(10):5327–5335. https://doi.org/10.1021/es400113y

Jha S, Pudake RN (2016) Molecular mechanism of plant–nanoparticle interactions. In: Plant nanotechnology. Springer International Publishing, Cham, pp 155–181

Joner EJ, Hartnik T, Amundsen CE (2008) Environmental fate and ecotoxicity of engineered nanoparticles. Norw Pollut Cont Authority 159

Karimzadeh F, Haddad R, Garoosi G, Khademian R (2019) Effects of nanoparticles on activity of lignan biosynthesis enzymes in cell suspension culture of linum usitatissimum L. Russ J Plant Physiol 66(5):756–762. https://doi.org/10.1134/S1021443719050078

Kasote DM, Lee JHJ, Jayaprakasha GK et al (2019) Seed priming with iron oxide nanoparticles modulate antioxidant potential and defense-linked hormones in watermelon seedlings. ACS Sustain Chem Eng 7(5):5142–5151. https://doi.org/10.1021/acssuschemeng.8b06013

Kaveh R, Li Y-S, Ranjbar S, Tehrani R, Brueck CL, Van Aken B (2013) Changes in Arabidopsis thaliana gene expression in response to silver nanoparticles and silver ions. Environ Sci Technol 47(18):10637–10644

Khot LR, Sankaran S, Maja JM et al (2012) Applications of nanomaterials in agricultural production and crop protection: A review. Crop Prot 35:64–70. https://doi.org/10.1016/j.cropro.2012.01.007

Ko JA, Furuta N, Lim HB (2019) New approach for mapping and physiological test of silica nanoparticles accumulated in sweet basil (Ocimum basilicum) by LA-ICP-MS. Anal Chim Acta 1069:28–35. https://doi.org/10.1016/j.aca.2019.04.033

Koelmel J, Leland T, Wang H, Amarasiriwardena D, Xing B (2013) Investigation of gold nanoparticles uptake and their tissue level distribution in rice plants by laser ablation-inductively coupled-mass spectrometry. Environ Poll 174:222–228. https://doi.org/10.1016/j.envpol.2012.11.026

Kolenčík M, Ernst D, Urík M et al (2020) Foliar application of low concentrations of titanium dioxide and zinc oxide nanoparticles to the common sunflower under field conditions. Nano 10(8):1619. https://doi.org/10.3390/nano10081619

Koul M, Thomas L, Karmakar K (2021) Functional aspects of solanaceae trichomes in heavy metal detoxification. Nord J Bot 39(5):n/a. https://doi.org/10.1111/njb.03171

Lahiani MH, Chen J, Irin F et al (2015) Interaction of carbon nanohorns with plants: Uptake and biological effects. Carbon 81:607–619. https://doi.org/10.1016/j.carbon.2014.09.095

Landa P (2021) Positive effects of metallic nanoparticles on plants: overview of involved mechanisms. Plant Physiol Biochem 161:12–24. https://doi.org/10.1016/j.plaphy.2021.01.039

Larue C, Castillo-Michel H, Sobanska S et al (2014) Foliar exposure of the crop Lactuca sativa to silver nanoparticles: Evidence for internalization and changes in Ag speciation. J Hazard Mater 264:98–106. https://doi.org/10.1016/j.jhazmat.2013.10.053

Le VN, Rui Y, Gui X et al (2014) Uptake, transport, distribution and bio-effects of SiO2 nanoparticles in Bt-transgenic cotton. J Nanobiotechnol 12(1):50. https://doi.org/10.1186/s12951-014-0050-8

Lee W, Kwak JI, An Y (2012) Effect of silver nanoparticles in crop plants Phaseolus radiatus and Sorghum bicolor: Media effect on phytotoxicity. Chemosphere 86(5):491–499. https://doi.org/10.1016/j.chemosphere.2011.10.013

Levard C, Hotze EM, Lowry GV et al (2012) Environmental transformations of silver nanoparticles: impact on stability and toxicity. Environ Sci Technol 46(13):6900–6914. https://doi.org/10.1021/es2037405

Li C, Wang P, van der Ent A et al (2019) Absorption of foliar-applied Zn in sunflower (Helianthus annuus): importance of the cuticle, stomata and trichomes. Ann Bot 123(1):57–68. https://doi.org/10.1093/aob/mcy135

Li J, Chang PR, Huang J et al (2013) Physiological Effects of Magnetic Iron Oxide Nanoparticles Towards Watermelon. J Nanosci Nanotechnol 13(8):5561–5567. https://doi.org/10.1166/jnn.2013.7533

Lin D, Tian X, Wu F, Xing B (2010) Fate and transport of engineered nanomaterials in the environment. J Environ Qual 39(6):1896–1908. https://doi.org/10.2134/jeq2009.0423

Lin YF, Aarts MGM (2012) The molecular mechanism of zinc and cadmium stress response in plants. Cell Mol Life Sci 69(19):3187–3206

Liu W, Zeb A, Lian J, Wu J, Xiong H, Tang J, Zheng S (2020) Interactions of metal-based nanoparticles (MBNPs) and metal-oxide nanoparticles (MONPs) with crop plants: a critical review of research progress and prospects. Environ Rev 28(3):294–310

López-Moreno ML, de la Rosa G, Hernández-Viezcas JA, Castillo-Michel H, Botez CE, Peralta-Videa JR, Gardea-Torresdey JL (2010) Evidence of the differential biotransformation and genotoxicity of ZnO and CeO2 nanoparticles on soybean (Glycine max) plants. Environ Sci Technol 44(19):7315–7320. https://doi.org/10.1021/es903891g

Lu K, Shen D, Liu X et al (2020) Uptake of iron oxide nanoparticles inhibits the photosynthesis of the wheat after foliar exposure. Chemosphere 259:127445. https://doi.org/10.1016/j.chemosphere.2020.127445

Lv J, Christie P, Zhang S (2019) Uptake, translocation, and transformation of metal-based nanoparticles in plants: recent advances and methodological challenges. Environ Sci Nano 6(1):41–59. https://doi.org/10.1039/C8EN00645H

Ma X, Yan J (2018) Plant uptake and accumulation of engineered metallic nanoparticles from lab to field conditions. Curr Opin Environ Sci Health 6:16–20. https://doi.org/10.1016/j.coesh.2018.07.008

Ma Y, He X, Zhang P, Zhang Z, Guo Z, Tai R, Xu Z, Zhang L, Ding Y, Zhao Y, Chai Z (2011) Phytotoxicity and biotransformation of La_2O_3 nanoparticles in a terrestrial plant cucumber (Cucumis sativus). Nanotoxicology 5(4):743–753. https://doi.org/10.3109/17435390.2010.545487

Mahakham W, Sarmah AK, Maensiri S et al (2017) Nanopriming technology for enhancing germination and starch metabolism of aged rice seeds using phytosynthesized silver nanoparticles. Sci Rep 7(1):8263–8221. https://doi.org/10.1038/s41598-017-08669-5

Mahdi KNM, Peters RJB, Klumpp E et al (2017) Silver nanoparticles in soil: aqueous extraction combined with single-particle ICP-MS for detection and characterization. Env Nanotechnol Monit Manag 7:24–33. https://doi.org/10.1016/j.enmm.2016.12.002

Malandrakis AA, Kavroulakis N, Avramidou M et al (2021) Metal nanoparticles: phytotoxicity on tomato and effect on symbiosis with the fusarium solani FsK strain. Sci Total Environ 787:147606. https://doi.org/10.1016/j.scitotenv.2021.147606

Mali SC, Raj S, Trivedi R (2020) Nanotechnology a novel approach to enhance crop productivity. Biochem Biophys Rep. https://doi.org/10.1016/j.bbrep.2020.100821. 24100821-S240558082030131X 100821

Manna I, Sahoo S, Bandyopadhyay M (2021) Effect of engineered nickel oxide nanoparticle on reactive oxygen species-nitric oxide interplay in the roots of Allium cepa L. Front Plant Sci 12:586509. https://doi.org/10.3389/fpls.2021.586509

Maurer-Jones MA, Gunsolus IL, Murphy CJ, Haynes CL (2013) Toxicity of engineered nanoparticles in the environment. Anal Chem 85(6):3036–3049. https://doi.org/10.1021/ac303636s

Mazumdar H, Ahmed GU (2011) Phytotoxicity effect of silver nanoparticles on Oryza sativa. Int J Chem Tech Res 3(3):1494–1500

Miao A, Schwehr KA, Xu C et al (2009) The algal toxicity of silver engineered nanoparticles and detoxification by exopolymeric substances. Environ Pollut 157(11):3034–3041. https://doi.org/10.1016/j.envpol.2009.05.047

Mirzajani F, Askari H, Hamzelou S et al (2013) Effect of silver nanoparticles on Oryza sativa L. and its rhizosphere bacteria. Ecotoxicol Environ Safety 88:48–54. https://doi.org/10.1016/j.ecoenv.2012.10.018

Mosa KA, El-Naggar M, Ramamoorthy K et al (2018) Copper nanoparticles induced genotoxicty, oxidative stress, and changes in Superoxide Dismutase (SOD) gene expression in cucumber (Cucumis sativus) plants. Front Plant Sci 9. https://doi.org/10.3389/fpls.2018.00872

Nair P, Chung I (2014) Impact of copper oxide nanoparticles exposure on Arabidopsis thaliana growth, root system development, root lignificaion, and molecular level changes. Environ Sci Pollut Res 21(22):12709–12722. https://doi.org/10.1007/s11356-014-3210-3

Pace HE, Rogers NJ, Jarolimek C et al (2012) Single particle inductively coupled plasma-mass spectrometry: a performance evaluation and method comparison in the determination of nanoparticle size. Environ Sci Technol 46(22):12272–12280. https://doi.org/10.1021/es301787d

Pariona N, Martínez AI, Hernandez-Flores H et al (2017) Effect of magnetite nanoparticles on the germination and early growth of Quercus macdougallii. Sci Total Environ 575:869–875. https://doi.org/10.1016/j.scitotenv.2016.09.128

Parveen A, Rao S (2015) Effect of Nanosilver on Seed Germination and Seedling Growth in Pennisetum glaucum. J Clust Sci 26(3):693–701. https://doi.org/10.1007/s10876-014-0728-y

Patlolla AK, Berry A, May L, Tchounwou PB (2012) Genotoxicity of silver nanoparticles in Vicia faba a pilot study on the environmental monitoring of nanoparticles. Int J Environ 9(5): 1649–1662. https://doi.org/10.3390/ijerph9051649

Peharec Štefanić P, Jarnević M, Cvjetko P, Biba R, Šikić S, Tkalec M, Cindrić M, Letofsky-Papst I, Balen B (2019) Comparative proteomic study of phytotoxic effects of silver nanoparticles and silver ions on tobacco plants. Environ Sci Pollut Res Int 26(22):22529–22550. https://doi.org/10.1007/s11356-019-05552-w

Perreault F, Samadani M, Dewez D (2014) Effect of soluble copper released from copper oxide nanoparticles solubilisation on growth and photosynthetic processes of Lemna gibba L. Nanotoxicology 8(4):374–382. https://doi.org/10.3109/17435390.2013.789936

Pokhrel LR (2013) Evaluation of colloidal stability and ecotoxicity of metal-based nanoparticles in the aquatic and terrestrial systems

Pozebon D, Scheffler GL, Dressler VL (2017) Recent applications of laser ablation inductively coupled plasma mass spectrometry (LA-ICP-MS) for biological sample analysis: a follow-up review. J Anal At Spectrom 32(5):890–919. https://doi.org/10.1039/C7JA00026J

Pradas del Real AE, Vidal V, Carrière M, Castillo-Michel H, Levard C, Chaurand P, Sarret G (2017) Silver nanoparticles and wheat roots: a complex interplay. Environ Sci Technol 51(10): 5774–5782. https://doi.org/10.1021/acs.est.7b00422

Pradhan S, Patra P, Das S et al (2013) Photochemical modulation of biosafe manganese nanoparticles on vigna radiata: a detailed molecular, biochemical, and biophysical study. Environ Sci Technol 47(22):13122–13131. https://doi.org/10.1021/es402659t

Rajput VD, Singh A, Singh VK et al (2021) Impact of nanoparticles on soil resource. In: Amrane A, Mohan D, Nguyen TA et al (eds) Nanomaterials for Soil Remediation. Elsevier, pp 65–85

Raliya R, Franke C, Chavalmane S et al (2016) Quantitative understanding of nanoparticle uptake in watermelon plants. Front Plant Sci 7:1288. https://doi.org/10.3389/fpls.2016.01288

Rastogi A, Zivcak M, Sytar O et al (2017) Impact of metal and metal oxide nanoparticles on plant: A critical review. Front Chem 5:78. https://doi.org/10.3389/fchem.2017.00078

Reddy PVL, Hernandez-Viezcas JA, Peralta-Videa JR et al (2016) Lessons learned: Are engineered nanomaterials toxic to terrestrial plants? Sci Total Environ 568:470–479. https://doi.org/10.1016/j.scitotenv.2016.06.042

Reith F, Etschmann B, Grosse C, Moors H, Benotmane MA, Monsieurs P, Grass G, Doonan C, Vogt S, Lai B, Martinez-Criado G, George GN, Nies DH, Mergeay M, Pring A, Southam G, Brugger J (2009) Mechanisms of gold biomineralization in the bacterium Cupriavidus metallidurans. PNAS 106(42):17757–17762. https://doi.org/10.1073/pnas.0904583106

Rossi L, Fedenia LN, Sharifan H et al (2019) Effects of foliar application of zinc sulfate and zinc nanoparticles in coffee (Coffea arabica L.) plants. Plant Physiol Biochem 135:160–166. https://doi.org/10.1016/j.plaphy.2018.12.005

Santner A, Calderon-Villalobos LIA, Estelle M (2009) Plant hormones are versatile chemical regulators of plant growth. Nat Chem Biol 5(5):301–307. https://doi.org/10.1038/nchembio.165

Sanzari I, Leone A, Ambrosone A (2019) Nanotechnology in plant science: to make a long story short. Front Bioeng Biotechnol 7:120. https://doi.org/10.3389/fbioe.2019.00120

Savithramma N, Ankanna S, Bhumi G (2012) Effect of nanoparticles on seed germination and seedling growth of Boswellia ovalifoliolata an endemic and endangered medicinal tree taxon. Nano Vision 2(1):2

Schwab F, Zhai G, Kern M et al (2016) Barriers, pathways and processes for uptake, translocation and accumulation of nanomaterials in plants - Critical review. Nanotoxicology 10(3):257–278. https://doi.org/10.3109/17435390.2015.1048326

Shahid M, Dumat C, Khalid S et al (2019) Foliar uptake of arsenic nanoparticles by spinach: an assessment of physiological and human health risk implications. Environ Sci Pollut Res Int 26(20):20121–20131. https://doi.org/10.1007/s11356-018-3867-0

Shaw AK, Ghosh S, Kalaji HM et al (2014) Nano-CuO stress induced modulation of antioxidative defense and photosynthetic performance of Syrian barley (Hordeum vulgare L.). Environ Exp Bot 102:37–47. https://doi.org/10.1016/j.envexpbot.2014.02.016

Shrivastava M, Srivastav A, Gandhi S, Rao S, Roychoudhury A, Kumar A, Singhal R, Jha SK, Singh S (2019) Monitoring of engineered nanoparticles in soil-plant system: a review. Environ Nanotechnol Monit Manag 11:100218. https://doi.org/10.1016/j.enmm.2019.100218

Slomberg DL, Schoenfisch MH (2012) Silica nanoparticle phytotoxicity to arabidopsis thaliana. Environ Sci Technol 46(18):10247–10254. https://doi.org/10.1021/es300949f

Sotelo-Boyás ME, Bautista-Baños S, Correa-Pacheco ZN et al (2016) Biological activity of chitosan nanoparticles against pathogenic fungi and bacteria. In: Bautista-Baños S, Gianfranco R, Antoinio J (eds) chitosan in the preservation of agricultural commodities. Elsevier Inc, pp 339–349. https://doi.org/10.1016/B978-0-12-802735-6.00013-6

Spielman-Sun E, Lombi E, Donner E, Howard D, Unrine JM, Lowry GV (2017) Impact of surface charge on cerium oxide nanoparticle uptake and translocation by wheat (Triticum aestivum). Environ Sci Technol 51(13):7361–7368. https://doi.org/10.1021/acs.est.7b00813

Stegemeier JP, Schwab F, Colman BP et al (2015) Speciation matters: bioavailability of silver and silver sulfide nanoparticles to alfalfa (Medicago sativa). Env Sci Technol 49(14):8451–8460. https://doi.org/10.1021/acs.est.5b01147

Stokes DJ (2008) Principles and practice of variable pressure, environmental scanning electron microscopy (VP-ESEM). Chichester, West Sussex, UK, Wiley

Sturikova H, Krystofova O, Huska D, Adam V (2018) Zinc, zinc nanoparticles and plants. J Hazard Mater 349:101–110. https://doi.org/10.1016/j.jhazmat.2018.01.040

Sundaria N, Singh M, Upreti P et al (2019) Seed Priming with iron oxide nanoparticles triggers iron acquisition and biofortification in wheat (Triticum aestivum L.) grains. J Plant Growth Regul 38(1):122–131. https://doi.org/10.1007/s00344-018-9818-7

Taylor AF, Rylott EL, Anderson CWN, Bruce NC (2014) Investigating the toxicity, uptake, nanoparticle formation and genetic response of plants to gold. PLoS One 9(4):e93793. https://doi.org/10.1371/journal.pone.0093793

Torney F, Trewyn BG, Lin VS et al (2007) Mesoporous silica nanoparticles deliver DNA and chemicals into plants. Nat Nanotechnol 2(5):295–300. https://doi.org/10.1038/nnano.2007.108

Tripathi DK, Singh S, Singh S et al (2017a) An overview on manufactured nanoparticles in plants: uptake, translocation, accumulation and phytotoxicity. Plant Physiol Biochem 110:2–12. https://doi.org/10.1016/j.plaphy.2016.07.030

Tripathi A, Liu S, Singh PK et al (2017b) Differential phytotoxic responses of silver nitrate (AgNO3) and silver nanoparticle (AgNps) in Cucumis sativus L. Plant Gene 11:255–264. https://doi.org/10.1016/j.plgene.2017.07.005

Ullah H, Li X, Peng L, Cai Y, Mielke HW (2020) In vivo phytotoxicity, uptake, and translocation of PbS nanoparticles in maize (Zea mays L.) plants. Sci Total Environ 737(139558). https://doi.org/10.1016/j.scitotenv.2020.139558

Vannini C, Domingo G, Onelli E, Prinsi B, Marsoni M, Espen L, Bracale M (2013) Morphological and proteomic responses of Eruca sativa exposed to silver nanoparticles or silver nitrate. PLoS One 8(7):e68752. https://doi.org/10.1371/journal.pone.0068752

Vannini C, Domingo G, Onelli E et al (2014) Phytotoxic and genotoxic effects of silver nanoparticles exposure on germinating wheat seedlings. J Plant Physiol 171(13):1142–1148. https://doi.org/10.1016/j.jplph.2014.05.002

Wang H, Kou X, Pei Z et al (2011) Physiological effects of magnetite (Fe3O4) nanoparticles on perennial ryegrass (Lolium perenne L.) and pumpkin (Cucurbita mixta) plants. Nanotoxicology 5(1):30–42. https://doi.org/10.3109/17435390.2010.489206

Wang M, Gao L, Dong S, Sun Y, Shen Q, Guo S (2017) Role of silicon on plant–pathogen interactions. Front Plant Sci 8:701. https://doi.org/10.3389/fpls.2017.00701

Wang T, Wu J, Xu S et al (2019) A potential involvement of plant systemic response in initiating genotoxicity of agnanoparticles in arabidopsis thaliana. Ecotoxicol Environ Saf 170:324–330. https://doi.org/10.1016/j.ecoenv.2018.12.002

Wang W, Liu J, Ren Y et al (2020) Phytotoxicity assessment of copper oxide nanoparticles on the germination, early seedling growth, and physiological responses in Oryza sativa L. Bull Environ Contam Toxicol 104(6):770–777. https://doi.org/10.1007/s00128-020-02850-9

Wang W, Tarafdar J, Biswas P (2013) Nanoparticle synthesis and delivery by an aerosol route for watermelon plant foliar uptake. J Nanopart Res 15(1):1–13. https://doi.org/10.1007/s11051-013-1417-8

Ward CS, Pan J, Colman BP et al (2019) Conserved microbial toxicity responses for acute and chronic silver nanoparticle treatments in wetland mesocosms. Environ Sci Technol 53(6): 3268–3276. https://doi.org/10.1021/acs.est.8b06654

Wen S-B, Mao X, Greif R, Russo RE (2007) Laser ablation induced vapor plume expansion into a background gas. II Experimental analysis J Appl Phys 101(2):023115. https://doi.org/10.1063/1.2431085

Wilschefski SC, Baxter MR (2019) inductively coupled plasma mass spectrometry: introduction to analytical Aspects. Clin Biochem Rev 40(3):115–133. https://doi.org/10.33176/aacb-19-00024

Xiong T, Dumat C, Dappe V et al (2017) Copper oxide nanoparticle foliar uptake, phytotoxicity, and consequences for sustainable urban agriculture. Environ Sci Technol 51(9):5242–5251. https://doi.org/10.1021/acs.est.6b05546

Xu L, Takemura T, Xu M, Hanagata N (2011) Toxicity of silver nanoparticles as assessed by global gene expression analysis. Mater Express 1(1):74–79. https://doi.org/10.1166/mex.2011.1010

Yan A, Chen Z (2018) Detection methods of nanoparticles in plant tissues. IntechOpen, London

Yan L, Li P, Zhao X et al (2020) Physiological and metabolic responses of maize (Zea mays) plants to Fe3O4 nanoparticles. Sci Total Environ 718:137400. https://doi.org/10.1016/j.scitotenv.2020.137400

Yanık F, Vardar F (2015) Toxic effects of aluminum oxide (Al2O3) nanoparticles on root growth and development in triticum aestivum. Water Air Soil Pollut 226(9):296. https://doi.org/10.1007/s11270-015-2566-4

Yasmeen F, Raja NI, Mustafa G, Sakata K, Komatsu S (2016) Quantitative proteomic analysis of post-flooding recovery in soybean root exposed to aluminum oxide nanoparticles. J Proteome 143:136–150. https://doi.org/10.1016/j.jprot.2016.03.014

Yasur J, Rani P (2013) Environmental effects of nanosilver: impact on castor seed germination, seedling growth, and plant physiology. Environ Sci Pollut Res 20(12):8636–8648. https://doi.org/10.1007/s11356-013-1798-3

Yusefi-Tanha E, Fallah S, Rostamnejadi A, Pokhrel LR (2020) Particle size and concentration dependent toxicity of copper oxide nanoparticles (CuONPs) on seed yield and antioxidant defense system in soil grown soybean (Glycine max cv. Kowsar). Sci Total Environ 715(136994). https://doi.org/10.1016/j.scitotenv.2020.136994

Zafar H, Ali A, Zia M (2017) CuO nanoparticles inhibited root growth from brassica nigra seedlings but induced root from stem and leaf explants. Appl Biochem Biotechnol 181(1):365–378. https://doi.org/10.1007/s12010-016-2217-2

Zari H, Babak P, Asad R (2015) The Effect of priming with nano-sliver on agronomic traits of safflower cultivars. J Essent oilbear Plants (Dehra Dun) 18(5):1148–1156. https://doi.org/10.1080/0972060X.2014.976664

Zhang B, Pan X, Cobb GP, Anderson TA (2006) Plant microRNA: a small regulatory molecule with big impact. Deve Biol 289(1):3–16. https://doi.org/10.1016/j.ydbio.2005.10.036

Zhang P, Ma Y, Zhang Z, He X, Zhang J, Guo Z, Tai R, Zhao Y, Chai Z (2012) Biotransformation of ceria nanoparticles in cucumber plants. ACS Nano 6(11):9943–9950. https://doi.org/10.1021/nn303543n

Zhang W-Y, Wang Q, Li M, Dang F, Zhou D-M (2019) Nonselective uptake of silver and gold nanoparticles by wheat. Nanotechnology 13(8):1073–1086. https://doi.org/10.1080/17435390.2019.1640909

Zhao L, Hernandez-Viezcas JA, Peralta-Videa JR, Bandyopadhyay S, Peng B, Munoz B, Keller AA, Gardea-Torresdey JL (2013) ZnO nanoparticle fate in soil and zinc bioaccumulation in corn plants (Zea mays) influenced by alginate. Environ Sci Processes Impacts 15(1):260–266. https://doi.org/10.1039/C2EM30610G

Zhao L, Lu L, Wang A et al (2020) Nano-biotechnology in agriculture: use of nanomaterials to promote plant growth and stress tolerance. J Agric Food Chem 68(7):1935–1947. https://doi.org/10.1021/acs.jafc.9b06615

Transcription Factors and Metal Stress Signalling in Plants

17

Parinita Agarwal, Anjali Chittora, Doddabhimappa R. Gangapur, and Pradeep K. Agarwal

Abstract

The sessile autotrophic plants possess versatile adaptive mechanisms to grow and complete their life cycle, exposed to dynamic range of environmental challenges. Soil constitutes the major source of heavy metals for plants, both essential and non-essential. Plants are greatly sensitive towards both low and high concentration of heavy metals and accordingly, explicit beneficial or adverse effect on plant growth and productivity. A number of plant transcriptional factors (TFs) are identified as regulators of metal stress signal transduction pathways. The MYB family of proteins with highly conserved MYB domain is vast and functions as transcription factors. NAC family of plant-specific transcription factor descended from three proteins: NAM (no apical meristem), ATAF 1/2 (*Arabidopsis* transcription activator factor 1/2), and CUC2 (cup-shaped cotyledon). WRKYs family possess highly conserved WRKYGQK heptapeptide at the N-terminus, WRKYs preferentially binds to W box *cis*-elements (having a TTGACC/T core sequence) in the promoters of downstream target genes. These families have been studied for different abiotic stresses including salinity and drought; however, only few studies report their role in heavy metal stress tolerance in plants. In the present chapter, we have highlighted the toxic effects of heavy metals on plants and the involvement of the MYB, NAC, and WRKY TFs with respect to different heavy metal stress.

P. Agarwal (✉) · A. Chittora · D. R. Gangapur · P. K. Agarwal
Division of Plant Omics, CSIR-Central Salt and Marine Chemicals Research Institute (CSIR-CSMCRI), Council of Scientific and Industrial Research (CSIR), Gijubhai Badheka Marg, Bhavnagar, Gujarat, India
e-mail: parinita.a@rediffmail.com

K. Kumar, S. Srivastava (eds.), *Plant Metal and Metalloid Transporters*,
https://doi.org/10.1007/978-981-19-6103-8_17

Keywords

Transcription factors · Heavy metals · MYB · NAC · WRKY

17.1 Introduction

Plants, being sessile, have to complete their life cycle at one place and therefore, are subjected to a myriad of abiotic and biotic factors (virus, bacteria, fungi, nematodes, etc.) alone and in combinations. The abiotic factors include salinity, drought, low or high temperature, light, and heavy metal stress. The increased anthropogenic activities, rapid urbanisation, industrialisation, and deployment of modern agricultural practices have caused heavy metal contamination in the environment, especially the terrestrial ecosystems. The application of fertilisers, sewage, contamination with industrial waste, and unhealthy irrigation have caused the agricultural soils to be polluted with heavy metal toxicity such as Cd, Cu, Zn, Ni, Co, Cr, Pb, and As. Heavy metals are present naturally in the earth's crust but due to natural and/or anthropogenic activities, they are released in excess, causing problem to living organisms. Based on the density (>5 g/cm^3), the 53 elements of the d-block are referred to as "heavy metals" (Järup 2003). In the periodic table of elements, they are grouped as transition elements; rare earth elements (lanthanides and actinides); and the lead group, which is a heterogeneous group that contain elements that form amphoteric oxides (Appenroth 2010). Only 19 elements like C, O, H, Mg, S, N, Ca, P, and K (macronutrients) and Cu, Zn, Mn, Fe, Mo, B, Ni, Co, Cl, and B (micronutrients) are required by angiosperms for basic metabolism (Ernst 2006). Certain elements, when present in extremely low quantities, function as

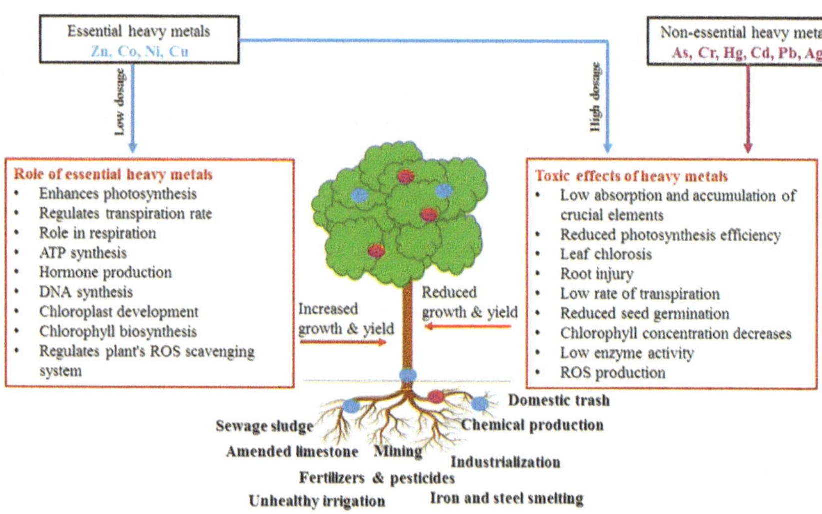

Fig. 17.1 Toxic effects of heavy metals on plants

micronutrients, facilitating different biochemical and physiological processes in living organisms; however, at concentrations higher than specific thresholds, they become toxic (Nagajyoti et al. 2010). The heavy metals comprise two groups (Fig. 17.1): (1) essential elements that support plant growth such as iron (Fe), magnesium (Mg), manganese (Mn), zinc (Zn), molybdenum (Mo), and copper (Cu) and (2) non-essential elements like cadmium (Cd), chromium (Cr), lead (Pb), aluminium (Al), and selenium (Se). Heavy metal pollution is caused by agricultural activities (irrigation, application of amended limestone, inorganic fertilisers, pesticides, and sewage sludge), the use of coal and oil to generate power, industrial activities (iron and steel smelting and chemical production), mining, and domestic trash (Fig. 17.1, Jaishankar et al. 2014). Heavy metal toxicity of soil imposes significant concerns, due to their high stability and lack of biodegradability (Singh and Prasad 2015); they are taken up by plant roots and subsequently translocated to other organs and thus enter the food chain (Shahid et al. 2016).

17.2 Heavy Metal Toxicity in Plants

Plant exposure to heavy metals adversely affects its morpho-physiological, biochemical, and molecular processes, including inhibition in growth and development, photosynthetic efficiency, ROS accumulation, and improper functioning of various metabolic pathways (Fig. 17.1). Depending on the toxicity levels, the heavy metals followtheorderas$As^{5+} < As^{3+} < Cr^{6+} < Co^{2+} < Zn^{2+} < Ni^{2+} < Cu^{2+} < Ti^{+} < Hg^{2+} < Cd^{2+}$- $< Ag^{+}$ (Van Assche and Clijsters 1990). Heavy metal toxicity in plants is thought to be caused by four different processes. These include the following: (1) similarities with nutritional cations, which result in absorption competition at the root surface, for example, As and Cd compete for absorption with P and Zn, respectively; (2) direct interaction of heavy metals with functional proteins' sulphhydryl groups (–SH), which destroys their function and structure, rendering them inactive; (3) displacement of essential cations from specific binding sites, resulting in functional collapse; and (4) generation of reactive oxygen species (ROS), which damages macromolecules (DalCorso et al. 2013). Some of the observed plant responses to heavy metals are changes in redox status, the level of signal molecules, the activity of antioxidant system enzymes, membrane permeability, cysteine, glutathione (GSH), phytochelatin (PCs), protein contents, expression of genes encoding pathogenesis-related (PR) proteins, genes encoding enzymes in the flavonoid biosynthesis pathway, and the level of phenolics (Morkunas et al. 2018).

Arsenic (As) is non-essential and toxic to plants, found naturally in all soils, and enters the farming via natural geochemical processes (Smedley and Kinniburgh 2002), use of As-based pesticides or irrigation. It enters human food chain via contaminated drinking water and crops like rice are identified as the major source (Meharg et al. 2009). The two forms of inorganic As, arsenate (AsV) and arsenite (AsIII), are taken up by roots. The AsV gets converted to the more toxic forms AsIII; however, both forms negatively impact plant metabolism. AsV is an analogue of inorganic phosphate (Pi), and AsV and Pi compete for uptake via the same transport mechanisms. Pi transporter proteins (PHT) carry AsV efficiently through the plasmalemma (Finnegan and Chen 2012). Also, AsIII is a thiol reactive molecule that can bind to sulphhydryl groups (Kitchin and Wallace 2006). As lowers fruit yield and decreases leaf fresh weight in tomato (*Lycopersicon esculentum*; Barrachina et al. 1995). As causes reduced growth, chlorosis, and wilting in canola (*Brassica napus*; Cox et al. 1996). Furthermore, As reduces seed germination, seedling height, leaf area, and dry matter production in rice (*Oryza sativa*; Marin et al. 1993; Abedin et al. 2002).

Chromium (Cr) is highly toxic and non-essential, found in two oxidation states in plants as trivalent (Cr^{3+}) and hexavalent (Cr^{6+}) species, with the former being less hazardous than the latter. Cr is toxic for agronomic plants at 5–100 mg g^{-1} in soil (Davies Jr et al. 2002). Cr accumulates via carrier ions such as sulphate or iron (Singh et al. 2013). Cr decreases seedling dry matter production and slows the development of stems and leaves during the early stages of plant growth. Cr toxicity prevents the cell division and elongation of plant roots, thus reducing the overall length of roots (Nematshahi et al. 2012; Shanker et al. 2005). Cr can hinder the absorption and accumulation of crucial elements including N, P, K, Fe, Mg, Mn, Mo, Zn, Cu, Ca, and B via altering the plasma membrane's uptake activity in root cells (Shanker et al. 2005). Cr^{6+} interferes with nitrogen metabolism, which causes a decrease in the levels of nitrate reductase, nitrite reductase, glutamine synthetase, glutamate dehydrogenase, and urease. It also reduces the photosynthesis efficiency in terms of CO_2 fixation, the transport of electrons, photophosphorylation, and photosynthetic enzyme activity (Shanker et al. 2009). Cr^{6+} affects chloroplast ultrastructure, causing the lamellar system to develop slowly and the mesophyll cells to disorganise. Plants exposed to Cr had lower levels of transpiration, stomatal conductance, and CO_2 assimilation (Schiavon et al. 2009).

Cobalt (Co) is beneficial for plants and naturally occurs in the earth's crust as cobaltite (CoAsS), erythrite ($Co_3(AsO_4)_2$), and smaltite ($CoAs_2$, Barceloux and Barceloux 1999). Normal Co concentrations in plants are as low as 0.1–10 mg kg^{-1} dry weight (Bakkaus et al. 2005). In chickpea, 50 mg kg^{-1} is a damage threshold level (Khan and Khan 2010). In cauliflower leaves, an excess of Co reduced the concentrations of Fe, chlorophyll, protein, and catalase activity. In addition, elevated levels of Co influenced the translocation of P, S, Mn, Zn, and Cu from cauliflower roots to tops. Water potential and transpiration rate were both considerably reduced by Co. When exposed to Co, diffusive resistance and relative water content in cauliflower leaves increased (Chatterjee and Chatterjee 2000). In radish (*Raphanus sativus*), shoot length, root length, and total leaf area decreased; chlorophyll concentration decreased; plant nutrient content and antioxidant enzyme activity decreased; plant sugar, amino acid, and protein levels have decreased (Jayakumar et al. 2007). Also, in tomato (*Lycopersicon esculentum*), reduction in plant nutrient content was observed (Jayakumar et al. 2013).

Zinc (Zn) is an essential heavy metal and its divalent state (Zn^{2+}) is the most common form found in soil and taken up by plants and soil pH is the most significant factor determining Zn availability (Broadley et al. 2007). The metal concentration varies in between 0.02 and 0.04 mg g^{-1} dry weight in soil in which plants grow and the concentrations above 0.2 mg g^{-1} dry matter develops phytotoxicity (Tsonev and Cebola Lidon 2012). Zn toxicity causes chlorosis in young leaves because of iron deficiency, purplish-red coloration in leaves due to phosphorus deficiency, and manganese and copper deficiencies in plant shoots that hinder transfer of these micronutrients from root to shoot. This indicates Zn^{2+} in excess may readily surpass other metals, especially those with equivalent ionic radii in enzymes or transporters (Fukao et al. 2011; Ebbs and Kochian 1997; Lee et al. 1996). The elevated level of Zn^{2+} causes decline in initial and maximum chlorophyll fluorescence that results in the repression of PSII activity (Tsonev and Cebola Lidon 2012). Excess Zn^{2+} in cells can generate ROS and affect membrane integration and permeability (Hosseini and Poorakbar 2013).

Nickel (Ni) is a micronutrient that exists in a variety of oxidative forms, and its divalent state (Ni^{2+}) is the most persistent in the environment and biological systems (Poonkothai and Vijayavathi 2012). Ni is essential for plants, but the required concentration is very low (0.05–10 mg kg^{-1} dry weight) in plant species. Higher Ni concentrations show toxic effects in plants (Bhalerao et al. 2015). In chickpea, 10 mg kg^{-1} was a damage threshold of Ni (Khan and Khan 2010). Excess Ni inhibits the action of enzymes such as amylase and protease, as well as disturbing the hydrolyzation of food storage in germinating seeds, which has a negative impact on the germination process and seedling growth traits of plants (Aydinalp and Marinova 2009; Sethy and Ghosh 2013). Ni toxicity has been associated with reduced lateral root growth and development in plants (Seregin et al. 2003). Furthermore, the agglomeration of Ni in the root apex substantially obstructs mitotic cell division in this organ, resulting in a decrease in growth (L'Huillier et al. 1996). The induction of ROS, due to Ni toxicity, is observed in *Jatropha curcas* L., which results in the impairment of cell membrane and enzymatic imbalance (Yan et al. 2008).

Copper (Cu) is an essential micronutrient that helps plants perform a variety of key physiological activities, such as redox reaction catalyser in mitochondria, chloroplasts, and cytoplasm and as an electron carrier during plant respiration (Fargasova 2004; Yruela 2009). Cu concentrations in soil generally range from 2 to 250 mg kg^{-1}, and healthy plants may absorb 20–30 mg kg^{-1}. Cu concentrations in cells must be kept low since, due to its strong redox properties, the element is very toxic (Azooz et al. 2012). The growth of rice plant was severely inhibited at 300–500 mg kg^{-1} or above soil Cu levels (Xu et al. 2006). The earliest symptom of Cu toxicity is a reduction in root elongation and growth. Chlorosis, necrosis, and leaf discoloration are some of the symptoms that follow (Tsay et al. 1995; Yruela 2009). Excess Cu can bind to sulphhydryl groups in cell membranes cause lipid peroxidation, resulting in membrane damage and the generation of free radicals in various plant organelles (Chen et al. 2000). At high Cu concentrations, there is production of ROS such as singlet oxygen (O^{2-}) and hydroxyl radical (HO·), which is triggered by the redox process causing damage to macromolecules such as DNA, RNA, lipids, carbohydrates, and proteins (Yurekli and Porgali 2006).

Titanium (Ti) occurs naturally in the form of titanium oxide (TiO_2). It has both beneficial and toxic effects, depending on a variety of experimental conditions (Cox et al. 2016). It causes growth inhibition, decreased photosynthetic efficiency, reduced plant biomass, decreased Na^+ K^+-ATPase activity, cell membrane changes, tissue and DNA damage, decrease in fertility and survival rate, induction of stress response gene expression, and ROS production (Hou et al. 2019).

Silver (Ag) is a toxic and non-essential metal, with concentrations in contaminated soil ranging from 3 to 42 mg kg^{-1} (Galazzi and Arruda 2018). In *Arabidopsis*, excessive levels of Ag nanoparticles (between 5 and 20 mg L^{-1}) caused toxic effects (Kaveh et al. 2013). Decrease in biomass, leaf area, chlorophyll, nutrient uptake, transpiration rate, root growth; inhibition of seed germination; variation in cell structure, cell division, and levels of phytohormone; and oxidative stress are all toxic effects of Ag nanoparticles (Yan and Chen 2019).

Mercury (Hg) is a highly toxic, non-essential heavy metal. It exists in a variety of forms, including HgS, Hg^{2+}, Hg°, and methyl-Hg. In agricultural soil, however, the ionic form (Hg^{2+}) predominates (Han et al. 2006). *Avena sativa* was affected at lower Hg concentration (below 20 mg kg^{-1}) and *Phaseolus vulgaris* was affected above 20 mg kg^{-1}. The study showed 0.36 mg kg^{-1} of Hg in soil as a critical concentration, and above this plant will be affected (Lima et al. 2019). Plant cells are severely harmed by high levels of Hg^{2+} that can cause apparent damage and physiological disorders (Zhou et al. 2007). Hg^{2+} binds to water channel proteins, causing leaf stomata to close and water flow to be physically obstructed in plants (Zhang and Tyerman 1999). High levels of Hg^{2+} inhibit mitochondrial function and cause oxidative stress by stimulating the production of ROS. Plants' bio-membrane lipids and cellular metabolism are disrupted as a result of this (Messer et al. 2005; Cargnelutti et al. 2006).

Cadmium (Cd) is a non-essential and toxic heavy metal that has a regulation limitation of 100 mg kg^{-1} soil in agricultural soil (Salt et al. 1995; Gill and Tuteja 2011). Seed germination is suppressed in plants exposed to hazardous amounts of Cd owing to a lack of water absorption, limiting the amount of water available for seed embryo development (Vijayaragavan et al. 2011). Water scarcity, as well as inactivation of starch mobilisation in the endosperm and poor transport of soluble carbohydrates to the embryonic axis, can all contribute to increased hunger on the embryonic axis (Kuriakose and Prasad 2008). Chlorosis, growth inhibition, browning of root tips, and eventually death are also the symptoms of a high Cd levels in plants (Baryla et al. 2001; Dai et al. 2006). Cd has been related to the usage, storage, and utilisation of a variety of elements (such as Ca, Mg, P, and K), as well as plant water uptake (Tran and Popova 2013). Cd toxicity also promotes the formation of ROS in the mitochondrial electron transport chain (Heyno et al. 2008). Cd-induced inhibition of iron (Fe^{3+}) reductase causes an iron (Fe^{2+}) shortage, which has a significant impact on photosynthesis and its apparatus (Hasan et al. 2009).

Lead (Pb) is a non-essential heavy metal, and it has become a significant environmental contaminant. Plants have a Pb threshold of about 2 mg kg^{-1}, whereas agricultural soils have a Pb threshold of 50–300 mg kg^{-1} (Zulfiqar et al. 2019). Pb affects photosynthesis, nutrient absorption, seedling development, enzyme activity, water balance, and membrane permeability in plants (Sharma and Dubey 2005;

Shahid et al. 2011; Kumar et al. 2012). Pb interferes with the normal functioning of chloroplasts at higher concentrations by inhibiting enzymes involved in chlorophyll biosynthesis, CO_2 fixation, and the pigment-protein complexation of photosystems (Sharma and Dubey 2005). PSII is more susceptible to Pb exposure than PSI since it is collapsed to a greater extent, resulting in negative effects on both the donor and acceptor sites, the oxygen evolving complex, and electron-transfer reactions (Pourrut et al. 2011; Romanowska et al. 2012). Pb increases the production of ROS in plant cells, which enhances processes such as lipid peroxidation and the proportion of saturated vs unsaturated fatty acids (which increases in content) in cell membranes (Malecka et al. 2001). Plants produce a variety of non-enzymatic, low molecular weight antioxidants, and enzymatic antioxidants to combat excess ROS and protect cells (Gratão et al. 2005).

Aluminium (Al) is a non-essential and toxic heavy metal. It is the third most prevalent metal in the earth's crust (Gallo-Franco et al. 2020). Al toxicity is a key component in restricting plant growth in the most acidic soils. As the pH of the soil lowers, Al becomes more soluble, and the proportion of phytotoxic Al ions in the soil increases. Under acidic circumstances, Al is liberated from soil as $Al(OH)_2^{+}$, Al $(OH)^{2+}$, and $Al(H_2O)^{3+}$, the latter usually referred to as Al^{3+} (Samac and Tesfaye 2003). When plants are exposed to Al, their leaves contract, curl along the margins, and appear chlorotic under the leaf margins. Due to decrease in mitotic activity, it has an effect on root and shoot growth. It influences the physical characteristics of the plasma membrane, reduces photosynthesis, and decreases chlorophyll content. Plants show DNA synthesis inhibition, ATP depletion, mitochondrial dysfunction, ROS generation, and programmed cell death (Imadi et al. 2016).

17.3 Signal Transduction Pathway in Response to Heavy Metal Stress

In plants, heavy metal stress activates a variety of signalling pathways, including calcium-dependent signalling, ROS signalling, hormone signalling, along with various signalling molecules that mediate signal transduction and boost the expression of stress-responsive genes as well as transcription factors (Fig. 17.2).

Increased ROS generation occurs as a result to heavy metal stress, known as "oxidative burst", which functions as an alarm signal and stimulates the gene expression of transcription factors and numerous defence-related genes (Sharma et al. 2012). Its production impairs cell activity and causes oxidative damage to macromolecules like DNA and proteins (Kapoor et al. 2015). The low ROS concentration functions as a signal, eliciting a plant response to stress, whereas the high ROS concentration offers a hazard to the plant cell, causing serious damage and cell death (Steffens 2014). There are two types of detoxification mechanisms: the primary detoxification mechanism includes increased antioxidant isoenzyme activity, which maintains ROS homeostasis by scavenging ROS, and the secondary detoxification mechanism involves heavy metals being chelated by peptides and sequestered into vacuoles (Kumar et al. 2015).

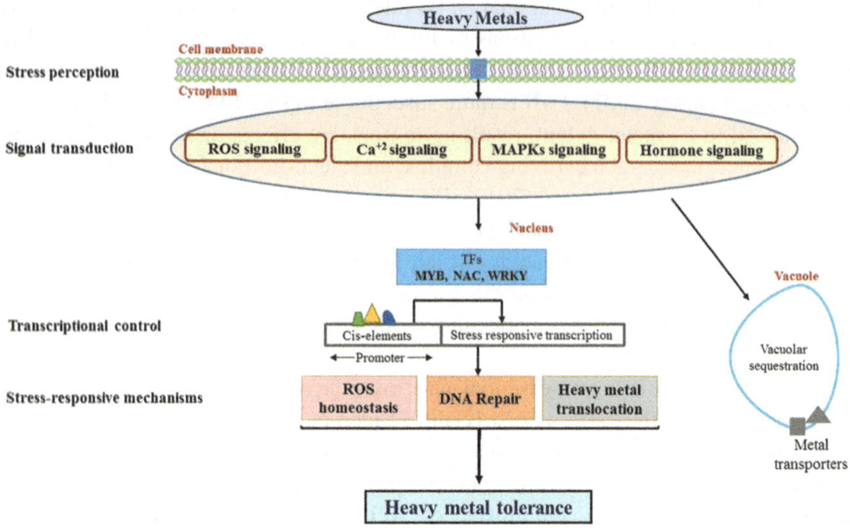

Fig. 17.2 Schematic representation of heavy metal-induced signal transduction pathways in plants

Ca^{2+} is a ubiquitous secondary messenger that plays a role in normal plant activity and the response to different metal stresses. Calmodulins (CaMs), CaM-like proteins (CMLs), calcineurin B-like proteins (CBLs), and Ca^{2+}-dependent protein kinases (CDPKs) are the calcium sensing proteins found in plants. They bind to Ca^{2+} and activate several downstream signalling pathways (Dodd et al. 2010).

Three tier components form the MAPK cascade. MAPKKKs, MAPKKs, and MAPKs mediate phosphorylation reactions from an upstream receptor to a downstream target (Hamel et al. 2006). MAPKs are known to be triggered by metal ligand perception as well as ROS molecules generated during metal stress (Smeets et al. 2013; Jalmi and Sinha 2015). In *Medicago sativa*, excessive heavy metals (Cu and Cd) activate MAPKs (Jonak et al. 2004). In rice, Cd activates MAPKs (OsMAPK2) as well as the MBP kinase gene (Yeh et al. 2007). The activation of MAPK cascades in *Brassica juncea* under AsIII stress transduces metal stress-mediated signals, resulting in increased resistance to As stress (Gupta et al. 2009).

Phytohormones have a role in signalling pathways in response to heavy metal stresses, and some of these are mentioned here. Auxin modulates auxin homeostasis, including auxin stability, transport, and redistribution, which has a direct impact on plant responses to metal stress (Potters et al. 2007). Cytokinin alleviated Cd's inhibition of photosynthetic pigment and chloroplast membranes, boosting photosynthetic capacity and primary metabolite levels (Piotrowska-Niczyporuk et al. 2012). The involvement of abscisic acid in mediating a gold stress response in *Arabidopsis* revealed alterations in glutathione-mediated detoxification and hormone biosynthesis pathways (Shukla et al. 2014). The transcription factors AP2 and ERF1 from *Medicago sativa*, as well as major ethylene synthesis genes from rice, were shown to be increased under Hg treatment (Chen et al. 2014; Montero-Palmero et al. 2014).

Heavy metal tolerance is enhanced by transport proteins. The ZIP gene family contains transporters that are involved in the transportation of a number of cations, including cadmium, iron, manganese, and zinc (Guerinot 2000). Metal ions can be transported by the NRAMP metal transporter. They are found in the plasma membrane as well as the tonoplast membrane of roots and shoots (Nevo and Nelson 2006). MRP and PDR are two subfamilies of the ABC transporter family, which are important in chelation and sequestering metal ions and transporting them to the vacuole (Manara 2012). Also, metal ions are transported from the cytoplasm to the vacuole, as well as to the apoplast and endoplasmic reticulum, via the CDF transporter family (also known as MTPs in plants) (Krämer et al. 2007).

17.4 Role of MYB, NAC, and WRKY Transcription Factors in Response to Heavy Metal Stress

Transcriptional control is a major mechanism facilitating gene expression. Transcription factors (TFs) and their *cis*-elements together regulate temporal and spatial expression of genes (Puranik et al. 2012). TFs constitute a major portion of the genome, the *Arabidopsis* genome possess 1510–1581 (20%) and rice 1611 TFs genes (Iida et al. 2005; Xiong et al. 2005). However, the single-celled yeast *Saccharomyces cerevisiae* contains only 2% of TFs (Mewes et al. 1997) suggesting that with evolution to complex life there is an increase in number of transcription regulators. Microarray analyses of root and shoot tissue in rice resulted in identification of MYB, NAC, and WRKY TFs (Ogawa et al. 2009). The analysis of Cd-treated maize and rice transcriptomes identified 880 ortholog groups, with a conserved function of Cd-responsive orthologs and paralogs, suggesting common molecular mechanisms of the plant response to Cd stress (Cheng et al. 2018).

17.5 MYB Transcription Factors

The MYB (myeloblastoma) TFs are found in all eukaryotes including animals, plants, fungi, and slime mould and forms a large family in plants with approximately 9% of total TF in *Arabidopsis* (Riechmann 2000). The MYBs are characterised by the presence of highly conserved MYB DNA-binding domain consisting of one to four imperfect amino acid sequence repeats (R) of about 52 amino acids, each forming three α-helices. These TFs are classified based on the number of repeats; 4R-MYB, containing either four R1 or R2 repeats, it forms the smallest group with only one member in many plant genomes (Dubos et al. 2010); R1R2R3-type MYB (3R-MYB), containing R1R2R3-type MYB (3R-MYB); R2R3-type MYB (2R-MYB), containing R2R3-type MYB (2R-MYB), constitutes the largest group of plant MYB family; 1R MYB has wide occurrence in plants and constitutes the second largest group of the MYB family. MYBs regulate different primary and

Table 17.1 Transcript regulation of MYB transcription factors towards metal stress treatments

S. no	Gene	Source plant	Metal stress	Response	References
1.	OsMYB45	*Oryza sativa*	Cd	Upregulation (roots)	Hu et al. (2017)
2.	OsARM1	*Oryza sativa*	As	Upregulation (roots, leaves, stem)	Wang et al. (2017)
3.	BjCdR12	*Brassica juncea*	Cd	Upregulation	Fusco et al. (2005)
4.	SbMYB15	*Salicornia brachiata*	Cd and Ni	Upregulation	Sapara et al. (2019)
5.	MYB72	*Arabidopsis thaliana*	Cd	Upregulation	van de Mortel et al. (2008)
		Thlaspi caerulescens			

Table 17.2 Metal stress tolerance potential of transgenics overexpressing MYB transcription factors

S. no	Gene	Transformed plant	Response	References
1.	OsARM1	*Oryza sativa*	Sensitive to As	Wang et al. (2017)
2.	SbMYB15	*Nicotiana tabacum*	Enhanced Cd and Ni tolerance	Sapara et al. (2019)

secondary metabolism, seed and floral development, cell fate and identity, and abiotic and biotic stress tolerance (Dubos et al. 2010). There exist limited studies towards the regulation and involvement of MYB TFs in response to heavy metals (Tables 17.1 and 17.2).

An *Arabidopsis MYB59* homologue (BjCdR12) regulated via Cd treatment was identified in *Brassica juncea* (Fusco et al. 2005). In *Oryza sativa* cv. *Nipponbare*, under Cd (100 μM) stress *OsMYB45* expression was induced in roots at 3 h and increased up to tenfold (Hu et al. 2017). Microarray analysis of Zn/Cd-hyperaccumulator *Thlaspi caerulescens* and the non-accumulator *Arabidopsis* revealed the regulation of MYB TFs (*MYB4*, *MYB10*, *MYB72*) on Cd and Zn exposure in *Arabidopsis*, while in *T. caerulescens*, MYB TF was regulated during Zn deficiency and high Cd exposure (van de Mortel et al. 2008). The binding of *JrMYB2* to the MYBCORE motif identified in the promoter region of Cd-inducible *JrVHAG1* suggest that *JrMYB2* might function as an upstream regulator of *JrVHAG1* to either regulate *JrVHAG1* or pair up with *JrVHAG1* to enhance plant $CdCl_2$ stress tolerance (Xu et al. 2018). Mutation of *OsMYB45* in rice caused hypersensitivity to Cd, and the quantity of H_2O_2 in mutant leaves increased, while catalase (CAT) activity reduced. In addition, the expression of CAT genes was reduced in mutant as compared to WT, suggesting *OsMYB45* plays a significant role in rice resistance to Cd stress (Hu et al. 2017). *SbMYB15* from *Salicornia brachiata*, a succulent halophyte, showed increased transcript accumulation with Cd and Ni stress. Furthermore, the transgenic tobacco overexpressing *SbMYB15* exhibited better growth, reduced heavy metal concentration, increased activity of antioxidative enzymes, and imparting tolerance towards Cd and Ni stress (Sapara et al. 2019).

In rice, As-associated transporter genes are controlled by the rice R2R3 MYB TF *OsARM1* (ARSENITE-RESPONSIVE MYB1). After AsIII treatment, *OsARM1* expression was rapidly induced in leaves, stems, basal regions, and roots, peaking after 1 h. Overexpression of *OsARM1* downregulated As transporters (*OsLsi1*, *OsLsi2*, and *OsLsi6*) leading to enhanced sensitivity to AsIII, while knockout of *OsARM1* downregulated the As transporters and improved resistance to AsIII, suggesting *OsARM1* regulates As absorption and root-to-shoot translocation in rice (Wang et al. 2017).

17.6 NAC Transcription Factors

The NAC TF is one of the largest plant-specific TF family, the nomenclature being derived from the initials of *Petunia* NAM (no apical meristem; Souer et al. 1996), ATAF1/2 (*Arabidopsis* transcription activation factor 1/2), and CUC2 (cup-shaped cotyledon 2) proteins (Aida et al. 1997).

NAC protein comprises a conserved 150–160 amino acid long N-terminal NAC domain, with five subdomains A–E, involved in DNA binding ability, nuclear localisation, and homo- or hetero-dimerisation and the C-terminal region, which is a variable transcriptional regulatory region, functioning as transcription activator or repressor (Tran et al. 2004). The NAC TFs possessing α-helical transmembrane motifs in their C-terminal regions are designated as NTLs; these motifs provide membrane anchoring to plasma membrane and endoplasmic reticulum (Tran et al. 2004). The NAC TFs regulate plant growth development and growth and also stress responses (Puranik et al. 2012). The transcript regulation of NAC TFs in response to heavy metal stress has been studied from *Aegilops*, *Oryza*, *Helianthus*, *Solanum*, and *Vigna* (Table 17.3).

The expression levels of NAC TF genes from *Helianthus annuus* showed differential expression in response to Cr and Cu stress alone and in combination (Yuce et al. 2019). Three NACs from *Aegilops markgrafii* (*AemNAC1*, *AemNAC2*, and *AemNAC3*) with Cd (100 μM) stress showed differential expression in shoot and root tissues. *AemNAC2* (fivefold) and *AemNAC3* (2.5-fold) were expressed highly under Cd stress (Du et al. 2020). A rice NAC family transcription factor, *OsNAC300*, with Cd (100 μM) stress was induced in roots in a dose-dependent manner (Hu et al. 2021). The overexpression of NAC TF enhances heavy metal tolerance in plants (Table 17.4). The wheat transgenics overexpressing *AemNAC2* showed low accumulation of Cd and high biomass as compared to untransformed wheat. *TaNRAMP5* and *TaHMA2* are involved in absorption and transportation of heavy metals in plant and are related to Cd sensitivity in wheat. The expression of *TaNRAMP5* and *TaHMA2* was inhibited in transgenics due to low Cd content. Thus, *AemNAC2* can enhance Cd tolerance in wheat (Du et al. 2020). The *osnac300* knockout mutants with 200 μM Cd showed necrosis and browning on the leaves, root elongation, and

Table 17.3 Transcript regulation of NAC transcription factors towards metal stress treatments

S. no	Gene	Source plant	Metal stress	Transcript response	References
1.	AemNAC2	*Aegilops markgrafii*	Cd	Upregulation (shoot and root)	Du et al. (2020)
2.	OsNAC5	*Oryza sativa*	Al	Upregulation (root)	Moreno-Alvarado et al. (2017)
3.	Han682	*Helianthus annuus*	Cr	Upregulation (root) Slight change (shoot)	Yuce et al. (2019)
			Cu	Downregulation (root, shoot)	
	Han2027	*Helianthus annuus*	Cr	Upregulation (root) Downregulation (shoot)	
			Cu	Upregulation (root) Downregulation (shoot)	
	Han2724	*Helianthus annuus*	Cr	Upregulation (root) Downregulation (shoot)	
			Cu	Slight change (root) Downregulation (shoot)	
4.	OsNAC300	*Oryza sativa*	Cd	Upregulation (root)	Hu et al. (2021)
5.	SlNAC	*Solanum lycopersicum*	Al	Upregulation (root)	Jin et al. (2020)
6.	VuNAR1	*Vigna umbellata*	Al	Upregulation (root)	Lou et al. (2020)
7.	SNAC1	*Oryza sativa*	Cd	Upregulation (shoot)	Wang et al. (2020)
8.	OsNAC5	*Oryza sativa*	Al	Upregulation (root, shoot)	Escobar-Sepúlveda et al. (2017)

Table 17.4 Metal stress tolerance potential of transgenics overexpressing NAC transcription factors

S. no	Gene	Transformed plant	Response	References
1.	AemNAC2	Wheat (cultivar bobwhite)	Enhanced Cd tolerance	Du et al. (2020)
2.	OsNAC300	*Oryza sativa*	Enhanced Cd tolerance	Hu et al. (2021)
3.	VuNAR1	*Arabidopsis thaliana*	Enhanced Al tolerance	Lou et al. (2020)

survival ratio were also low. Further, *OsNAC300* overexpression resulted in enhanced expression levels of *OsNAC300* and root elongation, suggesting that *OsNAC300* is required for Cd tolerance in rice (Hu et al. 2021).

The *SNAC1* and *OsNAC3* TFs were identified as Cd-tolerant rice genes from yeast cDNA library. *SNAC1* targets many Cd-tolerant genes (*OsMKK1*, *OsMKK6*, *OsMPK3*) that are important in the MAPK signalling cascade. Various networking genes closely related to *SNAC1*, such as several kinases and WRKY-type transcription factors, were highlighted on the gene network map. Under 500 μM Cd stress transient expression of the rice *SNAC1* gene in tobacco leaves showed decreased Cd-induced cell death and brown patches of H_2O_2 formation in the lesion sites, indicating involvement of *SNAC1* towards Cd tolerance (Wang et al. 2020). As rice is an Al-tolerant crop, withstanding two- to fivefold higher Al levels than wheat, sorghum, or maize (Famoso et al. 2010), therefore, the expression of 57 NAC genes in four Mexican rice cultivars (ssp. indica), Cotaxtla, Tres Ríos, Huimanguillo, and Temporalero, was analysed. Differential transcript expression of 25 genes was observed, with 21 genes in Cotaxtla, 19 in Tres Ríos, 18 in Huimanguillo, and 24 in Temporalero plants, suggesting that Al controls the expression of NAC transcription factors in rice. The expression of *OsNAC5* was induced in all four rice cultivars (Moreno-Alvarado et al. 2017). The rice NAC genes are Al-responsive and along with phytohormones play an important role in regulating growth (Escobar-Sepúlveda et al. 2017). Expression analyses of 19 (out of 93) *SlNACs* exhibited variable expression patterns with Al stress. Interestingly, 7 *SlNACs* showing substantially high expression were induced even in the absence of Al, by cycloheximide (CHX), indicating that in the absence of Al, a transcriptional repressor may inhibit the transcriptional activation of *SlNAC* TFs, and Al stress may induce the repressor to degrade. Therefore, *SlNAC* TFs are early genes in the tomato root apex that are implicated in the Al stress response (Jin et al. 2020). The *VuNAR1* (NAC-type Al responsive 1) from *Vigna umbellata* expression was enhanced in the root tip and basal root regions but not in the leaves when rice bean seedlings were exposed to Al. The expression increased in a dose-dependent manner (0–50 μM), and with 25 μM Al the expression initiated at 30 min and peaked at 2 h and dropped to control levels at 8 h (Lou et al. 2020). The *Arabidopsis* transgenics overexpressing *VuNAR1* showed better root growth and low accumulation of Al as compared to WT with Al stress. The reduced cell wall Al content can be attributed to direct transcriptional activation of cell wall associated receptor kinase 1 (*WAK1*) which regulates cell wall pectin metabolism (Lou et al. 2020).

17.7 WRKY Transcription Factors

The WRKY TFs are named based on the 60 amino acids long highly conserved WRKY domains, with a conserved WRKYGQK heptapeptide their N-terminal and a novel zinc finger-like motif at the C-terminal. The heptapeptide sequence and zinc finger-like motif play a prominent role towards the binding affinity of WRKY TFs to the *cis*-acting W box (TTGACT/C) elements present in the promoters. The WRKY

TFs are classified into three major groups: Group I comprise WRKY proteins with two WRKY domains, and Group II and Group III have only one WRKY domain (Eulgem et al. 2000). The Group I and Group II proteins show similar C2H2-type zinc finger motif (C–X4–5–C–X22–23–H–X1–H) and Group III possess the C2HC zinc finger motif (C–X7–C–X23–H–X–C). Group II WRKYs are further divided into subgroups IIa, IIb, IIc, IId, and IIe based on the amino acid sequence present outside the WRKY domain. The DNA-binding selectivity of WRKY TFs to the W box (TTGACC/T) is dependent on the sequences flanking the W box, also the interaction of WRKY TFs with other proteins forming higher-order protein DNA complexes, resulting in differential transcriptional activity (Ciolkowski et al. 2008). Although Group I have two WRKY domains, it has been reported that in SPF1, ZAP1, and PcWRKY1 proteins, the C-terminal WRKY domain facilitates sequence-specific binding to DNA. The N-terminal might increase affinity and specificity or provide support for protein–protein interactions. Also, mutations in the WRKYGQK sequence significantly reduces the DNA-binding activity, furthermore substituting the conserved cysteine and histidine residues in the C2H2-type zinc finger-like motif abolished the DNA-binding activity. WRKY TFs have been isolated from different plants including *Arabidopsis*, rice, mulberry (Baranwal et al. 2016), soybean (Schmutz et al. 2010), papaya, poplar, sorghum, *Physcomitrella patens* (Pandey and Somssich 2009), and *Jatropha* (Xiong et al. 2013). Furthermore, they are also reported from protists (*Giardia lamblia* and *Dictyostelium discoideum*) and green algae (*Chlamydomonas reinhardtii*), suggesting ancient origin of the gene family. The phylogenetic relations revealed the WRKY gene family arise during evolution through gene duplications and that the amplification of group III WRKY genes in rice is due to tandem and segmental gene duplications. The probability of an evolutionary pathway for WRKY–GCM1 superfamily via classical C2H2 fingers (Znf) is also suggested (Babu et al. 2006). In higher plants 20% of WRKYs belong to evolutionary important Group III which plays an important role in plant adaptation to environment (Rushton et al. 2010); this group is largely active in monocots with specific roles (Zhang and Wang 2005). WRKY TFs regulate stress signal pathways by interacting and regulating other TFs or genes. WRKY TFs show autoregulation, cross regulation, and also post translational modification like phosphorylation to regulate its activity (Rushton et al. 2010; Asai et al. 2002). WRKY TFs also provide tolerance against different metal toxicity.

The WRKYs TFs transcript response to different metal stress has been analysed (Table 17.5). In soybean, 29 WRKY genes were identified to be Cd-responsive, with 26 being upregulated and only 3 showed downregulation. Furthermore, the *GmWRKY142* expression was upregulated in soybean roots (5 days old) on Cd (10–50 μM Cd) treatment at different time durations. The maximum *GmWRKY142* expression was observed with 25 μM Cd at 4 h, followed by a reduction at the sixth hour (Cai et al. 2020). The *Arabidopsis WRKY13* gene expression was detected in all of the tissues, with the roots showing maximum accumulation with Cd stress (50 μM) at 1 h and then progressively declines (Sheng et al. 2019). Various Group I *WRKYs* are essential in abiotic stress responses as well as ABA signalling. After Cd (400 μM) treatment, ZmWRKY4 expression increased quickly, peaked at 60 min,

Table 17.5 Transcript regulation of WRKY transcription factors towards metal stress treatments

S. no	Gene	Source plant	Type	Metal stress	Transcript response	References
1.	WRKY12	*Arabidopsis thaliana*	II	Cd	Downregulation	Han et al. (2019)
2.	WRKY13	*Arabidopsis thaliana*	II	Cd	Upregulation (highly in roots)	Sheng et al. (2019)
3.	WRKY25	*Arabidopsis thaliana*	I	Zn	Upregulation	Wu et al. (2020)
4.	WRKY33	*Arabidopsis thaliana*	I	Zn		
5.	WRKY47	*Arabidopsis thaliana*	II	Al	No change	Li et al. (2020)
6.	*GmWRKY142*	*Glycine max*	III	Cd	Upregulation (roots)	Cai et al. (2020)
7.	MdWRKY11	*Malus domestica*	II	Cu	Upregulation (root, leaf)	Shi et al. (2020)
8.	OsWRKY22	*Oryza sativa*	III	Al	Upregulation (root tips, moderately in shoot)	Li et al. (2018)
9.	OsWRKY80	*Oryza sativa*	II	Fe	Upregulation (root, stem, leaf)	Ricachenevsky et al. (2010)
10.	ZmWRKY4	*Zea mays*	I	Cd	Upregulation	Hong et al. (2017)

Table 17.6 Metal stress tolerance potential of transgenics overexpressing WRKY transcription factors

S. no	Gene	Type	Transformed plant	Response	References
1.	WRKY12	II	*Arabidopsis thaliana*	Reduced Cd tolerance	Han et al. (2019)
2.	WRKY13	II	*Arabidopsis thaliana*	Enhanced Cd tolerance	Sheng et al. (2019)
3.	WRKY25	I	*Arabidopsis thaliana*	Enhanced Zn tolerance	Wu et al. (2020)
4.	WRKY33	I	*Arabidopsis thaliana*		
5.	WRKY47	II	*Arabidopsis thaliana*	Enhanced Al tolerance	Li et al. (2020)
6.	*GmWRKY142*	III	*Arabidopsis thaliana*	Enhanced Cd tolerance	Cai et al. (2020)
			Glycine max	Enhanced Cd tolerance	
7.	MdWRKY11	II	*Malus domestica*	Enhanced Cu tolerance	Shi et al. (2020)
8.	OsWRKY22	III	*Oryza sativa*	Enhanced Al tolerance	Li et al. (2018)
9.	ZmWRKY4	I	*Zea mays*	Enhanced Cd tolerance	Hong et al. (2017)

and then began to decrease in maize (Hong et al. 2017). Expression levels of *WRKY12* in *Arabidopsis* under Cd stress (50 μM) treatment showed that its transcription was inhibited by Cd stress. Expression pattern in stem was highest of all the examined tissues. Thus, *WRKY12* may have a role in regulation of Cd tolerance (Han et al. 2019). The transgenics overexpressing WRKY TFs show improved metal tolerance (Table 17.6). The overexpression of *GmWRKY142* in *Arabidopsis* transgenic plants reduced the degree of chlorosis in leaves and also promoted the growth of the plant height and fresh weights. The transgenics showed reduced Cd accumulation, which was attributed to the activated expression *ATCDT1* and *GmCDT1*-like genes, encoding Cd-binding Cys-rich proteins. Furthermore, in transgenics an array of downstream genes including TFs, metabolic regulators, metal stress-responsive, auxin-responsive, and cell wall structural genes were regulated (Cai et al. 2020). In *Arabidopsis*, overexpression of *WRKY13* reduced Cd accumulation and improved Cd tolerance, whereas loss-of-function of *WRKY13* increased Cd accumulation and sensitivity. Furthermore, when *Arabidopsis* was exposed to Cd (50 μM Cd), *WRKY13* directly triggered the expression of *PDR8*, which encodes a Cd^{2+} extrusion pump, reducing Cd accumulation and increasing Cd tolerance (Sheng et al. 2019). According to the study, *ZmWRKY4* is necessary for the ABA-induced increase in superoxide dismutase (SOD) and ascorbate peroxidase (APX) expression and activity. Overexpression of *ZmWRKY4* in maize protoplasts increased the expression and activity of antioxidant enzymes, but RNAi silencing of *ZmWRKY4* prevented

ABA-induced increases in antioxidant enzyme expression and activity. Therefore, *ZmWRKY4* upregulates SOD and APX, under Cd stress (Hong et al. 2017). *WRKY12* loss-of-function mutants (*wrky12–1* and *wrky12–2*) were given 0–75 μM Cd stress. The root length and fresh weights increased in mutants in Cd concentration-dependent manner, indicating that the mutants showed enhanced Cd tolerance. In addition, under Cd stress, *WRKY12*-OE plants showed increased Cd sensitivity. Also, Cd concentration was higher in wrky12 mutant and low in *WRKY12*-OE lines. Therefore, *WRKY12* specifically targets GSH1 and indirectly suppresses PC synthesis-related gene expression to negatively regulate Cd accumulation and tolerance in *Arabidopsis* (Han et al. 2019).

Microarray analysis in rice roots reveals differential regulation of 1613 genes, with 1050 being upregulated and 563 downregulated. *OsWRKY22* was upregulated and have a significant role in plant responses to Al stress. In rice, with Al stress, *OsWRKY22* expression was highly induced in root tips (25-fold), moderately induced in shoots (1.5-fold), and unchanged in basal roots. The *OsWRKY22* expression increases with increasing Al (0–50 μM) concentration, and it was maximum at 3 h duration (Li et al. 2018). The transgenic *oswrky22* homozygous plants showed inhibited root growth and high Al concentration in root tips under 25 μM Al stress for 5 days. Under identical Al treatment, results of transgenic complementation lines (*oswrky22*-Comp1 and 2) were similar to the WT but not to oswrky22. This demonstrates that loss of *OsWRKY22* function increases sensitivity to Al, but transgenic complementation of this loss of function restores tolerance to Al stress. Al-induced increase in *OsFRDL4* is regulated by *OsWRKY22* and *ART1*, and increase citrate secretion and Al tolerance in rice (Li et al. 2018). Under Al (30 μM) stress, in two distinct *WRKY47* loss-of-function mutant alleles (*wrky47–1* and *wrky47–2*), there was reduced relative root growth rate. The wrky*47–1* complementation line was able to restore the Al sensitivity of the wrky47–1 mutant and *WRKY47* overexpression lines promoted Al tolerance. *WRKY47* increases the expression of extensin-like protein (ELP) and xyloglucan endotransglucosylase-hydrolases 17 (XTH17), required for cell wall modification and Al tolerance by facilitating distribution of Al between root apoplast and symplast (Li et al. 2020).

The expression of *OsWRKY80* increased 1.6-fold after 6 days and 3-fold after 9 days with Fe stress (500 ppm, Ricachenevsky et al. 2010). Rice plants at the four-leaf stage (30 days) were given a 6-day Fe-excess (500 ppm) treatment. *OsWRKY80* expression was elevated in leaves, stems, and root tissues examined after exposure to excessive Fe. As a result, the responsiveness of *OsWRKY80* to Fe excess occurs throughout the plant. The upregulation was also observed after dark-induced senescence and drought stress, suggesting that *OsWRKY80* could be common stress response gene (Ricachenevsky et al. 2010).

The expression analysis of 29 *MdWRKYs* with 500 μM $CuSO_4$ reveals significant upregulation of *MdWRKY11* in both the roots and the leaves, indicating that this gene plays an essential role in the response to excess Cu. Transgenic apple plants overexpressing *MdWRKY11* were created to examine the potential role of *MdWRKY11* in Cu tolerance. WT leaves developed toxic symptoms after 30 days of treatment (500 μM $CuSO_4$). These toxic symptoms, on the other hand, were not detected in transgenic plants. Also, *MdWRKY11* improves Cu tolerance by increasing transcription of *MdHMA5*, a Cu transporter, functions in the storage of excess Cu in root cell walls and stems (Shi et al. 2020).

In Zn excess (10× Zn) condition, expression levels of *Arabidopsis WRKY33* and *WRKY25* decreased in *agb1–2* compared to control, highlighting that WRKY transcription factors mediate the AGB1-dependent nutrient responses. In both *Arabidopsis* and rice, the Gβ-null (*agb1*) mutant showed changed metal ion profiles and displayed severe growth inhibition and abnormal root waving under iron and zinc stress, whereas the Gα-null mutation reduced leaf chlorosis under iron deprivation. The expression of metal stress marker genes such as ion transporters and transcription factors were also examined in WT and *agb1–2* plants. Under zinc-deficient conditions, the zinc transporter genes *ZIP3* and *ZIP4* showed reduced induction in *agb1–2*. As a result, when the expression of these ion transporters is disrupted, poor nutrient absorption and nutrient depletion in the shoot tissues may occur (Wu et al. 2020).

17.8 Conclusion

Heavy metal stress is one of the key problems impacting agriculture. The situation is expected to worsen as a consequence of growing human activities, which have resulted in unfavourable changes in the environment. Some heavy metals cannot be decomposed and hence exist in the soil. Plants are the main staple and nourishment for the world's population. Heavy metals may be distributed from the roots to the stems, leaves, flowers, and seeds of plants, which can affect the consumers of these plants. Species of plants, heavy metal nature and amount, as well as charge density of heavy metals all influence their absorption, movement, and transportation within the plant. Heavy metals have the ability to disturb fundamental equilibrium in plant cells, shifting the balance, causing ROS production. Furthermore, heavy metal stress causes the replacement of necessary cations with poisonous heavy metal ions and their attachment to active groups of cofactors. It causes visible symptoms on plants like reduced growth, chlorosis, and wilting of leaves. TFs like MYB, WRKY, and NAC have also been found for providing heavy metal stress tolerance in plants. Plants have established numerous signalling mechanisms, including signal transduction, to counter metal stress. These pathways transfer signals, trigger the expression of defence-related genes, and offer tolerance to stress stimuli. Plant toxicity tolerance is determined by their capacity to inhibiting metal ions from entering the cell, complexing and chelating them in the extracellular space, and perhaps sequestering them in the vacuole. To accomplish this, plants must trigger defensive responses such as antioxidative enzyme production and activation, as well as mechanisms for avoiding oxidative stress abnormalities. Thus, a high-throughput analysis using different tools and techniques is required to understand the role of TFs, regulators of various components of signal transduction pathway. It would help to uncover the missing connections of the mechanisms for metal stress tolerance in plants.

References

Abedin MJ, Cotter-Howells J, Meharg AA (2002) Arsenic uptake and accumulation in rice (*Oryza sativa* L.) irrigated with contaminated water. Plant Soil 240(2):311–319

Aida M, Ishida T, Fukaki H (1997) Genes involved in organ separation in *Arabidopsis*: an analysis of the cup-shaped cotyledon mutant. Plant Cell 9:841–857

Appenroth KJ (2010) Definition of "heavy metals" and their role in biological systems. In: Soil heavy metals. Springer, Berlin, pp 19–29

Asai T, Tena G, Plotnikova J, Willmann MR, Chiu W-L, Gomez-Gomez L, Boller T, Ausubel FM, Sheen J (2002) MAP kinase signalling cascade in *Arabidopsis* innate immunity. Nature 415: 977–983

Aydinalp C, Marinova S (2009) The effects of heavy metals on seed germination and plant growth on alfalfa plant (*Medicago sativa*). Bulg J Agric Sci 15(4):347–350

Azooz MM, Abou-Elhamd MF, Al-Fredan MA (2012) Biphasic effect of copper on growth, proline, lipid peroxidation and antioxidant enzyme activities of wheat (*Triticum aestivum* cv. *Hasaawi*) at early growing stage. Aust J Crop Sci 6(4):688

Babu MM, Iyer LM, Balaji S, Aravind L (2006) The natural history of the WRKY–GCM1 zinc fingers and the relationship between transcription factors and transposons. Nucleic Acids Res 34(22):6505–6520

Bakkaus E, Gouget B, Gallien JP, Khodja H, Carrot F, Morel JL, Collins R (2005) Concentration and distribution of cobalt in higher plants: the use of micro-PIXE spectroscopy. Nucl Instrum Methods Phys Res Sect B 231(1–4):350–356

Baranwal VK, Negi N, Khurana P (2016) Genome-wide identification and structural, functional and evolutionary analysis of WRKY components of mulberry. Sci Rep 6(1):1–13

Barceloux DG, Barceloux D (1999) Cobalt. J Toxicol Clin Toxicol 37(2):201–216

Barrachina AC, Carbonell FB, Beneyto JM (1995) Arsenic uptake, distribution, and accumulation in tomato plants: effect of arsenite on plant growth and yield. J Plant Nutr 18(6):1237–1250

Baryla A, Carrier P, Franck F, Coulomb C, Sahut C, Havaux M (2001) Leaf chlorosis in oilseed rape plants (*Brassica napus*) grown on cadmium-polluted soil: causes and consequences for photosynthesis and growth. Planta 212(5):696–709

Bhalerao SA, Sharma AS, Poojari AC (2015) Toxicity of nickel in plants. Int J Pure Appl Biosci 3(2):345–355

Broadley MR, White PJ, Hammond JP, Zelko I, Lux A (2007) Zinc in plants. New Phytol 173(4): 677–702

Cai Z, Xian P, Wang H, Lin R, Lian T, Cheng Y, Ma Q, Nian H (2020) Transcription factor GmWRKY142 confers cadmium resistance by up-regulating the cadmium tolerance 1-like genes. Front Plant Sci 11:724

Cargnelutti D, Tabaldi LA, Spanevello RM, de Oliveira JG, Battisti V, Redin M, Linares CEB, Dressler VL, de Moraes Flores EM, Nicoloso FT, Morsch VM (2006) Mercury toxicity induces oxidative stress in growing cucumber seedlings. Chemosphere 65(6):999–1006

Chatterjee J, Chatterjee C (2000) Phytotoxicity of cobalt, chromium and copper in cauliflower. Environ Pollut 109(1):69–74

Chen LM, Lin CC, Kao CH (2000) Copper toxicity in rice seedlings: changes in antioxidative enzyme activities, H_2O_2 level, and cell wall peroxidase activity in roots. Bot Bull Acad Sin 41: 99–103

Chen YA, Chi WC, Trinh NN, Huang LY, Chen YC, Cheng KT, Huang TL, Lin CY, Huang HJ (2014) Transcriptome profiling and physiological studies reveal a major role for aromatic amino acids in mercury stress tolerance in rice seedlings. PLoS One 9(5):95163

Cheng D, Tan M, Yu H, Li L, Zhu D, Chen Y, Jiang M (2018) Comparative analysis of Cd-responsive maize and rice transcriptomes highlights Cd co-modulated orthologs. BMC Genomics 19(1):1–15

Ciolkowski I, Wanke D, Birkenbihl RP, Somssich IE (2008) Studies on DNA-binding selectivity of WRKY transcription factors lend structural clues into WRKY-domain function. Plant Mol Biol 68:81–92

Cox MS, Bell PF, Kovar JL (1996) Differential tolerance of canola to arsenic when grown hydroponically or in soil. J Plant Nutr 19(12):1599–1610

Cox A, Venkatachalam P, Sahi S, Sharma N (2016) Silver and titanium dioxide nanoparticle toxicity in plants: a review of current research. Plant Physiol Biochem 107:147–163

Dai LP, Xiong ZT, Huang Y, Li MJ (2006) Cadmium-induced changes in pigments, total phenolics, and phenylalanine ammonia-lyase activity in fronds of *Azolla imbricata*. Environ Toxicol Int J 21(5):505–512

DalCorso G, Manara A, Furini A (2013) An overview of heavy metal challenge in plants: from roots to shoots. Metallomics 5(9):1117–1132

Davies FT Jr, Puryear JD, Newton RJ, Egilla JN, Saraiva Grossi JA (2002) Mycorrhizal fungi increase chromium uptake by sunflower plants: influence on tissue mineral concentration, growth, and gas exchange. J Plant Nutr 25(11):2389–2407

Dodd AN, Kudla J, Sanders D (2010) The language of calcium signaling. Annu Rev Plant Biol 61: 593–620

Du X, He F, Zhu B, Ren M, Tang H (2020) NAC transcription factors from *Aegilops markgrafii* reduce cadmium concentration in transgenic wheat. Plant Soil 449:1–12

Dubos C, Stracke R, Grotewold E, Weisshaar B, Martin C, Lepiniec L (2010) MYB transcription factors in Arabidopsis. Trends Plant Sci 15(10):573–581

Ebbs SD, Kochian LV (1997) Toxicity of zinc and copper to *Brassica* species: implications for phytoremediation. J Environ Qual 26(3):776–781

Ernst WH (2006) Evolution of metal tolerance in higher plants. For Snow Landsc Res 80(3): 251–274

Escobar-Sepúlveda HF, Trejo-Téllez LI, García-Morales S, Gómez-Merino FC (2017) Expression patterns and promoter analyses of aluminum-responsive NAC genes suggest a possible growth regulation of rice mediated by aluminum, hormones and NAC transcription factors. PLoS One 12(10):0186084

Eulgem T, Rushton PJ, Robatzek S, Somssich IE (2000) The WRKY superfamily of plant transcription factors. Trends Plant Sci 5:199–206

Famoso AN, Clark RT, Shaff JE, Craft E, McCouch SR, Kochian LV (2010) Development of a novel aluminum tolerance phenotyping platform used for comparisons of cereal aluminum tolerance and investigations into rice aluminum tolerance mechanisms. Plant Physiol 153: 1678–1691

Fargasova A (2004) Toxicity comparison of some possible toxic metals (Cd, Cu, Pb, Se, Zn) on young seedlings of *Sinapis alba* L. Plant Soil Environ 50(1):33–38

Finnegan P, Chen W (2012) Arsenic toxicity: the effects on plant metabolism. Front Physiol 3(182): 00182

Fukao Y, Ferjani A, Tomioka R, Nagasaki N, Kurata R, Nishimori Y, Fujiwara M, Maeshima M (2011) iTRAQ analysis reveals mechanisms of growth defects due to excess zinc in *Arabidopsis*. Plant Physiol 155(4):1893–1907

Fusco N, Micheletto L, Dal Corso G, Borgato L, Furini A (2005) Identification of cadmium-regulated genes by cDNA-AFLP in the heavy metal accumulator *Brassica juncea* L. J Exp Bot 56(421):3017–3027

Galazzi RM, Arruda MAZ (2018) Evaluation of changes in the macro and micronutrients homeostasis of transgenic and non-transgenic soybean plants after cultivation with silver nanoparticles through ionomic approaches. J Trace Elem Med Biol 48:181–187

Gallo-Franco JJ, Sosa CC, Ghneim-Herrera T, Quimbaya M (2020) Epigenetic control of plant response to heavy metal stress: a new view on aluminum tolerance. Front Plant Sci 11:602625

Gill SS, Tuteja N (2011) Cadmium stress tolerance in crop plants: probing the role of sulfur. Plant Signal Behav 6(2):215–222

Gratão PL, Polle A, Lea PJ, Azevedo RA (2005) Making the life of heavy metal-stressed plants a little easier. Funct Plant Biol 32(6):481–494

Guerinot ML (2000) The ZIP family of metal transporters. Biochim Biophys Acta (BBA) Biomembr 1465(1–2):190–198

Gupta M, Sharma P, Sarin NB, Sinha AK (2009) Differential response of arsenic stress in two varieties of *Brassica juncea* L. Chemosphere 74(9):1201–1208

Hamel LP, Nicole MC, Sritubtim S, Morency MJ, Ellis M, Ehlting J, Beaudoin N, Barbazuk B, Klessig D, Lee J, Martin G (2006) Ancient signals: comparative genomics of plant MAPK and MAPKK gene families. Trends Plant Sci 11(4):192–198

Han FX, Su Y, Monts DL, Waggoner CA, Plodinec MJ (2006) Binding, distribution, and plant uptake of mercury in a soil from Oak Ridge, Tennessee, USA. Sci Total Environ 368(2–3): 753–768

Han Y, Fan T, Zhu X, Wu X, Ouyang J, Jiang L, Cao S (2019) WRKY12 represses GSH1 expression to negatively regulate cadmium tolerance in *Arabidopsis*. Plant Mol Biol 99(1–2): 149–159

Hasan SA, Fariduddin Q, Ali B, Hayat S, Ahmad A (2009) Cadmium: toxicity and tolerance in plants. J Environ Biol 30(2):165–174

Heyno E, Klose C, Krieger-Liszkay A (2008) Origin of cadmium-induced reactive oxygen species production: mitochondrial electron transfer versus plasma membrane NADPH oxidase. New Phytol 179(3):687–699

Hong C, Cheng D, Zhang G, Zhu D, Chen Y, Tan M (2017) The role of ZmWRKY4 in regulating maize antioxidant defense under cadmium stress. Biochem Biophys Res Commun 482(4): 1504–1510

Hosseini Z, Poorakbar L (2013) Zinc toxicity on antioxidative response in (*Zea mays* L.) at two different pH. J Stress Physiol Biochem 9(1):66–73

Hou J, Wang L, Wang C, Zhang S, Liu H, Li S, Wang X (2019) Toxicity and mechanisms of action of titanium dioxide nanoparticles in living organisms. J Environ Sci 75:40–53

Hu S, Yu Y, Chen Q, Mu G, Shen Z, Zheng L (2017) OsMYB45 plays an important role in rice resistance to cadmium stress. Plant Sci 264:1–8

Hu S, Shinwari KI, Song Y, Xia J, Xu H, Du B, Luo L, Zheng L (2021) OsNAC300 positively regulates cadmium stress responses and tolerance in rice roots. Agronomy 11(1):95

Iida K, Seki M, Sakurai T, Satou M, Akiyama K, Toyoda T, Konagaya A, Shinozaki K (2005) RARTF: database and tools for complete sets of *Arabidopsis* transcription factors. DNA Res 12: 247–256

Imadi SR, Waseem S, Kazi AG, Azooz MM, Ahmad P (2016) Aluminum toxicity in plants: an overview. Plant Metal Interact 2016:1–20

Jaishankar M, Tseten T, Anbalagan N, Mathew BB, Beeregowda KN (2014) Toxicity, mechanism and health effects of some heavy metals. Interdiscip Toxicol 7(2):60–72

Jalmi SK, Sinha AK (2015) ROS mediated MAPK signaling in abiotic and biotic stress-striking similarities and differences. Front Plant Sci 6:769

Järup L (2003) Hazards of heavy metal contamination. Br Med Bull 68(1):167–182

Jayakumar K, Jaleel CA, Vijayarengan P (2007) Changes in growth, biochemical constituents, and antioxidant potentials in radish (*Raphanus sativus* L.) under cobalt stress. Turk J Biol 31(3): 127–136

Jayakumar K, Rajesh M, Baskaran L, Vijayarengan P (2013) Changes in nutritional metabolism of tomato (*Lycopersicon esculantum* Mill.) plants exposed to increasing concentration of cobalt chloride. Int J Food Nutr Saf 4(2):62–69

Jin JF, Wang ZQ, He QY, Wang JY, Li PF, Xu JM, Zheng SJ, Fan W, Yang JL (2020) Genome-wide identification and expression analysis of the NAC transcription factor family in tomato (*Solanum lycopersicum*) during aluminum stress. BMC Genomics 21(1):1–14

Jonak C, Nakagami H, Hirt H (2004) Heavy metal stress. Activation of distinct mitogen-activated protein kinase pathways by copper and cadmium. Plant Physiol 136(2):3276–3283

Kapoor D, Sharma R, Handa N, Kaur H, Rattan A, Yadav P, Gautam V, Kaur R, Bhardwaj R (2015) Redox homeostasis in plants under abiotic stress: role of electron carriers, energy metabolism mediators and proteinaceous thiols. Front Environ Sci 3:13

Kaveh R, Li YS, Ranjbar S, Tehrani R, Brueck CL, Van Aken B (2013) Changes in *Arabidopsis thaliana* gene expression in response to silver nanoparticles and silver ions. Environ Sci Technol 47(18):10637–10644

Khan MR, Khan MM (2010) Effect of varying concentration of nickel and cobalt on the plant growth and yield of chickpea. Aust J Basic Appl Sci 4(6):1036–1046

Kitchin KT, Wallace K (2006) Dissociation of arsenite-peptide complexes: triphasic nature, rate constants, half-lives, and biological importance. J Biochem Mol Toxicol 20(1):48–56

Krämer U, Talke IN, Hanikenne M (2007) Transition metal transport. FEBS Lett 581(12): 2263–2272

Kumar A, Prasad MNV, Sytar O (2012) Lead toxicity, defense strategies and associated indicative biomarkers in *Talinum triangulare* grown hydroponically. Chemosphere 89(9):1056–1065

Kumar S, Dubey RS, Tripathi RD, Chakrabarty D, Trivedi PK (2015) Omics and biotechnology of arsenic stress and detoxification in plants: current updates and prospective. Environ Int 74:221–230

Kuriakose SV, Prasad MNV (2008) Cadmium stress affects seed germination and seedling growth in *Sorghum bicolor* (L.) Moench by changing the activities of hydrolyzing enzymes. Plant Growth Regul 54(2):143–156

Lee CW, Choi JM, Pak CH (1996) Micronutrient toxicity in seed geranium (*Pelargonium* × *hortorum* Bailey). J Am Soc Hort Sci 121(1):77–82

L'Huillier L, d'Auzac J, Durand M, Michaud-Ferrière N (1996) Nickel effects on two maize (*Zea mays*) cultivars: growth, structure, Ni concentration, and localization. Can J Bot 74(10): 1547–1554

Li GZ, Wang ZQ, Yokosho K, Ding B, Fan W, Gong QQ, Li GX, Wu YR, Yang JL, Ma JF, Zheng SJ (2018) Transcription factor WRKY 22 promotes aluminum tolerance via activation of Os FRDL 4 expression and enhancement of citrate secretion in rice (*Oryza sativa*). New Phytol 219 (1):149–162

Li CX, Yan JY, Ren JY, Sun L, Xu C, Li GX, Ding ZJ, Zheng SJ (2020) A WRKY transcription factor confers aluminum tolerance via regulation of cell wall modifying genes. J Integr Plant Biol 62(8):1176–1192

Lima FRD, Martins GC, Silva AO, Vasques ICF, Engelhardt MM, Cândido GS, Pereira P, Reis RHCL, Carvalho GS, Windmöller CC, Moreira FMS (2019) Critical mercury concentration in tropical soils: impact on plants and soil biological attributes. Sci Total Environ 666:472–479

Lou HQ, Fan W, Jin JF, Xu JM, Chen WW, Yang JL, Zheng SJ (2020) A NAC-type transcription factor confers aluminium resistance by regulating cell wall associated receptor kinase 1 and cell wall pectin. Plant Cell Environ 43(2):463–478

Malecka A, Jarmuszkiewicz W, Tomaszewska B (2001) Antioxidative defense to lead stress in subcellular compartments of pea root cells. Acta Biochim Pol 48(3):687–698

Manara A (2012) Plant responses to heavy metal toxicity. In: Plants and heavy metals. Springer, Dordrecht, pp 27–53

Marin AR, Pezeshki SR, Masschelen PH, Choi HS (1993) Effect of dimethylarsenic acid (DMAA) on growth, tissue arsenic, and photosynthesis of rice plants. J Plant Nutr 16(5):865–880

Meharg AA, Williams PN, Adomako A, Lawgali YY, Deacon C, Villada A, Cambell RCJ, Sun G, Zhu YG, Feldmann J, Rabb A, Zhao FJ, Islam R, Hossain S, Yanai J (2009) Geographical variation in total and inorganic arsenic content of polished (white) rice. Environ Sci Technol 43: 1612–1617

Messer RL, Lockwood PE, Tseng WY, Edwards K, Shaw M, Caughman GB, Lewis JB, Wataha JC (2005) Mercury(II) alters mitochondrial activity of monocytes at sublethal doses via oxidative stress mechanisms. J Biomed Mater Res Part B Appl Biomater 75(2):257–263

Mewes HW et al (1997) Overview of the yeast genome. Nature 387:7–65

Montero-Palmero MB, Martín-Barranco A, Escobar C, Hernández LE (2014) Early transcriptional responses to mercury: a role for ethylene in mercury-induced stress. New Phytol 201(1): 116–130
Moreno-Alvarado M, García-Morales S, Trejo-Téllez LI, Hidalgo-Contreras JV, Gómez-Merino FC (2017) Aluminum enhances growth and sugar concentration, alters macronutrient status and regulates the expression of NAC transcription factors in rice. Front Plant Sci 8:73
Morkunas I, Woźniak A, Mai VC, Rucińska-Sobkowiak R, Jeandet P (2018) The role of heavy metals in plant response to biotic stress. Molecules 23(9):2320
Nagajyoti PC, Lee KD, Sreekanth TVM (2010) Heavy metals, occurrence and toxicity for plants: a review. Environ Chem Lett 8(3):199–216
Nematshahi N, Lahouti M, Ganjeali A (2012) Accumulation of chromium and its effect on growth of (*Allium cepa* cv. Hybrid). Eur J Exp Biol 2(4):969–974
Nevo Y, Nelson N (2006) The NRAMP family of metal-ion transporters. Biochim Biophys Acta (BBA) Mol Cell Res 1763(7):609–620
Ogawa I, Nakanishi H, Mori S, Nishizawa NK (2009) Time course analysis of gene regulation under cadmium stress in rice. Plant Soil 325(1):97–108
Pandey SP, Somssich IE (2009) The role of WRKY transcription factors in plant immunity. Plant Physiol 150:1648–1655
Piotrowska-Niczyporuk A, Bajguz A, Zambrzycka E, Godlewska-Żyłkiewicz B (2012) Phytohormones as regulators of heavy metal biosorption and toxicity in green alga *Chlorella vulgaris* (Chlorophyceae). Plant Physiol Biochem 52:52–65
Poonkothai MVBS, Vijayavathi BS (2012) Nickel as an essential element and a toxicant. Int J Environ Sci 1(4):285–288
Potters G, Pasternak TP, Guisez Y, Palme KJ, Jansen MA (2007) Stress-induced morphogenic responses: growing out of trouble? Trends Plant Sci 12(3):98–105
Pourrut B, Shahid M, Dumat C, Winterton P, Pinelli E (2011) Lead uptake, toxicity, and detoxification in plants. Rev Environ Contam Toxicol 213:113–136
Puranik S, Sahu PP, Srivastava PS, Prasad M (2012) NAC proteins: regulation and role in stress tolerance. Trends Plant Sci 17:369–381
Ricachenevsky FK, Sperotto RA, Menguer PK, Fett JP (2010) Identification of Fe-excess-induced genes in rice shoots reveals a WRKY transcription factor responsive to Fe, drought and senescence. Mol Biol Rep 37(8):3735–3745
Riechmann JL (2000) Arabidopsis transcription factors: genome-wide comparative analysis among eukaryotes. Science 290:2105–2110
Romanowska E, Wasilewska W, Fristedt R, Vener AV, Zienkiewicz M (2012) Phosphorylation of PSII proteins in maize thylakoids in the presence of Pb ions. J Plant Physiol 169(4):345–352
Rushton PJ, Somssich IE, Ringler P, Shen QJ (2010) WRKY transcription factors. Trends Plant Sci 15:247–258
Salt DE, Blaylock M, Kumar NP, Dushenkov V, Ensley BD, Chet I, Raskin I (1995) Phytoremediation: a novel strategy for the removal of toxic metals from the environment using plants. Bio/Technology 13(5):468–474
Samac DA, Tesfaye M (2003) Plant improvement for tolerance to aluminum in acid soils—a review. Plant Cell Tissue Organ Culture 75(3):189–207
Sapara KK, Khedia J, Agarwal P, Gangapur DR, Agarwal PK (2019) SbMYB15 transcription factor mitigates cadmium and nickel stress in transgenic tobacco by limiting uptake and modulating antioxidative defence system. Funct Plant Biol 46(8):702–714
Schiavon M, Agostini G, Pittarello M, Dalla Vecchia F, Pastore P, Malagoli M (2009) Interactions between chromate and sulfate affect growth, photosynthesis and ultrastructure in *Brassica juncea* (L.) Czern. Sulfur metabolism in plants. Backhuys Publishers, Leiden
Schmutz J, Cannon SB et al (2010) Genome sequence of the palaeopolyploid soybean. Nature 463: 178–183
Seregin IV, Kozhevnikova AD, Kazyumina EM, Ivanov VB (2003) Nickel toxicity and distribution in maize roots. Russ J Plant Physiol 50(5):711–717

Sethy SK, Ghosh S (2013) Effect of heavy metals on germination of seeds. J Nat Sci Biol Med 4(2): 272

Shahid M, Pinelli E, Pourrut B, Silvestre J, Dumat C (2011) Lead-induced genotoxicity to *Vicia faba* L. roots in relation with metal cell uptake and initial speciation. Ecotoxicol Environ Saf 74(1):78–84

Shahid M, Dumat C, Khalid S, Niazi NK, Antunes PM (2016) Cadmium bioavailability, uptake, toxicity and detoxification in soil-plant system. Rev Environ Contam Toxicol 241:73–137

Shanker AK, Cervantes C, Loza-Tavera H, Avudainayagam S (2005) Chromium toxicity in plants. Environ Int 31(5):739–753

Shanker AK, Djanaguiraman M, Venkateswarlu B (2009) Chromium interactions in plants: current status and future strategies. Metallomics 1(5):375–383

Sharma P, Dubey RS (2005) Lead toxicity in plants. Braz J Plant Physiol 17(1):35–52

Sharma P, Jha AB, Dubey RS, Pessarakli M (2012) Reactive oxygen species, oxidative damage, and antioxidative defense mechanism in plants under stressful conditions. J Bot 2012:1–26

Sheng Y, Yan X, Huang Y, Han Y, Zhang C, Ren Y, Fan T, Xiao F, Liu Y, Cao S (2019) The WRKY transcription factor, WRKY13, activates PDR8 expression to positively regulate cadmium tolerance in *Arabidopsis*. Plant Cell Environ 42(3):891–903

Shi K, Liu X, Zhu Y, Bai Y, Shan D, Zheng X, Wang L, Zhang H, Wang C, Yan T, Zhou F (2020) MdWRKY11 improves copper tolerance by directly promoting the expression of the copper transporter gene MdHMA5. Horticult Res 7(1):1–10

Shukla D, Krishnamurthy S, Sahi SV (2014) Genome wide transcriptome analysis reveals ABA mediated response in *Arabidopsis* during gold ($AuCl^{-}{}_4$) treatment. Front Plant Sci 5:652

Singh A, Prasad SM (2015) Remediation of heavy metal contaminated ecosystem: an overview on technology advancement. Int J Environ Sci Technol 12(1):353–366

Singh HP, Mahajan P, Kaur S, Batish DR, Kohli RK (2013) Chromium toxicity and tolerance in plants. Environ Chem Lett 11(3):229–254

Smedley PL, Kinniburgh DG (2002) A review of the source, behaviour and distribution of arsenic in natural waters. Appl Geochem 17:517–568

Smeets K, Opdenakker K, Remans T, Forzani C, Hirt H, Vangronsveld J, Cuypers A (2013) The role of the kinase OXI1 in cadmium-and copper-induced molecular responses in *Arabidopsis thaliana*. Plant Cell Environ 36(6):1228–1238

Souer E, Van Houwelingen A, Kloos D (1996) The no apical meristem gene of petunia is required for pattern formation in embryos and flowers and is expressed at meristem and primordia boundaries. Cell 85:159–170

Steffens B (2014) The role of ethylene and ROS in salinity, heavy metal, and flooding responses in rice. Front Plant Sci 5:685

Tran TA, Popova LP (2013) Functions and toxicity of cadmium in plants: recent advances and future prospects. Turk J Bot 37(1):1–13

Tran LSP, Nakashima K, Sakuma Y (2004) Isolation and functional analysis of *Arabidopsis* stress-inducible NAC transcription factors that bind to a drought-responsive cis-element in the early responsive to dehydration stress 1 promoter. Plant Cell 16:2481–2498

Tsay CC, Wang LW, Chen YR (1995) Plant in response to copper toxicity. Taiwania 40(2): 173–181

Tsonev T, Cebola Lidon FJ (2012) Zinc in plants—an overview. Emir J Food Agric (EJFA) 24(4): 322–333

Van Assche F, Clijsters H (1990) Effects of metals on enzyme activity in plants. Plant Cell Environ 13:195–206

van de Mortel JE, Schat H, Moerland PD, Van Themaat EVL, Van Der Ent S, Blankestijn H, Ghandilyan A, Tsiatsiani S, Aarts MG (2008) Expression differences for genes involved in lignin, glutathione and sulphate metabolism in response to cadmium in *Arabidopsis thaliana* and the related Zn/Cd-hyperaccumulator *Thlaspi caerulescens*. Plant Cell Environ 31(3): 301–324

Vijayaragavan M, Prabhahar C, Sureshkumar J, Natarajan A, Vijayarengan P, Sharavanan S (2011) Toxic effect of cadmium on seed germination, growth and biochemical contents of cowpea (*Vigna unguiculata* L.) plants. Int Multidiscip Res J 1(5):1–6

Wang FZ, Chen MX, Yu LJ, Xie LJ, Yuan LB, Qi H, Xiao M, Guo W, Chen Z, Yi K, Zhang J (2017) OsARM1, an R2R3 MYB transcription factor, is involved in regulation of the response to arsenic stress in rice. Front Plant Sci 8:1868

Wang B, Zhang M, Zhang J, Huang L, Chen X, Jiang M, Tan M (2020) Profiling of rice Cd-tolerant genes through yeast-based cDNA library survival screening. Plant Physiol Biochem 155:429–436

Wu TY, Krishnamoorthi S, Goh H, Leong R, Sanson AC, Urano D (2020) Crosstalk between heterotrimeric G protein-coupled signaling pathways and WRKY transcription factors modulating plant responses to suboptimal micronutrient conditions. J Exp Bot 71(10): 3227–3239

Xiong Y, Liu T, Tian C, Sun S, Li J, Chen M (2005) Transcription factors in rice: a genome wide comparative analysis between monocots and eudicots. Plant Mol Biol 59:191–203

Xiong W, Xu X, Zhang L, Wu P, Chen Y, Li M, Wu G (2013) Genome-wide analysis of the WRKY gene family in physic nut (*Jatropha curcas* L.). Gene 524(2):124–132

Xu J, Yang L, Wang Z, Dong G, Huang J, Wang Y (2006) Toxicity of copper on rice growth and accumulation of copper in rice grain in copper contaminated soil. Chemosphere 62(4):602–607

Xu Z, Ge Y, Zhang W, Zhao Y, Yang G (2018) The walnut JrVHAG1 gene is involved in cadmium stress response through ABA-signal pathway and MYB transcription regulation. BMC Plant Biol 18(1):1–13

Yan A, Chen Z (2019) Impacts of silver nanoparticles on plants: a focus on the phytotoxicity and underlying mechanism. Int J Mol Sci 20(5):1003

Yan R, Gao S, Yang W, Cao M, Wang S, Chen F (2008) Nickel toxicity induced antioxidant enzyme and phenylalanine ammonia-lyase activities in *Jatropha curcas* L. cotyledons. Plant Soil Environ 54(7):294–300

Yeh CM, Chien PS, Huang HJ (2007) Distinct signalling pathways for induction of MAP kinase activities by cadmium and copper in rice roots. J Exp Bot 58(3):659–671

Yruela I (2009) Copper in plants: acquisition, transport and interactions. Funct Plant Biol 36(5): 409–430

Yuce M, Taspinar MS, Aydin M, Agar G (2019) Response of NAC transcription factor genes against chromium stress in sunflower (*Helianthus annuus* L.). Plant Cell Tissue Organ Culture (PCTOC) 136(3):479–487

Yurekli F, Porgali Z (2006) The effects of excessive exposure to copper in bean plants. Acta Biol Cracov Bot 48(2):7–13

Zhang WH, Tyerman SD (1999) Inhibition of water channels by $HgCl_2$ in intact wheat root cells. Plant Physiol 120(3):849–858

Zhang Y, Wang L (2005) The WRKY transcription factor superfamily: its origin in eukaryotes and expansion in plants. BMC Evol Biol 5:1–12

Zhou ZS, Huang SQ, Guo K, Mehta SK, Zhang PC, Yang ZM (2007) Metabolic adaptations to mercury-induced oxidative stress in roots of *Medicago sativa* L. J Inorg Biochem 101(1):1–9

Zulfiqar U, Farooq M, Hussain S, Maqsood M, Hussain M, Ishfaq M, Ahmad M, Anjum MZ (2019) Lead toxicity in plants: impacts and remediation. J Environ Manag 250:109557

Heavy Metal Transporters, Phytoremediation Potential, and Biofortification

18

S. Saharsha Reddy, Prasann Kumar, and Padmanabh Dwivedi

Abstract

For the normal growth and functioning, along with the macronutrients, plants also require essential micronutrients such as Cu, Zn, Mn, Fe, Ni, and Co, but they are needed in very small quantity; if the concentration and accumulation of these elements exceed the limits, it will become toxic to the plants. Here we can see the role of metal transporters; they aid in absorption, sequestration, and storage of these metals. This chapter focuses on the importance and the functions of the metal transport protein classes like CDF family, NRAMPs, ZIP family, ABC transporters, and CAX family in maintaining metal homeostasis. Another issue is that due to high concentration of heavy metals in the soil, these enter into the food chain which may pose threat to the human population; hence the context of phytoremediation becomes pertinent, which is a technique where hyperaccumulating plants/trees are grown for the removal or remediating the heavy metals present in the soil. We explain about various aspects taken into account to estimate the potential of the plant in sequestering the heavy metals from the soil, various ways of phytoremediation. Now-a-days, we are putting loads of pressure on land and resources to increase the yield of the product, but very little emphasis is given on improving (especially micronutrients) and conserving the nutritional qualities of the product; therefore, we can see that malnutrition is becoming very prominent in this era where about half of the world population suffers from the malnutrition of iron, zinc, and selenium; hence,

S. S. Reddy · P. Kumar
Department of Agronomy, School of Agriculture, Lovely Professional University, Jalandhar, Punjab, India

P. Dwivedi (✉)
Department of Plant Physiology, Institute of Agricultural Sciences, Banaras Hindu University, Varanasi, India

K. Kumar, S. Srivastava (eds.), *Plant Metal and Metalloid Transporters*,
https://doi.org/10.1007/978-981-19-6103-8_18

biofortification comes into account, in which plants can be fortified either by agronomic practices or through breeding or by using biotechnological approaches.

Keywords

Biofortification · Heavy metals · Metal transporters · Micronutrients · Phytoremediation

Abbreviations

ABC transporters	ATP-binding cassette transporter
Ag^{2+}	Gold
ATP	Adenosine triphosphate
BCF	Bioconcentration factor
Ca	Calcium
CAX	Cation exchanger
Cd^{2+}	Cadmium
CDF	Cation diffusion facilitators
CDF	Cation diffusion facilitators
Co^{2+}	Cobalt
COPT	Copper transporters
COPT	Copper transporters
Cr	Chromium
Cs	Cesium
Cu^{2+}	Copper
dgl	Deglycyrrhizinated licorice
DW	Dry weight
EDTA	Ethylenediamine tetra acetic acid
Fe^{2+}	Nickel
frd3	Ferric reductase defective 3
g	Gram
GLS	Glucosinolates
GM	Genetically modified
Hg^{2+}	Mercury
HM	Heavy metal
I	Iodine
Mg	Magnesium
mg	Milligrams
Mn^{2+}	Manganese
Mo	Molybdenum
MOT1	Molybdate transporter type 1
Ni^{2+}	Iron
NRAMP	Natural resistance-associated macrophage proteins

Pb^{2+}	Lead
Se	Selenium
Sr	Strontium
U	Uranium
YSL transporters	Yellow stripe-like proteins
ZIP	Zinc resistance transporter, Iron-resistance transporter-like proteins
Zn^{2+}	Zinc

18.1 Introduction

Along with the essential growth factors like water, sunlight, and minerals, plants also take certain heavy or toxic compounds through plant roots from the soil and water with the help of transporters for their growth and development (Chatterjee et al. 2013). Metals with an atomic mass of 20 with a specific gravity higher than 5 g/cm^3 are called heavy metals. Apart from the macromolecules, plants will also require micronutrients like copper (Cu), iron (Fe), manganese (Mn), nickel (Ni), molybdenum (Mo), cobalt (Co), and zinc (Zn) for their normal growth and functioning, and these are required in very small quantity; hence they are called as micronutrients, and these are involved in an enzyme-catalyzed redox reaction, electron transfer, and metalloprotein formations (Zenk 1996; Puig and Peñarrubia 2009; Rastgoo et al. 2011). When these metals/micronutrients are taken in higher concentrations, they get accumulated in the soil, and plant parts become toxic to plants (Peralta-Videaa et al. 2009). In plants, toxicity of these elements may cause chlorosis, necrosis, inhibition of root growth, and stunted growth (Marschner 2011; Van Assche and Clijsters 1990), and toxicity may also interrupt cell transport and may cause oxidative damage; hence the metal transporters play an important role in removing the excess of metal elements in plants, thereby maintaining metal homeostasis. There are many different classes of protein that are involved in metal uptake and translocation; some of them are heavy metal transporting ATPases: CPx-type ATPases, natural resistance-associated macrophage proteins (NRAMP), copper transporters (COPT), cation diffusion facilitators (CDF), ZIP transporters, and YSL transporters (Fig. 18.1). Some plants have the innate ability to regulate the uptake, accumulate, translocate these toxic elements, and detoxify them from the soil, plants, or environment; when this principle is involved in growing hyperaccumulating plants in the soil to detoxify the soil or water from the toxic or heavy metals, then this methodology is called as phytoremediation. The ability of the plants to take up the toxins or heavy metals present in the soil is called the phytoremediation potential, it varies from plant to plant, and these plants are tolerant to the high concentration of metals. These plants take up the heavy metals from the soil and store them in the aerial plant parts, 100–1000-folds higher than those found in non-hyperaccumulating species without exhibiting any symptoms of phytotoxicity (Reeves 2006). Biofortification is

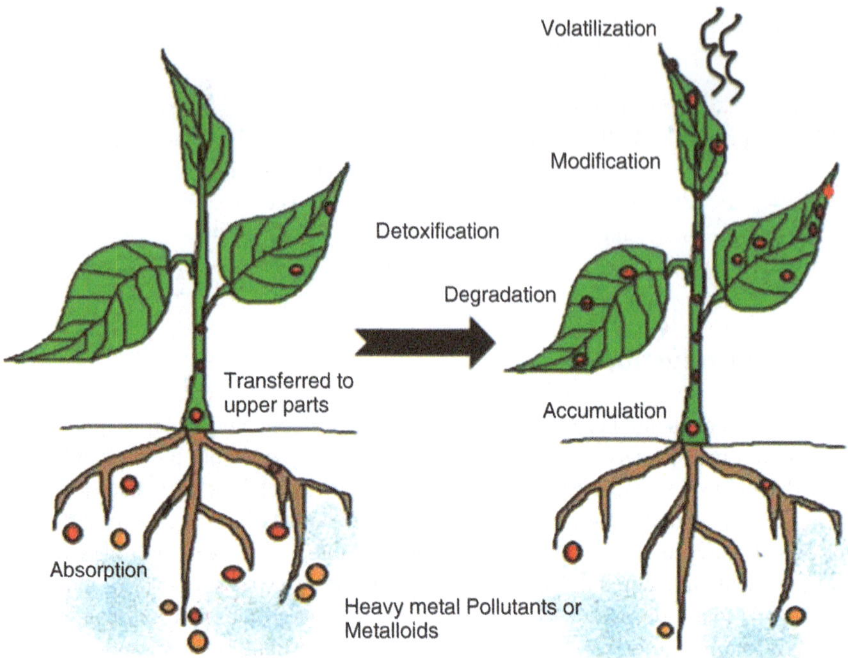

Fig. 18.1 Metal stress and binding protein

an agricultural process that increases the uptake and accumulation of specific nutrients (Rouached 2013); biofortification can be either through agronomic strategies, plant breeding, and biotechnology. The product obtained as a result of biofortification is rich in nutrients, and when this is fed to the people, it has the potential to correct their deficiencies and improve their health status.

18.2 Metal Transporters and Their Importance

Apart from abiotic factors such as temperature, humidity, sunlight, rainfall, and water, mineral nutrition is also one of the most essential factors for the normal functioning, growth, and development of the plants. Within the minerals, plants also require micronutrients such as copper (Cu^{2+}), zinc (Zn^{2+}), manganese (Mn^{2+}), iron (Fe^{2+}), nickel (Ni^{2+}), and cobalt (Co^{2+}), but when these elements are present in excess, then these and the other non-essential metal elements such as cadmium (Cd^{2+}), mercury (Hg^{2+}), gold (Ag^{2+}), and lead (Pb^{2+}) may be extremely toxic to the plants. For instance, copper (Cu^{2+}) is an essential trace element that helps plants in catalyzing several electron transport reactions during photosynthesis and respiration, and other elements such as zinc (Zn^{2+}) and manganese (Mn^{2+}) are important constituents of the enzymes or these are involved in enzyme activation (Marschner 2011); hence when these elements are not available, plants show deficiency symptoms. Similarly, when these elements are present in high concentration, they are very toxic to plants and the plants develop toxicity symptoms such as chlorosis, necrosis, inhibition of root growth, and stunted growth (Marschner 2011; Van Assche and Clijsters 1990); toxicity may also interrupt cell transport and may cause oxidative damage; hence the metal transporters play an important role in removing the excess of metal elements in plants, thereby maintaining metal homeostasis.

The heavy metals present in the soil may be essential or non-essential. Essential elements include copper, zinc, manganese, nickel, and iron; these metal elements help in many useful biological processes such as electron transfer proteins and co-factors for several enzymes (Fageria et al. 2009; Chaffai and Koyama 2011). Non-essential elements such as cadmium (Cd), mercury (Hg), and lead (Pb) have no role in biological processes, and their accumulation beyond the limit can be toxic to plants, apart from these non-essential metal elements by replacing the essential elements or by changing the structure of biomolecules and other stress regulatory proteins (Sarwar et al. 2010); hence to cope with such situation, plants adopt various tolerance mechanisms such as metal sequestration, compartmentalization in certain cell organelles, exclusion, and inactivation by exudation of organic ligands (Choppala et al. 2014).

18.3 Health Impacts of Heavy Metals

If the plants are unable to maintain metal homeostasis, then these chemicals may enter and accumulate in the product and may enter the entire food chain causing health issues in humans, animals, and other living organisms (Fig. 18.2). For instance, cadmium is carcinogenic, and it has a negative impact on calcium metabolism which leads to hypercalciuria and kidney dysfunctioning and may also cause anemia (Awofolu 2005). Zinc in high concentration causes fatigue and dizziness (Hess and Schmid 2002); a higher concentration of lead may cause renal failure, cardiovascular disease, short-term memory loss, coordination problem, and decreased learning ability in children; chromium toxicity leads to intense hair fall, a higher concentration of copper causes brain, kidney disorders, intestinal problems, and anemia (Salem et al. 2000); inhalation of nickel may make the person prone to lung cancer, its excess concentration may also lead to stomach cancers, and it is also toxic to the immune system and nervous system (Khan et al. 2007). To mitigate all these harmful impacts, plants have several classes of proteins that are involved in heavy metal transport and they are helpful in maintaining their homeostasis; some of the classes of proteins that are involved in metal transport are as follows:

18.3.1 Heavy Metal Transporting ATPases: CPx-Type ATPases

These are type metal ATPases; these are present in many organisms and play an important role in the transport of essential and toxic metals such as Cu^{+2}, Zn^{+2}, Cd^{+2},

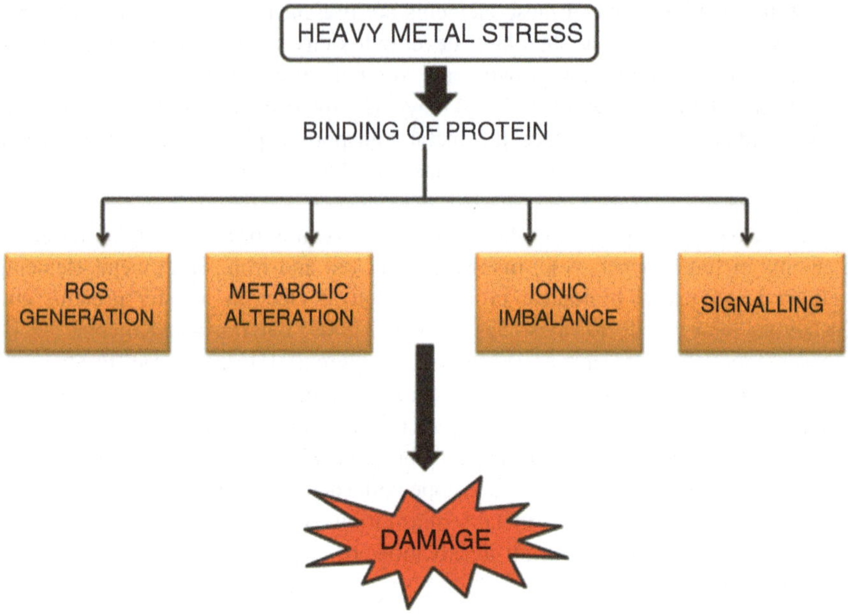

Fig. 18.2 Metal contamination and root surface

and Pb^{+2}; these use ATP to pump a variety of charged substrates across biological membranes and are distinguished by the formation of a phosphorylated intermediate during the reaction cycle (Axelsen and Palmgren 1998). CPx-ATPases, apart from supplying the required amount of heavy metals for the cell functioning, also remove the accumulated metal elements from the plant, thereby preventing its toxicity. These aid in the transport of copper, cadmium, and zinc in some cases (Beard et al. 1997; Thelwell et al. 1998).

18.3.1.1 Natural Resistance-Associated Macrophage Proteins (NRAMP)

There are three members of this NRAMP family—SMF1, SMF2, and SMF3—among these, SMF1 acts as a high-affinity transporter for manganese uptake and SMF2 acts as a high-affinity transporter which mostly prefers cobalt (Supek et al. 1996, 1997; Liu et al. 1997). Some NRAMP transporters aid in iron and manganese uptake by the root epidermis (Cailliatte et al. 2010). Apart from these functions, they are also involved in Fe and cadmium uptake and in maintaining homeostasis along with performing different physiological functions (Hall and Williams 2003).

18.3.1.2 Copper Transporters (COPT)

These are eukaryotic families of copper transporters (Eide 1998), and this class coordinates the copper during transmembrane transport (Petris 2004). In plants, they are involved in the uptake of copper from the soil to the root (Sancenón et al. 2004)

18.3.1.3 Cation Diffusion Facilitators (CDF)

CDF proteins are a family of heavy metal transporters that aid in the transport of zinc, cobalt, and cadmium (Paulsen and Saier Jr 1997). The members of this family are involved in metal uptake; some of them catalyze efflux, some of them are found in plasma membrane, while others are found in intracellular membranes (van der Zaal et al. 1999). They are involved in transporting and maintaining metal homeostasis and tolerance. They are involved in the efflux of Zn^{2+}, Cd^{2+}, Co^{2+}, Ni^{2+}, or Mn^{2+}, from the cytoplasm to the outside of the cell. The plant members of the CDF family are named as MTPs (Paulsen and Saier Jr 1997; Gaither and Eide 2001). Most of the CDFs play a role in metal detoxification, while others are involved in long-distance metal transport (Ricachenevsky et al. 2013). They are involved in the transport of zinc, cobalt, and cadmium and are found in bacteria and eukaryotes (Paulsen and Saier Jr 1997; Eide 1998; van der Zaal et al. 1999).

18.3.1.4 ZIP Transporters (Zinc Resistance Transporter, Iron-Resistance Transporter-Like Proteins)

These are transporters involved in metal homeostasis and mediate the influx of zinc, iron, cadmium, or manganese, from outside the cell into the cytoplasm (Guerinot 2000); they play an important role in the uptake of iron and zinc (Eide et al. 1996; Lin et al. 2009). They are also involved in the translocation of divalent cations such as Fe, Zn, Mn, and Cd across the membranes that differ in substrate range and specificity (Guerinot 2000; Mäser et al. 2001). They are also involved in the transport of Zn^{2+} and other metals from the extracellular or organelle lumen into the cytosol (Saier Jr 1999).

18.3.1.5 YSL Transporters (Yellow Stripe-Like Proteins)

YSL transporters play an important role in metal uptake from the soil in monocots and the distribution of these metals over a long distance in both monocots and dicots (Conte and Walker 2011).

18.3.1.6 MOT1 (Molybdate Transporter Type 1)

This class is involved in molybdenum uptake and transport in the plants (Fig. 18.3).

18.4 Phytoremediation

The word "phytoremediation" is derived from the Greek word Phyto meaning plant and the Latin word remedium meaning to correct or remove an evil (Ali et al. 2013). Phytoremediation is a technique where plants are employed in reducing or removing or degrading or immobilizing the harmful toxins or heavy metals from soil or the surrounding environment; this technique makes the land or the surrounding free from the target toxin and makes the land fit for further cultivation. There are several physical and biological techniques for remediation of metals and other toxins from the soil, but phytoremediation is a cost-effective method; the plants selected for this process should have that inherent capacity to uptake and accumulate heavy metals;

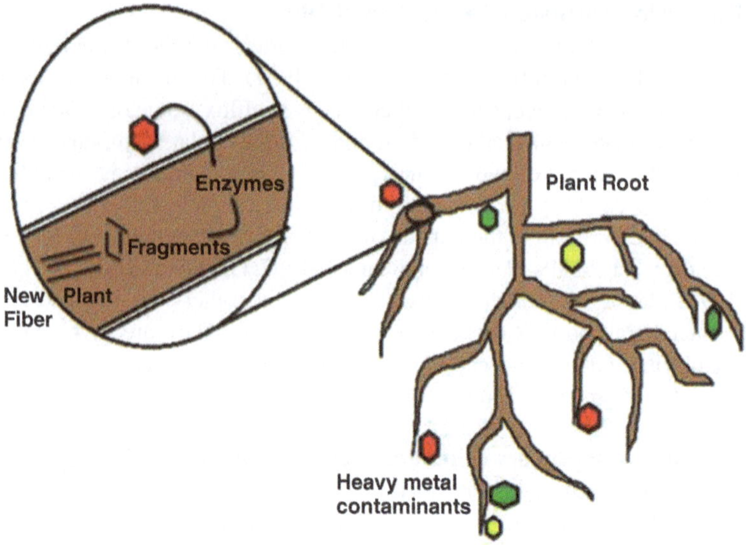

Fig. 18.3 Mode of phytoremediation

these plants are called hyperaccumulators; once the process of remediation is completed, the produce and residues of this hyperaccumulators can be safely disposed of or they can be collected and used for recovery of the elements in the areas where it is needed or where it deficient (Juwarkar et al. 2010).

18.5 Different Ways of Phytoremediation

18.5.1 Phytostabilization/Phytoimmobilization

In this method, the plants have the potential ability to reduce the mobility and bioavailability of the toxins or heavy metals either by preventing its leaching into the groundwater or by preventing its entry into the food chain by either including adsorption by roots, precipitation, and complexation in the root zone (Erakhrumen and Agbontalor 2007). Mahmoud and Abd El-Kader (2015) found that application of phosphogypsum immobilized Zn, Ni, Pb, and Cd from the soil. Through this method, there is no permanent removal of the heavy metals, but they are inactivated by converting them to unavailable forms.

18.5.2 Phytovolatilization

The plants can uptake the chemical, convert it into volatile form, and then release it into the atmosphere (Ghosh and Singh 2005). This method is most suitable for the elements like mercury where the mercuric ion is converted to a less toxic elemental form. This is a temporary measure because the elemental form released into the atmosphere may return to the soil through rainfall (Sarwar et al. 2017).

18.5.3 Phytoextraction

The plants can uptake heavy metals from the soil and store them in the aerial parts (Salt and Smith 1998; Lombi et al. 2002). The stored toxins or metals get accumulated in the aerial parts and are harvested. Phytoextraction can be done into two ways—they are continuous and induced; in continuous phytoextraction, the plants can uptake and store the heavy metals or toxins throughout their lifecycle, whereas in induced, phytoextraction chelates are used. For instance, EDTA is used in *Brassica juncea* (Indian mustard) and *Helianthus annuus* (sunflower); this helps in the mobilization of heavy elements like chromium (Cr), copper (Cu), nickel (Ni), and zinc (Zn) (Blaylock et al. 1997; Turgut et al. 2004).

18.5.4 Rhizodegradation

Some plants break down the contaminants into smaller units by the enzymes or chemicals released by the plant roots; these enzymes are produced by plants, which can degrade the pesticides (Hussain et al. 2009; Niti et al. 2013). The plant roots release certain secondary metabolites as root exudates which stimulate microbial activity; these microbes get actively involved in the degradation of organic pollutants (Pieper et al. 2004).

18.5.5 Rhizofiltration

As the name suggests, the plant filters and removes the toxins or heavy metals from the liquid media, and plants roots can adsorb or absorb the toxic chemicals. This method is used for the cleaning of groundwater. These plants are not directly planted; firstly plants are grown in clean water; once the roots are well established and large enough, they are placed in the water containing the toxic chemicals; now these plants take up toxic elements from the water; and at last, the produce from these plants is disposed safely. This method is mainly used for water bodies contaminated with a radionuclide (Niti et al. 2013).

18.6 Phytoremediation Potential

The phytoremediation potential of the plants can be calculated by analyzing three aspects:

18.6.1 Translocation Index

The translocation index indicates the potential ability of the plants to translocate the heavy metals from the soil through roots to the aerial plant parts (Zu 2005). The phytoremediation potential of the plant and the translocation index are directly related, high translocation index indicates the greater phytoremediation potential of the plant. Translocation index can be calculated by using the formula

$$\text{Translocation index} = \frac{\text{Target metal content in the leaves (mg/g DW)}}{\text{Target metal content in the leaves (control) (mg/g DW)}} \times 100$$

18.6.2 Tolerance Index

The tolerance index indicates the ability of the crops to grow well with high biomass in metal-contaminated conditions; this shows its tolerance level to that particular heavy metal.

If the tolerance index is greater than 1, then it shows that there is an increase in biomass and the plants have developed tolerance against the metal present in the soil; the tolerance level of 1 indicates that there is no difference when compared to non-heavy metal control treatment; tolerance index less than 1 indicates that there is a decrease in biomass and the plant is in stress condition (Wilkins 1957, 1978).

$$\text{Tolerance index} = \frac{\text{Biomass of the treated plants (g/plant)}}{\text{Biomass of the control plants (g/plant)}}$$

18.6.3 Bioconcentration Factor (BCF)

The bioconcentration factor indicates the uptake of metal, its mobilization into the plant tissues, and storage in the aerial plant parts (Newman and Unger 2003). It is defined as the ratio of the metal concentration in the plant to the metal concentration in the soil (Zayed et al. 1998). BCF values greater than 1 indicate that the plants are potential HM-hyperaccumulators (Zhang et al. 2002). The formula for calculating the BCF is

$$\text{Bioconcentration factor} = \frac{\text{Target metal content in plant tissue (mg/kg DW)}}{\text{Target metal content in roots (mg/kg DW)}}$$

The maximum ability of the hyperaccumulating plants to take up and store the toxins or heavy metals is called phytoremediation potential. *Brassica campestris* uptake copper, cadmium, zinc, iron, nickel, manganese, and lead (Chandra et al.

2009). *Brassica juncea*, *Brassica rapa*, and *Brassica napus* uptake cadmium, copper, cobalt, lead, and arsenic. Lead (Pb) accumulation is more in plants having high biomass; plants producing biomass of 20 tons could accumulate more than 1% of lead in its shoots (Mcgrath and Zhao 2003). The most frequently used plants for phytoremediation technology are *Brassica juncea*, *Helianthus annuus*, *Brassica napus*, and *Zea mays* (Szczygłowska et al. 2011). *Brassica juncea* has a high innate capacity to accumulate cadmium (Cd) in shoots; it can also remove metals like lead (Pb) giving 28% reduction and selenium (Se) giving 13–48% reduction. *Brassica* plants can remove a larger quantity of zinc from the soil than *Thlaspi caerulescens* because of their high biomass production (Gisbert et al. 2006). The plants used for phytoremediation exhibit slow growth, and the toxic elements taken by the plants are stored in their leaves, roots, and seeds or fruits. The plants accumulated with these toxic elements are harvested once they are matured. To reduce the problem of disposal, the plants are dried. Plants belonging to the Brassicaceae family are good accumulators of toxic chemicals. The *Brassica* species has a special advantage; when cultivated in soils with high residual chemicals and metals, these plants uptake these metals and accumulate in their plant parts (Van Ginneken et al. 2007). In *Brassica*, this accumulation of metals stimulates the production of glucosinolates (GLS); these are organic compounds containing sulfur, the byproducts formed on degradation contains isothiocyanates, and this substance has biocidal properties and been hence used for fumigation. These compounds can control parasites, bacteria, and fungi (Clarke 2010). Several aquatic plants such as *Eichhornia crassipes*, pennywort (*Hydrocotyle umbellata* L.), and duckweed (*Lemna minor* L.) can remove heavy metals from the water. Indian mustard has the potential to remove several heavy metals such as Cd, Cr, Cu, Ni, Pb, and Zn, and sunflower can remove Pb, U, 137Cs, and 90Sr from the solution used for hydroponics (Vara Prasad and de Oliveira Freitas 2003). Several species of *Thlaspi* have the potential to accumulate more than one metal; on growing *Thlaspi* in nickel-contaminated soils, it can accumulate about 3% of its dry matter; *Thlaspi caerulescens* can also accumulate Cd, Ni, Zn, and Pb; it has the potential to remove 8.5 kg cadmium per hectare (Robinson et al. 1998); and another species of *Thlaspi rotundifolium* can accumulate Ni, Pb, and Zn (Krämer et al. 1996). Buckwheat (*Fagopyrum esculentum*), the first known Pb hyperaccumulator species with high biomass, can accumulate up to 4.2 mg g^{-1} dry weight of Pb in the shoots (Tamura et al. 2005). The phytoextraction and the remediation potential is based on the biomass production and metal bioconcentration factor (plant-to-soil concentration ratio) (Zhao et al. 2003).

18.7 Biofortification

The method in which the uptake and accumulation of a particular element is increased is called biofortification (Bhatia et al. 2013; Rouached 2013; Lin et al. 2014).

It aims to increase the concentration of micronutrients in the edible parts of the plant either through agronomic, traditional breeding, and biotechnological biofortification; it is cost-effective and helps in reducing the problems of malnutrition, especially in the rural population. Biofortification is a sustainable approach where the concentration of a particular element in the food is increased; if safe on testing when this product is fed to people, it can supply a higher concentration of the particular essential element; hence biofortification serves best in overcoming malnutrition (Mortvedt 1996; Zhu et al. 2009).

A lot of pressure is put on the resources and inputs to increase the crop yield, but we do not focus on the quality of food; since the past, there is a decline in the mineral contents in the grains especially the micronutrients such as iron (Fe), zinc (Zn), selenium (Se), iodine (I), and copper (Cu) and the macronutrients like calcium (Ca) and magnesium (Mg) (World Health Organization 2002; White and Broadley 2005; Garvin et al. 2006; Fan et al. 2008).

18.7.1 Types of Biofortification

Biofortification can be done in three ways:

18.7.1.1 Agronomic Biofortification

Agronomic biofortification involves the application of nutrients to the plants for temporary improvement of the nutrient concentration of the produce, and consuming this product improves the nutritional status of humans (Cakmak and Kutman 2018). Agronomic biofortification is a simple, inexpensive method; apart from the use of fertilizers, we can also use plant growth-promoting microorganisms like *Bacillus*, *Pseudomonas*, *Rhizobium*, and *Azotobacter*, which will increase the availability of the minerals (Chew et al. 2012; Fajardo et al. 2017). This is a fertilizer-based approach that is based on the soil and/or foliar application of micronutrients either alone or in combination with other fertilizers. It is a short-time solution .In wheat, there was a minor increase in iron content with the use of Fe-EDTA and nitrogen. Selenium and zinc content in the product can be increased by agronomic biofortification (Cakmak and Kutman 2018). For instance, when canola plants were given plant growth-promoting rhizobacteria, viz., *Azospirillum brasilense* and *Azotobacter vinelandii* along with chemical fertilizers; this resulted in increased protein, oleic acid, and linoleic acid in the seed (Nosheen et al. 2011); in barley with the integrated use of biofertilizers, inorganic fertilizers and vermicompost have increased the zinc and iron content in the grain (Malekif et al. 2011).

18.7.1.2 Traditional Breeding

Biofortification through plant breeding is a cost-effective and sustainable approach that can enhance the nutritional properties of the food produced which on consumption improve the health status of low-income people globally (Bouis 2002; Bouis et al. 2011; Blancquaert et al. 2014). This approach benefits large populations ranging from urban to remote rural areas (Bouis 2002; Bouis et al. 2011; Saltzman et al. 2013). With one-time investment in breeding, the crop can be grown and multiplied for many years without any additional cost. For successful biofortification, there should be wide genetic diversity in the gene pool (White and Broadley 2009; Garg et al. 2018). Firstly, genotypes with superior characteristics which are high in micronutrients should be identified by screening a wide range of germplasms, then the selected genotypes are crossed, the promising lines should be isolated and tested in multiple locations, after the best ones should be submitted to government agencies for testing for agronomic performance, and if they are expressing well, they are tested across multiple locations over multiple seasons (Bouis and Saltzman 2017).

18.7.1.3 Biotechnological Approach

In a transgenic or biotechnological approach, biofortification is done by improving the mobilization from the soil, through roots, transport of these elements to shoot, and their accumulation in the harvestable produce. In non-graminaceous plants, iron uptake and its tolerance can be enhanced by overexpressing genes encoding Fe(III) reductases (Samuelsen et al. 1998; Connolly et al. 2003). In graminaceous plants, Fe content can be increased by synthesis and exudation of phytosiderophores (Takahashi et al. 2001). In pea (brz and dgl) and *Arabidopsis* (frd3, also known as man1), mutants with constitutive Fe(III) reductase activity accumulate not only Fe but also Zn, Ca, Mg, Cu, and Mn in shoots, but additional Fe-chelates are necessary for phloem transport to the seed (Grusak 2000; Wang et al. 2003; Rogers and Guerinot 2002). Due to lack of long-term effectiveness, sustainability of fertilizer emphasis was given in developing economical, long-lasting methodologies to increase micronutrient content in the plants. Genetic engineering technology is being used in corn, rice, wheat, and soybeans, but there are many controversies regarding the cultivation, commercialization, and use of GM crops. Major limitations of this method are time-consuming and require a high standard legal framework, and it is expensive (Winkler 2011; Inaba and Macer 2004; Watanabe et al. 2005); for instance, in the early 2000 golden rice was developed which has the potential to deliver more than 50% of vitamin A, but it is not commercialized in any country due to risk factors involved in the regulatory approval processes (Wesseler and Zilberman 2014; Bouis and Saltzman 2017).

18.8 Use of Phytoremediation Plant Materials for Biofortification

The plants which are hyperaccumulators of heavy metals that are used in phytoremediation can be harvested and used as green manure in growing specific crops based on the specific need for fortification; for instance, Se-enriched plant material from phytoremediation is used as green fertilizer in biofortification practice (Bañuelos et al. 2010; Yasin et al. 2015). This product can be further harvested and fed to animals; for instance, in San Joaquin Valley of Central California, Indian mustard was used to remediate selenium-contaminated soils, and the produce from this crop was fed to animals (Banuelos 2006; Bañuelos et al. 2010). Before the integration of phytoremediation and biofortification, the chemical composition test of the produce is of prime importance; on test only if it is safe, then it can be used further as animal feed.

18.9 Conclusion

The population is increasing exponentially, so in order to feed this growing population, we are intensively putting the land into cultivation with excessive use of chemical fertilizers (especially for supplying macronutrients) and pesticides; mining has destructed the physical, chemical, and biological properties of the soil, and this has led to increasing in concentration of metals in the soil and other resources like water. This is becoming a burning issue in both developed and developing countries; on the other hand, there is also the problem of deficiency of nutrients especially micronutrients in the soil, and thereby the food that is produced in these soils are also deficient; it has been estimated that about 25% of the world population is at risk of zinc deficiency. For instance, there is accumulation of cadmium in the fields due to contaminants in the fertilizers used. Both the micronutrients and the heavy metals can enter the food chain from the soil, and also when micronutrients are taken in excess than required, it may be toxic to the plants and the people who consume that food; hence metal transporters play a prominent role in maintaining metal homeostasis in the plants. Phytoremediation is one of the best ways that can counteract the problem of accumulation of heavy toxic chemicals or metals in the soil and the surrounding ecosystem; to make this successful, we need to have an idea about the metal transporters and their role. Before selecting the plants, we need to estimate their phytoremediation potential in order to know how efficient they are in remediating the heavy metals. Malnutrition is one such burning issue, and mostly micronutrients are deficient; hence more emphasis is given to enhancing the trace elements bioavailability in the root zone, their translocation from the roots to shoots, and their distribution to the grain or economic plant part. Biofortification is of different types but agronomic biofortification and fortification through breeding are the most eminent and successful both in terms of efficiency and cost. Biofortification, enhancing trace element bioavailability in the rhizosphere, translocation from roots to shoots, and redistribution toward grain tissues are the obvious targets. It would be best if we integrate phytoremediation and biofortification; new methodologies should be developed in order to increase the phytoremediation and accumulation of more than one element; for instance, there should be an accumulation of both the micronutrients selenium and zinc through a single procedure.

References

Ali H, Khan E, Sajad MA (2013) Phytoremediation of heavy metals—concepts and applications. Chemosphere 91(7):869–881

Awofolu OR (2005) A survey of trace metals in vegetation, soil and lower animal along some selected major roads in metropolitan city of Lagos. Environ Monit Assess 105(1):431–447

Axelsen KB, Palmgren MG (1998) Evolution of substrate specificities in the P-type ATPase superfamily. J Mol Evol 46(1):84–101
Banuelos GS (2006) Phyto-products may be essential for sustainability and implementation of phytoremediation. Environ Pollut 144(1):19–23
Bañuelos GS, Roche JD, Robinson J (2010) Developing selenium-enriched animal feed and biofuel from canola planted for managing Se-laden drainage waters in the Westside of Central California. Int J Phytoremediation 12(3):243–254
Beard SJ, Hashim R, Membrillo-Hernández J, Hughes MN, Poole RK (1997) Zinc (II) tolerance in Escherichia coli K-12: evidence that the zntA gene (o732) encodes a cation transport ATPase. Mol Microbiol 25(5):883–891
Bhatia P, Aureli FD, Amato M, Prakash R, Cameotra SS, Nagaraja TP, Cubadda F (2013) Selenium bioaccessibility and speciation in biofortified Pleurotus mushrooms grown on selenium-rich agricultural residues. Food Chem 140(1–2):225–230
Blancquaert D, De Steur H, Gellynck X, Van Der Straeten D (2014) Present and future of folate biofortification of crop plants. J Exp Bot 65(4):895–906
Blaylock MJ, Salt DE, Dushenkov S, Zakharova O, Gussman C, Kapulnik Y et al (1997) Enhanced accumulation of Pb in Indian mustard by soil-applied chelating agents. Environ Sci Technol 31(3):860–865
Bouis HE (2002) Plant breeding: a new tool for fighting micronutrient malnutrition. J Nutr 132(3): 491S–494S
Bouis HE, Saltzman A (2017) Improving nutrition through biofortification: a review of evidence from HarvestPlus, 2003 through 2016. Glob Food Sec 12:49–58
Bouis HE, Hotz C, McClafferty B, Meenakshi JV, Pfeiffer WH (2011) Biofortification: a new tool to reduce micronutrient malnutrition. Food Nutr Bull 32(1_Suppl 1):S31–S40
Cailliatte R, Schikora A, Briat JF, Mari S, Curie C (2010) High-affinity manganese uptake by the metal transporter NRAMP1 is essential for Arabidopsis growth in low manganese conditions. Plant Cell 22(3):904–917
Cakmak I, Kutman UÁ (2018) Agronomic biofortification of cereals with zinc: a review. Eur J Soil Sci 69(1):172–180
Chaffai R, Koyama H (2011) Heavy metal tolerance in Arabidopsis thaliana. Adv Bot Res 60:1–49
Chandra R, Bharagava RN, Yadav S, Mohan D (2009) Accumulation and distribution of toxic metals in wheat (Triticum aestivum L.) and Indian mustard (Brassica campestris L.) irrigated with distillery and tannery effluents. J Hazard Mater 162(2–3):1514–1521
Chatterjee S, Mitra A, Datta S, Veer V (2013) Phytoremediation protocols: an overview. In: Plant-based remediation processes. pp 1–18
Chew SC, Loh SP, Khor GL (2012) Determination of folate content in commonly consumed Malaysian foods. Int Food Res J 19(1)
Choppala G, Saifullah, Bolan N, Bibi S, Iqbal M, Rengel et al (2014) Cellular mechanisms in higher plants governing tolerance to cadmium toxicity. Crit Rev Plant Sci 33(5):374–391
Clarke DB (2010) Glucosinolates, structures and analysis in food. Anal Methods 2(4):310–325
Connolly EL, Campbell NH, Grotz N, Prichard CL, Guerinot ML (2003) Overexpression of the FRO2 ferric chelate reductase confers tolerance to growth on low iron and uncovers posttranscriptional control. Plant Physiol 133(3):1102–1110
Conte SS, Walker EL (2011) Transporters contributing to iron trafficking in plants. Mol Plant 4(3): 464–476
Eide DJ (1998) The molecular biology of metal ion transport in Saccharomyces cerevisiae. Annu Rev Nutr 18(1):441–469
Eide D, Broderiusn M, Fett J, Guerinot ML (1996) A novel iron-regulated metal transporter from plants identified by functional expression in yeast. Proc Natl Acad Sci U S A 93(11):5624–5628
Erakhrumen AA, Agbontalor A (2007) Phytoremediation: an environmentally sound technology for pollution prevention, control and remediation in developing countries. Educ Res Rev 2(7): 151–156

Fageria NK, Filho MB, Moreira A, Guimarães CM (2009) Foliar fertilization of crop plants. J Plant Nutr 32(6):1044–1064
Fajardo V, Alonso-Aperte E, Varela-Moreiras G (2017) Total folate content in ready-to-eat vegetable meals from the Spanish market. J Food Compos Anal 64:223–231
Fan MS, Zhao FJ, Fairweather-Tait SJ, Poulton PR, Dunham SJ, McGrath SP (2008) Evidence of decreasing mineral density in wheat grain over the last 160 years. J Trace Elem Med Biol 22(4): 315–324
Gaither LA, Eide DJ (2001) Eukaryotic zinc transporters and their regulation. In: Zinc biochemistry, physiology, and homeostasis. pp 65–84
Garg M, Sharma N, Sharma S, Kapoor P, Kumar A, Chunduri V, Arora P (2018) Biofortified crops generated by breeding, agronomy, and transgenic approaches are improving lives of millions of people around the world. Front Nutr 5:12
Garvin DF, Welch RM, Finley JW (2006) Historical shifts in the seed mineral micronutrient concentration of US hard red winter wheat germplasm. J Sci Food Agric 86(13):2213–2220
Ghosh M, Singh SP (2005) A review on phytoremediation of heavy metals and utilization of it's by products. Asian J Energy Environ 6(4):18
Gisbert C, Clemente R, Navarro-Avinó J, Baixauli C, Ginér A, Serrano R et al (2006) Tolerance and accumulation of heavy metals by Brassicaceae species grown in contaminated soils from Mediterranean regions of Spain. Environ Exp Bot 56(1):19–27
Grusak MA (2000) Strategies for improving the iron nutritional quality of seed crops: lessons learned from the study of unique iron-hyperaccumulating pea mutants. Pisum Genet 32:1–5
Guerinot ML (2000) The ZIP family of metal transporters. Biochim Biophys Acta (BBA) Biomembr 1465(1–2):190–198
Hall JÁ, Williams LE (2003) Transition metal transporters in plants. J Exp Bot 54(393):2601–2613
Hess R, Schmid B (2002) Zinc supplement overdose can have toxic effects. J Paediatr Haematol/ Oncol 24:582–584
Hussain S, Siddique T, Arshad M, Saleem M (2009) Bioremediation and phytoremediation of pesticides: recent advances. Crit Rev Environ Sci Technol 39(10):843–907
Inaba M, Macer D (2004) Policy, regulation and attitudes towards agricultural biotechnology in Japan
Juwarkar AA, Singh SK, Mudhoo A (2010) A comprehensive overview of elements in bioremediation. Rev Environ Sci Biotechnol 9(3):215–288
Khan MA, Ahmad I, Rahman IU (2007) Effect of environmental pollution on heavy metals content of Withaniasomnifera. J Chin Chem Soc 54(2):339–343
Krämer U, Cotter-Howells JD, Charnock JM, Baker AJ, Smith JAC (1996) Free histidine as a metal chelator in plants that accumulate nickel. Nature 379(6566):635–638
Lin YF, Liang HM, Yang SY, Boch A, Clemens S, Chen CC et al (2009) Arabidopsis IRT3 is a zinc-regulated and plasma membrane localized zinc/iron transporter. New Phytol 182(2): 392–404
Lin ZQ, Haddad S, Hong J, Morrissy J, Bañuelos GS, Zhang LY (2014) Use of selenium-contaminated plants from phytoremediation for production of selenium-enriched edible mushrooms. In: Selenium in the environment and human health. pp 124–126
Liu XF, Supek F, Nelson N, Culotta VC (1997) Negative control of heavy metal uptake by the Saccharomyces cerevisiae BSD2 gene. J Biol Chem 272(18):11763–11769
Lombi E, Tearall KL, Howarth JR, Zhao FJ, Hawkesford MJ, McGrath SP (2002) Influence of iron status on cadmium and zinc uptake by different ecotypes of the hyperaccumulator Thlaspicaerulescens. Plant Physiol 128(4):1359–1367
Mahmoud E, Abd El-Kader N (2015) Heavy metal immobilization in contaminated soils using phosphogypsum and rice straw compost. Land Degrad Dev 26(8):819–824
Malekif S, Chaichi MR, Mazaheri D, Tavakkol AR, Savaghebi G (2011) Barley grain mineral analysis as affected by different fertilizing systems and by drought stress
Marschner H (2011) Marschner's mineral nutrition of higher plants. Academic Press

Mäser P, Thomine S, Schroeder JI, Ward JM, Hirschi K, Sze H, Talke IN, Amtmann A, Maathuis FJ, Sanders D, Harper JF, Tchieu J, Gribskov M, Persans MW, Salt DE, Kim SA, Guerinot ML (2001) Phylogenetic relationships within cation transporter families of Arabidopsis. Plant Physiol 126(4):1646–1667
McGrath SP, Zhao FJ (2003) Phytoextraction of metals and metalloids from contaminated soils. Curr Opin Biotechnol 14(3):277–282
Mortvedt JJ (1996) Heavy metal contaminants in inorganic and organic fertilizers. In: Fertilizers and environment. Springer, Dordrecht, pp 5–11
Newman MC, Unger MA (2003) Fundamentals of ecotoxicology. Lewis Publishers, Boca Raton, FL, p 458
Niti C, Sunita S, Kamlesh K, Rakesh K (2013) Bioremediation: an emerging technology for remediation of pesticides. Res J Chem Environ 17:4
Nosheen A, Bano A, Ullah F (2011) Nutritive value of canola (Brassica napus L.) as affected by plant growth promoting rhizobacteria. Eur J Lipid Sci Technol 113(11):1342–1346
Paulsen IT, Saier MH Jr (1997) A novel family of ubiquitous heavy metal ion transport proteins. J Membr Biol 156(2):99–103
Peralta-Videaa JR, Lopeza ML, Narayana M, Saupea G, Gardea-Torresdeya J (2009) The biochemistry of environmental heavy metal uptake by plants: implications for the food chain. Int J Biochem Cell Biol 41:1665–1677
Petris MJ (2004) The SLC31 (Ctr) copper transporter family. Pflügers Archiv 447(5):752–755
Pieper DH, dos Santos VAM, Golyshin PN (2004) Genomic and mechanistic insights into the biodegradation of organic pollutants. Curr Opin Biotechnol 15(3):215–224
Puig S, Peñarrubia L (2009) Placing metal micronutrients in context: transport and distribution in plants. Curr Opin Plant Biol 12(3):299–306
Rastgoo L, Alemzadeh A, Afsharifar A (2011) Isolation of two novel isoforms encoding zinc-and copper-transporting P1B-ATPase from Gouan (Aeluropus littoralis). Plant Omics J 4(7): 377–383
Reeves RD (2006) Hyperaccumulation of trace elements by plants. In: Phytoremediation of metal-contaminated soils. Springer, Dordrecht. pp 25–52
Ricachenevsky FK, Menguer PK, Sperotto RA, Williams LE, Fett JP (2013) Roles of plant metal tolerance proteins (MTP) in metal storage and potential use in biofortification strategies. Fron Plant Sci 4:144
Robinson BH, Leblanc M, Petit D, Brooks RR, Kirkman JH, Gregg PE (1998) The potential of Thlaspicaerulescens for phytoremediation of contaminated soils. Plant Soil 203(1):47–56
Rogers EE, Guerinot ML (2002) FRD3, a member of the multidrug and toxin efflux family, controls iron deficiency responses in Arabidopsis. Plant Cell 14(8):1787–1799
Rouached H (2013) Recent developments in plant zinc homeostasis and the path toward improved biofortification and phytoremediation programs. Plant Signal Behav 8(1):e22681
Saier MH Jr (1999) Genome archeology leading to the characterization and classification of transport proteins. Curr Opin Microbiol 2(5):555–561
Salem HM, Eweida EA, Farag A (2000) Heavy metals in drinking water and their environmental impact on human health. In: Int conference on the environ hazards mitigation, Cairo Univ Egypt. pp 542–556
Salt DE, Smith RD (1998) Raskin. Phytoremediation 49:643–668
Saltzman A, Birol E, Bouis HE, Boy E, De Moura FF, Islam Y, Pfeiffer WH (2013) Biofortification: progress toward a more nourishing future. Glob Food Sec 2(1):9–17
Samuelsen AI, Martin RC, Mok DW, Mok MC (1998) Expression of the yeast FRE genes in transgenic tobacco. Plant Physiol 118(1):51–58
Sancenón V, Puig S, Mateu-Andrés I, Dorcey E, Thiele DJ, Peñarrubia L (2004) The Arabidopsis copper transporter COPT1 functions in root elongation and pollen development. J Biol Chem 279(15):15348–15355
Sarwar N, Malhi SS, Zia MH, Naeem A, Bibi S, Farid G (2010) Role of mineral nutrition in minimizing cadmium accumulation by plants. J Sci Food Agric 90(6):925–937

Sarwar N, Imran M, Shaheen MR, Ishaque W, Kamran MA, Matloob A et al (2017) Phytoremediation strategies for soils contaminated with heavy metals: modifications and future perspectives. Chemosphere 171:710–721

Supek F, Supekova L, Nelson H, Nelson N (1996) A yeast manganese transporter related to the macrophage protein involved in conferring resistance to mycobacteria. Proc Natl Acad Sci U S A 93(10):5105–5110

Supek F, Supekova L, Nelson H, Nelson N (1997) Function of metal-ion homeostasis in the cell division cycle, mitochondrial protein processing, sensitivity to mycobacterial infection and brain function. J Exp Biol 200(2):321–330

Szczygłowska M, Piekarska A, Konieczka P, Namieśnik J (2011) Use of brassica plants in the phytoremediation and biofumigation processes. Int J Mol Sci 12(11):7760–7771

Takahashi M, Nakanishi H, Kawasaki S, Nishizawa NK, Mori S (2001) Enhanced tolerance of rice to low iron availability in alkaline soils using barley nicotianamine aminotransferase genes. Nat Biotechnol 19(5):466–469

Tamura H, Honda M, Sato T, Kamachi H (2005) Pb hyperaccumulation and tolerance in common buckwheat (Fagopyrum esculentum Moench). J Plant Res 118(5):355–359

Thelwell C, Robinson NJ, Turner-Cavet JS (1998) An SmtB-like repressor from Synechocystis PCC 6803 regulates a zinc exporter. Proc Natl Acad Sci U S A 95(18):10728–10733

Turgut C, Pepe MK, Cutright TJ (2004) The effect of EDTA and citric acid on phytoremediation of Cd, Cr, and Ni from soil using Helianthus annuus. Environ Pollut 131(1):147–154

Van Assche F, Clijsters H (1990) Effects of metals on enzyme activity in plants. Plant Cell Environ 13(3):195–206

van der Zaal BJ, Neuteboom LW, Pinas JE, Chardonnens AN, Schat H, Verkleij JA, Hooykaas PJ (1999) Overexpression of a novel Arabidopsis gene related to putative zinc-transporter genes from animals can lead to enhanced zinc resistance and accumulation. Plant Physiol 119(3): 1047–1056

Van Ginneken L, Meers E, Guisson R et al (2007) Phytoremediation for heavy metal-contaminated soils combined with bioenergy production. J Environ Eng Landsc Manag 15(4):227–236

Vara Prasad MN, de Oliveira Freitas HM (2003) Metal hyperaccumulation in plants: biodiversity prospecting for phytoremediation technology. Electron J Biotechnol 6(3):285–321

Wang TL, Domoney C, Hedley CL, Casey R, Grusak MA (2003) Can we improve the nutritional quality of legume seeds? Plant Physiol 131(3):886–891

Watanabe KN, Sassa Y, Suda E, Chen CH, Inaba M, Kikuchi A (2005) Global political, economic, social and technological issues on transgenic crops. Plant Biotechnol 22(5):515–522

Wesseler J, Zilberman D (2014) The economic power of the Golden Rice opposition. Environ Dev Econ 19(6):724–742

White PJ, Broadley MR (2005) Biofortifying crops with essential mineral elements. Trends Plant Sci 10:586–593

White PJ, Broadley MR (2009) Biofortification of crops with seven mineral elements often lacking in human diets–iron, zinc, copper, calcium, magnesium, selenium and iodine. New Phytol 182(1):49–84

Wilkins DA (1957) A technique for the measurement of lead tolerance in plants. Nature 180(4575): 37–38

Wilkins DA (1978) The measurement of tolerance to edaphic factors by means of root growth. New Phytol 80(3):623–633

Winkler JT (2011) Biofortification: improving the nutritional quality of staple crops. In: Access not excess. Smith-Gordon. pp 100–112

World Health Organization (2002) The world health report 2002: reducing risks, promoting healthy life. World Health Organization

Yasin M, El Mehdawi AF, Jahn CE, Anwar A, Turner MF, Faisal M, Pilon-Smits EA (2015) Seleniferous soils as a source for production of selenium-enriched foods and potential of bacteria to enhance plant selenium uptake. Plant Soil 386(1):385–394

Zayed A, Gowthaman S, Terry N (1998) Phytoaccumulation of trace elements by wetland plants: I. Duckweed. J Environ Qual 27(3):715–721

Zenk MH (1996) Heavy metal detoxification in higher plants—a review. Gene 179(1):21–30

Zhang W, Cai Y, Tu C, Ma LQ (2002) Arsenic speciation and distribution in an arsenic hyperaccumulating plant. Sci Total Environ 300(1–3):167–177

Zhao FJ, Lombi E, McGrath SP (2003) Assessing the potential for zinc and cadmium phytoremediation with the hyperaccumulator Thlaspicaerulescens. Plant Soil 249(1):37–43

Zhu YG, Pilon-Smits EAH, Zhao FJ, Williams PN, Meharg AA (2009) Selenium in higher plants: understanding mechanisms for biofortification and phytoremediation. Trends Plant Sci 14:436–442

Zu YQ (2005) Hyperaccumulator of Pb, Zn, and Cd in herbaceous grown on lead-zinc mining area in Yunnan, China. Environ Int 31(5):755–762

Phytoremediation and Biofortification: Contrasting yet Similar Approaches of Manipulating Plant Metal(loid) Homeostasis for Societal Benefit

19

Anuska Sen, Binayak Das, Sanket Roy, and Surajit Bhattacharya

Abstract

Agricultural losses due to heavy metal(loid) stresses have been a major area of concern in the last few decades. Several remediation technologies have been in practice to remove or reduce the concentration of the toxic metal(loid)s from the contaminated lands, out of which phytoremediation has been one of the most favorable approaches due to its economic and eco-friendly nature. On the other hand, malnutrition of essential nutrients such as iron, zinc, selenium, vitamins, etc. is another rising problem leading to hidden hunger all around the globe. Although both phytoremediation and biofortification have several methods, their enhancement through the genetic engineering approach of combating heavy metal(loid) stress and malnutrition by the creation of suitable transgenics stands out as the most promising techniques in the near future. In this chapter, we discuss the different essential nutrients and their necessity in the plant and human body. We also discuss about the major toxic heavy metal(loid)s, their ill effects on plant and human life, and the phytoremediation techniques to mitigate these stresses. Further, the different biofortification approaches are also discussed with an emphasis on the genetic engineering methods to increase the bioavailability of essential nutrients. Thereafter, we deliberate on the need of employing different omics technologies to identify the genetic elements involved in maintaining metal (loid) homeostasis for efficient phytoremediation and/or biofortification applications. Lastly, we hypothesize the possibility of a combinatorial approach of phytoremediation and biofortification to create a transgenic "super-remediator" plant for simultaneous phytoremediation of harmful heavy metal(loid)s in non-edible parts, along with biofortification of essential nutrients into the edible

A. Sen · B. Das · S. Roy · S. Bhattacharya (✉)
Amity Institute of Biotechnology, Amity University Kolkata, Kolkata, West Bengal, India
e-mail: sbhattacharya1@kol.amity.edu

K. Kumar, S. Srivastava (eds.), *Plant Metal and Metalloid Transporters*,
https://doi.org/10.1007/978-981-19-6103-8_19

parts of the plants toward large-scale sustainable production of nutritious and safe food.

Keywords

Biofortification · Genetic engineering · Metal(loid) stress · Omics technologies · Phytoremediation · Super-remediator plant · Transgenic plant

19.1 Introduction

Nutrients are integral components for the proper functioning and development of any living organism, be it plants, microbes, or animals, including humans. Based on their requirement, nutrients are broadly classified into two categories, namely, macronutrients that are required in large quantities, and micronutrients, which are needed in comparatively small amounts. Macronutrients consist of elements such as nitrogen (N), phosphorus (P), potassium (K), magnesium (Mg), calcium (Ca), sulfur (S), etc. On the other hand, some of the major micronutrients are iron (Fe), copper (Cu), manganese (Mn), zinc (Zn), etc. Interestingly, a majority of these macronutrient and micronutrient elements are metals in terms of chemical properties, and the plants require these metal elements for a wide range of physiological processes that help in overall plant growth, development, and reproduction (Table 19.1). However, these quintessential elements that help the plants to maintain various biochemical and physiological functions are often associated with toxic effects when present beyond certain threshold concentrations (Table 19.1), especially for the consuming organisms (Bashir et al. 2019). The entry and accretion of these elements in different plants are regulated by the metal transporters (Table 19.1). Plants, being immobile organisms, have very limited mechanism against stress prevention and face-specific challenges by environmental metal ion concentrations while absorbing sufficient mineral micronutrients for their continued growth and development and at the same time have to often deal with toxic levels of both the essential and nonessential metal (loid)s. Since plants obtain both essential and beneficial elements from the soil, due to their faulty selectivity toward metal ion absorbance, plants tend to take up nonessential elements when those toxic elements are bioavailable (Clemens and Ma 2016). Since metal homeostasis in plants is associated with the mobilization, uptake, binding/chelation, trafficking, storage, both long and short-distance transport of metals, the regulation of these processes play a crucial role in the accumulation of these metal(loid)s, Further, these plant parts (roots, stems, grains, etc.) with excess accumulated toxic metal(loid)s upon consumption by human beings and other animals causes dietary toxicity and in turn pave way for the inevitable biomagnification of that specific toxic metal(loid). Thus, the borderline between the benefits and toxicity of metal nutrients is so fine and minute that strict regulation of their concentration in the living systems needs to be monitored. Therefore, this review aims to elucidate the importance of phytoremediation for reducing metal contamination in food supplies leading to improvement of the food nutritional value

Table 19.1 Tabular representation of essential macro- and micro-nutrients, their major role in plants, the safe limit for consumption, and the metal transporters that aid in their efflux-influx in cells

Type	Elements	Major physiological role	The safe limit for human consumption	Major transporters	References
Macro element	Nitrogen (N)	Starch synthesis in leaves; amino acids production; photosynthesis rate (chlorophyll content) enhancement; increase in crop yield	<10 mg/L in the form of nitrate (as per EPA)	Nitrate peptide transporter family (NPF); amino acid permease (AAP), etc.	Bassi et al. (2018), Kumar et al. (2021), Tegeder and Masclaux-Daubresse (2018), Ward et al. (2018)
	Phosphorus (P)	Carbohydrate metabolism; cell signaling; a component of DNA/RNA; mediates salt stress	Max. 4000 mg in daily food	Pht1 family (in the form of inorganic phosphate)	Kumar et al. (2021), NIH (2021b), Smith et al. (2003)
	Sulfur (S)	Maintenance of structure and function of enzymes; mediates heavy metal and oxidative stresses	Not found	SULTR family	Gigolashvili and Kopriva (2014), Zhao et al. (2008)
	Potassium (K)	Nitrogen metabolism; protein synthesis; sugar transport, stomatal opening, and closing; regulation of ATPase in the cell membrane	Max. 5000–6000 mg/day	AKT1 (*Arabidopsis* K+ transporter 1); HAK5 (high-affinity K+ transporter 5); etc.	Xu et al. (2020), Thor (2019), Turck et al. (2016)
	Calcium (Ca)	Cell signaling (as second messenger); structural component of cell wall and membrane	2000–2500 mg daily	Ca(2+)/H(+) antiporter; Ca (2+)-ATPase; etc.	Thor (2019), Yang and Jie (2005), Cormick and Belizán (2019)

(continued)

Table 19.1 (continued)

Type	Elements	Major physiological role	The safe limit for human consumption	Major transporters	References
	Magnesium (Mg)	Chloroplast formation, photosynthesis rate; regulation of cellular stress response; co-factor of certain enzymes	350 mg for adults; 65 mg for children (daily intake)	MGT6; Mg^{2+}/H^{+} exchangers	Hauer-Jákli and Tränkner (2019), NIH (2021a), Shaul (2002)
	Iron (Fe)	Maintenance of chloroplast structure; chlorophyll production; co-factors for cytochromes, catalases, etc.	0.8 mg/kg of body weight (safe limit)	ZIP family	Bashir et al. (2019), Kumar et al. (2021), WHO (1996)
	Copper (Cu)	Co-factor of enzymes involved in photosynthesis and respiration; lignin synthesis; carbohydrate and protein metabolism	1.3 mg/L (as per EPA)	COPT2; HMA family	Bashir et al. (2019), Taylor et al. (2020)
	Manganese (Mn)	Co-factor of the oxygen-evolving complex in photosynthesis; ROS scavenging; glycosylation	9–11 mg for adults; 2–3 mg for children (daily intake)	NRAMP family; CaCA family; BICAT family; MTP/CDF family; P-type ATPase; ZIP family	Alejandro et al. (2020), NIH (2021c)
	Zinc (Zn)	Carbohydrate metabolism; cell differentiation and proliferation; co-factor of all six classes of enzymes; maintains the structure of various proteins	34–40 mg for adults; 4–5 mg for children (daily intake)	ZIP family; HMA family	Bashir et al. (2019), Kumar et al. (2021), NIH (2021d)

and biofortification for the nutritional enhancement of the plant by manipulating the metal transporters in plant systems. Moreover, understanding the mechanisms by which both essential and nonessential metals are sequestered, stored, and detoxified in various plants may contribute to the optimization of a combinatorial biotechnological application for simultaneous phytoremediation and biofortification can be hypothesized for the sustainable benefit of human kind.

19.2 Metal(loid)s and Their Toxic Effects on Living Systems

In chemical terms, the metals are the elements capable of readily forming cations by donating electrons and demonstrating metallic bonds and are often described as a lattice of positive ions surrounded by delocalized electrons cloud. Generally, the metals are considered to be lustrous, ductile, malleable, and good conductors of electricity, whereas the non-metals are mostly brittle (for solid non-metals), lack luster, and are insulators. On the other hand, metalloids or semi-metals are characterized by intermediate physical and chemical properties and are metallic in appearance but are brittle being poor conductors of electricity. The term heavy metals especially in the biological sense have generally been used for those metals and semimetals or metalloids that have the potential for human or environmental toxicity (Tchounwou et al. 2012). Heavy metal(loid)s are frequently present naturally in soils, but many anthropogenic activities (e.g., mining, agriculture, sewage processing, the metal industry, and automobiles) increase their ubiquity in the environment resulting in concentrations that are toxic to both animals and plants. These non-essential metal(loid)s like arsenic (As), lead (Pb), mercury (Hg), chromium (Cr), and cadmium (Cd) cause phytotoxicity and severely affect plant physiology by inducing stress symptoms. Further, the removal of these metal(loid)s from the soil is particularly difficult not only because they are relatively immobile in the soil and can persist for longer periods (Martínez-Alcalá and Clemente 2020), but also as they are non-biodegradable and therefore cannot be easily broken down into simpler forms by soil microbes (Atieh et al. 2017). Since the bioaccumulation of these heavy metal(loid)s are associated with abundant negative effects on both plant and animal (including human) health, in the US Agency for Toxic Substances and Disease Registry (ATSDR) Priority List of Hazardous Substances in 2013, the elements As, Pb, Hg, and Cd have been ranked first, second, third, and seventh, respectively.

The most toxic heavy metal(loid)s, As is taken up by plant roots either as As(III) or as As(V) from the surrounding soil as per the availability by different transporters (Table 19.2). It affects the plant at the physiological, biochemical, and molecular levels, the most prominent effect being the generation of ROS molecules. This inhibits the germination rates of the plants and other cellular activities (Abbas et al. 2018). Since arsenic is highly prone to be transferred from the soil to food grain, it also affects animal (including human) life once these contaminated grains are consumed. It induces a plethora of diseases in humans by interfering with various biochemical pathways in the body. Even in low levels, chronic exposure to As via

Table 19.2 Common heavy metal(loid)s and transporters involved in their movement and their major negative impact on human health

Metal(loid)	Transporter used	Major effects on human health	References
Arsenic (As)	Inositol transporters (INTs); ABCC subfamily; MATE2; INT family; PTR family; ACR transporters	Diabetes, liver fibrosis, neurodegenerative issues, cardiovascular disorders, chromosomal aberrations, carcinogenesis, etc.	Engwa et al. (2019), Tang and Zhao (2020)
Cadmium (Cd)	HMA; CAX; NRAMP; ABCC; MTP	Oxidative stress generation (ROS production), apoptosis induction, DNA damage	Tchounwou et al. (2012), Zhang et al. (2018a)
Chromium (Cr)	Phosphate and sulfate transporters	Pulmonary sensitization; lung and nasal cancer; dermatitis and skin ulcers; cardiovascular disorders; certain genetic mutations	Srivastava et al. (2021), Yu (2013)
Lead (Pb)	HMA; NRAMPs;	Neurodegenerative disorders; reduces IQ; hypertension; renal damage	Wani et al. (2015)
Mercury (Hg)	ABCC subfamily	Nervous system damage; cardiomyopathy; anemia; leukemia; pulmonary fibrosis; bronchitis	Rice et al. (2014)

drinking water can cause problems such as diabetes, liver fibrosis, neurodegenerative issues, cardiovascular disorders, etc. (Table 19.2). It can also affect the nervous system, endocrine system, and gastrointestinal system, weaken the immune system, and cause epigenetic alterations (Engwa et al. 2019). When exposed to As in moderate levels for a longer period, disorders as grave as chromosomal aberrations and arsenic-induced carcinogenesis can occur (Table 19.2). As disrupts the DNA repair machinery in cells and thus multiple signaling pathways get disrupted, which further results in diseases such as melanoma, lung cancer, etc. (Biswas et al. 2020).

On the other hand, Pb mostly remains localized to the root cells of the plants as the negative charges of the cell walls of the root cells block its further transport to the upper parts of the part. Pb induces toxicity in the plant cells by triggering ROS production, in the way disrupting various essential proteins and enzymes needed for multiple biochemical pathways in the plant. It vehemently hinders plant developmental rates by interfering with seed germination processes, biogenesis, metabolism of lipid, chlorophyll production, etc. (Pourrut et al. 2011). Pb has the potential to affect almost every organ of the human body, and chronic exposure to lead is known to induce neurodegenerative disorders, reduce the intelligence quotient (IQ), and cause other behavioral issues (Table 19.2). Other conditions such as hypertension, renal damage, etc. might also be caused by long-term Pb exposure (Wani et al. 2015).

Hg is a heavy metal found to naturally exist in three forms: elemental, organic, and inorganic. It is considered to be a hazardous element and has a severe impact on human and plant health (Table 19.2). On entering a cell, mercury (Hg^{2+}) and methyl mercury (MeHg) start forming covalent bonds with cysteine residues and reduce the cellular antioxidants. Antioxidant enzymes act as cellular defenses against Hg (Valko et al. 2006). The interaction of mercury compounds triggers the production of oxidative damage through the accumulation of reactive oxygen species (ROS) which is normally eliminated by cellular antioxidants. Both organic and inorganic Hg is known to alter calcium homeostasis. Organic mercury (MeHg) is believed to increase the intracellular calcium concentration by increasing the amount of calcium influx from the extracellular medium and releasing intracellular stores. On the other hand, inorganic Hg (Hg^{2+}) increases calcium concentration in cells only by increasing the influx rate of calcium into the cells from the extracellular medium (Tchounwou et al. 2012). Mercury can also replace the central magnesium atom from chlorophyll, thus preventing them from properly utilizing the light for photosynthesis (Patra and Sharma 2000). In humans, Hg can impair neurological development in infants and causes nervous system damage in adults (Table 19.2). It can also affect the cardiovascular and pulmonary systems of the human body (Rice et al. 2014).

Another prominent heavy metal(loid) that causes multiple adverse effects on plant and human health is Cd (Table 19.2). Plants growing in Cd-rich soils turn pale due to decreased chlorophyll content and degradation of other vital proteins (Mohanpuria et al. 2007). They gradually show chlorosis, that is, browning or yellowing of leaves, and ultimately die out of nutrition deprivation and other biochemical pathway disruptions. The most significant implications are in oxidative stress generation (by the production of ROS), induction of apoptosis, and causing DNA damage (Engwa et al. 2019). Cd if inhaled or ingested affects severely the pulmonary and GI tract of humans (Table 19.2). Symptoms like nausea, vomiting, salivation, muscle cramps, and vertigo appear within 30 min (Table 19.2). Continued exposure to Cd reduces the levels of serotonin, norepinephrine, and acetylcholine (Tchounwou et al. 2012). Contrarily, Cr is found in varying oxidation states from Cr (II) to Cr(VI). This great range of oxidation states is a major factor in the toxicity of the element. In plants, Cr toxicity caused problems such as a decrease in root length, reduction in seed germination capability, and inhibited translocation of sugars across different plant parts (Nagajyoti et al. 2010). Humans can be exposed to increased levels of Cr(III) via different modes such as inhalation, ingestion, or skin contact with contaminated surfaces (Wilbur et al. 2012). On inhalation, Cr causes pulmonary sensitization and might induce lung and nasal cancer in the future. In the case of dermal contact, Cr(VI) can cause issues such as dermatitis and skin ulcers. When ingested in certain concentrations, chromium heavily affects the gastrointestinal mucosal tissue, triggers cardiovascular disorders, and also leads to genetic mutations at the molecular level, etc. (Table 19.2) (Yu 2013).

19.3 Plant Metal Transporters Regulating the Uptake, Transport, and Accumulation of Metals in the Plant System

During the uptake, the heavy metal(loid)s themselves are not capable of crossing the cell wall and membrane to enter the cytosol of the root cells. Various factors such as their size, charge, the permeability of the membrane, species, etc. hinder their entry into the cells. The metal transporters, like carrier proteins, channel proteins, and pumps, not only facilitate the entry and exit of metal ions into and out of the cells, but also aid in the translocation of these heavy metal(loid)s from the root to the shoot of the plant (DalCorso et al. 2019). Further, these transporters are often located in the membranes of different organelles, to allow the influx-efflux of metal(loid) ions into and out of the organelles, thereby maintaining cellular homeostasis of these metal (loid)s. Thus, the metal transporters perform three major roles concerning heavy metal(loid) transport: (1) uptake of the metal(loid)s via plant roots, (2) translocation of these metal(loid)s from the root to shoot system, and (3) redistribution to different plant parts (Jogawat et al. 2021). According to the homology of their sequences, metal transporters have been categorized into various families such as ZRT, IRT, ZIP, NRAMP, CDF/ MTP, HMA, ECA, MHX, DTX, etc. Some of them have shown differential expression (specifically upregulation) in certain hyperaccumulator species in the presence of specific metal(loid)s. This suggests that genetically engineering plants by altering the structure and function of such metal transporters might be a possibility for phytoremediation of toxic metals (Chaturvedi et al. 2016) as well as biofortification of essential metal(loid)s.

(a) **ZIP (ZRT-IRT-like proteins)**: ZIP family are integral membrane transporters that are mainly responsible for the transport of Zn^{2+} ions and maintenance of metal homeostasis in plants. They are also majorly involved in the uptake, translocation, and accumulation of metals and metalloids such as cadmium (Cd^{2+}), copper (Cu^{2+}), manganese (Mn^{2+}), iron (Fe^{2+}), nickel (Ni^{2+}), and cobalt (Co^{2+}) (Pedas and Husted 2009). ZIP transporters play a major role in zinc uptake, intracellular trafficking, sequestration, and detoxification. This property of ZIP transporters makes them a suitable candidate for the biofortification of crops with Zn (Krishna et al. 2020). A specific member of the family *OsZIP1* acts as a metal efflux transporter and prevents the accumulation of excess Zn^{2+}, Cu^{2+}, and Cd^{2+} in rice plants (Liu et al. 2019).

(b) **HMA (P_{1B}-type heavy metal ATPase) family**: HMA transporters play an integral role in the detoxification of metals such as zinc and cadmium in hyperaccumulator plants. They have diverse functions in plants. For instance, *OsHMA2* translocates Zn^{2+} and Cd^{2+} ions from the root to shoot and further into developing rice grains. Another member of the family, *OsHMA3* takes part in vacuolar sequestration of these metals in the root cells (Takahashi et al. 2012). Other members such as *AtHMA1*, *AtHMA3*, etc. help in the detoxification and accumulation of Zn and Cd in Arabidopsis, respectively (Kim et al. 2009; Morel et al. 2009). Alterations made to the structure and functioning of these transporters might be a promising remediation technique to remove heavy metal (loid) contamination in food crops such as rice.

(c) **MTP (metal tolerance protein) or CDFs (cation diffusion facilitators) family**: These transporters are responsible for the maintenance of homeostasis and heavy metal stress tolerance in plants (Gao et al. 2020). Due to the presence of a C-terminal cation efflux domain, they are mainly involved in the transport of divalent cations such as Co^{2+}, Fe^{2+}, Mn^{2+}, Zn^{2+}, Cu^{2+}, Cd^{2+}, etc. (Montanini et al. 2007). The increased presence of MTP transporters in hyperaccumulator plant species such as *Arabidopsis halleri* and *Noccaea caerulescens* proves their involvement in heavy metal stress tolerance (Ricachenevsky et al. 2013). MTP transporters have also been in use for the biofortification of certain micronutrients. A member of the family, MTP1 has been used for the biofortification of Zn (Palmgren et al. 2008). Alteration of these MTP transporters may help to increase the endosperm-specific overexpression of micronutrients in various plant tissues, which would be a giant leap in biofortification strategies.

(d) **NRAMP (naturally resistant associated macrophage protein) family**: NRAMP transporters are found almost across all the kingdoms, from the simplest organisms like bacteria to complex ones such as plants and animals. NRAMP family members can transport various metals and metalloids. For instance, AtNRAMPs have been found to transport both iron (Fe) as an essential micronutrient and also cadmium (Cd) which is toxic to the plant. A yeast NRAMP homolog, SMF1 functions as a manganese (Mn) transporter; a mammalian homolog known as *NRAMP2* is involved in the iron (Fe) transport system (Thomine et al. 2000). Another member, *HvNRAMP5* (identified in *Hordeum vulgare*) shows involvement in cadmium (Cd) and manganese (Mn) transport, but not in iron (Fe) transport (Wu et al. 2016).

19.4 Remediation of Soil Metal(loid) Contamination

The numerous ill effects of heavy metal(loid) toxicity make their remediation an indispensable act to save the biotic community. Several remediation techniques such as biosorption, chemical extraction, chlorination, electrokinetics, membrane separation, pyrometallurgical separation, phytoremediation, etc. have been in practice for decades. Some of these technologies rely on physical methods, while some are based on chemical separation, whereas some depend on biological organisms for remediation (Table 19.3). It has been seen that the side effects of most of the remediation techniques affect the environment in some way or the other. However, in comparison to the physical and chemical remediation techniques, the biological methods have demonstrated fewer negative effects. Apart from the individual techniques, various combinatorial remediation methods have also been adapted in recent times. For instance, the application of microbial treatment in combination with electrokinetics has proved to be an effective method in treating heavy metal(loid) contaminated soil (Huang et al. 2017). Similarly, an amalgamation of electrokinetics with

Table 19.3 The summarized form of the major remediation methods available for the removal of environmental heavy metal(loid) contamination

Remediation method	Brief description	Machinery used (major)	References
Adsorption-desorption	Heavy metal(loids) in the soil can be adsorbed as ions; the common interface used is inorganic colloids	Thermostated rotary shaker	Xie et al. (2018)
Biosorption	It has the same principle of adsorption, but here the functionally active adsorbent used is agricultural wastes such as seeds, fruit shells, etc. (ATP may or may not be involved)	For batch process: continuous stirred tank bioreactor; for continuous process: column bioreactor	Shamim (2018), Yan and Viraraghavan (2001)
Bioleaching (or biomining) method	Extraction of metals from contaminated soil with the help of microorganisms (works on the principle of microbe-mineral interaction)	Bioslurry reactor	Pathak et al. (2009)
Chemical extraction	Chemicals such as acids can be used to extract metals for wastewater. Such chemical treatments prevent the free movement of the contaminants in the soil/water and thus stop their spread	Chemical reactors (type as per requirement)	Stylianou et al. (2007)
Chlorination	Removal of heavy metal (loids) by the use of chlorine. It may be followed by thermal treatment at times	Laboratory scale rotary reactor	Fraissler et al. (2009), Nowak et al. (2010)
Electrokinetics (also called electro bioremediation)	Separation of contaminants (heavy metals or metalloids) in the presence of an applied direct current electric field	Electrodes (anode and cathode) are connected to either side of the soil compartment	Pamukcu and Wittle (1992), Virkutyte et al. (2002)
Ion-exchange	Ion-exchange resins (cationic or anionic type as per need) in permeable reactive barriers (PRBs) lead to the reduction of heavy metal(loid) concentration from the soil and groundwater	Ion exchanger	Lin and Kiang (2003), Vilensky et al. (2002)
Membrane separation	Making contaminated water free of heavy metal		

(continued)

Table 19.3 (continued)

Remediation method	Brief description	Machinery used (major)	References
	(loids) such as Cu^{2+} and Cd^{2+} by applying the principles of reverse osmosis (RO) and nanofiltration (NF)	Filtration membranes of suitable pore sizes are needed for RO and NF	Qdais and Moussa (2004)
Pyrometallurgical separation	Separation of metals from contaminated soils by the application of high temperatures	Industrial furnace (operated at very high temperatures)	Diels et al. (2003)
Solidification and stabilization	In solidification, the contaminants (metals) are encapsulated in a solid matrix. In stabilization, the contaminants are relatively immobilized by the formation of chemical bonds	Large-scale plants	Mulligan et al. (2001b)
Vitrification	Intensive heat energy is applied to the contaminated soil to first melt it down and then allow it to cool to transform into a glass-like solid. (Vitrification can be induced by electricity, heat, or gas plasma)	For electrical vitrification: electrodes For thermal vitrification: Microwave radiation For plasma vitrification: electrical discharge-induced gas plasma	Liu et al. (2018), Meuser (2013)
Phytoremediation	The use of plants (and associated microbes at times) to remove heavy metal(loids) toxicity from contaminated soils	Depends on the different types of phytoremediation techniques used (discussed in detail later)	Yan et al. (2020)

phytoremediation is a much more cost-effective method and allows metal recovery easily (Selvi et al. 2019). Likewise, an integrated approach involving membrane separation and microbial remediation helps to remove heavy metal(loid)s from contaminated water (Köber et al. 2005). Phytoremediation is a better approach than most of the others mentioned since it utilizes various natural sources like plants, solar energy, etc. in the remediation process, using inexpensive components with easy maintenance. Also, phytoremediation can reduce environmental contamination without the use of any chemical agents, which makes it an eco-friendly approach. It can be applied over large areas of contaminated soils without any significant risk of reduction of soil fertility (Yan et al. 2020). Further, the disposal of the extracted contaminants such as the heavy metal(loid)s is relatively easier, as the plant-microbe interactions take care of that.

19.4.1 Phytoremediation

Phytoremediation is the eradication of various contaminants occurring in the form of nutrients, biodegradable compounds, and non-biodegradable compounds like heavy metals and metalloids using plants and associated microorganisms from soil and water bodies (Chaney 1983; Chaney et al. 1997). It is undoubtedly one of the most economic, eco-friendly, and sustainable solutions to the problems of soil and water body contamination by toxic metal(loid)s. Therefore, improving the efficiency and effectivity of various phytotechnologies (phytoremediation techniques) has been the objective of extensive scientific research across the globe (Clemens 2001; Kawahigashi 2009; LeDuc and Terry 2005; Lone et al. 2008; Odjegba and Fasidi 2007; Saier and Trevors 2010; Sarma 2011; Suresh and Ravishankar 2004; Vithanage et al. 2012). The novelty of phytoremediation over other conventional, modern, and ultra-modern engineering or related remediation techniques like excavation, soil incineration, soil washing, flushing, solidification (Mulligan et al. 2001a), vitrification (Baker et al. 1994), pyrometallurgical separation (Diels et al. 2003), electrokinetics (Pamukcu and Wittle 1992), bioleaching (Sreekrishnan et al. 1993), biosorption (Yan and Viraraghavan 2001), biomethylation (Kosolapov et al. 2004), chemical treatments, biohydrometallurgical process (Veglio and Beolchini 1997), etc., is due to its cost-effectivity, high efficiency, and environmental sustainability (Wuana and Okieimen 2011; Ullah et al. 2015; Sarwar et al. 2017; Van Aken 2009).

The type of phytotechnology appropriate for carrying out remediation of a particular piece of land or aquatic body is largely dependent upon the bioavailability of the contaminants that are required to be removed or stabilized (Berti and Cunningham 2000; Lasat 2002; Violante et al. 2010). The most significant inorganic soil contaminants metal(loid)s can be broadly categorized based on their bioavailability into high (cadmium, arsenic, selenium, copper, zinc, and nickel), intermediate (cobalt, manganese, and iron), and low (lead, chromium, and uranium) (Prasad 2003), although the concentrations of each of these will vary with locations and habitats. Plants normally uptake metal(loid)s through their root system without harming the topsoil and in turn depict restorative capabilities by enriching the soil with organic matter, thus enhancing its overall fertility (Mench et al. 2010). However, their inherent ability to accumulate the insoluble metalloids that are either bound to soil particles or have precipitated (Sheoran et al. 2016) falls short for the purpose of remediation for most plants, with regard to the comparatively enormous bioavailability of the metal(loid)s. Thus, the objective of present-day technology and research is to enhance their phytoremediation capabilities by genetically engineering (incorporating) them with heterologous genes to create suitable hyperaccumulator transgenic plants. Plants that are naturally hyperaccumulatory in nature are ideally of great interest for such purposes due to their ability to grow on metalliferous soils and accumulate metal(loid)s and xenobiotics in the leaves and related parts in significantly high concentrations (Rascio and Navari-Izzo 2011). Strategies where the metal(loid)s are adsorbed onto or absorbed into the roots (rhizofiltration) or rhizosphere (phytostabilization) and others (strategies) discussed in the following sections

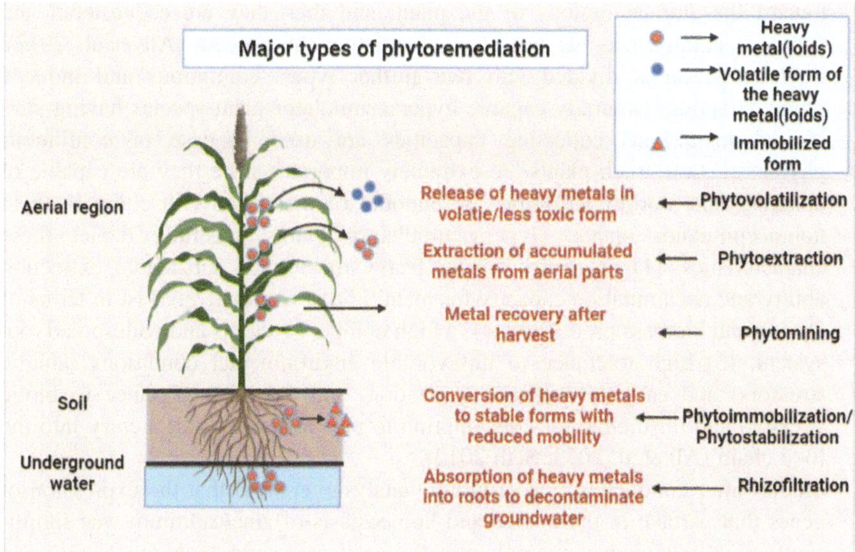

Fig. 19.1 The schematic diagram shows the common methods of phytoremediation (phytovolatilization, phytoextraction, phytomining, phytoimmobilization/phytostabilization, and rhizofiltration) in practice and the area of a plant where they are usually performed. The red circles represent the heavy metal(loid)s that need to be remediated, the blue circles represent the volatile forms of those heavy metal(loid)s, and the red triangles represent their immobilized forms

have their own approaches at solving the problem at hand from different angles (Fig. 19.1).

1. **Phytoextraction**, also known as phytoaccumulation or phytosequestration, is deemed to be not only the most effective but also the most challenging out of all the phytotechnologies. This method involves the uptake of metal(loid)s and their storage in the stems, roots, and leaves of a metal-tolerant plant. Land and aquatic plants have been explored for their bioaccumulative capacities to sieve out those with comparatively better accumulatory capabilities (hyperaccumulators) (Jacob et al. 2018; Salt et al. 1995). The criteria for being categorized as a hyperaccumulator is that the plant must at least accumulate 1000 mg/kg (dry weight) of a particular type of metal(loid) that is occurring in its soil. The process involves the cultivation of hyperaccumulators which can concentrate the metal(loid)s into their organs for harvesting at a later stage. The harvested, dried, and metal(loid)-rich plants are then industrially processed by either dumping in special areas or are smelted to further profit from the energy gained by burning the biomass (Krämer 2005; Salt et al. 1998; Vishnoi and Srivastava 2007). There are broadly four steps comprising phytoextraction. The first step is where the metal(loid)s are collected in the rhizosphere and then in the next step the plant, through its roots takes in the concentrated metal(loid)s. The third step takes place internally, where the metal(loid)s are translocated

toward the storage regions of the plant, and then they are sequestered and compartmentalized as the last step, completing the process (Ali et al. 2013). Phytoextraction is divided into two further types: continuous and induced phytoextraction. Naturally capable hyperaccumulator plant species having substantial metal(loid) collecting capacities are used in case of continuous phytoextraction. Such plants are extremely important since they are capable of uptaking and storing 100 times the amount of metal(loid)s in comparison to non-accumulator plants. Hyperaccumulatory plants generally depict these characteristics: (1) resistivity toward heavy metals, (2) substantial extraction ability and accumulation capacity for metal(loid)s, (3) relatively fast in terms of growth and biomass production, (4) a high number of shoots and widespread root system, (5) high tolerance to unfavorable environmental conditions (abiotic stressors) and easy cultivation-harvestation, and (6) high tolerance to biotic stressors and unsuitable for consumption to prevent metal(loid)'s entry into the food chain (Ali et al. 2013; Seth 2012).

Studies after conducting transcriptomic analysis reported that the expression of genes that partake in the uptake and homeostasis of zinc/cadmium was significantly high in the hyperaccumulators: *T. caerulescens* and *Arabidopsis halleri* in comparison to the non-hyperaccumulators *Thlaspi arvense* or *A. thaliana* (Hammond et al. 2006; Talke et al. 2006; van de Mortel et al. 2006). Two P_{1B}-type ATPases *HMA4* and *HMA2* were found to be playing a crucial role in the accumulatory ability to translocate Zn/Cd from the root to the aerial portions by mobilizing the metal(loid)s from the pericycle to the xylem (Hussain et al. 2004; Wong and Cobbett 2009). Silencing the two of them led to the decrease of Zn and Cd depositions in the shoots but increased concentrations in roots (Hanikenne et al. 2008). Another hyperaccumulator plant, *P. vittata* is known to accumulate about 2.3% of As by uptaking and sending it from the roots to the xylem through the formation of arsenite-phytochelatin complexes (Ma et al. 2001; Su et al. 2008; Zhao et al. 2009). Although the exact list of genes responsible for this process remains unknown, *PvACR2*, an arsenate reductase encoding gene, was suspected to be instrumental in the process (Ellis et al. 2006). Transgenic *A. thaliana* with a silenced *AtACR2* lead to a 10–12-fold increase in the accumulation of As in the shoots, with a decrease of the same in the roots, all in comparison to the wild type (Dhankher et al. 2006). However, another study reported the opposite phenotypic outcome (decreased As storage in roots of *AtACR2* silenced plant), making the exact role of *AtACR2* uncertain (Bleeker et al. 2006). Thus the *in planta* role of metal(loid) accumulating genes need further exploration to better understand the hyperaccumulator functions.

In case of *induced phytoextraction*, natural or synthetic chelating agents are added to metal(loid)-contaminated soils to increase their bioavailability, leading to a faster uptake and accumulation of the metal(loid)s in the plants' aerial regions. This is considered for scenarios where the hyperaccumulators are substantially high-biomass-producing in nature (Ebbs et al. 1997). The major drawback of this technique is that the increased solubility of the chelated metal(loid)s significantly increases the risk of their leaching into deeper cleaner soil and water levels, ultimately defeating the very purpose of phytoextraction (Neugschwandtner et al. 2008; Wenzel et al. 2003). This technique has been conducted on previous accounts for extracting Pb from lead-rich soils (Babaeian et al. 2016; Blaylock et al. 1997; Walker et al. 2003).

In terms of limitations of phytoextraction, this technique at times turns out as unfeasible within a realistic time frame in cases where the metal(loid) contamination of the soil is severe. According to mass-balance data, phytosequestration is feasible for only low or moderately contaminated soils. And for severe soils, phytostabilization (discussed in detail later on) is preferred to instead stabilize the metal(loid)s and prevent their seepage into deeper soil and water layers from erosion and leaching.

2. **Phytomining** is the term used to refer to the process of recovering metal(loid)s from the biomass of hyperaccumulatory plants after they have been harvested, in an economically feasible manner (Brooks et al. 1998; LaCoste et al. 2001; Nkrumah et al. 2016; Sheoran et al. 2009). This technique is considered profitable specifically for metal(loid)s like gold, thallium, cobalt, and nickel as their market price evaluation is on the higher side (Sheoran et al. 2009). Pyrolysis is one of the methods performed to extract the metal(loid)s from the toxic biomasses on an industrial scale, where the biomass is decomposed under anaerobic conditions to obtain pyrolytic fluid and coke. The latter being rich in metal(loid)s is further used as a viable substitute for smelting (Bridgwater et al. 1999)
3. **Phytovolatilization** is the process where the plants after uptaking the contaminants from the contaminated soil/water body convert the metal(loid)s into their volatile forms and then release them into the atmosphere by transpiration from leaves or other foliage regions. This process is considered in soils that are contaminated with metal(loid)s that can form volatile hydride and methyl compounds. Volatilization of heavy metals like Se, As, and Hg is possible (Jia et al. 2012; Sakakibara et al. 2010; Terry et al. 2000; Zhu and Rosen 2009). This process can be further divided into two types: direct and indirect phytovolatilization. In direct phytovolatilization, the contaminants are directly translocated from roots to the aerial regions (stems or leaves), from where volatilization of those compounds takes place (without them undergoing any changes or modifications). On the other hand, in the case of indirect phytovolatilization, the contaminant compounds transform while inside the aerial regions into lesser toxic compounds, before they get volatilized. For instance, mercury and selenium when absorbed by metallophytes get converted into their volatile forms mercuric oxide and dimethyl selenide, respectively, before escaping into the atmosphere (Ali et al. 2013). Other plants such as *Canna glauca* L., *Colocasia esculenta*, *Cyperus papyrus* L., *Azolla caroliniana*, *Arundo donax* L., *Pteris vittata*, *Brassica juncea*, *Lupinus* sp., *Liriodendron tulipifera*, and *Typha angustifolia* L. have also been studied for their ability to phytovolatilize heavy metals (Pilon-Smits et al. 1999a). Certain transgenic plants expressing mercuric reductase and bacterial organomercurial lyase (merb) are able to uptake both divalent and methylmercury but release into the atmosphere Hg(0) (Rahman et al. 2008). For instance, transgenic *Arabidopsis thaliana* and *Nicotiana tabacum*

expressing merC proteins depicted enhanced uptake of divalent mercury when compared to their wild types (Sasaki et al. 2006). The only advantage that this particular phytotechnology provides is that of removing metal(loid)s from the soil by dispersing the volatile contaminants (unchanged or less-toxic versions) into the atmosphere. But, although that qualifies as remediation, this advantage is flawed in reality as complete "removal" of contaminants does not take place. Not only is the volatilization process slower in comparison to other phytotechnologies (Tangahu et al. 2011), the overall phenomenon occurring is that of the contaminants being transferred from the soil to the atmosphere (Heaton et al. 1998). The compounds would then cause atmospheric pollution and are likely to return to the soil owing to rain (Mulligan et al. 2001b; Vangronsveld et al. 2009).

4. **Phytostabilization** or phytoimmobilization is the remediation technology that utilizes metallophytes to immobilize metal(loid)s occurring in the soil to decrease their bioavailability, to restrict them from entering into the food chain (Marques et al. 2009; Wong 2003). It is a viable phytotechnology suited for soils with highly severe contaminations, in which case other methods such as phytoaccumulation do not suffice to remove the contaminants as effectively. The objective of this technique is to de-functionalize the metal(loid) contaminants via concentrating them by inducing their precipitation, reduction in the rhizosphere, accumulation into the roots, or onto their surface (Gerhardt et al. 2017; Kumpiene et al. 2012). Plants by stabilizing the metal(loid)s in the soil not only prevent their entry into aquatic bodies but also into the atmosphere (Vangronsveld et al. 2009; Mench et al. 2010). The plants considered ideal to qualify for conducting phytostabilization must have well developed and widespread root system because they are critical in not only stabilizing the metal(loid)s but also the soil structure, thus preventing erosion. Studies have explored and identified multiple plant species that are suitable for phytostabilizing metal(loid)-contaminated soils (Burges et al. 2018), e.g., *Agrostis* spp., *Festuca* spp., *Deschampsia cespitosa*, *Piptatherum miliaceum*, *Zygophyllum fabago*, *Lygeum spartum*, *Bromus inermis*, *Phleum pratense*, *Festuca pratensis*, *Alopecurus pratensis*, *P. tremula*, *Picea abies*, etc. (García et al. 2003). The efficacy of this process is usually enhanced by the addition of various additives that alter metal (loid) states and adjust the pH of the soil in doing so. These in turn help by plummeting metal(loid) bioavailability (Alvarenga et al. 2009; Burges et al. 2018; Epelde et al. 2009). For instance, *Medicago sativa* L. with powdered marble as an additive was used to successfully remediate an acidic mining site (Midhat et al. 2018). Another study reported the phytostabilization of Cd using *Osmanthus fragrans*, *Ligustrum vicaryi* L., *Cinnamomum camphora*, *Loropetalum chinense* var. *rubrum*, and *Euonymus japonicus* Aureo-mar (EJ) in greenhouse conditions (Zeng et al. 2018). They found that growing these plants in the Cd-rich soil leads to an increase in the levels of urease and sucrose while also decreasing the levels of dehydrogenase. A different study utilizing a hydropic system explored the role of *Thalia geniculata*, *Cyperus alternifolius*, *Canna indica*, *Eichhornia crassipes*, and *Pistia stratiotes* (water lettuce) with *Vetiveria zizanioides* (vetiver) and then found *E. crassipes* and *P. stratiotes* to be containing translocation factors

necessary for phytostabilizing Cd and Zn. The aquatic plants depicted great ability for storage of metal(loid)s in their roots as well (Sricoth et al. 2018). The most substantial advantage that this phytoremediation technology offers is not having to deal with any contaminant-rich biomass or exploring options for the eco-friendly disposal of the same, which is applicable for phytoextraction (Wuana and Okieimen 2011). Another significant aspect of phytostabilization is that microorganisms such as bacteria and mycorrhiza that live in the rhizosphere can absorb metal(loid)s, produce chelators, and enhance the precipitation process. They also promote the growth of a better root system, expediting phytostabilization and at the same time filtering the toxic compounds during their translocation to the aerial regions of the plant (Göhre and Paszkowski 2006; Mastretta et al. 2009; Ma et al. 2011).

5. **Phytofiltration** is considered to be the next most effective phytoremediation technique after phytoextraction for the remediation of metal(loid)s (Raskin et al. 1997). This phytotechnology utilizes different plant parts to either adsorb or absorb and store the contaminants from soils and water bodies (mostly), thus restricting them from leaching into the groundwater and entry into the food chain, respectively. The technique is categorized into three subcategories based on the part of the plant in which the contaminants are to be stored, the types being (1) rhizofiltration, where the contaminants are adsorbed onto/absorbed into the roots; (2) caulofiltration, where the contaminants are stored in the shoot; and (3) blastofiltration, where the same is stored in the seedlings (Mesjasz-Przybyłowicz et al. 2004; Sarma 2011).

   Rhizofiltration utilizes the roots or other submerged plant regions of tolerant aquatic plants to absorb/adsorb metal(loid)s from the surface or underground water bodies. The efficacy of this process depends upon how large the surface area is for adsorption to occur, and how tolerant the plants are to hypoxia (Dushenkov et al. 1995). Reeds (macrophyte) were one of the most suitable species used for rhizofiltering wastewaters (Dürešová et al. 2014; Mohanty and Patra 2011). Plants to be used for rhizofiltration are grown hydroponically in greenhouses with water as substrate rather than soil. Once a widespread root system develops, the usual water source is gradually switched out with metal (loid)-rich wastewaters from contaminated water bodies. The fully grown plants are then harvested once the metal(loid)s get saturated in the roots. A study exploring feasible plant species for rhizofiltration reported that common sunflower (*Helianthus annuus*) plants depict the highest efficiency in eradicating Pb from contaminated water, in comparison to *Brassica juncea*, *Nicotiana tabacum*, *Secale cereale*, *Spinacia oleracea*, and *Zea mays* (Raskin and Ensley 2000). Indian mustard is also efficient in sequestering metal(loid)s and is thus a good candidate for rhizofiltration due to its long and hairy roots in comparison to other aquatic plants (Dhanwal et al. 2017; Rezania et al. 2016; Tomé et al. 2008). Apart from lead, this phytotechnology is also applicable for the remediation of cadmium, copper, zinc, chromium, and nickel (Dushenkov et al. 1995). Caulofiltration (shoot filtration; Latin *caulis*= shoot) of Pb was observed in the aquatic macrophyte *Ipomoea aquatica* Forsk. (Chanu and Gupta 2016). The study reported that although primary accumulation of Pb was observed in the root system, accumulation of excess metal in the lower stem tissues was also observed whenever the concentration of Pb exceeded 20 mg L^{-1}. Thus, the stem region would be considered as a secondary reservoir, after the roots. Another study reported the caulofiltration of the metal(loid)s—arsenic, copper, selenium, and aluminum—by excised *Stevia rebaudiana* stems (Hajar et al. 2014). On the other hand, blastofiltration is the process in which seedlings are used for remediation as a variety of seedlings are capable of adsorbing or absorbing substantial amounts of metal(loid)s due to the rapid increase in surface to volume occurring during germination (Krishna et al. 2012). Another study explored the blastofiltration potential in castor, melon, okra, and moringa seeds. The seedlings were introduced to lead- and cadmium-rich waters (60 ppm each) separately for 72 h, and about 96–99% of the metal content was reduced. The overall efficiency of blastofiltration was highest in okra and castor seeds, whereas moringa seeds were able to remove 100% of the cadmium (Udokop 2018). Another study exploring the blastofiltering potential of *Moringa oleifera* seeds reported it could uptake up to 95% for Cu, 93% for Pb, 76% for Cd, and 70% for Cr, from wastewaters (Ravikumar and Sheeja 2013). Other studies have indicated papaya seeds and mango seed powders to be effective in blastofiltering Zn at pH 5.0 (highest uptake at this pH) and Cu, Cd, and Pb at 85–100% efficiency, respectively (Ong et al. 2012; Parekh et al. 2002).

6. **Phytotransformation** also referred to as phytodegradation is described as the process by which the contaminants that are extracted from the contaminated water, sediment, or soil undergo chemical alterations inside the plant system (in both root and shoot). The transformed contaminants are inactivated, degraded, or immobilized (Pivetz 2001; Tangahu et al. 2011). This phenomenon also resembles phytovolatilization, where the metal(loid)s are transformed into lesser toxic volatile compounds of themselves. Phytotransformation is referred to as rhizodegradation when it is occurring in the roots (Bulak et al. 2014). The chemical transformation of the extracted toxic compounds is induced by various enzymes synthesized within the plant system, often with the assistance of microorganisms such as bacteria, yeasts, and fungi (Dobos and Puia 2010). Studies observing chromium toxicity in plants such as *Pistia stratiotes* reported that it converted Cr^{6+} to Cr^{3+} (lower mobility and toxicity in comparison) (Sen et al. 1987; Shanker et al. 2005).
7. **Phytodesalination** is one of the newer phytotechnologies in comparison to others (Zorrig et al. 2012). Phytodesalination, as the name suggests, is where plants are used to extract excess salts from overly saline soils so that plant growth can resume as usual (Manousaki and Kalogerakis 2011; Sakai et al. 2012). Salinity stress is a major contributor to abiotic stress that causes serious devastation of essential crops. A study explored three halophytes for their phytodesalination capacities and ranked them based on their efficiency for Na^{+} accumulation from saline water (Islam et al. 2019), the three plants being *Ipomoea aquatica*, *Alternanthera philoxeroides*, and *Ludwigia adscendens*. Among these three, *I. aquatica* depicted the highest capacity (130 kg Na^{+} ha^{-1}) for accumulation,

with *A. philoxeroides* (105 kg Na^+ ha^{-1}) and *L. adscendens* (80 kg Na^+ ha^{-1}) coming in second and third place, respectively. Similarly in another study, plants, *Suaeda maritima* and *Sesuvium portulacastrum*, were found to accumulate 504 kg and 474 kg of NaCl per hectare, respectively, over a span of 4 months (Ravindran et al. 2007). Hence, such plants can be used to extract salts from overly saline soils.

19.4.2 Methods to Enhance Phytoremediation

19.4.2.1 Chemical Induced Enhancement of Phytoremediation

Certain chemical compounds or surfactants have proven to catalyze the accumulation of meta(loid)s in plants by making the movement of the metal(loid)s easier in the soil by lowering their surface tension. The reagent (1,2-cyclohexylenedinitrilo) tetraacetic acid, (CDTA) when added in the amount of 5 mmol kg^{-1} to As-contaminated soils, increased the accumulation of As from 950 mm to 1400 ppm, without the help of any chelators (Bagga and Peterson 2001). Another study depicted that the addition of chelating agent ethylenediaminetetraacetic acid, or EDTA, to Pb-contaminated soils lead to the enhancement in the sequestration of Pb into the aerial regions of plants. The accumulated Pb increased from 24.23–680.56 mg/kg to 29.07–1905.57 mg/kg in the storage regions of the plants (Wang et al. 2007). Although qualifying as a phytoremediator, the addition of EDTA and the formed EDTA-Pb complexes are highly soluble in the soil while also being difficult to biodegrade. This leads to an increase in environmental risks as significant concentrations of EDTA are fatal to plants (Epelde et al. 2008; Hovsepyan and Greipsson 2005; Wu et al. 2004). On the other hand, ethylenediamine-*N,N*-′-disuccinic acid (EDDS) is another chelating agent, but one that is easily biodegradable and is considered as a viable, environment-friendly alternative to EDTA (Epelde et al. 2008). The only compromise here is that the accumulatory efficiency of EDDS, although viable, is still considerably lesser in comparison to EDTA. This was depicted in an experiment where both these chelating agents were added to Pb-contaminated soils separately, and the phytoextraction results were compared in the plant *Cynara cardunculus*. It was deduced that the addition of 1 g of EDTA kg^{-1} of soil lead to a value of 1332 mg Pb kg^{-1} DW shoot. And the addition of 1 g of EDDS kg^{-1}, the value obtained was a mere 310 mg Pb kg^{-1} DW. It was also seen that EDDS being highly degradable did less harm to the soil microbiota in comparison to EDTA in uncontaminated soils. As for Pb-contaminated soils, the additions of EDDS lead to better basal and substrate-induced respiration (Epelde et al. 2008).

19.4.2.2 Genetic Engineering Mediated Enhancement of Phytotechnologies

It has been observed that the phytoremediation abilities or the accumulation capacities of various metal(loid)-tolerant (e.g., halophytes and metallophytes) and hyperaccumulatory plants could be significantly enhanced through the application of genetic engineering. In this method, the nucleotide sequences encoding the gene

(s) of interest are inserted into the host's genome, which then is inherited during DNA recombination, leading to the manifestation of the desired traits. In this manner, the efficiency of phytoremediation could be enhanced by overexpressing genes encoding metal(loid) chelators, metal(loid) transporters, phytochelatins (PCs), and metallothioneins (MTs), leading to improved accumulation, translocation, and sequestration (Ibañez et al. 2015; Das et al. 2016; Mani and Kumar 2014). This method allows us to utilize sexually incompatible species of plants and other organisms such as bacteria also, which is otherwise unachievable through conventional breeding methods (Berken et al. 2002; Marques et al. 2009). Hence, the total production of biomass and the accumulatory capabilities of hyperaccumulatory plant species can be substantially enhanced through this method (Gisbert et al. 2003; Heaton et al. 1998).

Oxidative stress caused by overproduction of ROS due to metal(loid) stress is fairly common among plants growing in metal(loid)-contaminated soils or aquatic bodies. Hence, improving the plant's antioxidant system is one of the primary focuses for bestowing effective metal(loid) tolerance (Koźmińska et al. 2018). Some of such genes have been identified and targeted for genetic engineering that are responsible for the translocation of heavy metals and metal(loid)s including ZRT-IRT-like proteins (ZIP), multidrug and toxic compound extrusion (MATE), heavy metal associated (HMA), and metal tolerance proteins (MTP) (Das et al. 2016, 2018; He et al. 2020; Liu et al. 2019). Other studies have shown that overexpressing genes responsible for the production of metal chelators and metal transporters also improve their abilities to uptake and translocate metal(loid)s (Wu et al. 2010). Multiple studies have indicated successful phytoremediation in terms of increased metal(loid) accumulation and distribution in various parts of transgenic plants (Gomes et al. 2016, refer to Table 19.2 for details). Studies have depicted enhanced phytovolatilization capabilities for the phytoremediation of Hg-contaminated sites using genetically engineered (transgenic) plants expressing *merA* (mercuric ion reductase) and *merB* (organomercury lyase) genes taken from *Escherichia coli* (Krämer 2005; Meagher and Heaton 2005). Another study adopted the same approach for creating transgenic lines of *Spartina alterniflora* (aquatic plant species) expressing both *merA* and *merB* genes for the phytoremediation of highly Hg-contaminated swamplands (Czakó et al. 2006).

Apart from using genetic engineering to enhance uptake, translocation, and sequestration, a different approach in using genetic engineering to improve metal (loid) phytoremediation has also been explored in recent times. This approach encompasses rendering morphological amendments to induce improved metal (loid) uptake. This was observed in transgenic tobacco plants with hairy roots which could efficiently accumulate Cu from Cu-contaminated soils. And the same held true for transgenic *Arabidopsis* plants also having hairy roots, which expressed the CopC protein (Pérez-Palacios et al. 2017). Although genetic engineering-mediated enhancement of phytoremediation has potential, there are limitations in seeking this approach as well. The primary cause of which is due to the lack of research into identifying genes and revealing their functionalities, and the roles of various associated proteins and enzymes that are also involved in morphological changes and the translocation-sequestration of the extracted metal(loid)s. As a result, hairy roots are unable to adapt to fluctuations in environmental cues such as varying metal(loid) concentrations, hydraulic conditions, and unwanted microbes (Khandare and Govindwar 2015). The second limitation is the long-existing controversy on the ethicality of the usage of genetically modified crops. A large group of independent scientists, environmentalists, farmers, and consumers are against the use of GMOs due to the potential rise in superweeds and superbugs (Maghari and Ardekani 2011).

19.4.2.3 Microbe-Mediated Enhancement of Phytotechnologies

Enhancing phytoremedial abilities through the collaboration of plants and microorganisms has been a domain of exploration for many years now. Numerous studies have successfully demonstrated successful remediations of multiple toxic metal(loid)s through the introduction of rhizospheric microbial communities that co-operate with host plants in extracting the contaminants more efficiently (Fasani et al. 2018; Gupta et al. 2013; Jentschke and Godbold 2000; Upadhyay et al. 2018). The plant-microbe partnerships have proven to influence various factors such as plant growth, yield, and tolerance to abiotic and biotic stresses; hence their relationship can be manipulated and applied for phytoremediation (De-la-Peña and Loyola-Vargas 2014; Rout and Southworth 2013). Microorganisms can effectively extract metal(loid)s because of the presence of anionic configurations on their negatively charged surfaces. They are also able to lower the soil pH, secrete chelators, and amend redox potentials, hence enhancing phytoimmobilization of metal(loid)s as well (Gadd 2004; Kidd et al. 2009). Some of the microorganisms that have been identified in this regard are from the genera *Pseudomonas*, *Aspergillus*, *Bacillus*, and *Micrococcus* spp. (Aboudrar et al. 2013; Dursun et al. 2003; Gupta and Keegan 1998; Limcharoensuk et al. 2015). Plant growth-promoting microorganisms (PGPMs) comprising a range of microorganisms such as bacteria, cyanobacteria, algae, and fungi help in plant growth while also helping minimize the lethal effects of pathogenic and abiotic stresses (Ma et al. 2016; Mishra et al. 2017; Upadhyay et al. 2016). A study created transgenic *Rhizobium leguminosarum* by inserting the As(III) S-adenosylmethionine methyltransferase gene from *Chlamydomonas reinhardtii (CrarsM)*. They found that the transgenic *R. leguminosarum* was able to methylate As(III) into other species of As and together with red clover plants lead to a total phytoextraction of 42.4% and phytovolatilization of 0.01–0.02% of the methylated As species (Zhang et al. 2017). Another study demonstrated enhanced fungal stability in contaminated regions, induced from chlamydospores of *Trichoderma asperellum*. The study reported that both plant growth and As accumulation had increased in the plant *Ipomoea aquatic* due to the symbiotic relationship between the two. Thus they inferred that this combination could be applied for enhancing the phytoremediation capacities of plants (Su et al. 2017). Apart from the use of bacteria and fungi, microalgae or algal biomass also allows for an environment-friendly, inexpensive, and efficient alternative for enhancing phytoremediation. The degree of sorption of metal(loid)s is largely based on the type of functional group (e.g., carboxyl, phosphoryl, sulfuryl, hydroxyl, etc.) occurring on the cell surfaces of algae (Kaplan 2013). The production of phytochelatins

(PCs) and phytochelatin (PCs)-metal(loid) complexes are instrumental in the translocation of the contaminants. This was indeed the case in another study where the production of (As)-PC complexes was deemed as a crucial step in the process of As remediation using *Chlorella vulgaris* (Munoz et al. 2014). These PCs and PC-metal (loid) complexes of reductive and oxidative nature prevent potential cellular damage from oxidative stress by inhabiting the role of a redox buffer system. Another study used *C. vulgaris* and *Nannochloropsis* sp. to explore their potentials in displaying phytoremedial roles for rice plants that were treated with doses of As as metal(loid) stress. As a result, the detrimental effects of the metalloid were reported as reduced, along with a decrease in the sequestration of the same in both the roots and shoots (Upadhyay et al. 2016).

19.5 Biofortification

Biofortification, also called biological fortification, refers to the process of enhancing the bioavailable contents of essential nutrients in food crops to combat the increasing problem of micronutrient malnutrition across the globe, especially in developing and under-developed countries (Singh et al. 2016a, b). The primary aim of biofortification is to provide nutritionally enriched food in sufficient quantity to mitigate "hidden hunger." It must be safe for consumption for the majority of people, and the process should preferably be sustainable (Saltzman et al. 2014). It employs either conventional plant breeding methods or agronomic interventions or transgenic techniques (genetic selection or engineering) to create nutritionally fortified crops (Fig. 19.2) that simultaneously produce high yields and profit to the agriculture industry (Bouis and Welch 2010).

Biofortification is a sustainable technology to combat the increasing malnutrition in the world, especially among people of the lower-income group. It is beneficial to the world at three major levels, the agricultural sector, economic, and population level. In the agricultural sector, it helps to enhance the resistance ability of plants to several stress factors (both biotic and abiotic). Biofortification reduces crops' dependency on chemical fertilizers and also ensures higher yield due to adequate or greater nutrient concentrations (Campos-Bowers and Wittenmyer 2007). It is a cost-effective method as compared to other techniques to mitigate nutritional deficiencies in people (Marques et al. 2021). The development of a biofortified crop indeed requires quite some costly laboratory instruments, involvement of specialized researchers in the field, and significant expenses during research and development. However, once the desired cell line of the biofortified crop has been prepared, then it requires a minimalistic cost to maintain the line; making it a one-time investment (Bouis 2002). Again, if the produced modified crop gains industrial success and acceptance among people, then it gives much greater financial gains than the initial investment during its development. Further, people need not change their regular diet to be benefitted from the nutritionally enhanced food. Thus people in the rural areas can consume the biofortified version of their simple foods such as rice, wheat, vegetables, etc., and get the nutritional benefits (Bouis 1996). In the case of

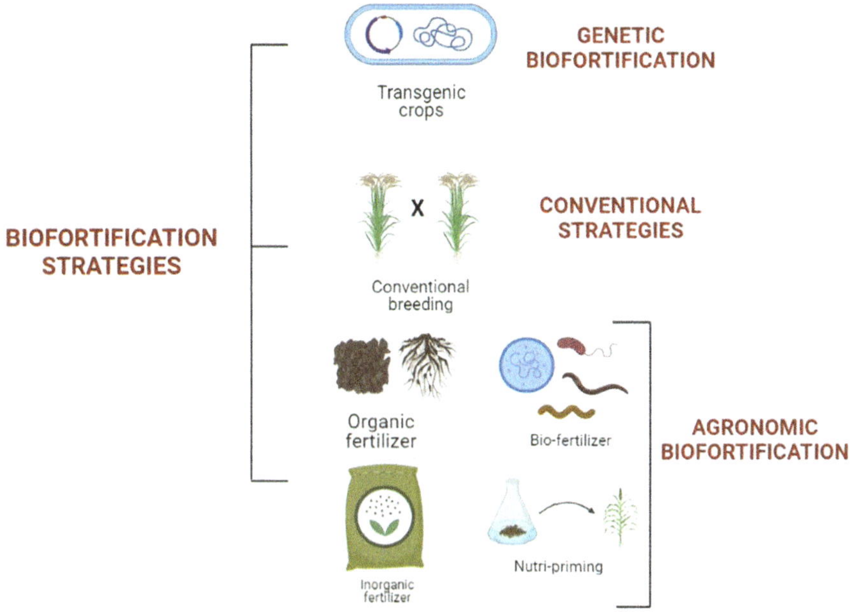

Fig. 19.2 The schematic diagram representing the three major biofortification strategies in practice, i.e., genetic/transgenic biofortification, conventional strategies, and agronomic biofortification (involving the use of organic and inorganic fertilizers, bio-fertilizers, and nutripriming methods)

biofortification, the fortificants (nutrients to be enriched with) are added to the crops during the production phase, that is at the time the crops are grown in lands. On the other hand, other existing fortification techniques add fortificants during the post-harvest processing phase of the crops. This is one of the major advantages of biofortification over other methods (Yadav et al. 2020). Therefore, over the years several strategies have been attempted for biofortification of different plants for societal benefits (Table 19.4).

19.5.1 Different Approaches Adapted for Plant Biofortification

1. **Genetic biofortification** relies on the genetic selection of superior crop varieties or creating transgenics capable of producing such nutritionally enriched food crops. This approach is mainly used in the case of crops that do not have much genetic diversity among their different varieties (Zhu et al. 2007). A great benefit of transgenic biofortification is that the gene(s) of interest, once identified or prepared, can be further used in case of multiple crop fortification (Garg et al. 2018). Transgenic biofortification is used to increase the desired nutrient concentration in a crop, enhance its bioavailability, and reduce the concentration of anti-nutrients (such as phytate, tannins, oxalate, etc.) that hinder the bioavailability of the desired micronutrient to be fortified (Welch and Graham 2004). Transgenic

Table 19.4 The tabular representation of the list of various crops, their types, the nutrients biofortified, and the various strategy (or strategies) used to biofortify them

Name of crops	Type of crop	Type of biofortification (nutrient biofortified) and the strategy(ies)
Pea	Legumes/ Pulses	**By agronomic biofortification**: Zinc
Grapes	Fruit	**By conventional breeding:** Antioxidants
Millets	Cereal	**By conventional breeding:** Iron and zinc (pearl millets)
Mango	Fruit	**By conventional breeding:** Beta-carotene, vitamin C
Apple	Fruit	**By genetic biofortification:** Stilbenes (natural phenolic defense compounds)
Linseed/ flax	Oil seed	**By genetic biofortification:** Flavonoids, PUFAs, carotenoids, essential amino acids
Alfalfa	Fodder	**By genetic biofortification:** Iso-flavonoids, methionine, phytase
Banana	Fruit	**By conventional breeding:** Provitamin A **By genetic biofortification:** Beta-carotene
Cauliflower	Vegetable	**By conventional breeding:** Beta-carotene **By genetic biofortification:** Beta-carotene
Cassava	Vegetable	**By conventional breeding:** Vitamin A, iron, carotene **By genetic biofortification:** Beta-carotene, provitamin A, iron, proteins
Carrot	Vegetable	**By agronomic biofortification**: Iodine, selenium **By genetic biofortification:** Calcium
Mustard	Oil seed	**By agronomic biofortification**: Selenium **By genetic biofortification:** γ-linolenic acid
Barley	Cereal	**By agronomic biofortification**: Iron, zinc **By genetic biofortification:** Zinc, phytase, lysine, beta glucan, resistant starch, PUFAs, human lactoferrin
Soybean	Legumes/ pulses	**By agronomic biofortification**: Selenium **By genetic biofortification:** Beta-carotene, vitamin E, cysteine, methionine, linoleic acid, linolenic acid, stearidonic acid, oleic acid, arachidonic acid, flavonoids, ω-3 fatty acids
Lettuce	Vegetable	**By agronomic biofortification**: Iodine, selenium **By genetic biofortification:** Iron
Tomato	Fruit	**By conventional breeding:** Anthocyanin (deep purple tomatoes under the name of Sun Black and Black Galaxy) **By agronomic biofortification**: Iodine **By genetic biofortification:** Folate, lycopene, beta-carotene, phytoene, carotenoids, provitamin A, ascorbate, antioxidant anthocyanins
Potato	Vegetable	**By conventional breeding:** Zinc, iron, antioxidants, copper, manganese **By agronomic biofortification**: Zinc, selenium **By genetic biofortification:** Beta-carotene, zeaxanthin, ascorbate, amino acids (methionine), cyclodextrins, anthocyanins, phenolic acid, fructan, inulin, enhanced amylopectin content in starch grains
Sweet potato	Vegetable	**By conventional breeding:** Vitamin A (orange sweet potato), beta-amylase

(continued)

Table 19.4 (continued)

Name of crops	Type of crop	Type of biofortification (nutrient biofortified) and the strategy(ies)
		By agronomic biofortification: Beta-carotene **By genetic biofortification:** Beta-carotene, antioxidants
Sorghum	Cereal	**By conventional breeding:** Zinc, iron, beta-carotene **By agronomic biofortification**: nutrient uptake, nutrient and metabolic profile, increased grain yield **By genetic biofortification:** Provitamin A, lysine
Wheat	Cereal	**By conventional breeding:** Zinc, iron, carotene, lutein, anthocyanins (colored wheat) **By agronomic biofortification**: Iron, zinc, selenium **By genetic biofortification:** Provitamin A, carotenoids, iron, phytic acid, amino acids, anthocyanin, amylose content
Rice	Cereal	**By conventional breeding:** Zinc, iron **By agronomic biofortification**: Iron, zinc, selenium **By genetic biofortification**: Phytoene (precursor of beta-carotene), folate, iron, phytic acid, zinc, alpha-linolenic acid, flavonoids and antioxidants, resistant starch, human lactoferrin
Maize	Cereal	**By conventional breeding:** Vitamin A (orange maize), lysine and tryptophan (quality protein maize), provitamin A, carotenoids, vitamin E, phenolic compounds, anthocyanins, fatty acids **By agronomic biofortification**: Zinc, selenium **By genetic biofortification:** Provitamin A, carotenoids, multivitamins, phytase, ferritin, amino acids (lysine, tryptoplan, methionine), human lactoferrin
Common bean	Legumes/ pulses	**By conventional breeding:** Zinc, iron **By agronomic biofortification**: Zinc **By genetic biofortification:** Methionine

approaches have been able to successfully fortify crops with nutrients such as vitamins, minerals, essential fatty acids, amino acids, etc. However, transgenic biofortification faces significant challenges due to its low acceptance rate among farmers and other common people. The regulatory rules and criteria for the release of biofortified crops differ among countries, which is the other major limitation. The transgenic method generally targets different genes responsible for the transport and accumulation of the desired micronutrients across the plant. However, it must be ensured that the nutrient enhancement takes place in the edible parts of the plant only and not in any other part (Garg et al. 2018).

2. **Conventional strategies** of biofortification rely on various crop breeding programs and are known to be the most widely accepted approach for the nutritional enrichment of various crops. This method is a good alternative to the transgenic approach if sufficient genetic diversity exists among the varieties of a crop to be fortified. Parent lines containing enriched nutritional levels are cross-bred with the recipient lines that contain the desired agronomic traits. Genetic variation is usually limited within the gene pool of the parent and recipient varieties, which can sometimes be increased by crossing distantly related plant

varieties (Garg et al. 2018). Using conventional breeding techniques, zinc- and iron-fortified varieties of rice, namely, BRRIdhan 62, BRRIdhan 64, and BRRIdhan 74, have been released in Bangladesh. The PBW1Zn variety of wheat fortified with Zn by the Punjab Agricultural University has been released in India. Other zinc-fortified varieties of wheat under the HarvestPlus program, named BHU 1, BHU 3, BHU 5, BHU 6, BHU 17, and BHU 18, have been released in India. Black-grained wheat fortified with anthocyanin pigments has been released for use in China (Havrlentová et al. 2014). Iron- and zinc-fortified millet varieties Barimasur-4, Barimasur-5, Barimasur-6, Barimasur-7, and Barimasur-8 (in Bangladesh); Khjura-1, Khajura-2, Shital, Shekhar, and Simal (in Nepal); and L4704 and Pusa Vaibhav (in India) have been produced by the International Center for Agricultural Research in the Dry Areas ICARDA and HarvestPlus programs (Govindaraj et al. 2019).

3. **Agronomic biofortification** employs fertilizers rich in the required micronutrient. These fertilizers can be organic, inorganic, or biologically derived. These nutrient-rich fertilizers can be applied either to the soil where the crop grows or can be directly sprayed to the leaves of the crop (foliar application) (de Valença et al. 2017). Foliar application of nutrients is a better approach as compared to the soil application method, as there is a risk of nutrient immobilization in the soil, which is not present in the foliar method. It ensures more efficient and rapid uptake of the nutrients by the edible parts of the plant since they have to be translocated for comparatively less distance (leaf to grain as compared to the soil to grain) (Lawson et al. 2015). However, in the case of foliar application, there is a chance of the nutrients being easily washed away due to rain. Certain studies have also shown that a combinatorial application of soil and foliar techniques can provide better results (Phattarakul et al. 2012). Nowadays, fertilizers rich in nitrogen, phosphorus, and potassium compounds (NPK fertilizers) are in great use, owing to their increasing success in enriching nutrient bioavailability and yield of grains (Prasad et al. 2014). Nutri-priming methods are also being practiced, where the seeds are soaked in nutrient-rich media, before plantation in the soil. This enhances seed germination, induces faster growth of the plant, gives better yield, and also increases the nutrient content in the crops. Further, some studies report that seed priming helps plants to combat various abiotic stress factors such as salinity stress (Johnson and Puthur 2021).

19.5.2 Application of Genetic Engineering in Biofortification

The major advantage of employing genetic biofortification strategies is that it can utilize an unlimited pool of genetic characters that can be transferred and expressed (overexpressed or knocked out) in the desirable plant with enhanced nutritional properties.

19.5.2.1 Mineral Biofortification in Plants

Iron-Fortified Crops

Altered expression of transporters *AtNRAMP3* and *AtNRAMP4* in *Arabidopsis* was found to increase the storage of iron in the vacuoles, which further led to enhanced iron content in the seeds (Lanquar et al. 2005). Overexpression of the barley nicotianamine synthase gene *HvNAS1* caused as much as a threefold increase in the iron content of the transgenic rice plants created. This was due to the overproduction of nicotinamine in the roots, shoots, and seeds, which facilitated greater translocation of Fe into the rice grains (Masuda et al. 2009). The constitutive expression of the NAS1 gene in *Arabidopsis* (*AtNAS1*) along with endosperm-specific ferritin expression showed a considerable increase in Fe concentration (Wirth et al. 2009). *OsYSL2*, a metal-nicotinamine transporter in rice, is involved in the transport of iron over long distances. The transgenic created by the expression of the *OsYSL2* gene showed increased iron efflux into rice grains (Ishimaru et al. 2010). Further studies reported that the introduction of multiple genes responsible for iron-homeostasis showed a greater impact on the increase in iron content in the transgenic rice plants as compared to the introduction of a single iron-responsive gene. A transgenic was created by targeting three iron fortification approaches. Firstly, iron accumulation in the rice seeds was increased by the expression of a Fe-storage protein Ferritin via endosperm-specific promoters. Secondly, overproduction of a metal chelator nicotinamine was used to increase Fe translocation, and thirdly, the expression of the Fe(II)-NA transporter *OsYSL2* under the control of an endosperm-specific promoter and sucrose transporter promoter was regulated to increase Fe flux into the endosperm. The cumulative effect of all three genetic engineering approaches showed ~6-fold enhancement in the iron content of the rice seeds (Masuda et al. 2012). The stress-responsive gene in rice, *OsNAC5*, was found to be involved in metal re-mobilization. Hence, overexpression of the gene could lead to the enhanced iron content in rice grains (Ricachenevsky et al. 2013). A study reported that under the control of an endosperm-specific promoter, the overexpression of a vacuolar iron transporter gene in wheat, *TaVIT2*, showed a twofold increase in the Fe content (Connorton et al. 2017). Another study mentioned that the overexpression of the *AtZIP1* gene in roots led to a significant enhancement in iron content in barley (Ramesh et al. 2004).

Zinc-Fortified Crops

Overexpression of the exogenous barley gene, *HvNAS1*, in *Arabidopsis* and tobacco plants led to a significant increase in the zinc content in the seeds (Kim et al. 2005). The Zn responsive nature of NAS can be understood by the fact that genes that code for NAS are differentially regulated in response to metals such as Fe and Zn in plants such as *Arabidopsis* (Mizuno et al. 2003). Also, the overexpression of the *HvNAS1* gene in rice has shown ~1.5-fold enhancement in the Zn content in the transgenic rice grains (Masuda et al. 2009). The constitutive overexpression of the three genes, *OsNAS1*, *OsNAS2*, and *OsNAS3*, showed increased Fe and Zn content in rice endosperm, out of which *OsNAS2* caused the greatest enhancement (Johnson et al.

2011). A member of the NAC transcription factor family, *NAM-B1*, has been reported to increase the zinc content in wheat grains by ~12% (Distelfeld et al. 2007).

Iodine-Fortified Crops

Although not much data exists about the transporters responsible for the transport and accumulation of iodine in the plant body, certain studies report their role in biofortification. A study reported the involvement of halide ion methyltransferase (*HMT*) and halide/thiol methyltransferase (*HTMT*) in the volatilization of iodine in the form of methyl halides (mainly methyl iodide) in plants such as wheat, paddy rice, and daikon radish (Itoh et al. 2009). Another study conducted on *Arabidopsis* found that the iodine content in *Arabidopsis* seeds can be enhanced by facilitating its uptake via the overexpression of the *NIS* (human sodium-iodide symporter) gene or via the reduction of its volatilization by knocking out a halide methyltransferase gene, *HOL-1*, which in turn encodes for *HMT* enzyme (Landini et al. 2012).

Selenium-Fortified Crops

A study reported that the overexpression of an *Arabidopsis* gene, *APS1*, encoding the enzyme ATP sulfurylase (responsible for selenate to selenite transition) caused increased selenate reduction in Indian mustard (*B. juncea*). The APS1 overexpression caused about a 2–2.5-fold increase in the ATP sulfurylase enzyme content in the plants. The transgenics thus created could tolerate and accumulate more selenium (~2–3-fold enhancement in Se accumulation in the form of methyl-SeCys was noticed) (Pilon-Smits et al. 1999a, b). Another study showed that the overexpression of the selenocysteine methyltransferase (*SMT*) gene in *Arabidopsis* and mustard plants showed increased selenium accumulation in the form of methyl-SeCys (LeDuc et al. 2004). Since the methyl-SeCys form of selenium is more suitable for biofortification, such genetic engineering approaches of creating APS and SMT transgenics are some of the successful attempts of selenium biofortification of crops (Malagoli et al. 2015). Further studies have shown that double-transgenic mustard plants created by the overexpression of both APS and SMT caused up to nine-fold increase in Se accumulation (in the form of methyl-SeCys) as compared to wild-type plants (LeDuc et al. 2006). Also, transgenic Indian mustard plants created by the overexpression of the cystathionine-gamma-synthase (*CGS*) gene demonstrated an enhanced selenium volatilization rate, reduced shoot Se levels, and increased Se tolerance ability as compared to the wild-type plants (Huysen et al. 2004). Another study showed that the expression of the genes *HMT*, *S3H*, and *SAMT* led to an increased iodine content in the leaves and fruits of the transgenic tomato plant created (Halka et al. 2019).

19.5.2.2 Vitamin Biofortification in Plants

Vitamin A-Biofortified Plants

A study conducted on cassava (*Manihot esculenta*) showed that the expression of the phytoene synthase (*PSY*) gene causes increased accumulation of the colored provitamin A carotenoids in the plant roots. This is because the expressed *PSY* gene enhances carbon flux via carotenogenesis and thus overexpression of the *PSY* transgene produced yellow-fleshed cassava roots with enhanced provitamin A content (Welsch et al. 2010). An earlier study had reported that the expression of a *PSY* gene from daffodil, under the regulation of seed-specific promoter of glutenin, enhances the β carotene (precursor to vitamin A) content in the endosperm of transgenic rice (*japonica* variety) plants (Burkhardt et al. 1997). The biofortification of β carotene in rice grains was also performed by the insertion of three novel genes, *Erwinia uredovora* phytoene desaturase gene (*crtI*), daffodil phytoene synthase (*psy1*) gene, and lycopene β-cyclase (*β-lcy*) from daffodil gene under the control of specific promoters in the rice genome (Ye et al. 2000). The introduction of the two genes *psy* and *β-lcy* were able to enhance the β carotene content in rice grains, which led to the creation of a successful biofortified product named "Golden rice" (Beyer et al. 2002). Further, a study had demonstrated that the simultaneous introduction of the *psy* and *crtI* genes could increase the β carotene accumulation in rice plants by ~23-folds and produced another transgenic named "Golden Rice 2" (Paine et al. 2005). Experiments performed on maize (following the basic concepts of β carotene biofortification in Golden rice) showed that the overexpression of both *crtB* and *crtI* genes is necessary to enhance provitamin A content in maize grains (Aluru et al. 2008). Transformation of the maize *PSY1* gene and *CrtI* gene, under the regulation of specific promoters in wheat, showed a ~10.8-fold increase in β carotene levels in the transgenic wheat (Cong et al. 2009). Another study conducted on wheat demonstrated ~8-fold enhancement in β carotene content in the transgenic wheat seeds by the co-expression of the bacterial *CrtB* and *CrtI* genes (Wang et al. 2014). The expression of the *Erwinia uredovora crtB* gene encoding phytoene synthase, in the tubers of the potato, showed a ~6.25-fold increase in carotenoid accumulation in the tubers of the transgenic potato plants. However, despite the increase in carotenoid accumulation levels, no considerable upregulation of significant carotenoid biosynthetic genes was noticed in the transgenic potato tubers (Ducreux et al. 2005). Several studies in potato have reported that the knock-out of certain genes such as the lycopene ε-cyclase (*ε-LCY*) gene and the non-heme β-carotene hydroxylase (*CHY*) gene causes a significant increase in the β carotene content of the transgenic potato tubers (Diretto et al. 2006, 2007).

Vitamin B-Biofortified Plants

A study has shown that the overexpression of the genes *PDX1* and *PDX2* in *Arabidopsis* caused the vitamin B6 content of the transgenic *Arabidopsis* seeds to double (Chen and Xiong 2009). Enhanced content of folic acid in tomatoes can be caused by the overexpression of GTP cyclohydrolase I specifically in the fruit (de la Garza et al. 2004). Researchers have demonstrated an approximately 100-fold increase in vitamin B9 content in rice seeds by the overexpression of two *Arabidopsis thaliana* genes, *Glb-1* and *GluB1*, belonging to the pterin and para-aminobenzoate branches of the folate biosynthetic pathway (Storozhenko et al. 2007).

Vitamin C-Biofortified Plants

Several studies have shown that the overexpression of the GDP-l-galactose phosphorylase (GGP or VTC2) gene, under the regulation of specific promoter, leads to enhanced ascorbate (vitamin C) content in plants such as tomato, strawberry, and potato (Bulley et al. 2012). Again, another study reported that when the GalUR gene (codes for NADPH-dependent D-galacturonate reductase) of strawberry was overexpressed in *Arabidopsis*, it led to a two- to three-fold increase in the vitamin C content of the transgenic *Arabidopsis* seeds (Agius et al. 2003). A transgenic corn created by the expression of rice dehydroascorbate reductase (dhar) under the control of the barley D-hordein promoter showed up to sixfold enhanced vitamin C content (Naqvi et al. 2009).

Vitamin E-Biofortified Plants

Multiple research studies have reported that the overexpression of the genes, *HGGT* (homogentisate geranylgeranyl transferase) and *HPT* (homogentisate phytyltransferase) encoding for the respective enzymes, leads to enhanced vitamin E content in plants such as *Arabidopsis*, tobacco, and corn (Dolde and Wang 2011; Tanaka et al. 2015; Yang et al. 2011).

19.5.3 Challenges of Biofortification Methods

Firstly, the nutritional enrichment should be in the edible parts of the plant. For instance, in the case of crops such as rice and wheat, the only consumed part is the food grain. Hence, the fortification strategies should aim at increasing the nutrient content of the endosperm. Only then will the process be successful in its aim to deliver the increased nutrient content to humans consuming the crops. Secondly, apart from achieving high nutrient density in the crops, the cost-effectiveness, production rates, and accessibility of the methods have to be assessed. Thirdly, sufficient clinical trials need to be held to validate a certain fortified crop to assure that it shows improved nutrition status in humans consuming them. Fourthly, the products of biofortification should be cheap enough for large masses of the population to adopt them, so that it can bring a global impact on the nutritional status of people across the world (Bouis and Welch 2010). Thus, the success of biofortification techniques on a large scale majorly depends on the acceptance of the method and the fortified crops by the producers and consumers, meaning the farmers and the common people (Singh et al. 2016a, b).

19.6 Utilizing Omics Technologies to Accelerate Phytoremediation and Biofortification

The accumulation process of nutrients in leaves, seeds, and roots is highly regulated by a subset of genes that controls the concentration of minerals in the whole plant body. Advancement in "omics" technologies and gene analysis machineries has given a boost in the identification of these genes at the molecular level so as to assess their functionality in different conditions (Conn et al. 2012). Germplasm screening is one such method that identifies variations in the germplasm among and within a given plant species. This can lead to the identification of nutrient homeostasis responsive gene discovery (Karley and White 2009). In the presence of different nutrients and heavy metal(loid)s, the expression of responsive (differentially expressed) genes is regulated. Detection of these changes requires a long list of procedures consisting of the study of naturally grown population analysis or laboratory-made mutant plants, followed by the identification of individual genes and their functions. Pioneering development in transcriptomics, the study of the whole set of RNA transcripts that are synthesized from the genome, has accelerated the process of gene discovery and their expression pattern analysis in the presence of different nutrients and heavy metals (Carvalho and Vasconcelos 2013). High-throughput RNA sequencing has provided a non-targeted, transcriptome-wide expression analysis of all the genes, at a given time in the presence of one or more than one nutrient (Carvalho and Vasconcelos 2013). 454 pyrosequencing, microarray, SAGE (serial analysis of gene expression), SSH (suppression subtractive hybridization), and macroarray technology are the major parts of transcriptomics that helped in the identification of nutrient responsive genes in various crops such as wheat, soybean, rice, barley, etc. (Lai et al. 2012; Narayanan et al. 2007; O'Rourke et al. 2009; Sperotto et al. 2009; White et al. 2006).

Bioactive food compounds and nutrients can regulate the expression of genes. Associating the genomic information with high-throughput omics technologies helped in unraveling the nutrient-gene interactions related to genotypic characteristics (Riaz et al. 2020). The involvement of proteomic analysis in the identification and abundance of proteins has helped in understanding their contributions to plant nutrient homeostasis. Genome and transcriptome influence the regulation of protein levels in the plant body. Hence, fluctuations in protein expressions in the presence of different minerals can easily be monitored using various proteomic techniques such as isobaric tags for relative and absolute quantification (iTRAQ), matrix-assisted laser desorption/ionization time-of-flight mass spectrometry (MALDI-TOF/TOF MS), and two-dimensional differential gel electrophoresis (2D-DIGE). A total of 521 microsomal zinc-responsive proteins (including membrane proteins and transporters) have been discovered in *Arabidopsis* roots using iTRAQ (Fukao et al. 2011). Proteomic studies majorly identify enzymes and transporters involved in mineral uptake and translocation. Out of 92 identified elements, 17 are essential for the optimal growth of a plant (Karley and White 2009). Ionomic studies help to identify these essential elements and their involvements in genetic and environmental responses (Williams and Salt 2009). Additionally, extraction and detection of plant metabolites can also help in the better understanding of various mechanisms involved in the regulation of relevant nutritional metabolites in our food (Hall et al. 2008). Along with other omics technologies (genomics, transcriptomics, nutrigenomics, proteomics, etc.), metabolomics techniques such as gas chromatography-mass spectrometry (GCMS), capillary electrophoresis (CE), nuclear magnetic resonance (NMR), and high-performance liquid chromatography (HLPC) help in the detection of diverse plant metabolites and their presence in a different range of concentrations (Carvalho and Vasconcelos 2013; De Vos et al. 2007). Taken together, the genomic, transcriptomic, proteomic, nutrigenomic, and metabolomic analyses have provided a database of various genes and transporters which are involved in the transport and accumulation of one or more nutrients that maintain the mineral homeostasis in plants (Winkel et al. 2012); further exploitation of these valuable genetic elements and dataset can help us devise novel strategies to manipulate the metal(loid) homeostasis for efficient phytoremediation and/or biofortification applications.

19.7 Conclusion and Future Prospects

Careful analysis of the approaches of phytoremediation and biofortification clearly indicates these two to be contrasting yet similar approaches of manipulating the metal homeostasis in cells for differing outcomes. However, it can be said that for practical applications, a combination of phytoremediation and biofortification will be a giant leap toward the large-scale production of nutritious and safe food sustainably. It is usually seen that the difference in the maximum safe limit and toxic concentration of a micronutrient is very less. Plant biomass produced after phytoremediation that is rich in certain micronutrients (such as selenium, zinc, iron, etc.) can be used to produce nutritionally enriched food products. Hence, the two techniques—phytoremediation and biofortification—can be considered to be the two sides of a coin and can be successfully integrated for both environmental and agricultural benefits (Yin et al. 2012). The two methods can be closely related because both of them rely on the same basic processes such as metal uptake from the soil, their accumulation, and translocation. Both aim for the transfer of metal (loid)s from soil to plant (further to humans in biofortification).

However, a correct choice has to be made regarding which plants will be selected for both phytoremediation and biofortification. Only such plants can be chosen which accumulate the nutrients in their edible parts, that too in safe concentrations. For instance, hyperaccumulator species such as *Thlaspi caerulescens* (zinc accumulator) and *Stanleya pinnata* (selenium accumulator) have increased efficiency of their respective metal uptake and accumulation in the edible parts of the plant (Cappa and Pilon-Smits 2014). Thus, they can be used as perfect candidates for an integrated phytoremediation and biofortification approach. Again it has to be taken care of that the edible portion of the plant should not store too much toxic metal(loid)s, which will then defeat the entire process of phytoremediation and biofortification. This gives rise to a major problem that the plants used for remediation of heavy metal (loid)-contaminated lands can in no way be directly utilized for the production of biofortified food crops (Yin et al. 2012).

Therefore, a possible approach to integrate phytoremediation and biofortification is by the development of transgenic plants by altering the structure and expression of different plant metal(loid) transporters and metal(loid) responsive genes (Fig. 19.3). The genetically modified "super remediator plant" thus created will be capable of

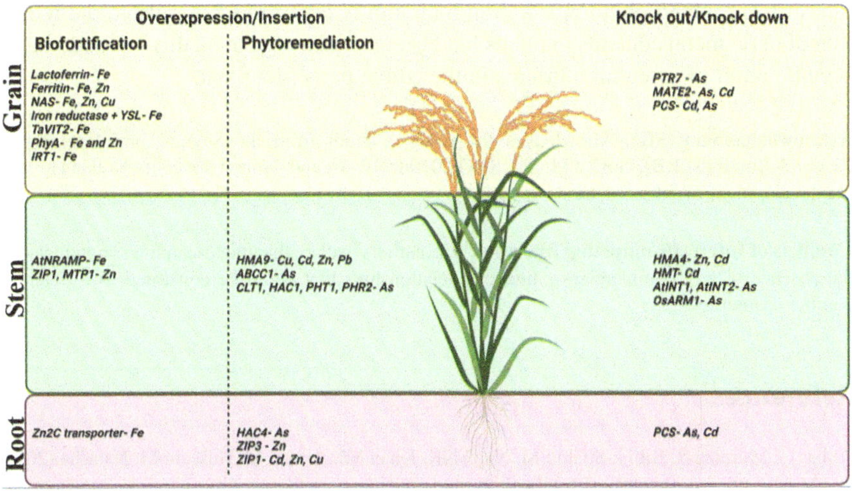

Fig. 19.3 The schematic diagram depicts a hypothetical model of a genetically modified "super remediator" plant for simultaneous phytoremediation of harmful heavy metal(loid)s in non-edible parts, along with selective biofortification of essential nutrients into the edible parts

tissue-specific remediation along with simultaneous selective fortification of essential metals. An ideal transgenic as hypothesized must have the ability to significantly uptake more than one metal or metalloid from the soil, out of which at least one should be an essential micronutrient and one, a toxic heavy metal(loid). The purpose of this transgenic will be to limit the translocation and accumulation of the heavy metal(loid) to the plant roots, while the essential nutrient will be stored in higher concentrations in the upper parts of the plants, such as the crops (edible part). This can be achieved by the manipulation of different metal transporters and genes at specific tissue of the plant. For instance, arsenic translocation from root to shoot can be stopped by knocking out the *HMA4* transporters present in the plant. Besides that, overexpression of *HMA9*, *ABCC1*, and *PHT1* transporters in the shoot region can enhance the accumulation of non-essential toxic metal(loid)s such as Cu, Cd, Pb, As, etc. (Zhang et al. 2018a, b). Selective knockout of *PCS* genes in the root can inhibit the translocation of carcinogenic heavy metal(loid)s such as Cd and As. This will be the first level of defense to reduce heavy metal(loid) concentration in the plant body. At the same time, the essential nutrient transporters in the root (such as *CAX1*, *Zn2C*, etc.) should be overexpressed to effectively translocate more of the nutrient to the shoot. In the shoot, genes responsible (*NRAMP1*, *ZIP1*, *MTP1*, etc.) for the remobilization of the essential nutrients to the grain are differentially expressed to ensure nutrient accumulation in the edible part of the plant (Liu et al. 2019). Along with that, the overexpression of nutrient binding genes (*ferritin*, *lactoferrin*, etc.), transporters (*IRT1*, *TaVIT2*, etc.), and other metal responsive genes (*PhyA*, *Iron reductase*, etc.) can help in the enhancement of grain unloading of essential nutrients in plant body (Malik and Maqbool 2020). In this way, it can be hypothesized that biofortification and phytoremediation may jointly contribute not only to the reduction of toxic metal contents in plants but also in enhancing the quality of food via the accumulation of essential nutrients in the edible parts of a plant.

Acknowledgement SB acknowledges the research grant from the Science and Engineering Research Board (SERB), Govt. of India (SRG/2019/001247), and Amity University Kolkata toward the development of infrastructural support.

Conflicts of Interest/Competing Interests The authors declare that the research was conducted in the absence of any commercial or financial relationships that could be construed as a potential conflict of interest.

References

Abbas G, Murtaza B, Bibi I, Shahid M, Niazi NK, Khan MI, Amjad M, Hussain M, Natasha (2018) Arsenic uptake, toxicity, detoxification, and speciation in plants: physiological, biochemical, and molecular aspects. Int J Environ Res Public Health 15(1):59

Aboudrar W, Schwartz C, Morel JL, Boularbah A (2013) Effect of nickel-resistant rhizosphere bacteria on the uptake of nickel by the hyperaccumulator Noccaea caerulescens under controlled conditions. J Soils Sediments 13(3):501–507

Agius F, González-Lamothe R, Caballero JL, Muñoz-Blanco J, Botella MA, Valpuesta V (2003) Engineering increased vitamin C levels in plants by overexpression of a D-galacturonic acid reductase. Nat Biotechnol 21(2):177–181

Alejandro S, Höller S, Meier B, Peiter E (2020) Manganese in plants: from acquisition to subcellular allocation. Front Plant Sci 11:300

Ali H, Khan E, Sajad MA (2013) Phytoremediation of heavy metals—concepts and applications. Chemosphere 91(7):869–881

Aluru M, Xu Y, Guo R, Wang Z, Li S, White W, Wang K, Rodermel S (2008) Generation of transgenic maize with enhanced provitamin A content. J Exp Bot 59(13):3551–3562

Alvarenga P, Gonçalves AP, Fernandes RM, de Varennes A, Vallini G, Duarte E, Cunha-Queda AC (2009) Organic residues as immobilizing agents in aided phytostabilization: (I) Effects on soil chemical characteristics. Chemosphere 74(10):1292–1300

Atieh MA, Ji Y, Kochkodan V (2017) Metals in the environment: toxic metals removal. Bioinorg Chem Appl 2017:1–2

Babaeian E, Homaee M, Rahnemaie R (2016) Chelate-enhanced phytoextraction and phytostabilization of lead-contaminated soils by carrot (Daucus carota). Arch Agron Soil Sci 62(3):339–358

Bagga DK, Peterson S (2001) Phytoremediation of arsenic-contaminated soil as affected by the chelating agent CDTA and different levels of soil pH. Remediat J 12(1):77–85

Baker AJM, McGrath SP, Sidoli CMD, Reeves RD (1994) The possibility of in situ heavy metal decontamination of polluted soils using crops of metal-accumulating plants. Resour Conserv Recycl 11(1–4):41–49

Bashir K, Seki M, Nishizawa NK (2019) The transport of essential micronutrients in rice. Mol Breed 39(12):168

Bassi D, Menossi M, Mattiello L (2018) Nitrogen supply influences photosynthesis establishment along the sugarcane leaf. Sci Rep 8(1):2327

Berken A, Mulholland MM, LeDuc DL, Terry N (2002) Genetic engineering of plants to enhance selenium phytoremediation. Crit Rev Plant Sci 21(6):567–582

Berti WR, Cunningham SD (2000) Phytostabilization of metals. In: Phytoremediation of toxic metals: using plants to clean-up the environment. Wiley, New York. pp 71–88

Beyer P, Al-Babili S, Ye X, Lucca P, Schaub P, Welsch R, Potrykus I (2002) Golden rice: introducing the beta-carotene biosynthesis pathway into rice endosperm by genetic engineering to defeat vitamin A deficiency. J Nutr 132(3):506S–510S
Biswas JK, Warke M, Datta R, Sarkar D (2020) Is arsenic in rice a major human health concern? Curr Pollut Rep 6(2):37–42
Blaylock MJ, Salt DE, Dushenkov S, Zakharova O, Gussman C, Kapulnik Y, Ensley BD, Raskin I (1997) Enhanced accumulation of Pb in Indian mustard by soil-applied chelating agents. Environ Sci Technol 31(3):860–865
Bleeker PM, Hakvoort HWJ, Bliek M, Souer E, Schat H (2006) Enhanced arsenate reduction by a CDC25-like tyrosine phosphatase explains increased phytochelatin accumulation in arsenate-tolerant Holcus lanatus. Plant J 45(6):917–929
Bouis H (1996) Enrichment of food staples through plant breeding: a new strategy for fighting micronutrient malnutrition. Nutr Rev 54(5):131–137
Bouis HE (2002) Plant breeding: a new tool for fighting micronutrient malnutrition. J Nutr 132(3): 491–494
Bouis HE, Welch RM (2010) Biofortification—a sustainable agricultural strategy for reducing micronutrient malnutrition in the global south. Crop Sci 50:S20–S32
Bridgwater AV, Meier D, Radlein D (1999) An overview of fast pyrolysis of biomass. Org Geochem 30(12):1479–1493
Brooks RR, Chambers MF, Nicks LJ, Robinson BH (1998) Phytomining. Trends Plant Sci 1:359–362
Bulak P, Walkiewicz A, Brzezińska M (2014) Plant growth regulators-assisted phytoextraction. Biol Plant 58(1):1–8
Bulley S, Wright M, Rommens C, Yan H, Rassam M, Lin-Wang K, Andre C, Brewster D, Karunairetnam S, Allan AC, Laing WA (2012) Enhancing ascorbate in fruits and tubers through over-expression of the L-galactose pathway gene GDP-L-galactose phosphorylase. Plant Biotechnol J 10(4):390–397
Burges A, Alkorta I, Epelde L, Garbisu C (2018) From phytoremediation of soil contaminants to phytomanagement of ecosystem services in metal contaminated sites. Int J Phytoremediation 20(4):384–397
Burkhardt PK, Beyer P, Wünn J, Klöti A, Armstrong GA, Schledz M, von Lintig J, Potrykus I (1997) Transgenic rice (Oryza sativa) endosperm expressing daffodil (*Narcissus pseudonarcissus*) phytoene synthase accumulates phytoene, a key intermediate of provitamin A biosynthesis. Plant J 11(5):1071–1078
Campos-Bowers MH, Wittenmyer BF (2007) Biofortification in China: policy and practice. Health Res Policy Syst 5(1):10
Cappa JJ, Pilon-Smits EAH (2014) Evolutionary aspects of elemental hyperaccumulation. Planta 239(2):267–275
Carvalho SMP, Vasconcelos MW (2013) Producing more with less: strategies and novel technologies for plant-based food biofortification. Food Res Int 54(1):961–971
Chaney RL (1983) Plant uptake of inorganic waste constituents. Land treatment of hazardous wastes. Noyes Data Corporation, Park Ridge, pp 50–76
Chaney RL, Malik M, Li YM, Brown SL, Brewer EP, Angle JS, Baker AJ (1997) Phytoremediation of soil metals. Curr Opin Biotechnol 8(3):279–284
Chanu BL, Gupta A (2016) Phytoremediation of lead using Ipomoea aquatica Forsk. in hydroponic solution. Chemosphere 156:407–411
Chaturvedi R, Varun M, Paul MS (2016) Phytoremediation: uptake and role of metal transporters in some members of Brassicaceae. In: Phytoremediation. Springer, Cham. Chapter 16. pp 453–468
Chen H, Xiong L (2009) Enhancement of vitamin B(6) levels in seeds through metabolic engineering. Plant Biotechnol J 7(7):673–681
Clemens S (2001) Molecular mechanisms of plant metal tolerance and homeostasis. Planta 212(4): 475–486

Clemens S, Ma JF (2016) Toxic heavy metal and metalloid accumulation in crop plants and foods. Annu Rev Plant Biol 67:489–512
Cong L, Wang C, Chen L, Liu H, Yang G, He G (2009) Expression of phytoene synthase1 and carotene desaturase crtI genes result in an increase in the total carotenoids content in transgenic elite wheat (*Triticum aestivum* L.). J Agric Food Chem 57(18):8652–8660
Conn SJ, Berninger P, Broadley MR, Gilliham M (2012) Exploiting natural variation to uncover candidate genes that control element accumulation in Arabidopsis thaliana. New Phytol 193(4): 859–866
Connorton JM, Jones ER, Rodríguez-Ramiro I, Fairweather-Tait S, Uauy C, Balk J (2017) Wheat vacuolar iron transporter TaVIT2 transports Fe and Mn and is effective for biofortification. Plant Physiol 174(4):2434–2444
Cormick G, Belizán JM (2019) Calcium intake and health. Nutrients 11(7):1606
Czakó M, Feng X, He Y, Liang D, Márton L (2006) Transgenic Spartina alterniflora for phytoremediation. Environ Geochem Health 28(1–2):103–110
DalCorso G, Fasani E, Manara A, Visioli G, Furini A (2019) Heavy metal pollutions: State of the art and innovation in phytoremediation. Int J Mol Sci 20(14):3412
Das N, Bhattacharya S, Maiti MK (2016) Enhanced cadmium accumulation and tolerance in transgenic tobacco overexpressing rice metal tolerance protein gene *OsMTP1* is promising for phytoremediation. Plant Physiol Biochem 105:297–309
Das N, Bhattacharya S, Bhattacharyya S, Maiti MK (2018) Expression of rice MATE family transporter OsMATE2 modulates arsenic accumulation in tobacco and rice. Plant Mol Biol 98(1–2):101–120
de la Garza RD, Quinlivan EP, Klaus SM, Basset GJ, Gregory JF III, Hanson AD (2004) Folate biofortification in tomatoes by engineering the pteridine branch of folate synthesis. Proc Natl Acad Sci U S A 101(38):13720–13725
de Valença AW, Bake A, Brouwer ID, Giller KE (2017) Agronomic biofortification of crops to fight hidden hunger in sub-Saharan Africa. Glob Food Sec 12:8–14
De Vos RC, Moco S, Lommen A, Keurentjes JJ, Bino RJ, Hall RD (2007) Untargeted large-scale plant metabolomics using liquid chromatography coupled to mass spectrometry. Nat Protoc 2(4):778–791
De-la-Peña C, Loyola-Vargas VM (2014) Biotic interactions in the rhizosphere: a diverse cooperative enterprise for plant productivity. Plant Physiol 166(2):701–719
Dhankher OP, Rosen BP, McKinney EC, Meagher RB (2006) Hyperaccumulation of arsenic in the shoots of Arabidopsis silenced for arsenate reductase (ACR2). Proc Natl Acad Sci U S A 103(14):5413–5418
Dhanwal P, Kumar A, Dudeja S, Chhokar V, Beniwal V (2017) Recent advances in phytoremediation technology. In: Advances in environmental biotechnology. Springer, Singapore. pp 227–241
Diels L, Spaans PH, Van Roy S, Hooyberghs L, Ryngaert A, Wouters H, Walter E, Winters J, Macaskie L, Finlay J, Pernfuss B, Woebking H, Pümpel T, Tsezos M (2003) Heavy metals removal by sand filters inoculated with metal sorbing and precipitating bacteria. Hydrometallurgy 71(1–2):235–241
Diretto G, Tavazza R, Welsch R, Pizzichini D, Mourgues F, Papacchioli V, Beyer P, Giuliano G (2006) Metabolic engineering of potato tuber carotenoids through tuber-specific silencing of lycopene epsilon cyclase. BMC Plant Biol 6:13
Diretto G, Welsch R, Tavazza R, Mourgues F, Pizzichini D, Beyer P, Giuliano G (2007) Silencing of beta-carotene hydroxylase increases total carotenoid and beta-carotene levels in potato tubers. BMC Plant Biol 7:11
Distelfeld A, Cakmak I, Peleg Z, Ozturk L, Yazici AM, Budak H, Saranga Y, Fahima T (2007) Multiple QTL-effects of wheat Gpc-B1 locus on grain protein and micronutrient concentrations. Physiol Plant 129(3):635–643
Dobos L, Puia C (2010) The most important methods for depollution of hydrocarbons polluted soils. Bull Univ Agric Sci Vet Med Cluj-Napoca Agric 66(1)

Dolde D, Wang T (2011) Oxidation of crude corn oil with and without elevated tocotrienols. J Am Oil Chem Soc 88(9):1367–1372
Ducreux LJ, Morris WL, Hedley PE, Shepherd T, Davies HV, Millam S, Taylor MA (2005) Metabolic engineering of high carotenoid potato tubers containing enhanced levels of beta-carotene and lutein. J Exp Bot 56(409):81–89
Dürešová Z, Šuňovská A, Horník M, Pipíška M, Gubišová M, Gubiš J, Hostin S (2014) Rhizofiltration potential of Arundo donax for cadmium and zinc removal from contaminated wastewater. Chem Pap 68(11):1452–1462
Dursun AY, Uslu G, Cuci Y, Aksu Z (2003) Bioaccumulation of copper(II), lead(II) and chromium (VI) by growing Aspergillus niger. Process Biochem 38(12):1647–1651
Dushenkov V, Kumar PB, Motto H, Raskin I (1995) Rhizofiltration: the use of plants to remove heavy metals from aqueous streams. Environ Sci Technol 29(5):1239–1245
Ebbs SD, Lasat MM, Brady DJ, Cornish J, Gordon R, Kochian LV (1997) Phytoextraction of cadmium and zinc from a contaminated soil. J Environ Qual 26(5):1424–1430
Ellis DR, Gumaelius L, Indriolo E, Pickering IJ, Banks JA, Salt DE (2006) A novel arsenate reductase from the arsenic hyperaccumulating fern Pteris vittata. Plant Physiol 141(4): 1544–1554
Engwa GA, Ferdinand PU, Nwalo FN, Unachukwu MN (2019) Mechanism and health effects of heavy metal toxicity in humans. Poisoning in the modern world—new tricks for an old dog? IntechOpen
Epelde L, Hernández-Allica J, Becerril JM, Blanco F, Garbisu C (2008) Effects of chelates on plants and soil microbial community: comparison of EDTA and EDDS for lead phytoextraction. Sci Total Environ 401(1–3):21–28
Epelde L, Becerril JM, Mijangos I, Garbisu C (2009) Evaluation of the efficiency of a phytostabilization process with biological indicators of soil health. J Environ Qual 38(5): 2041–2049
Fasani E, Manara A, Martini F, Furini A, DalCorso G (2018) The potential of genetic engineering of plants for the remediation of soils contaminated with heavy metals: transgenic plants for phytoremediation. Plant Cell Environ 41(5):1201–1232
Fraissler G, Jöller M, Mattenberger H, Brunner T, Obernberger I (2009) Thermodynamic equilibrium calculations concerning the removal of heavy metals from sewage sludge ash by chlorination. Chem Eng Process Process Intensif 48(1):152–164
Fukao Y, Ferjani A, Tomioka R, Nagasaki N, Kurata R, Nishimori Y, Fujiwara M, Maeshima M (2011) iTRAQ analysis reveals mechanisms of growth defects due to excess zinc in Arabidopsis. Plant Physiol 155(4):1893–1907
Gadd GM (2004) Microbial influence on metal mobility and application for bioremediation. Geoderma 122(2–4):109–119
Gao Y, Yang F, Liu J, Xie W, Zhang L, Chen Z, Peng Z, Ou Y, Yao Y (2020) Genome-wide identification of metal tolerance protein genes in Populus trichocarpa and their roles in response to various heavy metal stresses. Int J Mol Sci 21(5):1680
García G, Faz Á, Conesa HM (2003) Selection of autochthonous plant species from SE Spain for soil lead phytoremediation purposes. Water Air Soil Pollut Focus 3(3):243–250
Garg M, Sharma N, Sharma S, Kapoor P, Kumar A, Chunduri V, Arora P (2018) Biofortified crops generated by breeding, agronomy, and transgenic approaches are improving lives of millions of people around the world. Front Nutr 5:12
Gerhardt KE, Gerwing PD, Greenberg BM (2017) Opinion: taking phytoremediation from proven technology to accepted practice. Plant Sci 256:170–185
Gigolashvili T, Kopriva S (2014) Transporters in plant sulfur metabolism. Front Plant Sci 5
Gisbert C, Ros R, De Haro A, Walker DJ, Pilar Bernal M, Serrano R, Navarro-Aviñó J (2003) A plant genetically modified that accumulates Pb is especially promising for phytoremediation. Biochem Biophys Res Commun 303(2):440–445
Göhre V, Paszkowski U (2006) Contribution of the arbuscular mycorrhizal symbiosis to heavy metal phytoremediation. Planta 223(6):1115–1122

Gomes MAC, Hauser-Davis RA, de Souza AN, Vitória AP (2016) Metal phytoremediation: general strategies, genetically modified plants and applications in metal nanoparticle contamination. Ecotoxicol Environ Saf 134:133–147

Govindaraj M, Yadav OP, Srivastava RK, Gupta SK (2019) Conventional and molecular breeding approaches for biofortification of pearl millet. In: Quality breeding in field crops. pp 85–107

Gupta G, Keegan B (1998) Bioaccumulation and biosorption of lead by poultry litter microorganisms. Poult Sci 77(3):400–404

Gupta DK, Vandenhove H, Inouhe M (2013) Role of phytochelatins in heavy metal stress and detoxification mechanisms in plants. In: Heavy metal stress in plants. pp 73–94

Hajar EWI, Sulaiman AZB, Sakinah AMM (2014) Assessment of heavy metals tolerance in leaves, stems and flowers of Stevia rebaudiana plant. Procedia Environ Sci 20:386–393

Halka M, Smoleń S, Czernicka M, Klimek-Chodacka M, Pitala J, Tutaj K (2019) Iodine biofortification through expression of HMT, SAMT and S3H genes in *Solanum lycopersicum* L. Plant Physiol Biochem 144:35–48

Hall RD, Brouwer ID, Fitzgerald MA (2008) Plant metabolomics and its potential application for human nutrition. Physiol Plant 132(2):162–175

Hammond JP, Bowen HC, White PJ, Mills V, Pyke KA, Baker AJM, Whiting SN, May ST, Broadley MR (2006) A comparison of the Thlaspi caerulescens and Thlaspi arvense shoot transcriptomes. New Phytol 170(2):239–260

Hanikenne M, Talke IN, Haydon MJ, Lanz C, Nolte A, Motte P, Kroymann J, Weigel D, Krämer U (2008) Evolution of metal hyperaccumulation required cis-regulatory changes and triplication of HMA4. Nature 453(7193):391–395

Hauer-Jákli M, Tränkner M (2019) Critical leaf magnesium thresholds and the impact of magnesium on plant growth and photo-oxidative defense: a systematic review and meta-analysis from 70 years of research. Front Plant Sci 10:766

Havrlentová M, Pšenáková I, Žofajová A, Rückschloss Ľ, Kraic J (2014) Anthocyanins in wheat seed—a mini review. Nova Biotechnol t Chim 13(1):1–12

He G, Qin L, Tian W, Meng L, He T, Zhao D (2020) Heavy metal transporters-associated proteins in Solanum tuberosum: genome-wide identification, comprehensive gene feature, evolution and expression analysis. Genes 11(11):1269

Heaton ACP, Rugh CL, Wang N, Meagher RB (1998) Phytoremediation of mercury- and methylmercury-polluted soils using genetically engineered plants. J Soil Contam 7(4):497–509

Hovsepyan A, Greipsson S (2005) EDTA-enhanced phytoremediation of lead-contaminated soil by corn. J Plant Nutr 28(11):2037–2048

Huang T, Peng Q, Yu L, Li D (2017) The detoxification of heavy metals in the phosphate tailing-contaminated soil through sequential microbial pretreatment and electrokinetic remediation. Soil Sediment Contam Int J 26(3):308–322

Hussain D, Haydon MJ, Wang Y, Wong E, Sherson SM, Young J, Camakaris J, Harper JF, Cobbett CS (2004) P-type ATPase heavy metal transporters with roles in essential zinc homeostasis in Arabidopsis. Plant Cell 16(5):1327–1339

Huysen VT, Terry N, Pilon-Smits EA (2004) Exploring the selenium phytoremediation potential of transgenic Indian mustard overexpressing ATP sulfurylase or cystathionine-gamma-synthase. Int J Phytoremediation 6(2):111–118

Ibañez SG, Paisio CE, Wevar Oller AL, Talano MA, González PS, Medina MI, Agostini E (2015) Overview and new insights of genetically engineered plants for improving phytoremediation. In: Phytoremediation. pp 99–113

Ishimaru Y, Masuda H, Bashir K, Inoue H, Tsukamoto T, Takahashi M, Nakanishi H, Aoki N, Hirose T, Ohsugi R, Nishizawa NK (2010) Rice metal-nicotianamine transporter, OsYSL2, is required for the long-distance transport of iron and manganese: OsYSL2 is required for Fe and Mn transport to the endosperm. Plant J 62(3):379–390

Islam MS, Hosen MML, Uddin MN (2019) Phytodesalination of saline water using Ipomoea aquatica, Alternanthera philoxeroides and Ludwigia adscendens. Int J Environ Sci Technol 16(2):965–972

Itoh N, Toda H, Matsuda M, Negishi T, Taniguchi T, Ohsawa N (2009) Involvement of S-adenosylmethionine-dependent halide/thiol methyltransferase (HTMT) in methyl halide emissions from agricultural plants: Isolation and characterization of an HTMT-coding gene from Raphanus sativus (Daikon radish). BMC Plant Biol 9(1):116

Jacob JM, Karthik C, Saratale RG, Kumar SS, Prabakar D, Kadirvelu K, Pugazhendhi A (2018) Biological approaches to tackle heavy metal pollution: a survey of literature. J Environ Manag 217:56–70

Jentschke G, Godbold DL (2000) Metal toxicity and ectomycorrhizas. Physiol Plant 109(2): 107–116

Jia Y, Huang H, Sun G-X, Zhao F-J, Zhu Y-G (2012) Pathways and relative contributions to arsenic volatilization from rice plants and paddy soil. Environ Sci Technol 46(15):8090–8096

Jogawat A, Yadav B, Chhaya, Narayan OP (2021) Metal transporters in organelles and their roles in heavy metal transportation and sequestration mechanisms in plants. Physiol Plant 13370:1–17

Johnson R, Puthur JT (2021) Seed priming as a cost effective technique for developing plants with cross tolerance to salinity stress. Plant Physiol Biochem 162:247–257

Johnson AAT, Kyriacou B, Callahan DL, Carruthers L, Stangoulis J, Lombi E, Tester M (2011) Constitutive overexpression of the osnas gene family reveals single-gene strategies for effective iron- and zinc-biofortification of rice endosperm. PLoS One 6(9):e24476

Kaplan D (2013) Absorption and adsorption of heavy metals by microalgae. In: Handbook of microalgal culture. pp 602–611

Karley AJ, White PJ (2009) Moving cationic minerals to edible tissues: potassium, magnesium, calcium. Curr Opin Plant Biol 12:291–298

Kawahigashi H (2009) Transgenic plants for phytoremediation of herbicides. Curr Opin Biotechnol 20(2):225–230

Khandare RV, Govindwar SP (2015) Phytoremediation of textile dyes and effluents: current scenario and future prospects. Biotechnol Adv 33(8):1697–1714

Kidd P, Barceló J, Bernal MP, Navari-Izzo F, Poschenrieder C, Shilev S, Clemente R, Monterroso C (2009) Trace element behaviour at the root–soil interface: Implications in phytoremediation. Environ Exp Bot 67(1):243–259

Kim S, Takahashi M, Higuchi K, Tsunoda K, Nakanishi H, Yoshimura E, Mori S, Nishizawa NK (2005) Increased nicotianamine biosynthesis confers enhanced tolerance of high levels of metals, in particular nickel, to plants. Plant Cell Physiol 46(11):1809–1818

Kim YY, Choi H, Segami S, Cho HT, Martinoia E, Maeshima M, Lee Y (2009) AtHMA1 contributes to the detoxification of excess Zn(II) in Arabidopsis. Plant J 58(5):737–753

Köber R, Daus B, Ebert M, Mattusch J, Welter E, Dahmke A (2005) Compost-based permeable reactive barriers for the source treatment of arsenic contaminations in aquifers: column studies and solid-phase investigations. Environ Sci Technol 39(19):7650–7655

Kosolapov DB, Kuschk P, Vainshtein MB, Vatsourina AV, Wießner A, Kästner M, Müller RA (2004) Microbial processes of heavy metal removal from carbon-deficient effluents in constructed wetlands. Eng Life Sci 4(5):403–411

Koźmińska A, Wiszniewska A, Hanus-Fajerska E, Muszyńska E (2018) Recent strategies of increasing metal tolerance and phytoremediation potential using genetic transformation of plants. Plant Biotechnol Rep 12(1):1–14

Krämer U (2005) Phytoremediation: novel approaches to cleaning up polluted soils. Curr Opin Biotechnol 16(2):133–141

Krishna R, Fulekar MH, Pathak B (2012) Rhizofiltration: a green technology for remediation of heavy metals. Int J Innov Biosci 2(4):193–199

Krishna ATP, Maharajan T, Victor Roch G, Ignacimuthu S, Antony Ceasar S (2020) Structure, function, regulation and phylogenetic relationship of ZIP family transporters of plants. Front Plant Sci 11:662

Kumar S, Kumar S, Mohapatra T (2021) Interaction between macro- and micro-nutrients in plants. Front Plant Sci 12:665583

Kumpiene J, Fitts JP, Mench M (2012) Arsenic fractionation in mine spoils 10 years after aided phytostabilization. Environ Pollut 166:82–88

LaCoste C, Robinson B, Brooks R (2001) Uptake of thallium by vegetables: its significance for human health, phytoremediation, and phytomining. J Plant Nutr 24(8):1205–1215

Lai K, Duran C, Berkman PJ, Lorenc MT, Stiller J, Manoli S, Hayden MJ, Forrest KL, Fleury D, Baumann U, Zander M, Mason AS, Batley J, Edwards D (2012) Single nucleotide polymorphism discovery from wheat next-generation sequence data. Plant Biotechnol J 10(6):743–749

Landini M, Gonzali S, Kiferle C, Tonacchera M, Agretti P, Dimida A, Vitti P, Alpi A, Pinchera A, Perata P (2012) Metabolic engineering of the iodine content in Arabidopsis. Sci Rep 2(1):338

Lanquar V, Lelièvre F, Bolte S, Hamès C, Alcon C, Neumann D, Vansuyt G, Curie C, Schröder A, Krämer U, Barbier-Brygoo H, Thomine S (2005) Mobilization of vacuolar iron by AtNRAMP3 and AtNRAMP4 is essential for seed germination on low iron. EMBO J 24(23):4041–4051

Lasat MM (2002) Phytoextraction of toxic metals: a review of biological mechanisms. J Environ Qual 31(1):109–120

Lawson PG, Daum D, Czauderna R, Meuser H, Härtling JW (2015) Soil versus foliar iodine fertilization as a biofortification strategy for field-grown vegetables. Front Plant Sci 6:450

LeDuc DL, Terry N (2005) Phytoremediation of toxic trace elements in soil and water. J Ind Microbiol Biotechnol 32(11–12):514–520

LeDuc DL, Tarun AS, Montes-Bayon M, Meija J, Malit MF, Wu CP, AbdelSamie M, Chiang CY, Tagmount A, deSouza M, Neuhierl B, Böck A, Caruso J, Terry N (2004) Overexpression of selenocysteine methyltransferase in Arabidopsis and Indian mustard increases selenium tolerance and accumulation. Plant Physiol 135(1):377–383

LeDuc DL, AbdelSamie M, Móntes-Bayon M, Wu CP, Reisinger SJ, Terry N (2006) Overexpressing both ATP sulfurylase and selenocysteine methyltransferase enhances selenium phytoremediation traits in Indian mustard. Environ Pollut 144(1):70–76

Limcharoensuk T, Sooksawat N, Sumarnrote A, Awutpet T, Kruatrachue M, Pokethitiyook P, Auesukaree C (2015) Bioaccumulation and biosorption of Cd(2+) and Zn(2+) by bacteria isolated from a zinc mine in Thailand. Ecotoxicol Environ Saf 122:322–330

Lin SH, Kiang CD (2003) Chromic acid recovery from waste acid solution by an ion exchange process: equilibrium and column ion exchange modeling. Chem Eng J 92(1–3):193–199

Liu L, Li W, Song W, Guo M (2018) Remediation techniques for heavy metal-contaminated soils: principles and applicability. Sci Total Environ 633:206–219

Liu XS, Feng SJ, Zhang BQ, Wang MQ, Cao HW, Rono JK, Chen X, Yang ZM (2019) OsZIP1 functions as a metal efflux transporter limiting excess zinc, copper and cadmium accumulation in rice. BMC Plant Biol 19(1):283

Lone MI, He ZL, Stoffella PJ, Yang XE (2008) Phytoremediation of heavy metal polluted soils and water: progresses and perspectives. J Zhejiang Univ Sci B 9(3):210–220

Ma LQ, Komar KM, Tu C, Zhang W, Cai Y, Kennelley ED (2001) A fern that hyperaccumulates arsenic. Nature 409(6820):579–579

Ma Y, Prasad MNV, Rajkumar M, Freitas H (2011) Plant growth promoting rhizobacteria and endophytes accelerate phytoremediation of metalliferous soils. Biotechnol Adv 29(2):248–258

Ma Y, Oliveira RS, Freitas H, Zhang C (2016) Biochemical and molecular mechanisms of plant-microbe-metal interactions: relevance for phytoremediation. Front Plant Sci 7:918

Maghari BM, Ardekani AM (2011) Genetically modified foods and social concerns. Avicenna J Med Biotechnol 3(3):109–117

Malagoli M, Schiavon M, dall'Acqua S, Pilon-Smits EA (2015) Effects of selenium biofortification on crop nutritional quality. Front Plant Sci 6:280

Malik KA, Maqbool A (2020) Transgenic crops for biofortification. Front Sustain Food Syst 4: 571402

Mani D, Kumar C (2014) Biotechnological advances in bioremediation of heavy metals contaminated ecosystems: an overview with special reference to phytoremediation. Int J Environ Sci Technol 11(3):843–872

Manousaki E, Kalogerakis N (2011) Halophytes present new opportunities in phytoremediation of heavy metals and saline soils. Ind Eng Chem Res 50(2):656–660

Marques APGC, Rangel AOSS, Castro PML (2009) Remediation of heavy metal contaminated soils: phytoremediation as a potentially promising clean-up technology. Crit Rev Environ Sci Technol 39(8):622–654

Marques E, Darby HM, Kraft J (2021) Benefits and limitations of non-transgenic micronutrient biofortification approaches. Agronomy 11(3):464

Martínez-Alcalá I, Clemente R (2020) Phytoremediation of metalloid-contaminated soil. Metalloids Plants 1:27–46

Mastretta C, Taghavi S, van der Lelie D, Mengoni A, Galardi F, Gonnelli C, Barac T, Boulet J, Weyens N, Vangronsveld J (2009) Endophytic bacteria from seeds of Nicotiana tabacum can reduce cadmium phytotoxicity. Int J Phytoremediation 11(3):251–267

Masuda H, Usuda K, Kobayashi T, Ishimaru Y, Kakei Y, Takahashi M, Higuchi K, Nakanishi H, Mori S, Nishizawa NK (2009) Overexpression of the barley nicotianamine synthase gene hvnas1 increases iron and zinc concentrations in rice grains. Rice 2(4):155–166

Masuda H, Ishimaru Y, Aung MS, Kobayashi T, Kakei Y, Takahashi M, Higuchi K, Nakanishi H, Nishizawa NK (2012) Iron biofortification in rice by the introduction of multiple genes involved in iron nutrition. Sci Rep 2(1):543

Meagher RB, Heaton AC (2005) Strategies for the engineered phytoremediation of toxic element pollution: mercury and arsenic. J Ind Microbiol Biotechnol 32(11–12):502–513

Mench M, Lepp N, Bert V, Schwitzguébel J-P, Gawronski SW, Schröder P, Vangronsveld J (2010) Successes and limitations of phytotechnologies at field scale: outcomes, assessment and outlook from COST Action 859. J Soils Sediments 10(6):1039–1070

Mesjasz-Przybyłowicz J, Nakonieczny M, Migula P, Augustyniak M, Tarnawska M, Reimold W, Koeberl C, Przybyłowicz W, Elz A, Głowacka B (2004) Uptake of cadmium, lead, nickel and zinc from soil and water solutions by the nickel hyperaccumulator Berkheya coddii. Acta Biol Cracov Ser Bot 46:75–85

Meuser H (2013) Soil remediation and rehabilitation: treatment of contaminated and disturbed land. Springer

Midhat L, Ouazzani N, Hejjaj A, Bayo J, Mandi L (2018) Phytostabilization of polymetallic contaminated soil using Medicago sativa L. in combination with powdered marble: sustainable rehabilitation. Int J Phytoremediation 20(8):764–772

Mishra J, Singh R, Arora NK (2017) Alleviation of heavy metal stress in plants and remediation of soil by rhizosphere microorganisms. Front Microbiol 8:1706

Mizuno D, Higuchi K, Sakamoto T, Nakanishi H, Mori S, Nishizawa NK (2003) Three nicotianamine synthase genes isolated from maize are differentially regulated by iron nutritional status. Plant Physiol 132(4):1989–1997

Mohanpuria P, Rana NK, Yadav SK (2007) Cadmium induced oxidative stress influence on glutathione metabolic genes of Camellia sinensis (L.) O. Kuntze. Environ Toxicol 22(4): 368–374

Mohanty M, Patra HK (2011) Attenuation of chromium toxicity in mine waste water using water hyacinth. J Stress Physiol Biochem 7(4):335–346

Montanini B, Blaudez D, Jeandroz S, Sanders D, Chalot M (2007) Phylogenetic and functional analysis of the cation diffusion facilitator (CDF) family: improved signature and prediction of substrate specificity. BMC Genomics 8:107

Morel M, Crouzet J, Gravot A, Auroy P, Leonhardt N, Vavasseur A, Richaud P (2009) AtHMA3, a P1b-ATPase allowing Cd/Zn/Co/Pb vacuolar storage in Arabidopsis. Plant Physiol 149(2): 894–904

Mulligan CN, Yong RN, Gibbs BF (2001a) An evaluation of technologies for the heavy metal remediation of dredged sediments. J Hazard Mater 85(1–2):145–163

Mulligan CN, Yong RN, Gibbs BF (2001b) Remediation technologies for metal-contaminated soils and groundwater: an evaluation. Eng Geol 60(1–4):193–207

Munoz LP, Purchase D, Jones H, Feldmann J, Garelick H (2014) Enhanced determination of As–phytochelatin complexes in Chlorella vulgaris using focused sonication for extraction of water-soluble species. Anal Methods 6(3):791–797

Nagajyoti PC, Lee KD, Sreekanth TVM (2010) Heavy metals, occurrence and toxicity for plants: a review. Environ Chem Lett 8(3):199–216

Naqvi S, Zhu C, Farre G, Ramessar K, Bassie L, Breitenbach J, Perez Conesa D, Ros G, Sandmann G, Capell T, Christou P (2009) Transgenic multivitamin corn through biofortification of endosperm with three vitamins representing three distinct metabolic pathways. Proc Natl Acad Sci U S A 106(19):7762–7767

Narayanan NN, Vasconcelos MW, Grusak MA (2007) Expression profiling of Oryza sativa metal homeostasis genes in different rice cultivars using a cDNA macroarray. Plant Physiol Biochem 45(5):277–286

Neugschwandtner RW, Tlustoš P, Komárek M, Száková J (2008) Phytoextraction of Pb and Cd from a contaminated agricultural soil using different EDTA application regimes: Laboratory versus field scale measures of efficiency. Geoderma 144(3–4):446–454

NIH (National Institutes of Health) (2021a) Office of dietary supplements—magnesium

NIH (National Institutes of Health) (2021b) Office of dietary supplements—phosphorus

NIH (National Institutes of Health) (2021c) Office of dietary supplements—manganese

NIH (National Institutes of Health) (2021d) Office of dietary supplements—zinc

Nkrumah PN, Baker AJ, Chaney RL, Erskine PD, Echevarria G, Morel JL, van der Ent A (2016) Current status and challenges in developing nickel phytomining: an agronomic perspective. Plant Soil 406(1–2):55–69

Nowak B, Pessl A, Aschenbrenner P, Szentannai P, Mattenberger H, Rechberger H, Hermann L, Winter F (2010) Heavy metal removal from municipal solid waste fly ash by chlorination and thermal treatment. J Hazard Mater 179(1–3):323–331

O'Rourke JA, Nelson RT, Grant D, Schmutz J, Grimwood J, Cannon S, Vance CP, Graham MA, Shoemaker RC (2009) Integrating microarray analysis and the soybean genome to understand the soybeans iron deficiency response. BMC Genomics 10:376

Odjegba VJ, Fasidi IO (2007) Phytoremediation of heavy metals by Eichhornia crassipes. Environmentalist 27(3):349–355

Ong S-T, Yip S-P, Keng P-S, Lee S-L, Hung Y-T (2012) Papaya (Carica papaya) seed as a low-cost sorbent for zinc removal. Afr J Agric Res 7(5):810–819

Paine JA, Shipton CA, Chaggar S, Howells RM, Kennedy MJ, Vernon G, Wright SY, Hinchliffe E, Adams JL, Silverstone AL, Drake R (2005) Improving the nutritional value of Golden Rice through increased pro-vitamin A content. Nat Biotechnol 23(4):482–487

Palmgren MG, Clemens S, Williams LE, Krämer U, Borg S, Schjørring JK, Sanders D (2008) Zinc biofortification of cereals: problems and solutions. Trends Plant Sci 13(9):464–473

Pamukcu S, Wittle JK (1992) Electrokinetic removal of selected heavy metals from soil. Environ Prog 11(3):241–250

Parekh D, Patel J, Sudhakar P, Koshy VJ (2002) Removal of trace metals with mango seed powder. Indian J Chem Technol 9:540–542

Pathak A, Dastidar MG, Sreekrishnan TR (2009) Bioleaching of heavy metals from sewage sludge: a review. J Environ Manag 90(8):2343–2353

Patra M, Sharma A (2000) Mercury toxicity in plants. Bot Rev 66(3):379–422

Pedas P, Husted S (2009) Zinc transport mediated by barley ZIP proteins are induced by low pH. Plant Signal Behav 4(9):842–845

Pérez-Palacios P, Agostini E, Ibáñez SG, Talano MA, Rodríguez-Llorente ID, Caviedes MA, Pajuelo E (2017) Removal of copper from aqueous solutions by rhizofiltration using genetically modified hairy roots expressing a bacterial Cu-binding protein. Environ Technol 38(22): 2877–2888

Phattarakul N, Rerkasem B, Li LJ, Wu LH, Zou CQ, Ram H, Sohu VS, Kang BS, Surek H, Kalayci M, Yazici A, Zhang FS, Cakmak I (2012) Biofortification of rice grain with zinc through zinc fertilization in different countries. Plant Soil 361(1–2):131–141

Pilon-Smits EAH, de Souza MP, Hong G, Amini A, Bravo RC, Payabyab ST, Terry N (1999a) Selenium volatilization and accumulation by twenty aquatic plant species. J Environ Qual 28(3): 1011–1018

Pilon-Smits EA, Hwang S, Mel Lytle C, Zhu Y, Tai JC, Bravo RC, Chen Y, Leustek T, Terry N (1999b) Overexpression of ATP sulfurylase in Indian mustard leads to increased selenate uptake, reduction, and tolerance. Plant Physiol 119(1):123–132

Pivetz BE (2001) Ground water issue: phytoremediation of contaminated soil and ground water at hazardous waste sites. EPA, Washington, DC, pp 1–36

Pourrut B, Shahid M, Dumat C, Winterton P, Pinelli E (2011) Lead uptake, toxicity, and detoxification in plants. Rev Environ Contam Toxicol 213:113–136

Prasad MNV (2003) Phytoremediation of metal-polluted ecosystems: hype for commercialization. Russ J Plant Physiol 50(5):686–701

Prasad R, Shivay YS, Kumar D (2014) Agronomic biofortification of cereal grains with iron and zinc. Adv Agron 125:55–91

Qdais HA, Moussa H (2004) Removal of heavy metals from wastewater by membrane processes: a comparative study. Desalination 164(2):105–110

Rahman RAA, Abou-Shanab RA, Moawad H (2008) Mercury detoxification using genetic engineered Nicotiana tabacum. Glob Nest J 10(3):432–438

Ramesh SA, Choimes S, Schachtman DP (2004) Over-expression of an Arabidopsis zinc transporter in Hordeum vulgare increases short-term zinc uptake after zinc deprivation and seed zinc content. Plant Mol Biol 54(3):373–385

Rascio N, Navari-Izzo F (2011) Heavy metal hyperaccumulating plants: how and why do they do it? And what makes them so interesting? Plant Sci 180(2):169–181

Raskin I, Ensley BD (2000) Phytoremediation of toxic metals: using plants to clean up the environment. Wiley, New York

Raskin I, Smith RD, Salt DE (1997) Phytoremediation of metals: using plants to remove pollutants from the environment. Curr Opin Biotechnol 8(2):221–226

Ravikumar K, Sheeja AK (2013) Heavy metal removal from water using Moringa oleifera seed coagulant and double filtration. Int J Sci Eng Res 4(5):10–13

Ravindran KC, Venkatesan K, Balakrishnan V, Chellappan KP, Balasubramanian T (2007) Restoration of saline land by halophytes for Indian soils. Soil Biol Biochem 39(10):2661–2664

Rezania S, Taib SM, Md Din MF, Dahalan FA, Kamyab H (2016) Comprehensive review on phytotechnology: heavy metals removal by diverse aquatic plants species from wastewater. J Hazard Mater 318:587–599

Riaz U, Aziz H, Anum W, Mehdi SM, Murtaza G, Jamil M (2020) Biofortification technologies used in agriculture in relation to micronutrients. Plant micronutrients. Springer, Cham, pp 225–239

Ricachenevsky FK, Menguer PK, Sperotto RA, Williams LE, Fett JP (2013) Roles of plant metal tolerance proteins (MTP) in metal storage and potential use in biofortification strategies. Front Plant Sci 4:144

Rice KM, Walker EM Jr, Wu M, Gillette C, Blough ER (2014) Environmental mercury and its toxic effects. J Prev Med Public Health 47(2):74–83

Rout ME, Southworth D (2013) The root microbiome influences scales from molecules to ecosystems: The unseen majority. Am J Bot 100(9):1689–1691

Saier MH, Trevors JT (2010) Phytoremediation. Water Air Soil Pollut 205(S1):61–63

Sakai Y, Ma Y, Xu C, Wu H, Zhu W, Yang J (2012) Phytodesalination of a salt affected soil with four halophytes in China. J Arid Land Stud 22:17–20

Sakakibara M, Watanabe A, Inoue M, Sano S, Kaise T (2010) Phytoextraction and phytovolatilization of arsenic from as-contaminated soils by Pteris vittata. Proc Annu Int Conf Soils Sediments Water Energy 12(1):26

Salt DE, Blaylock M, Kumar NPBA, Dushenkov V, Ensley BD, Chet I, Raskin I (1995) Phytoremediation: a novel strategy for the removal of toxic metals from the environment using plants. Nat Biotechnol 13(5):468–474

Salt DE, Smith RD, Raskin I (1998) Phytoremediation. Annu Rev Plant Physiol Plant Mol Biol 49(1):643–668
Saltzman A, Birol E, Bouis HE, Boy E, De Moura FF, Islam Y, Pfeiffer WH (2014) Biofortification: progress toward a more nourishing future. Glob Food Sec 2(1):9–17
Sarma H (2011) Metal hyperaccumulation in plants: a review focusing on phytoremediation technology. J Environ Sci Technol 4(2):118–138
Sarwar N, Imran M, Shaheen MR, Ishaque W, Kamran MA, Matloob A, Rehim A, Hussain S (2017) Phytoremediation strategies for soils contaminated with heavy metals: modifications and future perspectives. Chemosphere 171:710–721
Sasaki Y, Hayakawa T, Inoue C, Miyazaki A, Silver S, Kusano T (2006) Generation of mercury-hyperaccumulating plants through transgenic expression of the bacterial mercury membrane transport protein merc. Transgenic Res 15(5):615–625
Selvi A, Rajasekar A, Theerthagiri J, Ananthaselvam A, Sathishkumar K, Madhavan J, Rahman PKSM (2019) Integrated remediation processes toward heavy metal removal/recovery from various environments—a review. Front Environ Sci 7:66
Sen AK, Mondal NG, Mandal S (1987) Studies of uptake and toxic effects of Cr(VI) on Pistia stratiotes. Water Sci Technol 19(1–2):119–127
Seth CS (2012) A review on mechanisms of plant tolerance and role of transgenic plants in environmental clean-up. Bot Rev 78(1):32–62
Shamim S (2018) Biosorption of heavy metals. Biosorption. InTech
Shanker A, Cervantes C, Lozatavera H, Avudainayagam S (2005) Chromium toxicity in plants. Environ Int 31(5):739–753
Shaul O (2002) Magnesium transport and function in plants: the tip of the iceberg. Biometals 15(3): 307–321
Sheoran V, Sheoran AS, Poonia P (2009) Phytomining: a review. Miner Eng 22(12):1007–1019
Sheoran V, Sheoran AS, Poonia P (2016) Factors affecting phytoextraction: a review. Pedosphere 26(2):148–166
Singh SS, Hazra KK, Praharaj CS Singh U (2016a) Biofortification: pathway ahead and future challenges. In: Biofortification of food crops. Springer, India. pp 479–492
Singh U, Praharaj CS, Chaturvedi SK, Bohra A (2016b) Biofortification: introduction, approaches, limitations, and challenges. In: Biofortification of food crops. Springer, India. pp 3–18
Smith FW, Mudge SR, Rae AL, Glassop D (2003) Phosphate transport in plants. Plant Soil 248(1): 71–83
Sperotto RA, Ricachenevsky FK, Duarte GL, Boff T, Lopes KL, Sperb ER, Grusak MA, Fett JP (2009) Identification of up-regulated genes in flag leaves during rice grain filling and characterization of OsNAC5, a new ABA-dependent transcription factor. Planta 230(5):985–1002
Sreekrishnan TR, Tyagi RD, Blais JF, Campbell PGC (1993) Kinetics of heavy metal bioleaching from sewage sludge—I. Effects of process parameters. Water Res 27(11):1641–1651
Sricoth T, Meeinkuirt W, Saengwilai P, Pichtel J, Taeprayoon P (2018) Aquatic plants for phytostabilization of cadmium and zinc in hydroponic experiments. Environ Sci Pollut Res Int 25(15):14964–14976
Srivastava D, Tiwari M, Dutta P, Singh P, Chawda K, Kumari M, Chakrabarty D (2021) Chromium stress in plants: toxicity, tolerance and phytoremediation. Sustainability 13(9):4629
Storozhenko S, De Brouwer V, Volckaert M, Navarrete O, Blancquaert D, Zhang GF, Lambert W, Van Der Straeten D (2007) Folate fortification of rice by metabolic engineering. Nat Biotechnol 25(11):1277–1279
Stylianou MA, Kollia D, Haralambous KJ, Inglezakis VJ, Moustakas KG, Loizidou MD (2007) Effect of acid treatment on the removal of heavy metals from sewage sludge. Desalination 215(1–3):73–81
Su YH, McGrath SP, Zhu YG, Zhao FJ (2008) Highly efficient xylem transport of arsenite in the arsenic hyperaccumulator Pteris vittata. New Phytol 180(2):434–441

Su S, Zeng X, Bai L, Williams PN, Wang Y, Zhang L, Wu C (2017) Inoculating chlamydospores of *Trichoderma asperellum* SM-12F1 changes arsenic availability and enzyme activity in soils and improves water spinach growth. Chemosphere 175:497–504

Suresh B, Ravishankar GA (2004) Phytoremediation—a novel and promising approach for environmental clean-up. Crit Rev Biotechnol 24(2–3):97–124

Takahashi R, Bashir K, Ishimaru Y, Nishizawa NK, Nakanishi H (2012) The role of heavy-metal ATPases, HMAs, in zinc and cadmium transport in rice. Plant Signal Behav 7(12):1605–1607

Talke IN, Hanikenne M, Krämer U (2006) Zinc-dependent global transcriptional control, transcriptional deregulation, and higher gene copy number for genes in metal homeostasis of the hyperaccumulator Arabidopsis halleri. Plant Physiol 142(1):148–167

Tanaka H, Yabuta Y, Tamoi M, Tanabe N, Shigeoka S (2015) Generation of transgenic tobacco plants with enhanced tocotrienol levels through the ectopic expression of rice homogentisate geranylgeranyl transferase. Plant Biotechnol 32(3):233–238

Tang Z, Zhao FJ (2020) The roles of membrane transporters in arsenic uptake, translocation and detoxification in plants. Crit Rev Environ Sci Technol 1–36

Tangahu BV, Sheikh Abdullah SR, Basri H, Idris M, Anuar N, Mukhlisin M (2011) A review on heavy metals (As, Pb, and Hg) uptake by plants through phytoremediation. Int J Chem Eng 2011:1–31

Taylor AA, Tsuji JS, Garry MR, McArdle ME, Goodfellow WL, Adams WJ, Menzie CA (2020) Critical review of exposure and effects: implications for setting regulatory health criteria for ingested copper. Environ Manag 65(1):131–159

Tchounwou PB, Yedjou CG, Patlolla AK, Sutton DJ (2012) Heavy metal toxicity and the environment. Exp Suppl 101:133–164

Tegeder M, Masclaux-Daubresse C (2018) Source and sink mechanisms of nitrogen transport and use. New Phytol 217(1):35–53

Terry N, Zayed AM, de Souza MP, Tarun AS (2000) Selenium in higher plants. Annu Rev Plant Physiol Plant Mol Biol 51(1):401–432

Thomine S, Wang R, Ward JM, Crawford NM, Schroeder JI (2000) Cadmium and iron transport by members of a plant metal transporter family in Arabidopsis with homology to Nramp genes. Proc Natl Acad Sci U S A 97(9):4991–4996

Thor K (2019) Calcium—nutrient and messenger. Front Plant Sci 10:440

Tomé VF, Rodríguez BP, Lozano JC (2008) Elimination of natural uranium and 226Ra from contaminated waters by rhizofiltration using Helianthus annuus L. Sci Total Environ 393(2–3): 351–357

Turck D, Bresson J, Burlingame B, Dean T, Fairweather-Tait S, Heinonen M, Hirsch-Ernst KI, Mangelsdorf I, McArdle H, Neuhäuser-Berthold M, Nowicka G, Pentieva K, Sanz Y, Siani A, Sjödin A, Stern M, Tomé D, Van Loveren H et al (2016) EFSA Panel on Dietetic Products, Nutrition and Allergies (NDA). Dietary reference values for potassium. EFSA J 14(10)

Udokop U (2018) Studies on the blastofiltration potentials of selected plant seeds for the treatment of lead and cadmium contaminated water samples. Afribary

Ullah A, Heng S, Munis MFH, Fahad S, Yang X (2015) Phytoremediation of heavy metals assisted by plant growth promoting (PGP) bacteria: a review. Environ Exp Bot 117:28–40

Upadhyay AK, Singh NK, Singh R, Rai UN (2016) Amelioration of arsenic toxicity in rice: Comparative effect of inoculation of Chlorella vulgaris and Nannochloropsis sp. on growth, biochemical changes and arsenic uptake. Ecotoxicol Environ Saf 124:68–73

Upadhyay MK, Yadav P, Shukla A, Srivastava S (2018) Utilizing the potential of microorganisms for managing arsenic contamination: a feasible and sustainable approach. Front Environ Sci 6: 24

Valko M, Rhodes CJ, Moncol J, Izakovic M, Mazur M (2006) Free radicals, metals and antioxidants in oxidative stress-induced cancer. Chem Biol Interact 160(1):1–40

Van Aken B (2009) Transgenic plants for enhanced phytoremediation of toxic explosives. Curr Opin Biotechnol 20(2):231–236 van de Mortel JE, Almar Villanueva L, Schat H, Kwekkeboom J, Coughlan S, Moerland PD, Loren V, van Themaat E, Koornneef M, Aarts MGM (2006) Large expression differences in genes for iron and zinc homeostasis, stress response, and lignin biosynthesis distinguish roots of Arabidopsis thaliana and the related metal hyperaccumulator Thlaspi caerulescens. Plant Physiol 142(3):1127–1147

Vangronsveld J, Herzig R, Weyens N, Boulet J, Adriaensen K, Ruttens A, Thewys T, Vassilev A, Meers E, Nehnevajova E, van der Lelie D, Mench M (2009) Phytoremediation of contaminated soils and groundwater: lessons from the field. Environ Sci Pollut Res 16(7):765–794

Veglio F, Beolchini F (1997) Removal of metals by biosorption: a review. Hydrometallurgy 44(3): 301–316

Vilensky MY, Berkowitz B, Warshawsky A (2002) In situ remediation of groundwater contaminated by heavy- and transition-metal ions by selective ion-exchange methods. Environ Sci Technol 36(8):1851–1855

Violante A, Cozzolino V, Perelomov L, Caporale AG, Pigna M (2010) Mobility and bioavailability of heavy metals and metalloids in soil environments. J Soil Sci Plant Nutr 10(3):268–292

Virkutyte J, Sillanpää M, Latostenmaa P (2002) Electrokinetic soil remediation—critical overview. Sci Total Environ 289(1–3):97–121

Vishnoi SR, Srivastava PN (2007) Phytoremediation-green for environmental clean. In: Proceedings of Taal2007: the 12th World lake conference. 1016:1021

Vithanage M, Dabrowska BB, Mukherjee AB, Sandhi A, Bhattacharya P (2012) Arsenic uptake by plants and possible phytoremediation applications: a brief overview. Environ Chem Lett 10(3): 217–224

Walker DJ, Clemente R, Roig A, Bernal MP (2003) The effects of soil amendments on heavy metal bioavailability in two contaminated Mediterranean soils. Environ Pollut 122(2):303–312

Wang HQ, Lu SJ, Li H, Yao ZH (2007) EDTA-enhanced phytoremediation of lead contaminated soil by Bidens maximowicziana. J Environ Sci (China) 19(12):1496–1499

Wang C, Zeng J, Li Y, Hu W, Chen L, Miao Y, Deng P, Yuan C, Ma C, Chen X, Zang M, Wang Q, Li K, Chang J, Wang Y, Yang G, He G (2014) Enrichment of provitamin A content in wheat (*Triticum aestivum* L.) by introduction of the bacterial carotenoid biosynthetic genes CrtB and CrtI. J Exp Bot 65(9):2545–2556

Wani AL, Ara A, Usmani JA (2015) Lead toxicity: a review. Interdiscip Toxicol 8(2):55–64

Ward MH, Jones RR, Brender JD, de Kok TM, Weyer PJ, Nolan BT, Villanueva CM, van Breda SG (2018) Drinking water nitrate and human health: an updated review. Int J Environ Res Public Health 15(7):1557

Welch RM, Graham RD (2004) Breeding for micronutrients in staple food crops from a human nutrition perspective. J Exp Bot 55(396):353–364

Welsch R, Arango J, Bär C, Salazar B, Al-Babili S, Beltrán J, Chavarriaga P, Ceballos H, Tohme J, Beyer P (2010) Provitamin A accumulation in cassava (Manihot esculenta) roots driven by a single nucleotide polymorphism in a phytoene synthase gene. Plant Cell 22(10):3348–3356

Wenzel WW, Unterbrunner R, Sommer P, Sacco P (2003) Chelate-assisted phytoextraction using canola (*Brassica napus* L.) in outdoors pot and lysimeter experiments. Plant Soil 249(1):83–96

White J, Pacey-Miller T, Crawford A, Cordeiro G, Barbary D, Bundock P, Henry R (2006) Abundant transcripts of malting barley identified by serial analysis of gene expression (Sage). Plant Biotechnol J 4(3):289–301

Wilbur S, Abadin H, Fay M, Yu D, Tencza B, Ingerman L, Klotzbach J, James S (2012) Toxicological profile for chromium. Agency for Toxic Substances and Disease Registry (US)

Williams L, Salt DE (2009) The plant ionome coming into focus. Curr Opin Plant Biol 12(3): 247–249

Winkel LH, Johnson CA, Lenz M, Grundl T, Leupin OX, Amini M, Charlet L (2012) Environmental selenium research: from microscopic processes to global understanding. Environ Sci Technol 46:571–579

Wirth J, Poletti S, Aeschlimann B, Yakandawala N, Drosse B, Osorio S, Tohge T, Fernie AR, Günther D, Gruissem W, Sautter C (2009) Rice endosperm iron biofortification by targeted and synergistic action of nicotianamine synthase and ferritin. Plant Biotechnol J 7(7):631–644
Wong MH (2003) Ecological restoration of mine degraded soils, with emphasis on metal contaminated soils. Chemosphere 50(6):775–780
Wong CKE, Cobbett CS (2009) HMA P-type ATPases are the major mechanism for root-to-shoot Cd translocation in Arabidopsis thaliana. New Phytol 181(1):71–78
World Health Organization (WHO) (1996) Iron in drinking water. WHO guidelines for drinking water quality, vol 2. 2nd edn
Wu LH, Luo YM, Xing XR, Christie P (2004) EDTA-enhanced phytoremediation of heavy metal contaminated soil with Indian mustard and associated potential leaching risk. Agric Ecosyst Environ 102(3):307–318
Wu G, Kang H, Zhang X, Shao H, Chu L, Ruan C (2010) A critical review on the bio-removal of hazardous heavy metals from contaminated soils: issues, progress, eco-environmental concerns and opportunities. J Hazard Mater 174(1–3):1–8
Wu D, Yamaji N, Yamane M, Kashino-Fujii M, Sato K, Feng Ma J (2016) The HvNRAMP5 transporter mediates uptake of cadmium and manganese, but not iron. Plant Physiol 172(3): 1899–1910
Wuana RA, Okieimen FE (2011) Heavy metals in contaminated soils: a review of sources, chemistry, risks and best available strategies for remediation. Int Sch Res Notices Ecol 2011: 1–20
Xie S, Wen Z, Zhan H, Jin M (2018) An experimental study on the adsorption and desorption of Cu (II) in silty clay. Geofluids 2018:1–12
Xu X, Du X, Wang F, Sha J, Chen Q, Tian G, Zhu Z, Ge S, Jiang Y (2020) Effects of potassium levels on plant growth, accumulation and distribution of carbon, and nitrate metabolism in apple dwarf rootstock seedlings. Front Plant Sci 11:904
Yadav DN, Bansal S, Tushir S, Kaur J, Sharma K (2020) Advantage of biofortification over fortification technologies. In: Wheat and barley grain biofortification. Elsevier. pp 257–273
Yan G, Viraraghavan T (2001) Heavy metal removal in a biosorption column by immobilized M. rouxii biomass. Bioresour Technol 78(3):243–249
Yan A, Wang Y, Tan SN, Yusof MLM, Ghosh S, Chen Z (2020) Phytoremediation: a promising approach for revegetation of heavy metal-polluted land. Front Plant Sci 11:359
Yang HQ, Jie YL (2005) Uptake and transport of calcium in plants. J Plant Physiol Mol Biol 31(3): 227–234
Yang W, Cahoon RE, Hunter SC, Zhang C, Han J, Borgschulte T, Cahoon EB (2011) Vitamin E biosynthesis: functional characterization of the monocot homogentisate geranylgeranyl transferase. Plant J 65(2):206–217
Ye X, Al-Babili S, Klöti A, Zhang J, Lucca P, Beyer P, Potrykus I (2000) Engineering the provitamin A (beta-carotene) biosynthetic pathway into (carotenoid-free) rice endosperm. Science 287(5451):303–305
Yin X, Yuan L, Liu Y, Lin Z (2012) Phytoremediation and biofortification: two sides of one coin. Springer
Yu D (2013) Chromium (Cr) toxicity: what are the physiologic effects of chromium exposure? Agency for Toxic Substances and Disease Registry-Environmental Health and Medicine Education
Zeng P, Guo Z, Cao X, Xiao X, Liu Y, Shi L (2018) Phytostabilization potential of ornamental plants grown in soil contaminated with cadmium. Int J Phytoremediation 20(4):311–320
Zhang J, Xu Y, Cao T, Chen J, Rosen BP, Zhao FJ (2017) Arsenic methylation by a genetically engineered Rhizobium-legume symbiont. Plant Soil 416(1–2):259–269
Zhang J, Martinoia E, Lee Y (2018a) Vacuolar transporters for cadmium and arsenic in plants and their applications in phytoremediation and crop development. Plant Cell Physiol 59(7): 1317–1325

Zhang Y, Chen K, Zhao F-J, Sun C, Jin C, Shi Y, Sun Y, Li Y, Yang M, Jing X, Luo J, Lian X (2018b) OsATX1 interacts with heavy metal P1B-type ATPases and affects copper transport and distribution. Plant Physiol 178(1):329–344

Zhao F, Tausz M, De Kok LJ (2008) Role of sulfur for plant production in agricultural and natural ecosystems. Sulfur metabolism in phototrophic organisms, 27. Springer, Dordrecht. pp 417–435

Zhao FJ, Ma JF, Meharg AA, McGrath SP (2009) Arsenic uptake and metabolism in plants. New Phytol 181(4):777–794

Zhu Y-G, Rosen BP (2009) Perspectives for genetic engineering for the phytoremediation of arsenic-contaminated environments: from imagination to reality? Curr Opin Biotechnol 20(2): 220–224

Zhu C, Naqvi S, Gomez-Galera S, Pelacho AM, Capell T, Christou P (2007) Transgenic strategies for the nutritional enhancement of plants. Trends Plant Sci 12(12):548–555

Zorrig W, Rabhi M, Ferchichi S, Smaoui A, Abdelly C (2012) Phytodesalination: a solution for salt-affected soils in arid and semi-arid regions. J Arid Land Stud 22:299–302

The manufacturer's authorised representative in the EU is Springer Nature Customer Service Centre GmbH, Europaplatz 3, 69115 Heidelberg, Germany. If you have any concerns regarding our products, please contact ProductSafety@springernature.com

Printed and bound by CPI Group (UK) Ltd, Croydon, CR0 4YY

30/09/2026

02225242-0018